Petar Sabev Varbanov, Jiří Škorpík, Jiří Pospíšil, Jiří Jaromír Klemeš
Sustainable Utility Systems

Also of interest

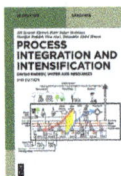

Sustainable Process Integration and Intensification.
Saving Energy, Water and Resources
Klemeš, Varbanov, Wan Alwi, Manan, 2018
ISBN 978-3-11-053535-8, e-ISBN 978-3-11-053536-5

Basic Process Engineering Control
Agachi, Cristea, Makhura, 2020
ISBN 978-3-11-064789-1, e-ISBN 978-3-11-064793-8

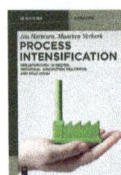

Process Intensification.
Breakthrough in Design, Industrial Innovation Practices,
and Education
Harmsen, Verkerk, 2020
ISBN 978-3-11-065734-0, e-ISBN 978-3-11-065735-7

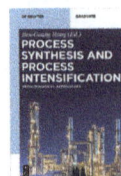

Process Synthesis and Process Intensification.
Methodological Approaches
Rong, 2017
ISBN 978-3-11-046505-1, e-ISBN 978-3-11-046506-8

Product and Process Design
Harmsen, Swinkels, 2018
ISBN 978-3-11-046772-7, e-ISBN 978-3-11-046774-1

Petar Sabev Varbanov, Jiří Škorpík,
Jiří Pospíšil, Jiří Jaromír Klemeš

Sustainable Utility Systems

Modelling and Optimisation

DE GRUYTER

Authors

Dr. Petar Sabev Varbanov
Sustainable Process Integration Lab.
SPIL, NETME Centre
Faculty of Mechanical Engineering
Brno University of Technology
Technická 2896/2
616 00 Brno
Czech Republic
varbanov@fme.vutbr.cz

Dr. Jiří Škorpík
Energy Institute
Faculty of Mechanical Engineering
Brno University of Technology
Technická 2896/2
616 00 Brno
Czech Republic
skorpik@fme.vutbr.cz

Dr. Jiří Pospíšil
Sustainable Process Integration Lab.
SPIL, NETME Centre
Faculty of Mechanical Engineering
Brno University of Technology
Technická 2896/2
616 00 Brno
Czech Republic
pospisil.j@fme.vutbr.cz

Prof. Jiří Jaromír Klemeš
Sustainable Process Integration Lab.
SPIL, NETME Centre
Faculty of Mechanical Engineering
Brno University of Technology
Technická 2896/2
616 00 Brno
Czech Republic
klemes@fme.vutbr.cz

ISBN 978-3-11-063004-6
e-ISBN (PDF) 978-3-11-063009-1
e-ISBN (EPUB) 978-3-11-063013-8

Library of Congress Control Number: 2020948254

Bibliographic information published by the Deutsche Nationalbibliothek
The Deutsche Nationalbibliothek lists this publication in the Deutsche Nationalbibliografie;
detailed bibliographic data are available on the Internet at http://dnb.dnb.de.

© 2021 Walter de Gruyter GmbH, Berlin/Boston
Cover image: PassionStudio/iStock/Getty Images Plus
Typesetting: Integra Software Services Pvt. Ltd.
Printing and binding: CPI books GmbH, Leck

www.degruyter.com

Foreword

Most chemical processes operate in the context of an existing site in which a number of processes are linked to the same utility system. Whilst different utilities are necessary for the site to function, the most prominent are those to provide the heating and cooling utilities necessary for the processes to maintain their energy balance. Heating is most often supplied via a steam system that also cogenerates power. The efficiency and reliability of this steam and cogeneration system is a crucial factor in the success of site operations. On most processing sites, the steam system is the largest energy consumer on the site. The steam system must be efficient and flexible enough to accommodate changes in the operation of the processes. Key to the efficient operation is the ability to understand both the individual units involved and the complex interactions that occur on the site. The integration of the processing and utility units is most often poorly understood. Not only is there typically integration within each processing unit through heat recovery systems, but there is also integration between the units through the utility system. On large sites, steam is generated from waste heat in the processing units and fed into the steam system. Other units draw this steam from the steam system. This means that heat recovery is taking place between processing units through the steam system. Thus, there are complex interactions between the processing units and the utility system and between processing units through the utility system. Tools are required to achieve an understanding and be able to optimise the performance of the utility system. In order to achieve such understanding, it is necessary to be able to model and optimise these interactions to maximise the system performance.

Whilst operating cost is a prominent issue for commercial enterprises, as society moves to greater sustainability of manufacturing systems, environmental releases, resource depletion and societal issues need to be also included in the evaluation of performance. Particularly prominent for utility systems is the emission of greenhouse gases from the burning of fossil fuels. Such fossil fuels will need to be replaced by more sustainable energy sources. This will place great challenges to the development of manufacturing systems in the process industries. To address these challenges effectively requires a holistic approach that recognises all of the complex interactions and constraints imposed by the move to sustainable manufacturing.

This book provides a practical guide to modelling and optimisation of steam and cogeneration systems and the basis of how to develop an understanding of how they might be integrated more efficiently. Complex steam and cogeneration utility systems usually feature many important degrees of freedom to be optimised. This book addresses the challenges to provide guidance to practising engineers, as well as undergraduate and postgraduate students of chemical and mechanical engineering.

Professor Robin Smith, FREng, FIChemE
Centre for Process Integration
University of Manchester

https://doi.org/10.1515/9783110630091-203

Contents

Part 2: Components of utility networks

Part 3: **Utility networks as a whole – modelling
          and optimising utility systems**

1 Introduction

Energy is essential for the economies to prosper. Statistics and research publications have reported a strong correlation between the Gross Domestic Product (GDP) per capita and the energy consumption per capita (Brown et al., 2011). A similar strong correlation has also been found between the Human Development Index (HDI) and the energy consumption (Šlaus and Jacobs, 2011). Naturally, the countries with higher per capita energy consumption feature higher GDP and HDI levels. Although the correlations alone do not establish causality, the linkage itself is apparent. Energy is needed in all sectors: industrial, domestic, commercial, transportation and power generation. Fossil fuel has been the primary source of energy since decades. As a result, even several decades after the start of the sustainable development ideas and efforts, the picture is essentially the same – at the current economic, societal and technological set-up, the higher level of development (e.g. HDI) comes at the expense of higher ecological footprint, as clearly detected by Holden et al. (2014).

Contribution of various activities and processes to the sustainability of human development is difficult to quantify precisely. Within the overall economy, every activity or process produces a combination of environmental impacts. These are often expressed as footprints (Klemeš, 2015), of which the most widely known are the greenhouse gas footprint and the water footprint. Understanding the footprints of various processes is a critical activity, which enables decision making for effective management of the economy in general and industry in particular. The intrinsic links among the various footprints and the used resources are complicated due to the complex networks that convert the resource inputs into products, side products and waste.

The strategic evaluation of the resources and footprints of industrial and business processes can be performed taking energy consumption as a basis due to the fact that any material and transportation flows inevitably require energy to run. An interesting indication of energy consumption nation-wide is the example of the US energy consumption evaluation, performed annually by the Lawrence Livermore National Laboratory (2020) in the form of energy flow charts as shown in Fig. 1.1. The chart indicates above 67% energy losses and only a little above 31% useful energy delivered for services. The sourced primary energy is also surveyed, and the statistic shows about 11% renewables (including biomass, wind, geothermal and hydro), as well as about 8% supply from nuclear power, the rest being supplied by fossil fuels.

Taking that structure of the energy supplies, useful energy services and energy losses as a representative indicator of the economic performance leads to a couple of essential conclusions for the US economy contribution to sustainability.

https://doi.org/10.1515/9783110630091-001

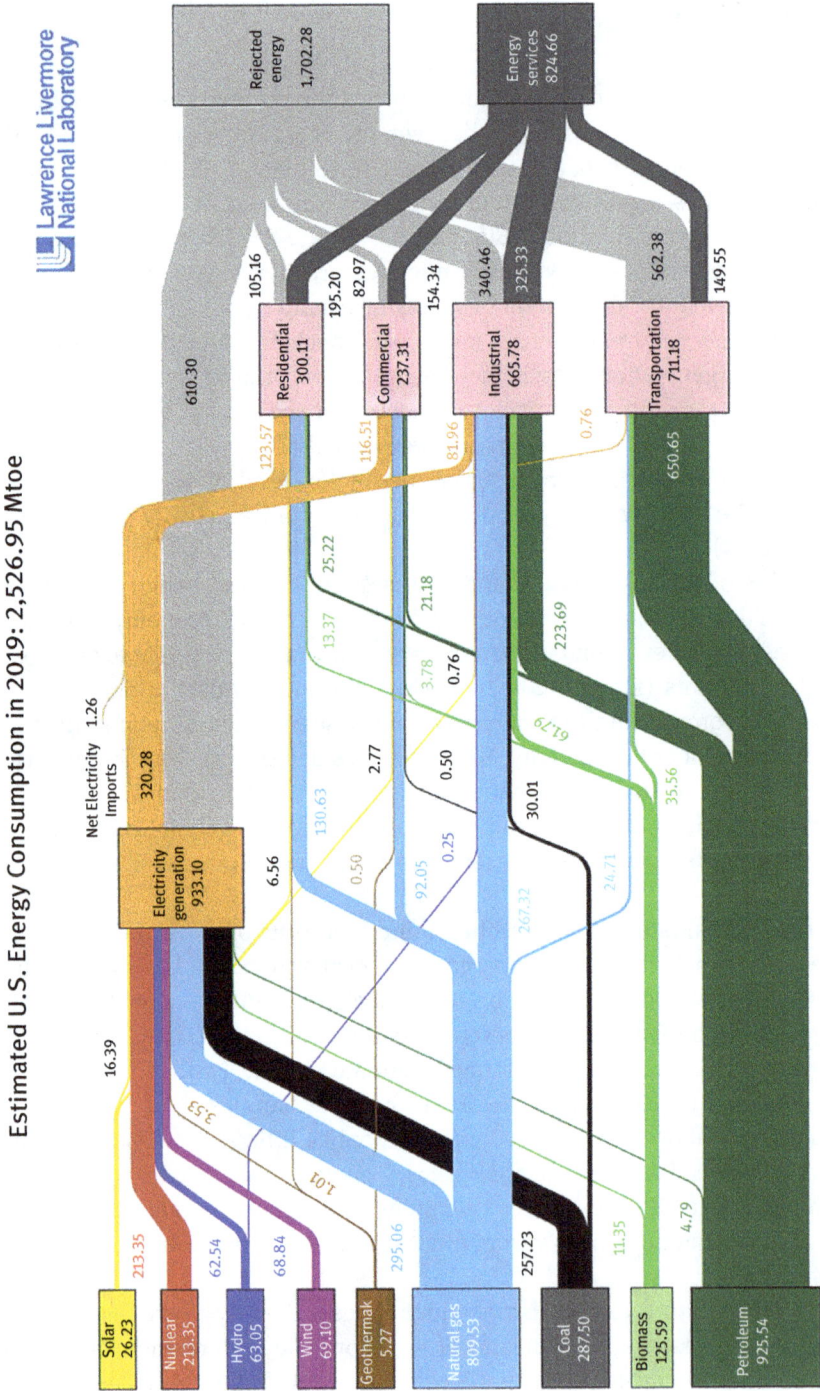

Fig. 1.1: Energy consumption of the US Economy in 2018 – resource extraction, conversion, use and waste (amended from Lawrence Livermore National Laboratory, 2020).

A positive achievement is the 11% share of renewables in the primary energy mix. This certainly helps the US economy in lowering the environmental impact. However, the footprint reductions are not automatic. The use of renewables is still associated with specific environmental footprints (Klemeš, 2015), including greenhouse gases, water and nitrogen footprints, which need further research for allowing sufficient quantification. Analysing the outputs of the economy, nearly 2/3 of the extracted primary energy was lost, and only 1/3 has reached the final users as services.

It can be seen that although the goal of increasing the supply of energy from renewable sources is an essential component in ensuring sustainable development, the structure of the outputs indicates an alarmingly low efficiency of the economy in utilising the overall pool of primary energy supply, making the overall task of supplying renewable energy and emission reduction more challenging, to the extent that its feasibility may be disputed. This is because the core of the energy problems – the low efficiency – has not been resolved yet.

Having established the main issue, a follow-up question arose. How would one go about reducing energy use and energy waste in the economy? The usual troubleshooting strategy, when confronted with multiple actors on the market, is to reveal the most significant potential for energy, waste and emission reduction. Energy is used by many economy sectors, including residential, commercial, industrial and transportation. Fuel consumption for electricity generation in the United States approximately rated as high as 39,341 PJ in 2017, according to the Annual Energy Outlook survey (AEO, 2018) by the Energy Information Administration of the Department of Energy. According to that source, the 2017 fuel consumption in the main sector groups was: industry 32,869 PJ, transportation 29,326 PJ, residential 20,849 PJ and commercial 19,011 PJ. The projections of future consumption are shown in Fig. 1.2a. It should be noted that among the sectors where power generation is not the purpose for fuel consumption, the industrial sector has been the most significant energy consuming one, closely followed by transportation. As such, the overall industry is responsible for large quantities of CO_2 emissions, which, despite the past and current efforts, are projected by the Energy Information Administration to level off and eventually slightly increase until 2050 (Fig. 1.2b).

Within an industrial site, the utility system is where most of the fuel and other resources enter the site, and most of the energy conversion takes place. The utility system supplies the heating, cooling and power/drive services to the site processes. This makes the utility system the key area of any industrial site. The usual metaphor of the engineers is that this is the "heart" of the site. For this reason, optimising the design and the operation of utility systems is of crucial importance, bearing the potential for significant energy savings directly, as well as pointing to areas of the site for further energy savings to be made in an economically viable way.

(a)

(b)

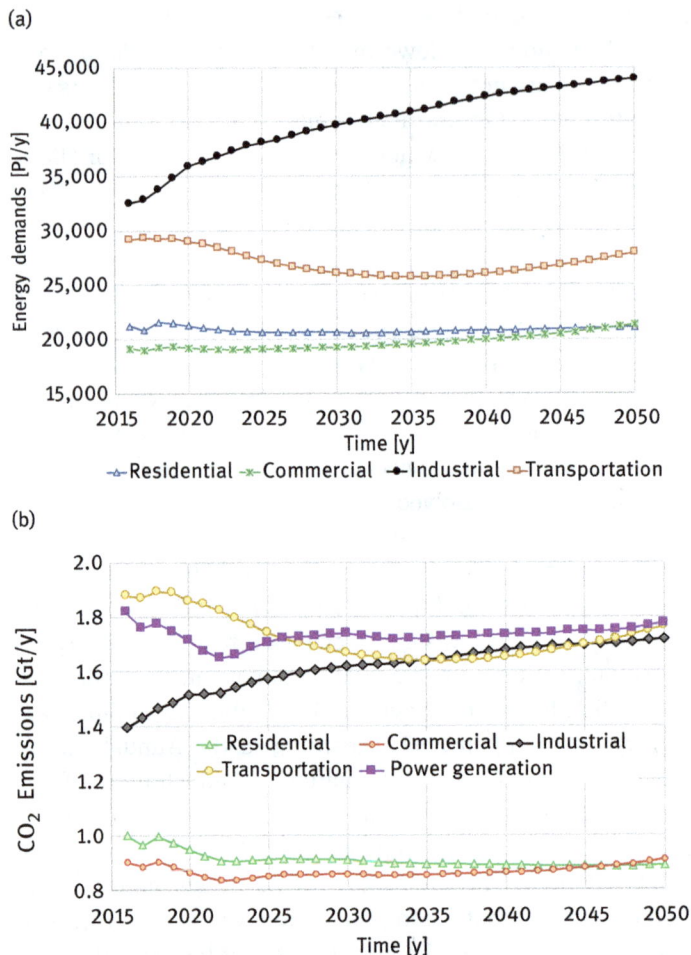

Fig. 1.2: Trend projections for the energy consumption (a) and GHG emissions (b) for the United States in the period until 2050 (adapted from AEO, 2018).

1.1 Introduction to site utility systems

Utility systems perform a variety of functions, depending on the nature of the sites they serve. The dominant function is to provide heating, cooling and power to the site processes. Another function, always implied and rarely recognised explicitly, is that it serves as a kind of a marketplace for exchanging process heat at various temperature levels, as is discussed further in the book.

1.1.1 Utility systems in the site context

A typical chemical or another industrial site usually consists of a number of production and auxiliary processes. These processes require the supply of various utilities in order to carry out their functions. Such utilities are:

- **Process heating**. Steam is usually the preferred heating medium, because of its high specific heat content in the form of latent heat and superheat. High-temperature processes, however, may require heating with hot oil or directly with flue gas in furnaces.
- **Process cooling**. This is typically performed by using cooling water, ambient air or refrigeration.
- **Power demands**. These arise from the need for driving process equipment such as pumps, compressors and mills and also lighting and electric heating for processes where high precision and responsiveness are necessary.
- **Water supply and disposal**. This is also a critical utility. It includes mainly the supply of freshwater, eventually treated to satisfy the water quality requirements, as well as the wastewater treatment, recycling and disposal.

In this book, the utility system is considered in a more specific context. Here, it has been defined as the system supplying the site processes with heating and cooling as well as satisfying their power demands. The focus of the book is on the heat and power cogeneration and emission reduction. The cooling is assumed to be provided by a separate site sub-system at a given cost.

A typical utility system configuration is shown in Fig. 1.3. It interacts with the site processes in several ways. It supplies steam for heating or generates some steam from high-temperature process cooling. The residual cooling demands are met by using direct utility cooling (usually with water). Also, the utility system satisfies the power demands of different processes.

The utility system generates some amounts of useful heat (in the form of steam) and power on-site, burning fuels. The on-site power generation may result in a power balance, deficit or excess with respect to the power demands of the site (coming from processes and internal needs). Any power deficit is covered by import from the central grid, resulting in individual cost for the imported power. Regarding the power excess, this is a particular case. If the regulations and the company contracts allow, the excess may be exported to the central grid, resulting in revenue for the site.

1.1.2 Operational and economic features

The economics of a site utility system is determined by the energy consumption in the form of fuel or power import, the potential export of power or steam from the site, as well as by the capital cost of the installed equipment. All these system

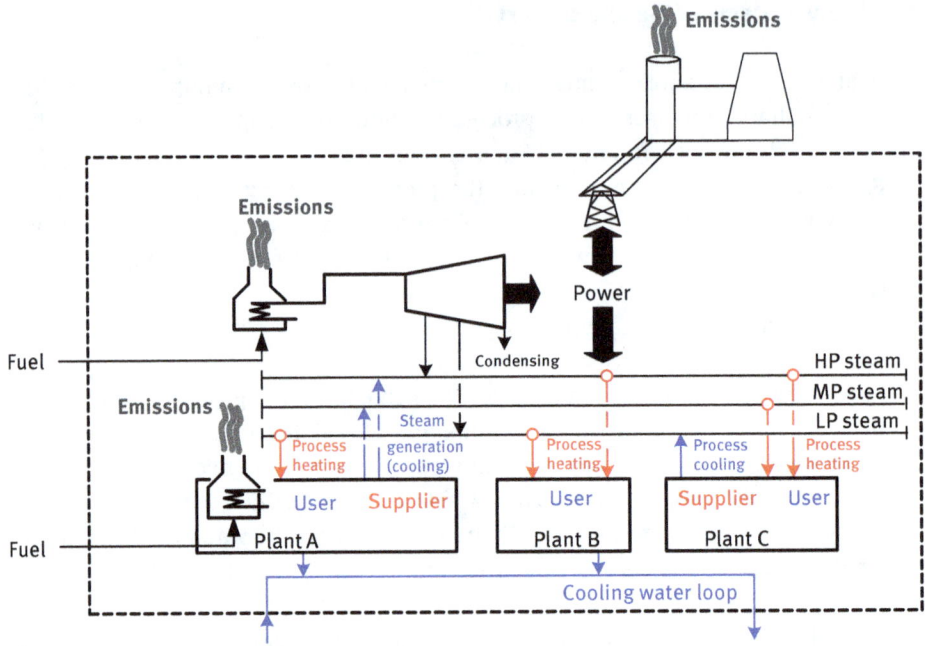

Fig. 1.3: Site utility system as a marketplace with the processes viewed as heat users or suppliers (adapted from Klemeš et al., 2018).

properties are interconnected and form several fundamental trade-offs. The main economic trade-offs are:
- Fuel consumption versus power import/export. This is an energy–energy trade-off.
- More capital for on-site power generation versus more power import. This is a capital–energy trade-off.
- Larger gas turbines versus larger steam turbines. This is a capital–capital trade-off.
- Larger gas turbines versus larger steam boilers. This is a capital–capital trade-off, as well as an energy–energy trade-off.

When a processing site is considered, the economic conditions, such as the market prices of fuels and electricity, usually vary with time. In parallel, there are also variations in the market demands for the site products, feedstock compositions, fuel compositions, the temperatures of the ambient and the different auxiliary input streams such as the make-up water. This variability of the site parameters introduces another dimension into the trade-off between the capital spent for power generation equipment and the power import. It becomes apparent that the efficiency and economic performance of the site vary together with the site parameters.

In the research literature, the most popular approach to deal with the variability of system specifications is the so-called multiperiod optimisation. The development of the concept within the mathematical programming and process systems engineering communities dates back to the 1970s – an example is the work by Williams (1978) treating general process optimisation models – a mining investment and a multiperiod blending problem. The method has been implemented in a software tool for energy planning – MARKAL (Fishbone and Abilock, 1981) – based on multiperiod Linear Programming (LP). Floudas and Grossmann (1986) applied the technique to the synthesis of Heat Exchanger Networks (HEN) under flexibility requirements.

In the field of utility systems, the first comprehensive models relied on LP, assuming a single steady state, as the model by Papoulias and Grossmann (1983). This was followed by the extension of the Heat Integration concept from process to the site level, giving rise to the Total Site Profiles (Dhole and Linnhoff, 1993), bringing the Pinch Analysis insights to the site-level heat recovery and reuse. That was followed by the development of the targeting method for cogeneration (Klemeš et al., 1997). The further developments for utility system optimisation include the development of a multiperiod Mixed-Integer LP (MILP) method for optimising the design and the operation of steam turbine networks by Mavromatis and Kokossis, published in two parts. The first part (Mavromatis and Kokossis, 1998a) developed the general steam turbine and the targeting models and provided a procedure for optimal selection of the steam pressure levels. The second part (Mavromatis and Kokossis, 1998b) dealt with the optimisation of the steam network design and operation. This line of method development has been further continued (Shang and Kokossis, 2004) with the development of a transhipment MILP model for the optimal selection of the steam pressure levels from a set of predefined candidates. This was further followed by a complete overhaul and formulation of the mathematical models of all essential utility system components and an integrated method for utility system operational optimisation (Varbanov et al., 2004) and for multiperiod synthesis (Varbanov et al., 2005). A variation of the operational optimisation includes the operational planning for a single step in changing process energy demands (Velasco-Garcia et al., 2011).

Longer term variations bound to process energy demands and device efficiency pose further modelling challenges of practical significance. This has been tackled by Zhao and You (2019), who presented a hybrid modelling framework by introducing the operating data into a mechanism to account for the changes of device efficiency and operating conditions and reflect them on the model used for optimising utility system operation. The illustrative case study, provided by the authors, clearly demonstrated that not accounting for the variability of the conditions is prone to lead to overly optimistic optimisation solutions not attainable in practice. The reported cost reduction resulting from the traditional deterministic optimisation was 14.5 % with respect to the initial operating point, while the robust optimisation accounting for the uncertainties lead to 10.1 % cost reduction.

1.2 Sustainability and industrial contribution to it

There have been numerous definitions of sustainability, many of them cite the "Brundtland Report" (World Commission on Environment and Development, 1987) and his address to the United Nations (Brundtland, 1992), while others simply take the main phrases without any quotations. Regardless of the specific phrasing, the main principle of sustainable development has been the same, as formulated in that report: to meet the needs of the present without compromising the ability of the future generations to meet their own needs. This has been the basis for all sustainability-related research, mainly aiming at the preservation of natural resource storages, ecosystems and their services, societal and economic components of the human life.

Within the industrial context, contributions to sustainable development are mainly viewed as pollution minimisation, applied at the smallest possible costs, in order to enable the industry to fulfil its function of delivering products and services to the society efficiently. A full discussion of these issues from the perspective of engineers can be found in Klemeš (2015). Applied to energy, the primary energy resources, currently used in industry, are fossil fuels. These generate vast amounts of emissions of oxides of carbon, sulphur and nitrogen. These gases are the major cause of the greenhouse effect and global climate change (global warming). This is a problem of prime importance to the sustainable development of the world civilisation.

The greenhouse gas (GHG) emissions from industrial utility systems can be reduced using different options, involving exploitation of renewable primary energy resources and energy efficiency measures. It is one of the goals of the current book to discuss the economics of these emission reduction options and to incorporate the issue into the procedure of utility system optimisation.

1.3 Historical development of site utility and cogeneration modelling

Historically, like all process synthesis problems, the early methods for utility system optimisation and synthesis have been based mainly on heuristics such as that by Nishio et al. (1980), and combinatorial algorithms such as the one by Petroulas and Reklaitis (1984). These were further developed by introducing more process insights with the thermodynamic approach by Chou and Shih (1987).

The next stage of development is the introduction of mathematical programming methods. Most earlier works employed MILP procedures based on linearised utility system models. These include the already mentioned (Mavromatis and Kokossis, 1998b) dealing with the steam network optimisation, the design of flexible utility systems optimising economics at pre-specified steam pressure levels (Shang and Kokossis, 2005) and the full synthesis procedure and model featuring unit models for all essential utility system components (Varbanov et al., 2005).

There was also an attempt to improve the modelling accuracy, which led to the Mixed-Integer Non-Linear Programming (also referred to as MINLP) procedure developed by Bruno et al. (1998).

The further developments include adding the Global Warming Potential (GWP) as a second criterion in the utility system optimisation (Papandreou and Shang, 2008), who showed a Pareto-front diagram clearly mapping the cost increases to potential reductions of the GWP. Further interesting works in this line include the utility system optimisation model embedding models of steam turbines with multiple extractions (Luo et al., 2011), trigeneration utility systems using biomass (Andiappan et al., 2015) and joint planning of operation and cleaning of utility systems (Zulkafli and Kopanos, 2017).

1.3.1 Targeting of utility systems

As mentioned earlier, Dhole and Linnhoff (1993) brought the Process Integration thinking to the domain of utility system evaluation and optimisation. To develop a good understanding of Total Site energy systems, they formulated a graphical method based on the concept of the **Site Heat Source and Heat Sink Profiles**. An example of constructing the Total Site Profiles is shown in Fig. 1.4. The method

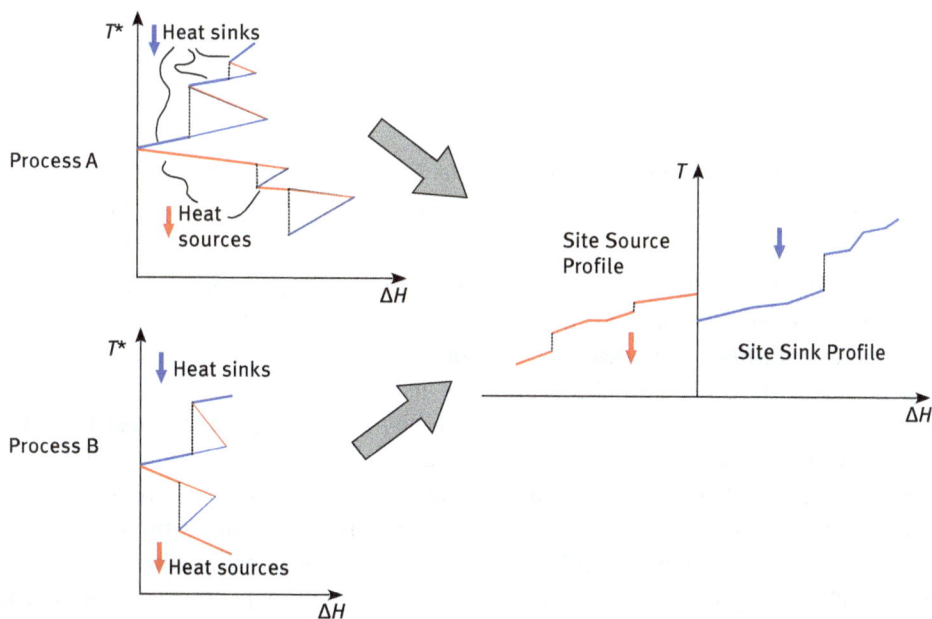

Fig. 1.4: Construction of Total Site Profiles from individual process Grand Composite Curves (adapted from Klemeš et al., 1997).

allows a target to be set for the Total Site heat recovery before any system design or
retrofit is started. The process heat recovery data are first converted to Grand
Composite Curves (GCCs), following the well-known Pinch Analysis procedure
(Linnhoff et al., 1994). The pockets on the GCCs that represent the scope for process
heat recovery are removed. The remaining GCC segments are combined to form a
Site Heat Source Profile and a **Site Heat Sink Profile**. Further, the Sink and
Source Profiles are superimposed on the steam header saturation temperatures.
From the resulting graph, the Composite Curves of steam generation and usage are
constructed, accounting for feasible heat transfer from the heat source profile to the
Steam Generation Composite Curve (SGCC) and from the **Steam Usage
Composite Curve (SUCC)** to the Heat Sink Profile (see Fig. 1.5).

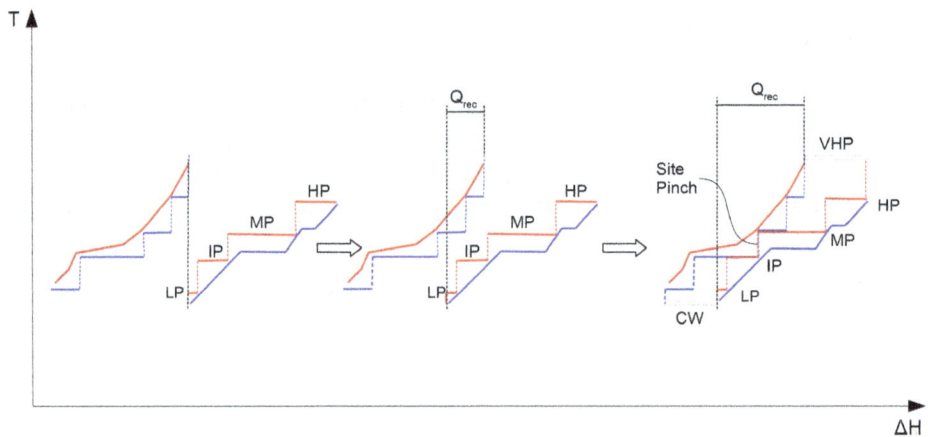

Fig. 1.5: Constructing Total Site Composites from Total Site Profiles (adapted from Klemeš et al.,
1997).

The Steam Composite Curves in Total Site Heat Integration (Klemeš et al., 2018) are
analogous to the Composite Curves for individual process Heat Integration
(Linnhoff et al., 1994). The heat recovery for the Total Site is indicated by the overlap between the SGCC and the SUCC. The maximum amount of heat recovery,
achieved through the steam system, is limited by the **Site Pinch** that restricts any
further overlap of these two curves. The steam mains, located at the Site Pinch, feature opposite net steam loads. This means that the steam main above the Site Pinch
is a net steam user and the one below the Site Pinch is a net steam supplier.

The work by Dhole and Linnhoff (1993) has been developed further by Klemeš
et al. (1997). Using the Total Site Profiles and the Steam Composite Curves allowed
them to set thermodynamic targets for cogeneration of heat and power and for fuel
consumption for the cases of maximum heat recovery and minimum cost of utilities.
The case of maximum heat recovery leads to minimum boiler (Very High Pressure)

steam requirement, achieved through steam recovery. However, for this case, the power generation by steam turbines is also minimal, which results in maximum power import. This scenario can be represented by the Total Site Profiles shifted to a position of maximum overlap as shown for the last stage in Fig. 1.5. It is important to emphasise that this target represents the thermodynamic limitation on the system efficiency and not a mandatory specification that must be achieved. The case of minimum utility cost is established by exploring the trade-off between steam recovery and power generation by steam turbines. If followed as a design guideline, it usually leads to a design that is different from the minimum fuel case.

Another direction in utility system targeting is to explore its cogeneration efficiency for a specific range of site power demands at constant heat demand. Such a procedure would assess the thermodynamics of the utility system with respect to a varying power-to-heat ratio. The latter is also referred to as **R-ratio**. The book by Kenney (1984) established the power-to-heat ratio as an analytical tool for utility system targeting. He studied various utility system configurations assuming a particular steam system for varying R-ratio values, building the so-called **fuel utilisation curves**. This idea has been further developed by Kimura and Zhu (2000), who distinguish between **"Ideal R-Curve"** and **"Actual R-Curve"**. The former is constructed for an ideal utility system and the latter – for an existing utility system. The R-Curve analysis and the power-to-heat ratio are of significant importance to practising engineers. This can be tracked by the continued use of this tool for renewable energy integration (Baniassadi et al., 2016) and desalination (Salimi and Amidpour, 2017) applications. Therefore, special attention is paid to this concept in Chapter 10.

1.3.2 Optimisation and synthesis of utility systems and standalone power stations

Much work has been published on the design and optimisation of utility systems. While some researchers advocate the use of heuristics and thermodynamic insights, others propose mathematical optimisation. The heuristic method of Nishio et al. (1980) for designing utility systems is based on the estimation of thermodynamic losses and the irreversibility in the system. The heuristics were used to minimise the exergy losses from the system. That was combined with a simple LP algorithm for steam turbine allocation to driver demands.

Petroulas and Reklaitis (1984) addressed the synthesis of plant utility systems by applying a decomposition strategy. They considered the utility system only as a source of mechanical energy for direct drives. The method decomposed the design task into two sub-problems – header selection and driver allocation. They introduced a discretisation technique to handle the choice of pressure levels for steam headers using a linear model. Their approach uses dynamic programming and LP techniques. This is, effectively, a combinatorial optimisation approach.

This method considers the utility system as a whole. It outlines that there exist complex interactions among its subsystems. It also recognises the importance of selecting the optimal number and pressure levels of the steam headers. The provided model defines a power loss-based function as the optimisation objective. The set of discrete candidate pressures for the steam headers is formed from the supply and target temperatures of the process heat sources and sinks. For a fixed number of headers, with regard to the minimisation of the objective function, it is shown that there is a very high probability that the optimal pressure levels belong to the set of candidate pressures derived from the starting and final temperatures of the heating and cooling demands of the site processes. The limitations of this approach are related mainly to the underlying assumptions. In the part of selecting the steam header conditions, it tries to satisfy the site power demand entirely by on-site generation, achieving a power balance on the site. Further, the steam turbine efficiency is considered constant with the load, which introduces a significant inaccuracy, and the model may often end up optimising a fictional system instead of the right one. Finally, the objective function clearly incorporates only energy cost information, and no attempt is made to account for the capital costs. The latter usually constitute a substantial part of the overall site expenses.

Chou and Shih (1987) proposed a methodology for utility system design, based on thermodynamic insights, with an objective to minimise fuel consumption and capital investment or maximise the efficiency of the utility system. For given heat and power loads, a preliminary configuration is assumed, and the aim is to satisfy the process heat demand first. In the case, if the power demand is not met, additional condensing turbines are considered to meet the remaining power demand. The method is based on the minimisation of the inefficiencies throughout the system, using heuristics. The approach develops a good understanding of the utility system thermodynamics. The emphasis on the thermodynamic component tends to minimise fuel consumption. Some limitations of this approach are that the model does not take into account the complex interactions between the utility subsystems and that the turbine/boiler efficiencies are assumed constant. Although the utility system configurations obtained would feature low fuel consumption and, most likely – low operating cost – the method does not account for the energy-capital trade-off. Finally, this method also aims to bring sites to power balance, which may not be the most economical mode.

Peterson and Mann (1985) considered many practical aspects of utility system design. The paper gives a comprehensive overview of the issues to account for during the design process. They give an analysis of the construction of system steam balances, based on a thorough enumeration of nominal and extreme operation modes and what-if scenarios. Regarding the high-level utility system configuration, the authors emphasise on the importance to establish the selection and sizes of the major equipment items such as boilers and gas turbines early during the design process in order to account for the long periods of manufacturing and delivery of such facilities.

Another approach, using MILP, has been suggested by Papoulias and Grossman (1983) for constant process heat and power demand for a site. Iyer and Grossmann (1997) developed a multiperiod optimisation model, further extended to combined synthesis and operation optimisation (Iyer and Grossmann, 1998) by introducing multiperiod operation into the utility system model. They used a superstructure-based approach that accounted for all possible interconnections within the utility system and subjected it to structure–parameter optimisation. These papers prove the strength of mixed-integer method in optimising the structure and the operation of the utility system simultaneously. In this sense, mathematical programming is superior to purely heuristic methodologies. However, the emphasis on the simultaneous structure–parameter optimisation introduces a strong bias towards the solution technique, and the thermodynamic insights have lower priority. This situation leaves open several very important issues. First, the number of steam headers is not optimised but postulated. Second, the selection of the pressure levels for each steam header is not based on a systematic approach or underlying physical insight. Third, there is no consideration of the degree of steam recovery and the fact that steam may be generated by using process cooling. Finally, the variation of the efficiency of steam turbines with their size and current load are not taken into account.

The mentioned work by Bruno et al. (1998) presented an MINLP design model for utility systems, based on the approach of Papoulias and Grossmann (1983), discussed earlier. They used the same type of superstructure. Improvements were made with regard to complex steam turbines and their efficiency as a function of the load and more precise estimation of the steam properties. The advantage of the method is the simplicity of the synthesis procedure. It recognises the inherent non-linearity of the problem and subjects the superstructure to an explicit MINLP optimisation. However, similar to the base method it extends, this one also assumes fixed predefined number and pressure levels for the steam headers. The power import is not considered as an option for the site. Also, the steam turbine efficiency is represented merely by a second-order polynomial, which does not have any process-related basis.

The model of power generation by steam turbines applied to targeting, optimisation and synthesis of steam turbine networks that have been developed by Mavromatis and Kokossis (1998a) has its regression coefficients obtained for backpressure turbines only. The model applied to the design of steam turbine networks (Mavromatis and Kokossis, 1998b) embeds the cogeneration targeting procedure from the first part. The steam turbine network synthesis is based on MILP optimisation. This method also provides a post-optimisation procedure for merging component (i.e. single-stage) turbines into complex machines, observing thermodynamic and scheduling equivalence of the configurations. The technique is based on Chou and Shih (1987).

Most methods for optimising utility system and power plants help to reduce the flue gas emissions from a site indirectly through eventually reduced fuel consumption. There are some direct techniques for reducing flue gas emissions through end-of-pipe treatment or decrease of the emission formation and release (e.g. low SO_x

and NO_x burners) that are installed close to the source of the emissions. If approached in isolation, the projects determined may have no return on the capital investment. In that regard, the Total Site Integration method was developed by Singh et al. (1998), which accounts for flue gas emission minimisation and regulations. Besides the direct abatement approach, other options are also considered, including gas turbine integration, choice of fuels, changes to the utility system and production processes with their HENs (Fig. 1.6). The simultaneous optimisation of these options provides synergetic effects in reducing capital investment and operation cost while satisfying emissions regulation at the same time. To reduce the complexity and the size of the optimisation problem, all the initially formulated options are evaluated, eliminating those options that are technically infeasible and economically non-viable. The remaining options are optimised to obtain an optimal set that meets the objective of reducing flue gas emissions with minimum cost.

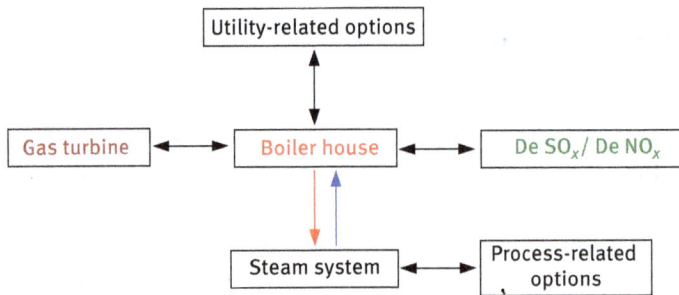

Fig. 1.6: The utility subsystems classification performed by Singh et al. (1998).

The significance of this work is in the demonstration that a systematic approach to the problem can effectively account for emission reduction and that it suggests a practical way to deal with existing regulations. However, it has to be treated with caution, as there are several drawbacks. The steam levels are not chosen systematically. Next, the screening of the design options is performed in isolation from each other, which introduces a contradiction with the overall philosophy to take the interactions into account. The latter may prevent the procedure from exploring the right solutions. It also relies on the simplistic turbine models from Mavromatis and Kokossis (1998a). Finally, this method produces solutions for a single period of operation and emphasises on minimisation of SO_x and NO_x emissions.

The steam turbine model first developed by Mavromatis and Kokossis (1998a) has been further extended by Shang (2000) to include also the modelling of condensing steam turbines. However, these steam turbine models have a number of shortcomings. First, they do not account for all factors influencing the steam turbine performance. For instance, they account for the pressure of the inlet steam via its saturation temperature, but not for the exhaust pressure. Second, the part-load modelling assumptions are too simplistic, introducing only the notion of the turbine isentropic efficiency and identifying the Willans Line intercept with the energy losses from the steam turbines, while there is no physical basis for this assumption.

In Shang (2000), hardware models for fired steam boilers and for gas turbines were also developed. The method also established a framework for steam levels selection, superstructure construction and subsequent optimisation of the resulting utility system configuration. The number of steam headers on the site is treated as a specification. Having this number, the temperature boundaries of the process demands for steam usage and generation are defined as level candidates for the various steam headers. After this initialisation stage, an MILP transhipment formulation is used to obtain the optimal steam header pressure levels. The formulation features one steam turbine per expansion zone. This approach to header selection extends the previously discussed technique, suggested by Petroulas and Reklaitis (1984). The difference from that earlier technique is that it specifies where to obtain the candidate pressures, based on the temperature boundaries of the site-level heat cascade.

There are several drawbacks to this approach. First, the number of steam headers is assumed fixed, and the system cost is not assessed for sensitivity with regard to the number of headers. Next, there is no systematic criterion for the decision to assign which temperature boundary to which header. The configuration of the steam and power generation equipment is optimised and decided after the selection of the steam levels rather than being optimised simultaneously with header selection. Also, in order to reduce the superstructure, the methodology applies thermodynamic-based equipment screening. This assessment is performed for each option alone, and the inefficient options are left out. This type of screening procedure tends to neglect very important trade-offs like that of the on-site power generation with the power import–export, the fuel–capital trade-off and fuel–power import–export trade-off.

A two-level hierarchical methodology for design and retrofit of standalone power plants has been proposed by Manninen and Zhu (2001). This methodology relies on a superstructure formulation and interaction between the design levels. One of the main modelling developments offered is the very convenient linear regression model of gas turbine performance and capital costs. It is based on ISO specifications, provided by gas turbine manufacturers. This model accounts only for the full-load performance (i.e. the design points). Another important feature of this work is the introduction of a hierarchy of two synthesis procedures (for high-level design and detailed design), which may interact with each other. This comes as a recognition of the fact that initially very little is known about the system being designed, so the first stage employs a simpler system model to establish the basic flowsheet and operating parameters and after that, the second procedure uses a more detailed model and optimises the operation and the lower level structural features.

The state of the art in utility system optimisation has developed steadily after 2004. The design of flexible utility systems has been revisited by Aguilar et al. –

first, by refining the turbine models (Aguilar et al., 2007a) and then updating the overall method (Aguilar et al., 2007b) emphasising the selection of direct-drive steam turbines. The GHG emissions from utility systems have been considered as a separate criterion in Papandreou and Shang (2008), as opposed to the GHG emission costing and inclusion into the cost function, applied in Varbanov et al. (2005). There have been further refinements, such as the inclusion of multiple-extraction steam turbine models in to the overall formulation (Luo et al., 2011), optimal scheduling of power plants (Mitra et al., 2013), combining the utility generation with cleaning schedules (Zulkafli and Kopanos, 2017).

1.4 The main issues and approaches followed

The problem of utility system optimisation is a subclass of general process optimisation. It has non-linear features arising from the energy balances, the estimation of the part-load performance of steam and gas turbines and the estimation of steam properties. Also, the problem features discrete structural and operating decisions such as equipment and fuel selection as well as operating schedules of the facilities and the power import/export mode.

The synthesis task, which is a special case of the system optimisation, involves an optimal choice among a number of different alternative flowsheets. In the majority of the known synthesis methods, the set of alternative flowsheets is represented by a common superstructure, which is subjected to structure–parameter optimisation.

The discussed modelling needs require a robust optimisation framework, based on component models that are just enough complex, to provide sufficient precision, but also simple enough, to provide seamless incorporation into the overall network optimisation models.

1.4.1 Modelling issues

The analysis of the existing approaches to the synthesis of thermal systems and industrial utility systems shows a significant development in the modelling and computational frameworks. The heuristic and combinatorial approaches are simple and easy to implement, but they often miss good design options and are inappropriate for large-scale systems and multiperiod operations.

The MILP approaches introduce a more comprehensive framework based on superstructure optimisation. However, they require a number of linearisation assumptions to be made. This, in many cases, significantly changes the problem being solved and produces inaccurate results. The MINLP methods feature better accuracy but lack robustness and reliability, although this is gradually changing with the development of novel global optimisation solvers, for example BARON ('BARON | The Optimization Firm' 2019). The problems are generally caused by the computational difficulties related to the non-convexity of the resulting optimisation model. A common drawback of the mathematical programming-based approaches to the network optimisation is that they produce a single solution. The practice of process synthesis, however, shows that there are rather sets of three to ten suboptimal solutions with the total cost varying within 5–10 %. This, compared with the low precision at this early stage of design, means that such solutions would be equivalent and they all have to be considered further at more detailed design stages. A possible remedy to this is the P-graph framework for process optimisation (Friedler et al., 2015), which has a multitude of proposed optimal solutions as a design feature. An example of such P-graph application is the work of Walmsley et al. (2018).

Another important issue is the availability of adequate models for the various equipment types, involved in the utility systems. The required balance of precision and complexity is usually achieved by several techniques – linearisation and discretisation being the most popular. These issues are covered in detail in Chapter 6.

Another key problem is the analysis of existing utility systems. Solving it can provide valuable information about the inherent limitations of an existing utility system and, consequently, the true economic value of potential utility savings as a result of energy efficiency improvements in the site production processes. The latter is usually done via retrofit of the process HENs or modifications to the processes. Retrofits require an adequate and sufficiently precise estimation of the potential financial and emission savings to perform a truly valuable process improvement campaign, in order to assess the payback and other economic parameters of the retrofit projects.

1.4.2 Reduction of greenhouse gas emissions

A general diagram of the material and energy flows around an industrial site is shown in Fig. 1.7. Currently, industrial sites burn fuels and import or export electrical power, satisfying their energy needs. The fuels burnt are primarily fossil fuels, although recently some smaller sites appear to turn their attention to burning biomass or other biofuels. There are several other principal sources of primary energy, such as solar irradiation, geothermal energy and wind energy. One other source can be nuclear power generation, which is usually done in standalone power plants only and is beyond the scope of this book.

Other fundamental ways to reduce the release of CO_2 and other GHG into the atmosphere are also shown in Fig. 1.7. These include two basic options. One is to capture and store the CO_2, which is termed sequestration. The other option is simply

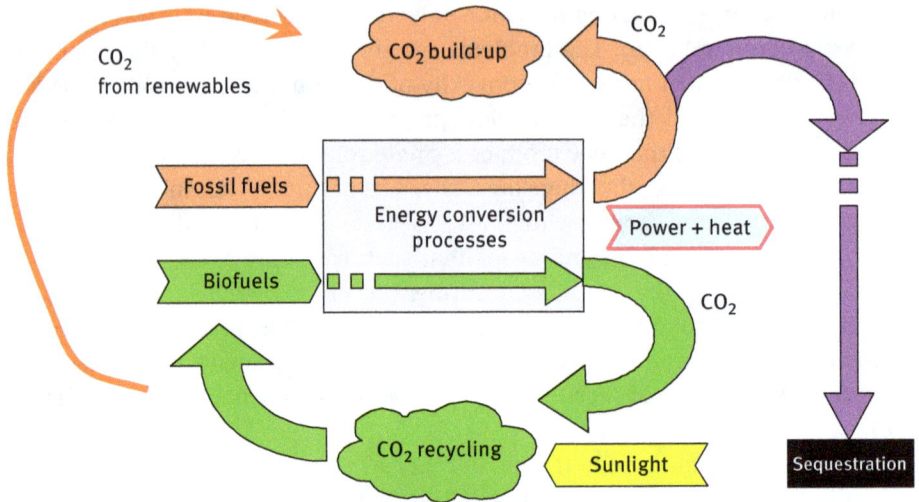

Fig. 1.7: Energy conversion, the carbon cycle and the hierarchy of measures (adapted from Varbanov et al., 2018).

to close the carbon cycle by using biofuels. The latter are usually produced from different types of biomass, generated by absorbing solar energy. The last option, capable of reducing the CO_2 emissions, which can be derived from Fig. 1.7, is the improvement of the energy efficiency of the site. This may include efficiency improvements to the utility system or to the production processes. Following the opening discussion of this chapter, centred around Fig. 1.1, this latter option has to play a significant or even dominant role in the efforts for reducing the environmental impacts from utility systems.

The options for CO_2 emission reduction, their applicability to industrial energy conversion to power and heat and the implications of adopting them as climate stabilisation measures can be summarised as follows.

Nuclear power. One option is to increase the capacity for nuclear power generation. This method of energy conversion applies solely to standalone power plants.

Wind, solar and geothermal energy. Another interesting possibility is the direct utilisation of solar and wind energy resources. This is usually done with wind turbines and photovoltaic panels in the case of power generation (Wang et al., 2017) and solar thermal collectors (Martínez-Rodríguez et al., 2019) for heating. These technologies produce relatively little GHG emissions during exploitation, but the production and installation of the facilities are bound to significant emissions. For the potential use of geothermal energy, the locations where this resource is available should be sufficiently close to the industrial sites. In the context of industrial utility systems, there are several major obstacles to the use of wind and solar energy:

- The flows of the primary resources – wind and sunlight – are unstable and quite volatile. This makes the utilisation of these energy sources extremely difficult. For instance, certain buffering capacities could be introduced in order to smooth the flow variations. There are ongoing research and development efforts in this area, with certain good achievements. A good example is the Tesla Megapack (O'Kane, 2019), claimed to be capable of up to 1 GWh storage arrays.
- The order of magnitude of the possible power generation using these sources is sometimes significantly smaller than the usual site demands for mechanical and electrical power, due to the distributed nature of such resources (Lam et al., 2011). For larger generation capacities, extended energy collection areas would need to be employed.

Capture and sequestration of carbon dioxide. There are processes that allow capturing the carbon dioxide carried with the flue gases and further storage into different sinks. In most studies, CO_2 capture process uses amine solvents (Oh et al., 2018). An absorption/regeneration process based on amines as sorbents can be used to capture CO_2 from flue gases. It is also possible to use indirect capture methods such as air separation before combustion (Tang and You, 2018), or fuel reforming combined with CO_2 separation and further burning of the remaining hydrogen. As a result, the flue gas after combustion would consist of almost pure CO_2 and water. The current estimation of the social cost of carbon includes the costs associated with climatic phenomena causing devastation and the recovery of infrastructure and private property assets from those disasters. According to the US-based "Environmental Defence Fund" ('The True Cost of Carbon Pollution', 2019), this is estimated to 40 USD/t CO_2.

Regarding geological sequestration, there is intensive ongoing research. For example, one can refer to the review concerning cement industry (Jang et al., 2016) and to a more general review on capture and storage technologies (Leung et al., 2014). A critical issue – energy waste during capture – is still extensively debated and evaluated (Yoro et al., 2019). The cost estimates depend very much on the considered context. The order of magnitude estimates is available on the Internet. Some estimates are based on what is known about natural gas production since geological sequestration is intended to fill in empty gas wells and similar geological formations under the sea bed or underground. A study exploring just the CO_2 capture was performed by Klemeš et al. (2005). The study showed a considerable variability of the estimates, in which coal-fired power plant reached 65 USD/t for the monoethanol amine as solvent, referring to the USD in 2004. A more recent study by Rubin et al. (2015) arrived at figures 46–99 USD/t for the combination of capture and storage processes. Sequestration into minerals has also been investigated theoretically. Lackner et al. (1997) performed a preliminary estimation of such processes. According to these sources, the likely cost of sequestration was of the order of 30–60 USD/t CO_2 avoided. It should be noted that the referenced cost estimates are reported to be optimistic lower bounds.

A third option is the biological capture of CO_2 and further possible utilisation or storage of the grown biomass. While specific cost estimates are difficult to find, there are already studies for combined CO_2 capture, wastewater cleaning and electricity production using microbial carbon capture cells (Wang et al., 2010), having the biomass and diesel as side products.

Biofuels. Another possibility is to use biofuels. These close the carbon cycle, and indirectly utilise the sunlight. The latter is, practically, abundant and free. The potential problem here lies in the capacity for biofuel production (Lam et al., 2011), since the growing and harvesting large amounts of biomass would ultimately compete with the food-producing agriculture processes. A possible challenge may be to combine food and biofuel production, sharing their primary resource base and costs.

Improving energy efficiency. After considering all the apparent emission reduction options, easily shown in Fig. 1.7, one should turn to the real picture of the problem in Fig. 1.1, which clearly shows the proportions and the possible changes when the various options are exploited. This leads to the option having the largest scope for improvement. Naturally, having all other conditions equal, an increase in the energy efficiency of some of the site processes or the utility system itself would cause a reduction in the fuel and power import requirements which is, in turn, an indirect way to reduce the CO_2 emissions. This is a part of an energy–capital cost trade-off.

1.5 Assumed prior knowledge

This book discusses the network-level modelling of utility systems. This includes balancing, simulation, optimisation and strategic analysis. For understanding these activities and to be able to apply the knowledge further, fundamental understanding of several disciplines is required:

– Mathematical modelling of engineering and business processes. This involves the ability to describe a process formally and to derive the relevant mathematical relations from the description, formulating simulation and optimisation models. Good sources to learn are (Williams, 2013) and (Klemeš et al., 2010).
– Process systems modelling and chemical process modelling. This means to perform mass and enthalpy balances, trace and combine process units via their connections. Good sources to refer to are (Towler and Sinnott, 2013) for the chemical engineering issues and (Smith, 2016).

- Thermodynamics – main thermodynamic properties and relationships can be learned from Smith et al. (2005) for a general text and from Pavelka et al. (2018) for more advanced concepts.
- Process Integration fundamentals – HENs, heat recovery, Heat Integration, Water Integration, process network synthesis and/or optimisation. Good sources for this are the textbook on *Sustainable Process Integration and Intensification* (Klemeš et al., 2018) and the thorough reference *Process Integration Handbook* (Klemeš, 2013).

1.6 Book scope, structure and intended readers

This book discusses and presents the components of utility systems for heat and power generation. Chapter 2 proceeds to an overview of the fundamentals of utility system modelling and thermodynamics. Chapter 3 presents an overview of the utility systems – their structure and the method developments. Chapter 4 provides a summary of the primary energy resources from the viewpoint of their relevance to industrial utility systems. Chapters 5, 6 and 7 deal with the models of steam generators, steam turbines and gas turbines, derived and tailored specifically for use in a large-scale mathematical model for utility network optimisation.

Chapter 8 presents the analysis and model building techniques for utility system modelling as a whole, analysing the degrees of freedom and the fundamental trade-offs. Chapter 9 uses the knowledge presented in the previous chapters to present a step-by-step example of how a utility system is balanced using MS-Excel and with a process simulator. Chapter 10 describes the basics of macro-type analyses of utility systems: Top-Level Analysis for revealing the most valuable steam savings and R-curve analysis for screening equipment selection for combined heat and power applications within utility systems. Chapter 11 summarises the Total Site Heat Integration developments, and Chapter 12 discusses the software tools available for utility system modelling and optimisation. Chapter 13 draws the conclusions and points to sources for further study and information.

The intended readers for this book include undergraduate and graduate students in mechanical and chemical engineering, as well as experts from related fields. The latter can be practising engineers and lecturers, preparing lecture courses on energy supply and conversion, as well as process system optimisation.

Nomenclature

Symbol	Measurement unit	Description
CW	–	Cooling water
GCC	–	Grand Composite Curve
GDP	EUR, USD	Gross Domestic Product
GHG	–	Greenhouse gas
GWP	–	Global Warming Potential
H (ΔH)	MW	Enthalpy flow or change of enthalpy flow
HDI	(1)	Human Development Index
HP (steam)	–	High Pressure
IP (steam)	–	Intermediate Pressure
ISO	–	International Standards Organisation
LP	–	Linear Programming
LP (steam)	–	Low Pressure
MILP	–	Mixed-Integer Linear Programming
MP (steam)	–	Medium Pressure
NO_x	–	Nitrogen oxides
SGCC	–	Steam Generation Composite Curve
SO_x	–	Sulphur oxides
SUCC	–	Steam Usage Composite Curve
T	°C	Temperature
T^*	°C	Shifted temperature

References

AEO. 2018. Annual Energy Outlook 2018 with Projections to 2050. February 2018. https://www.eia.gov/outlooks/aeo/pdf/AEO2018.pdf, accessed 13/08/2020.

Aguilar, O., Perry, S.J., Kim, J.-K., and Smith, R. 2007a. Design and Optimization of Flexible Utility Systems Subject to Variable Conditions: Part 1: Modelling Framework. Chemical Engineering Research and Design 85(8): 1136–48. https://doi.org/10.1205/cherd06062.

Aguilar, O., Perry, S.J., Kim, J.-K., and Smith, R. 2007b. Design and Optimization of Flexible Utility Systems Subject to Variable Conditions: Part 2: Methodology and Applications. Chemical Engineering Research and Design 85(8): 1149–68. https://doi.org/10.1205/cherd06063.

Andiappan, V., Tan, R.R., Aviso, K.B., and Ng, D.K.S. 2015. Synthesis and Optimisation of Biomass-Based Tri-Generation Systems with Reliability Aspects. Energy 89 (September): 803–18. https://doi.org/10.1016/j.energy.2015.05.138.

Baniassadi, A., Momen, M., Shirinbakhsh, M., and Amidpour, M. 2016. Application of R-Curve Analysis in Evaluating the Effect of Integrating Renewable Energies in Cogeneration Systems. Applied Thermal Engineering 93 (January): 297–307. https://doi.org/10.1016/j.applthermaleng.2015.09.101.

'BARON | The Optimization Firm'. 2019. 2019. https://www.minlp.com/baron, accessed 13/08/2020.

Brown, J.H., Burnside, W.R., Davidson, A.D., DeLong, J.P., Dunn, W.C., Hamilton, M.J., Mercado-Silva, N., et al. 2011. Energetic Limits to Economic Growth. BioScience 61(1): 19–26. https://doi.org/10.1525/bio.2011.61.1.7.

Brundtland, G.H. 1992. Our Common Future, Chapter 2: Towards Sustainable Development – A/42/427 Annex, Chapter 2 – UN Documents: Gathering a Body of Global Agreements. 1992. http://www.un-documents.net/ocf-02.htm, accessed 13/08/2020.

Bruno, J.C., Fernandez, F., Castells, F., and Grossmann, I.E. 1998. A Rigorous MINLP Model for the Optimal Synthesis and Operation of Utility Plants. Chemical Engineering Research and Design, Techno-Economic Analysis, 76(3): 246–58. https://doi.org/10.1205/026387698524901.

Chou, C.C. and Shih, Y.S. 1987. A Thermodynamic Approach to the Design and Synthesis of Plant Utility Systems. Industrial & Engineering Chemistry Research 26(6): 1100–1108. https://doi.org/10.1021/ie00066a009.

Dhole, V.R. and Linnhoff, B. 1993. Total Site Targets for Fuel, Co-Generation, Emissions, and Cooling. Computers & Chemical Engineering 17 (January): S101–9. https://doi.org/10.1016/0098-1354(93)80214-8.

Fishbone, L.G. and Abilock, H. 1981. Markal, a Linear-Programming Model for Energy Systems Analysis: Technical Description of the Bnl Version. International Journal of Energy Research 5 (4): 353–75. https://doi.org/10.1002/er.4440050406.

Floudas, C.A. and Grossmann, I.E. 1986. Synthesis of Flexible Heat Exchanger Networks for Multiperiod Operation. Computers & Chemical Engineering 10(2): 153–68. https://doi.org/10.1016/0098-1354(86)85027-X.

Friedler, F., Varbanov, P.S., and Fan, L.T. 2015. Applications of P-Graphs for Enhancing Sustainability of Industrial Plants. In: *Proceedings of the 4th International Congress on Sustainability Science and Engineering, ICOSSE 2015*, 406–8.

Holden, E., Linnerud, K., and Banister, D. 2014. Sustainable Development: Our Common Future Revisited. Global Environmental Change 26 (May): 130–39. https://doi.org/10.1016/j.gloenvcha.2014.04.006.

Iyer, R.R. and Grossmann, I.E. 1997. Optimal Multiperiod Operational Planning for Utility Systems. Computers & Chemical Engineering 21(8): 787–800. https://doi.org/10.1016/S0098-1354(96)00317-1.

Iyer, R.R. and Grossmann, I.E. 1998. Synthesis and Operational Planning of Utility Systems for Multiperiod Operation. Computers & Chemical Engineering 22(7): 979–93. https://doi.org/10.1016/S0098-1354(97)00270-6.

Jang, J.G., Kim, G.M., Kim, H.J., and Lee, H.K. 2016. Review on Recent Advances in CO2 Utilization and Sequestration Technologies in Cement-Based Materials. Construction and Building Materials 127 (November): 762–73. https://doi.org/10.1016/j.conbuildmat.2016.10.017.

Kenney, W.F. 1984. Energy Conservation in the Process Industries. Energy Science and Engineering. Orlando, United States: Academic Press.

Kimura, H. and Zhu, X.X. 2000. R – Curve Concept and Its Application for Industrial Energy Management. Industrial & Engineering Chemistry Research 39(7): 2315–35. https://doi.org/10.1021/ie9905916.

Klemeš, J., Bulatov, I., and Cockerill, T. 2005. Techno-Economic Modelling and Cost Functions of CO2 Capture Processes. Computer Aided Chemical Engineering, Vol. 20: 295–300, https://doi.org/10.1016/S1570-7946(05)80171-3.

Klemeš, J., Varbanov, P.S., Alwi, S.R.W., and Manan, Z.A. 2018. Sustainable Process Integration and Intensification: Saving Energy, Water and Resources, 2nd Ed, Walter de Gruyter GmbH, Berlin, Germany, ISBN:978-3-11-053535-8.

Klemeš, J.J., ed. 2013. Handbook of Process Integration (PI): Minimisation of Energy and Water Use, Waste and Emissions Woodhead Publishing Limited. Cambridge, UK ISBN: 978-0-85709-593-0.

Klemeš, J.J. 2015. Assessing and Measuring Environmental Impact and Sustainability. Oxford; Waltham, MA: Butterworth-Heinemann/Elsevier ISBN: 978-0-12-799968-5.

Klemeš, J.J., Dhole, V.R., Raissi, K., Perry, S.J., and Puigjaner, L. 1997. Targeting and Design Methodology for Reduction of Fuel, Power and CO_2 on Total Sites. Applied Thermal Engineering 17(8–10): 993–1003. https://doi.org/10.1016/S1359-4311(96)00087-7.

Klemeš, J.J., Friedler, F., Bulatov, I., and Varbanov, P.S. 2010. Sustainability in the Process Industry: Integration and Optimization. New York, USA: McGraw-Hill Education, ISBN: 978-0-07-160554-0.

Lackner, K.S., Butt, D.P., and Wendt, C.H. 1997. Progress on Binding CO_2 in Mineral Substrates. Energy Conversion and Management, 38 (January): S259–64. https://doi.org/10.1016/S0196-8904(96)00279-8.

Lam, H.L., Varbanov, P.S., and Klemeš, J.J. 2011. Regional Renewable Energy and Resource Planning. Applied Energy 88(2): 545–50. https://doi.org/10.1016/j.apenergy.2010.05.019.

Lawrence Livermore National Laboratory. 2020. Energy Flow Charts: Charting the Complex Relationships among Energy, Water, and Carbon. 21 April 2020. https://flowcharts.llnl.gov/content/assets/images/energy/us/Energy_US_2018.png, accessed 13/08/2020.

Leung, D.Y.C., Caramanna, G., and Maroto-Valer, M.M. 2014. An Overview of Current Status of Carbon Dioxide Capture and Storage Technologies. Renewable and Sustainable Energy Reviews 39 (November): 426–43. https://doi.org/10.1016/j.rser.2014.07.093.

Linnhoff, B., Townsend, D.W., Boland, D., Thomas, B.E.A., Guy, A.R., and Marsland, R.H. 1994. A User Guide on Process Integration for the Efficient Use of Energy. Revised First Edition. Rugby, UK: Institution of Chemical Engineers.

Luo, X., Zhang, B., Chen, Y., and Mo., S. 2011. Modeling and Optimization of a Utility System Containing Multiple Extractions Steam Turbines. Energy 36(5): 3501–12. https://doi.org/10.1016/j.energy.2011.03.056.

Manninen, J. and Zhu, X.X. 2001. Level-by-Level Flowsheet Synthesis Methodology for Thermal System Design. AIChE Journal 47(1): 142–59. https://doi.org/10.1002/aic.690470114.

Martínez-Rodríguez, G., Fuentes-Silva, A.L., Lizárraga-Morazán, J.R., and Picón-Núñez., M. 2019. Incorporating the Concept of Flexible Operation in the Design of Solar Collector Fields for Industrial Applications. Energies 12(3): 570. https://doi.org/10.3390/en12030570.

Mavromatis, S.P. and Kokossis, A.C. 1998a. Conceptual Optimisation of Utility Networks for Operational Variations – I. Targets and Level Optimisation. Chemical Engineering Science 53 (8): 1585–1608. https://doi.org/10.1016/S0009-2509(97)00431-4.

Mavromatis, S.P. and Kokossis, A.C. 1998b. Conceptual Optimisation of Utility Networks for Operational Variations – II. Network Development and Optimisation. Chemical Engineering Science 53(8): 1609–30. https://doi.org/10.1016/S0009-2509(97)00432-6.

Mitra, S., Sun, L., and Grossmann, I.E. 2013. Optimal Scheduling of Industrial Combined Heat and Power Plants under Time-Sensitive Electricity Prices. Energy 54 (June): 194–211. https://doi.org/10.1016/j.energy.2013.02.030.

Nishio, M., Itoh, J., Shiroko, K., and Umeda, T. 1980. A Thermodynamic Approach to Steam-Power System Design. Industrial & Engineering Chemistry Process Design and Development 19(2): 306–12. https://doi.org/10.1021/i260074a019.

O'Kane, S. 2019. Tesla's Megapack Battery Is Big Enough to Help Grids Handle Peak Demand. The Verge. 29 July 2019. https://www.theverge.com/2019/7/29/20746170/tesla-megapack-battery-pge-storage-announced, accessed 13/08/2020.

Oh, S.-Y., Yun, S., and Kim, J.-K. 2018. Process Integration and Design for Maximizing Energy Efficiency of a Coal-Fired Power Plant Integrated with Amine-Based CO_2 Capture Process. Applied Energy 216 (April): 311–22. https://doi.org/10.1016/j.apenergy.2018.02.100.

Papandreou, V. and Shang, Z. 2008. A Multi-Criteria Optimisation Approach for the Design of Sustainable Utility Systems. Computers & Chemical Engineering 32(7): 1589–1602. https://doi.org/10.1016/j.compchemeng.2007.08.006.

Papoulias, S.A. and Grossmann, I.E. 1983. A Structural Optimization Approach in Process Synthesis – I: Utility Systems. Computers & Chemical Engineering 7(6): 695–706. https://doi.org/10.1016/0098-1354(83)85022-4.

Pavelka, M., Klika, V., and Grmela, M. 2018. Multiscale Thermo-Dynamics: Introduction to GENERIC. Berlin, Germany; Boston USA: De Gruyter.

Peterson, J.F. and Mann, W.L. 1985. Steam System Design: How It Evolves. Chemical Engineering 92 (21): 62–74.

Petroulas, T. and Reklaitis, G.V. 1984. Computer-Aided Synthesis and Design of Plant Utility Systems. AIChE Journal 30(1): 69–78. https://doi.org/10.1002/aic.690300112.

Rubin, E.S., Davison, J.E., and Herzog, H.J. 2015. The Cost of CO_2 Capture and Storage. International Journal of Greenhouse Gas Control 40 (September): 378–400. https://doi.org/10.1016/j.ijggc.2015.05.018.

Salimi, M. and Amidpour, M. 2017. Investigating the Integration of Desalination Units into Cogeneration Systems Utilizing R-Curve Tool. Desalination 419 (October): 49–59. https://doi.org/10.1016/j.desal.2017.06.008.

Shang, Z. 2000. Analysis and Optimisation of Total Site Utility Systems. PhD Thesis, Manchester, UK: University of Manchester Institute of Science and Technology.

Shang, Z. and Kokossis, A. 2004. A Transhipment Model for the Optimisation of Steam Levels of Total Site Utility System for Multiperiod Operation. Computers & Chemical Engineering 28(9): 1673–88. https://doi.org/10.1016/j.compchemeng.2004.01.010.

Shang, Z. and Kokossis, A. 2005. A Systematic Approach to the Synthesis and Design of Flexible Site Utility Systems. Chemical Engineering Science 60(16): 4431–51. https://doi.org/10.1016/j.ces.2005.03.015.

Singh, H., Smith, R., and Zhu, X.X. 1998. Economic Achievement of Environmental Regulation in Chemical Process Industries. Computers & Chemical Engineering 22 (March): S917–20. https://doi.org/10.1016/S0098-1354(98)00180-X.

Šlaus, I. and Jacobs, G. 2011. Human Capital and Sustainability. Sustainability 3(1): 97–154. https://doi.org/10.3390/su3010097.

Smith, J.M., Van Ness, H.C., and Abbott, M.M. 2005. Introduction to Chemical Engineering Thermodynamics. 7th ed. McGraw-Hill Chemical Engineering Series. Boston USA: McGraw-Hill.

Smith, R. 2016. Chemical Process Design and Integration. 2nd ed. Chichester, West Sussex, United Kingdom: Wiley.

Tang, Y. and You, F. 2018. Life Cycle Environmental and Economic Analysis of Pulverized Coal Oxy-Fuel Combustion Combining with Calcium Looping Process or Chemical Looping Air Separation. Journal of Cleaner Production 181 (April): 271–92. https://doi.org/10.1016/j.jclepro.2018.01.265.

Towler, G.P. and Sinnott, R.K. 2013. Chemical Engineering Design: Principles, Practice, and Economics of Plant and Process Design. 2nd ed. Boston, MA, USA: Butterworth-Heinemann.

'The True Cost of Carbon Pollution'. 2019. Environmental Defense Fund. 2019. https://www.edf.org/true-cost-carbon-pollution, accessed 28/10/2020.

Varbanov, P., Perry, S., Klemeš, J., and Smith, R. 2005. Synthesis of Industrial Utility Systems: Cost-Effective de-Carbonisation. Applied Thermal Engineering 25(7): 985–1001. https://doi.org/10.1016/j.applthermaleng.2004.06.023.

Varbanov, P.S., Doyle, S., and Smith, R. 2004. Modelling and Optimization of Utility Systems. Chemical Engineering Research and Design 82(5): 561–78. https://doi.org/10.1205/026387604323142603.

Varbanov, P.S., Jia, X.X., Kukulka, D.J., Liu, X., and Klemeš, J.J. 2018. Emission Minimisation by Improving Heat Transfer, Energy Conversion, CO_2 integration and Effective Training. Applied Thermal Engineering 131: 531–39. https://doi.org/10.1016/j.applthermaleng.2017.12.001.

Velasco-Garcia, P., Varbanov, P.S., Arellano-Garcia, H., and Wozny, G. 2011. Utility Systems Operation: Optimisation-Based Decision Making. Applied Thermal Engineering 31(16): 3196–3205. https://doi.org/10.1016/j.applthermaleng.2011.05.046.

Walmsley, T.G., Jia, X., Philipp, M., Nemet, A., Liew, P.Y., Klemes, J.J., and Varbanov, P.S. 2018. Total Site Utility System Structural Design Using P-Graph. Chemical Engineering Transactions 63: 31–36. https://doi.org/10.3303/CET1863006.

Wang, X., El-Farra, N.H., and Palazoglu, A. 2017. Optimal Scheduling of Demand Responsive Industrial Production with Hybrid Renewable Energy Systems. Renewable Energy 100 (January): 53–64. https://doi.org/10.1016/j.renene.2016.05.051.

Wang, X., Feng, Y., Liu, J., Lee, H., Li, C., Li, N., and Ren, N. 2010. Sequestration of CO_2 Discharged from Anode by Algal Cathode in Microbial Carbon Capture Cells (MCCs). Biosensors and Bioelectronics 25(12): 2639–43. https://doi.org/10.1016/j.bios.2010.04.036.

Williams, H.P. 1978. The Reformulation of Two Mixed Integer Programming Problems. Mathematical Programming 14(1): 325–31. https://doi.org/10.1007/BF01588974.

Williams, H.P. 2013. Model Building in Mathematical Programming. 5th ed. Hoboken, N.J, USA: Wiley.

World Commission on Environment and Development, 1987. Our Common Future. Oxford Paperbacks. Oxford, UK; New York, USA: Oxford University Press.

Yoro, K.O., Sekoai, P.T., Isafiade, A.J., and Daramola, M.O. 2019. A Review on Heat and Mass Integration Techniques for Energy and Material Minimization during CO_2 Capture. International Journal of Energy and Environmental Engineering 10(3): 367–87. https://doi.org/10.1007/s40095-019-0304-1.

Zhao, L. and You, F. 2019. A Data-Driven Approach for Industrial Utility Systems Optimization under Uncertainty. Energy 182 (September): 559–69. https://doi.org/10.1016/j.energy.2019.06.086.

Zulkafli, N.I. and Kopanos, G.M. 2017. Integrated Condition-Based Planning of Production and Utility Systems under Uncertainty. Journal of Cleaner Production 167 (November): 776–805. https://doi.org/10.1016/j.jclepro.2017.08.152.

Part 1: **Basics**

2 Fundamentals for utility system modelling

This chapter reviews the knowledge necessary for understanding the follow-up content on utility system modelling and optimisation. It starts with an overview of the heat carriers in industry, followed by a concise presentation of the main thermodynamic concepts – the laws of thermodynamics, the concepts of state, energy, enthalpy, reversibility, entropy, followed by the fundamental thermodynamic diagrams used in the analysis of energy systems. The discussion is then provided of the ideal thermal engine – the Carnot engine, how to model the Carnot engine and a real heat engine.

The main unit of temperature used in this chapter is [K], using the absolute Kelvin scale. The reason for this is that the key definitions and other equations in thermodynamics assume that the temperature is measured in [K]. Any further engineering calculations need to use the conversion between the Kelvin scale and the particular temperature scale used by the engineers.

2.1 Heat carriers used in industry

Large amounts of thermal energy are used to perform heating in various branches of the industry. Examples of this can be found in crude oil preheating before distillation, preheating of feed flows to chemical reactors or even heat addition to carry out endothermic chemical reactions. Similarly, some processes like condensation, exothermic chemical reactions or product finalisation require heat to be extracted from them, resulting in process cooling.

2.1.1 Utility heating options

For utility process heating, there are many optional heat carriers. The most popular are:
- **Steam**. This is by far the most popular heating agent due to its very high specific heating value in the form of latent heat. On industrial sites, it is usually available at several different pressure levels. However, this way of utility heating has its inherent limitation arising from the limits on the pressures allowed in industrial equipment. The largest practical pressure for steam generation, and eventually process heating, is around 100 to 120 bar(a). This limits the highest possible steam saturation temperature to about 311–325 °C. However, the most common steam distribution pressure is around 40 bara, which results in an effective heating temperature of 250 °C. Referring to Engineering ToolBox (2003), the convective heat transfer coefficient for liquid can reach 3,000 W/(m$^2 \times$°C)

https://doi.org/10.1515/9783110630091-002

for free convection and much higher when forced convection is applied. For steam condensation or generation, the convective heat transfer coefficient is of the order of 10,000–12,000 W/(m^2 ×°C). Detailed instructions for obtaining the values of heat transfer coefficients can be found in handbooks, such as *Perry's Chemical Engineers Handbook* (Green, 2018).

– **Hot oil circuit.** This is another popular choice of heat carrier. Hot oil can be used for heating at temperatures up to 400 °C.

– **Direct fired heating.** This method is used when the desired heating temperatures are very high, up to 600–700 °C or higher.

It should be noted, however, that the heat transfer coefficients become lower in the order of enumeration of the heating media above. In the case of the direct-fired heating, this fact is partially compensated by the potentially higher temperature differences between the flue gases and the process streams being heated, as well as by the radiation mode of the heat transfer.

2.1.2 Utility cooling options

For utility process cooling, the most common options are (Merritt, 2016):

– **Water cooling.** This is the most popular choice of utility cooling wherever there is sufficient water supply at a reasonably low cost. It is implemented in a cooling water circuit, involving cooling heat exchangers, cooling tower and water make-up system (to compensate the water losses in the cooling tower, the exchangers and the piping). The make-up water is subjected to mechanical and chemical treatment to remove suspended and dissolved solids. Water cooling enables coolers with relatively smaller heat exchange area because of the high partial heat transfer coefficient on the waterside. However, it incurs water treatment and pumping costs, which in some cases may become quite significant. This method usually provides cooling down to temperatures around 30 °C, with possibility sometimes down to 10–15 °C, depending mostly on the ambient temperature.

– **Air cooling.** This is another popular way of cooling, which utilises the ambient air as a cooling agent. With regard to the heat exchange area, it is much less efficient. However, the cooling airflow incurs sometimes only relatively small costs for running air blowers (fans) and in most cases can be considered virtually free. In terms of cooling temperature achievable, it depends on the ambient temperature directly. For the climates in continental Europe, typical temperatures achievable with air cooling would be 5–10 °C in the winter and 25–45 °C in the summer. For a milder climate such that in the UK, the achievable cooling temperatures would be in the range of 10–15 in the winter and 25–35 in the summer.

– **Refrigeration**. This cooling option allows reaching very low temperatures – starting from chilling at 5 to 10 °C and dropping down to −20 to −30 °C and even lower. The process is usually based on compressor-driven or absorption-chilling refrigeration systems. Both processes incur significant capital and operating costs. The compressor-driven systems require relatively more moderate capital costs mainly for the compressors, heat exchangers and piping. They feature very significant power consumption, usually associated with high cost. Regarding the absorption chilling systems, they are capable of using low-temperature heat as energy input, which is either cheap or free, but feature relatively high capital cost for the absorbers and other auxiliary equipment. Therefore, they are economically attractive in cases where large amounts of low-temperature waste heat are available.

2.2 Fundamentals of thermodynamics and heat engines

This section consists of several main parts: an overview of the thermodynamics fundamentals, followed by the derivation of essential definition of efficiencies and description of thermodynamic visualisation tools, such as the temperature-entropy diagram and several well-known thermodynamic cycles, used for evaluation and benchmarking of real energy conversion processes.

2.2.1 Technical thermodynamics

Thermodynamics is the part of physics that deals with the state of matter, the change of these states (to distinguish them from biological, chemical, nuclear, and so on, we call them thermodynamic changes) and the consequences of those changes. The pillars of thermodynamics are its four laws (Atkins, 2007). A possible formulation is given below. It has to be noted that the wording of the formulations can vary, while they remain equivalent.

2.2.1.1 Zeroth law of thermodynamics
If thermodynamic system A is in thermal equilibrium with thermodynamic system B, and simultaneously thermodynamic system B is in thermal equilibrium with the thermodynamic system C, then thermodynamic system C is in thermal equilibrium with the thermodynamic system A (Balmer, 2011).

2.2.1.2 The first law of thermodynamics (1847 Herman von Helmholtz)

The total sum of the energy content of the thermodynamic systems that interact with each other remains constant. This is also known as the *law of conservation of energy* – providing the simplest direct statement of the law as *energy is conserved* (Balmer, 2011).

2.2.1.3 The second law of thermodynamics (1850 Rudolf Clausius)

The Second Law stipulates that the entropy of an isolated thermodynamic system increases during each spontaneous change. The Clausius formulation of the law reads as Potter and Somerton (2014):

It is impossible to construct a device which operates in a cycle and whose sole effect is the transfer of heat from a cooler body to a hotter body.

2.2.1.4 The third law of thermodynamics (1906 Walther Nernst)

The third law defines a reference point for thermodynamic calculations and for applying the other laws. The third law can be formulated as

The entropy of a pure substance is zero at absolute zero temperature.

The original formulation by Nernst only stated that the entropy of a pure substance is constant at the absolute zero, while the current definition is a result of a convention added by Planck, defining the value of that constant entropy as zero (Balmer, 2011).

These laws are based on observing the behaviour of large volumes of molecules or atoms and form the fundamentals of thermodynamics, often referred to as "classical thermodynamics". Statistical thermodynamics (Balmer, 2011) deals with the behaviour of average individual particles within a large cluster and links the laws of thermodynamics with quantum mechanics. Another part of thermodynamics deals with biological systems and is called bio-thermodynamics (Balmer, 2011).

2.2.2 Temperature

The zeroth law of thermodynamics allows to introduce a physical quantity that compares the state of molecules and atoms (here referred to as "particles") in individual systems, named *temperature*.

One is not able to sense or perceive the temperature of a single molecule, but only the vast clusters of particles that can be seen in the form of objects – solids or liquids, or that of the surrounding gases. It is possible to distinguish between a warmer and colder state of some observed system in a specific temperature range.

There are a number of temperature scales. One of the most popular is the Celsius scale (Balmer, 2011, p. 9) which is defined as (Fig. 2.1):

- The 0 °C point is given by the melting/freezing point of pure water at a pressure of 1 atm = 101.325 kPa.
- The boiling point of pure water at a pressure of 1 atm = 101.325 kPa defines the temperature of 100 °C.
- The parts of the scale between these two points form degrees, each degree taking 1 % of the distance between the freezing and the boiling points of water at 101.325 kPa.

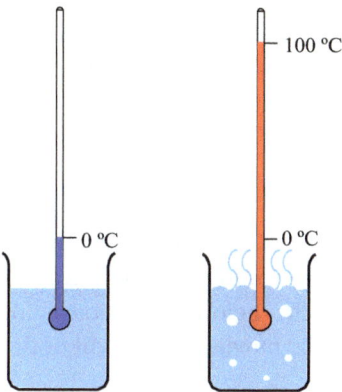

Fig. 2.1: Calibrating the measurement according to the Celsius temperature scale: (a) a vessel with freezing water and (b) a vessel with boiling water.

In thermodynamics, the Kelvin temperature scale is the standard and much more used than the Celsius scale. This is mainly due to the fact that many derivations are related to the Kelvin scale. The magnitudes of the units in both scales have been selected equal, that is, 1 K expresses the same temperature change as 1 °C. However, 0 K is referred to as "the absolute zero", which corresponds to −273.15 °C. Beyond these two scales, there are others that are widely used in some countries related to the British engineering traditions – for example the Fahrenheit scale is widespread in the United States. The conversions from the Celsius to the Kelvin and Fahrenheit scales can be expressed as follows:

$$T[\text{K}] = t[°\text{C}] + 273.15; \quad TT[\text{F}] = \frac{9}{5}t[°\text{C}] + 32 \tag{2.1}$$

where t is the temperature in the Celsius scale, T – Kelvin, TT – Fahrenheit.

At the molecular level, the temperature of a substance manifests itself by movement or vibration. The higher the temperature of the substance, the faster move the atoms and molecules that constitute it. In the case of gases and liquids, the movements include displacement (change of position), vibrations (oscillations) and rotation of molecules. For solids, molecules and atoms do not move within the occupied volume, but they retain their place in the crystal lattice, and higher temperature indicates greater vibration amplitude (the lattice may collapse at some temperature). Since temperature determines the state of the particles in the examined volume, it is referred to as a state variable.

2.2.3 Other state variables

The temperature alone is not sufficient to determine the state of the system being examined. Other state variables are, for example, volume, pressure and composition (number of moles of each substance present in the system). The pressure is defined as the force exerted by an object per unit of the area. This force is created by the fact that the particles of the fluid impinging on the surface examined and change their momentum.

2.2.4 Work

The movement of the gas particles in an enclosed volume, as in the system shown in Fig. 2.2, can create such a force that it can hold, for example, a weight-loaded piston or even lift it if the momentum of the gas particles below the piston is increased. In the latter case, the gas would perform work by moving the weight of the piston against the force of gravity.

   The momentum of the particles can be increased by increasing the gas temperature below the piston. For example, this can be achieved by heating the container of warm water at the bottom of the piston container (Fig. 2.2). By placing a vessel with warm water under the air vessel, the air begins to heat, as its temperature is lower than that of the hot water. By heating the air, the movement of the particles is intensified and the force acting on the piston is increased via the increased pressure, thereby moving the piston upwards. As the momentum of the gas particles increases, the piston with the weight starts to move from position 0 to position x, until the temperatures in both vessels become equal, and the values of the air state variables change from state 0 to state x.

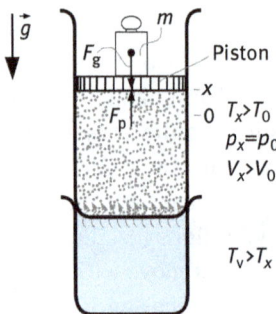

Fig. 2.2: Gas heating and expansion to perform work. Legend: $\vec{g}$ [m/s²], acceleration due to gravity as a vector; m [kg], mass; F_g [N], the force exerted by the weight on the piston; V_x [m³], the volume of gas below the piston; p_x [Pa], gas pressure below piston; F_p [N], the force of the gas under the piston by its pressure $F_p = F_g$; T_V [K], water temperature.

The work that the gas performs can be determined from the magnitude of the force applied to the piston and the length of the piston displacement. The displacement length is directly proportional to the change of the air volume under the piston, the force of the pressure and the piston area, from which the closed-gas operation can be derived as follows, reflecting the work done by the gas by moving the piston from state 0 to state x:

$$dA = p \times dV \tag{2.2}$$

$$A = \int_0^x p \times dV \tag{2.3}$$

2.2.5 Internal energy

Referring back to the example in Fig. 2.2, to change the condition of the air under the piston, it can be heated. During the heating of the air, the water cools down. For the current thought experiment, the effect of heat loss from the vessels to the ambient air is neglected. If a substance is capable of performing work by changing its state, it is said that it contains energy. Energy content in measured in "Joules" (SI symbol J), and energy, in general, is usually denoted by the symbol E. Since, in this case, this form of energy is enclosed within the discussed system or material, in the form of molecular motion, it is called "internal energy". Traditionally, in thermodynamics, it is denoted by U. The internal energy of a substance can change if its state changes, it is also a state variable. The overall internal energy of a system is an extensive quantity and is additive – for identical conditions, a larger system has more internal energy than a smaller one.

The internal energy of a substance can change, only by exchanging (passing or gaining) some of its energy with its surroundings. In the considered example (Fig. 2.2), the air in the upper vessel gains internal (thermal) energy via heat transfer from the hot water in the vessel but, at the same time, part of this internal energy obtained is immediately transformed into the work A done for lifting the piston. This work increases the potential energy of the piston.

Obviously, the internal energy of a substance is related to its temperature. If its temperature was equal to the absolute zero, it would be unable to do the work or transfer heat to the surrounding objects. After the work is performed by the gas on the piston, the gas volume is increased, which also results in a reduction of the density, while the pressure remains the same and the piston moves to a new (higher) position.

Since the gas density reduces, the momentum of the gas particles (temperature) is increased, to maintain the pressure. This means that the internal energy reduction of the water vessel is larger than the work done by the gas because some of this energy is needed to increase the internal energy of the gas for maintaining the pressure. From this point, it is possible to write the basic energy balance of the whole process, expressing the first law of thermodynamics for closed systems. The change in the internal energy of the water vessel is equal to the heat transfer to the gas vessel, and the heat gained by the gas vessel is distributed between the work done on the piston and the increase of the internal energy of the gas:

$$\Delta U_V = \Delta Q \tag{2.4}$$

$$\Delta Q = \Delta U_{VZ} + A \tag{2.5}$$

In differential terms, the relationship is expressed as

$$dQ = dU_{VZ} + dA = dU + p \times dV \tag{2.6}$$

In Eqs. (2.4) to (2.6), U [J] denotes the internal energy of the substance; ΔU [J] is a change of the internal energy; Q [J] is the heat transfer rate. The expression in Eq. (2.6) is a mathematical formulation of the *first law of thermodynamics* and it applies to any closed volume and any substance. The indices v and vz denote water and air. Heat does not refer to energy or work directly. Instead, the term heat denotes a flow – transfer of thermal energy from one body to another – which takes place over a temperature difference. Heat flows from the body with a higher temperature to that having a lower temperature.

Equation (2.6) suggests how the internal thermal energy of a system can be determined, using its change and the change of the heat transfer rate. If the examined substance is enclosed in a vessel with the constant volume during heating or cooling, the work performed will be zero because the volume differential will be zero

$$dA = p \times dV, \ dV = 0 \tag{2.7}$$

In such case, all heat added to or removed from the considered volume would only cause a change of the internal energy of the substance in the vessel. This, in turn, would change the temperature of the system:

$$V = \text{const} \Rightarrow \Delta Q = \Delta U \tag{2.8}$$

A process that takes place at constant volume is termed *isochoric*. The heat required to heat or cool a unit amount of a given substance is called *specific heat capacity at constant volume*. Mathematically, this can be written as

$$\Delta U|_V = m \times C_V \times \Delta T; \ dU|_V = m \times C_V \times dT \tag{2.9}$$

where C_V [J/kg/K] is the specific heat capacity of the substance at constant volume. This equation is also called *calorimetric*. C_V is a variable that is a function of the material composition and the state of the system. However, with small changes in temperature and pressure (without change of the state), this variable is often very close to a constant – the exact trend depends on the type and condition of the substance. For each chemical component, the values of C_V are measured or varied states. The measured values are tabulated in reference books containing data on thermodynamic properties of substances. Such a reference book is Poling et al.'s (2001).

It is important to note that the internal energy of a system is always measured as the difference from a chosen reference state because internal energy is referred to as "unobservable" (Balmer, 2011). Since the magnitude of U is related to the system temperature, the choice of zero point of the temperature scale also determines the zero point for internal energy. If the Celsius temperature is used, then $U = 0$ when $T = 0$ °C. However, in most thermodynamic treatments, the Kelvin scale is used, which sets $U = 0$ when $T = 0$ K.

2.2.6 Specific work and other quantities

The work per 1 kg of a substance in the system is called specific work, measured in kJ/kg or J/kg. In a similar way are defined specific heat content, specific internal energy, and *specific volume* (the volume occupied by 1 kg of working fluid [m³/kg]). The advantage of using specific quantities is that it is not necessary to know the overall weight of the working substance in the calculation. The equation of the first law of thermodynamics per 1 kg of working substance has the following form:

$$dq = du + p \times dv \qquad (2.10)$$

where

$$v = \frac{V}{m} = \frac{1}{\rho}; \; q = \frac{Q}{m}; \; u = \frac{U}{m}; \; a = \frac{A}{m} = p \times dv \qquad (2.11)$$

In Eqs. (2.10) and (2.11), the notation includes v [m³/kg] – specific volume of the working substance; m [kg] – mass of the working substance in the volume under test; ρ [kg/m³] – density; q [J/kg] – specific heat transfer; u [J/kg] – specific internal energy of the working substance; a [J/kg] – specific work. Typically, the symbols for the specific quantities are written in lowercase to distinguish them from absolute units of volume, energy and so on.

2.2.7 Equation of state

The state of the working fluid in a machine can be determined by measuring the state quantities. For many gases, the state can be evaluated using the equation of state for an ideal gas:

$$p \times v_M = R \times T \qquad (2.12)$$

In the equations of state, such as Eq. (2.12), the variables are usually given per unit of substance [mol]. In this case, the specific volume v_M is on a molar basis [m³/mol], the pressure p is in [N/m²], $R = 8.3145$ J/ mol K is the universal gas constant and T [K] is the temperature. Equation (2.12) is valid for ideal gas – that is for gas where the volume of the molecules is insignificant compared with the volume that the gas occupies as a whole, and the intermolecular interaction forces are negligible. This was formulated in 1820 by combining the works of Boyle, Charles and Gay-Lussac (Balmer, 2011). If the gas is very much compressed or undercooled and the volume of molecules is not negligible, then it is necessary to use other equations of state, that are applicable to real gases. Such equations have been developed following the ideal gas equation and include the Clausius and Van der Waals equations (Balmer, 2011) in the late nineteenth century.

With the fast development of the industry during the twentieth century, the need for more sophisticated process calculations and more comprehensive thermodynamic modelling has led to the development of a number of advanced equations of state, incorporated in the thermodynamic packages of process simulators. These equations include the Peng-Robinson equation of state (Peng and Robinson, 1976) with its variations discussed in a recent review (Lopez-Echeverry et al., 2017), the Soave–Redlich–Kwong equation (Soave, 1972), the UNIQUAC (Faramarzi et al., 2009) and UNIFAC (Fredenslund et al., 1975) models.

2.2.8 Energy transformations and conservation of energy

A system may also contain other types of energy beside thermal. These can be pressure energy, kinetic energy, nuclear energy and chemical energy. The total energy content is the sum of the individual energy components. Pressure, potential and kinetic energy components are referred to as *"mechanical energy"* types because they can be transformed into work directly.

Any type of energy can be transformed directly or indirectly into heat completely. For example, if a solid body is dropped into a vessel with liquid, its potential energy in the Earth's gravitational field is transformed into kinetic energy by motion. The kinetic energy of the body, moving in the liquid, is converted to heat by the friction

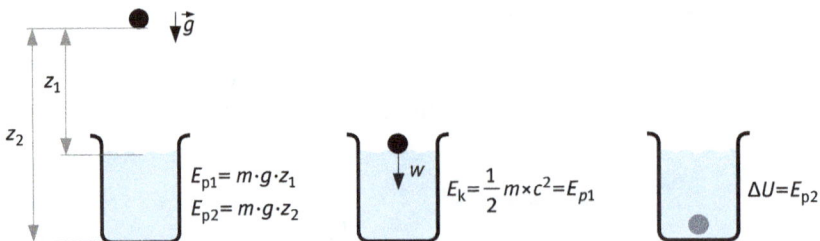

$$E_{p1} = m \cdot g \cdot z_1$$
$$E_{p2} = m \cdot g \cdot z_2$$

$$E_k = \frac{1}{2} m \times c^2 = E_{p1}$$

$$\Delta U = E_{p2}$$

Fig. 2.3: Transformation of potential energy into kinetic and internal heat energies.

of the solid boundary against the surrounding liquid until the body completely stops to move. The process is illustrated in Fig. 2.3.

The sphere with mass m [kg] hangs above the water surface in the vessel at height z_1 [m] and has potential energy E_{p1} [J] relative to the surface of the liquid in the vessel. After being released, by free fall, the sphere develops velocity c [m/s] equivalent to kinetic energy E_k [J], when reaching the liquid surface. When hitting the water surface, it slows down transferring part of its kinetic energy to the liquid. The rest of its kinetic and potential energies of the ball is further transferred to the internal energy of the liquid during the fall of the ball within the liquid environment, via friction of the ball surface and the surrounding liquid. In this way, the kinetic energy E_k of the ball is transformed into the internal energy of the liquid, which increases by ΔU [J]. The resistance of the media to the falling ball outside the liquid water and level change in the vessel when immersing the ball were neglected in the description.

Many transformations between the various energy types are possible. From those options, the transformation of other energy types into heat flow and internal energy is always spontaneous. For example, electrical energy can be directly transformed into the internal thermal energy of the wire in an electric heater. The wire of the electric heater, in turn, heats the air in the room with the help of a fan for forced air circulation. In this process, both parts of the used electricity (the part powering the wire and that powering the fan) end up transformed to the internal energy of the surrounding air.

Other examples can be given too. The energy released in the process of nuclear fission (De Sanctis et al., 2016), when the atomic nuclei break down, is transformed into internal thermal energy of the nuclear fuel and other content in the nuclear reactor, ending up, after the power generation, into the cooling water. The energy of a photon absorbed by the surface of the material results in heating up the material.

Transformation to other types of energy besides thermal is also possible. Different kinds of energy can be transformed not only into internal heat energy but also into motion, mechanical work, electricity and chemical energy. In all these transformations, energy is not created or destroyed but is conserved. This statement is widely known as another formulation of the *law of conservation of energy*. It states that if an isolated system contains various types of energy, their total sum remains the same regardless of the energy transformations that take place within the system. This formulation is equivalent to the first law of thermodynamics (mentioned earlier).

2.2.9 Enthalpy

The sum of the internal energy and the pressure energy of the fluids is denoted with the term *enthalpy*. The overall enthalpy of a system [Eq. (2.13)] is a quantity measured in [kJ] and other derivative units. The specific enthalpy (Eq. (2.14)) is defined per unit mass – often [kJ/kg] – or per unit amount of substance – often in [kJ/kmol]:

$$H = U + p \times V \tag{2.13}$$

$$h = u + p \times v \tag{2.14}$$

Especially in the case of gases, the supply or removal of heat to/from the working fluid is often associated with a change in both internal and pressure energy components when the gas changes its volume and pressure. For example, gas in a volume under investigation is heated under constant pressure, and it expands to displace the surrounding gas or increase the gas pressure. In such a case, the heat is absorbed heating up the gas and increasing its pressure energy. A similar situation occurs with hot air around the heater in a house. The air heats up and increases its volume under constant pressure, which is accompanied by an equivalent reduction of its density. Therefore, enthalpy is sometimes said to represent the thermal content of a substance.

Having the definitions for enthalpy and specific enthalpy allows rendering another expression of the first law of thermodynamics for a closed system, as follows:

$$u = h - p \times v \Rightarrow du = dh - p \times dv - v \times dp \tag{2.15}$$

$$dq = du + p \times dv \Rightarrow dq = dh - p \times dv - v \times dp + p \times dv \tag{2.16}$$

$$dq = dh - v \times dp \tag{2.17}$$

The expression of the first law, given in Eq. (2.17), is suitable for modelling the energy balance of open systems, which are part of a more extensive closed system. For example, it can be used to derive the amount of heat regenerated within a Stirling engine cycle (Organ, 2007), where the working volume of the Stirling engine is enclosed between two pistons and the regenerator is inside that assembly, that is, it is open to other engine volumes.

The ideal gas enthalpy can be determined relatively easily because the specific pressure energy of an ideal gas is, according to the equation of state, a function of the temperature only, and the specific internal energy also depends only on the temperature. The expression is as follows:

$$dh = C_P \times dT \tag{2.18}$$

where the quantity C_P [kJ/kg/K] is the specific heat capacity of an ideal gas at constant pressure. Further, the relationship between the specific heat capacities of an ideal gas at constant pressure and constant volume can be written as

$$C_P = C_V + r \ [kJ/kg/K] \tag{2.19}$$

where $r = R/M_{gas}$ is the individual gas constant obtained from the universal gas constant R by division over the molar weight pf, the gas M_{gas}. The derivations and a detailed discussion of these equations can be found in Balmer (2011).

The C_P is a variable for real gases because it is a function of the type and state of the gas. However, under small changes in the temperature and the pressure, without a change in state, this variable can be approximated as a constant. The precision of the approximation varies depending on the type and the condition of the substance. For each substance, the values of C_P are measured and tabulated. The readers can find data and equations for these in reference books such as Poling et al. (2001).

Changing the specific enthalpy of the fluid at constant pressure is approximately equal to the change in the specific internal energy of the heat ($dh \approx du$) because the volume change due to temperature change is very small $dv \approx 0$. In the case of other thermodynamic transitions, it is necessary to add the change in the pressure energy.

Similar to internal energy, enthalpy is also measured as a difference from a reference state, which depends on the used temperature scale – °C or K. In the case of temperature on the Celsius scale, the zero enthalpy point is at 0 °C.

2.2.10 Thermodynamic cycles, an example for a volume machine

The previously mentioned example of a weight-piston shows how the heat affects the state of the substance and how work can be done in a cylinder-piston system. The principle of heat transformation into work is illustrated using the example of a thermal machine. Permanently operating thermal piston machines are characterised by cyclic heating and cooling of the working substance, bringing the system, through a sequence of intermediate states, into the initial state. Such transition sequences are referred to as thermal (thermodynamic) cycles.

The principle of heat transformation to work applies to any substance, taking into account that such a transformation can only be partial and is limited by the temperatures of the available heat source and that of the ambient. However, it is realistic to use only the gaseous state of the substance, or a combination of the gaseous and liquid states, since the ability of the working fluid to change its volume is essential for the functioning of the machine. All substances have the effects of volume change with temperature (Kittel, 2005), including solid, liquid and gaseous substances. Solid materials can either expand (Bird and Ross, 2015) or contract – for example Fe_3Pt (Liu et al., 2017) – with raising their temperature. However, for a practical thermodynamic cycle implementation, such volume variations are insignificant, as can be established by any data book – for example *Perry's Chemical Engineers Handbook* (Green, 2018). The thermal expansion effects of solids and liquids are used in temperature measurement (Bird and Ross, 2015) – for example mercury thermometers and bimetallic thermometers.

The change of the state variables is manifested by cooling, heating, compressing and expanding the working substance, which is enclosed within the machine. Between the parts of the machine and the working substance, there is a force effect, which varies depending on the change of the state variables of the working substance, where the interactions produce useful work. An example of such a machine, implementing a thermal cycle, is a piston-cylinder (Fig. 2.4), containing an amount of gas as a working substance. The necessary change of the gas state variables is produced by its controlled heating and cooling, resulting in timed expansion and contraction of the working gas and forces acting on the piston.

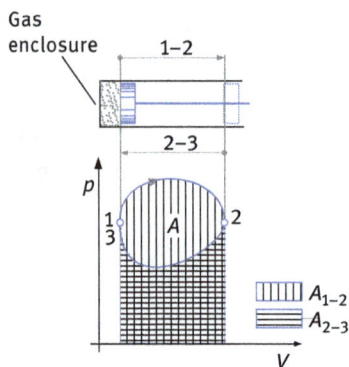

Fig. 2.4: Temperature–pressure diagram expressing the thermal cycle of a piston machine.
Transition 1–2: the gas is heated by supplying heat to the cylinder;
Transition 2–3: the gas is cooled by heat dissipation.

The state transitions denoted in Fig. 2.4 are:
- 1–2: The gas is heated by supplying heat from an external source.
- 2–3: The gas is cooled by heat removal/release from the cylinder.
- The transitions are marked with a circle, drawn in the clockwise direction, indicating a positive sign of the performed work.

The work performed by the gas in the cycle can be expressed as follows:

$$A = \oint (p \times dV) = A_{1-2} + A_{2-3} + \int_1^2 (p \times dV) + \int_2^3 (p \times dV) \tag{2.20}$$

where $A_{1-2} > 0;\ A_{2-3} < 0$

$$\oint \frac{dU}{} = 0 \Rightarrow \oint dq = A \tag{2.21}$$

In Eq. (2.20), A_{1-2} [kJ] is the work performed by the gas during the movement of the piston from position 1 to position 2; A_{2-3} [kJ] is the gas work during piston movement from position 2 to position 3. At position 3, the gas inside the cylinder has cooled down, so that its state is identical to that of point 1, closing the cycle. The cyclic integral of the internal thermal energy change must equal to zero because it represents an integration of the state variable and the working substance for the system that returns to its original thermodynamic state at the end of the cycle. For the time being, it is not specified how the gas is heated, how the heating is stopped and the gas cooling is started.

Thermal cycles can also be applied in the reverse direction. In that case, work can be applied to the working gas via the piston, whereby heat would be only removed from the cylinder, as a result of the conversion of the mechanical work to heat. In such reverse heat cycles, the work has a negative sign, and the cycle indication would be indicated in the $p-V$ diagram by a loop with an arrow in counterclockwise direction.

This example from Fig. 2.4 is strictly theoretical. A practical implementation of a cycle combining heating and cooling can be given with the Stirling engine (Organ, 2007), which can come in single- and in double-cylinder designs. The more wide-spread internal combustion engines feature open cycles, where the working gas is exchanged with the atmosphere.

2.2.11 Flow machines and their thermodynamic cycles

Unlike cylinder-piston machines, the flow machines always have at least one inlet and one outlet, all of which are constantly open. For example, an ordinary pipe can be considered a flow machine in which the pressure energy is transformed into internal heat due to fluid friction inside the pipe. This transformation is manifested by a pressure drop between the inlet and outlet of the pipe, also referred to as a "pressure loss".

Flow machines do not always have to perform work (Kim et al., 2019). For example, jet machines only transform mechanical energy between pressure and kinetic forms, as it is in ejectors and injectors. Another group of flow machines are blade-based, which usually transform energy of various forms directly or indirectly into work and vice versa. They are usually referred to as turbomachines and include turbines (water, steam), turbo-compressors, fans and propellers. Flow machines, almost without exception, use fluids as working substances. Depending on the type, they are capable of transforming internal heat energy, kinetic energy, potential energy and pressure energy of the working substance into work or vice versa.

The kinetic energy in a machine can be transformed based on the change in momentum of the working substance, resulting in the force acting on a moving machine part. In a turbomachine, the moving part (rotor) contains blades, which in the case of a turbine receive energy from the moving fluid and in the case of a compressor pass mechanical energy to the fluid. In both cases, the energy transfer is described by Euler's turbomachine equation (Dick, 2015). The potential energy of the working substance is transformed into work indirectly, for example, by first transforming it into the kinetic or pressure energy of the working substance, which then is used to drive the machine rotor and to perform work.

Flow machines form so-called open thermodynamic systems. This is because the energy flow with the produced work (power) is continuously generated, while assuming steady state, the content of the working substance in the machine is constant, but the substance flows through it.

If the substance in a turbomachine performs work, according to the first law of thermodynamics for open systems, that work is equal to the difference between the fluid energy flows at the inlet and the outlet of the machine. Of course, not all of that work reaches the shaft or the electrical generator, due to the mechanical losses incurred. The readers are referred to Dick (2015) and to Chapters 6 and 7 of this book for further details on turbines.

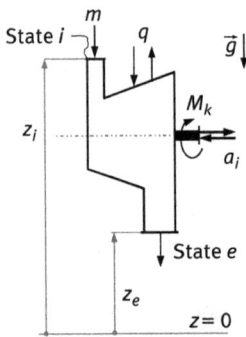

Fig. 2.5: Diagram for modelling the specific work of a flow machine at steady state.

The energy transfer and transformation in a turbine is illustrated in Fig. 2.5. The energy balance for the shown system accounts for the power generated and the energy contents in the working fluids at the inlet and outlets (kinetic, potential, pressure and internal thermal energy), the transmitted heat, neglecting the heat losses. In Fig. 2.5, the index i denotes the input and the index e denotes the output (exit) of the fluid. The variable denoted with a_i [kJ/kg] is the specific internal work performed by the working fluid; M_K [N × m] is the torque on the rotor shaft; q [kJ/kg] is the specific heat exchanged with the environment via the working fluid (positive value means heat is supplied to the machine; negative value means heat release to the environment); z [m] denotes the elevation of the inlet and outlet orifices. The specific work can be modelled as follows:

$$da_i = dq - du - d(p \times v) - \frac{dc^2}{2} - g \times dz \qquad (2.22)$$

where

$$du + d(p \times v) = dh \qquad (2.23)$$

Integration of the differential in Eq. (2.22) yields the following:

$$a_i = \int_i^e da_i = q + (u_i - u_e) + (v_i \times p_i - v_e \times p_e) + \frac{c_i^2 - c_e^2}{2} + g \times (z_i - z_e)$$

$$= q + \left(h_i + \frac{c_i^2}{2}\right) - \left(h_e + \frac{c_e^2}{2}\right) + g \times (z_i - z_e) \tag{2.24}$$

{heat exchange} {flow energy} {potential energy}

In Eqs. (2.22)–(2.24), in addition to the notation already introduced in Fig. 2.5, the symbols include
- u: specific internal energy [kJ/kg]
- p: pressure [kPa]
- v: specific volume of the flow [m³/kg] (inverse of density)
- c: linear velocity of the flow [m/s]
- g: acceleration due to gravity, 9.80665 m/s² (ESA, 2020)
- h: specific enthalpy of the flow [kJ/kg]

This equation can also be applied to any control volume within the machine. If the terms for the potential energy change and changes of kinetic energy are neglected (assumed insignificant), the differential of the specific work can be written as

$$da_i = dq - dh \tag{2.25}$$

In addition to purely flow-through machines, there are machines operating alternately as a closed and open system. A typical example is a piston steam engine, also including the variation known as Willans Engine (Hills, 1995). All flow machines work directly or indirectly in some heat-powered cycle. For example, a steam turbine is a device within a steam cycle. In addition to turbines, this cycle contains other essential equipment such as steam boilers, condensers and pumps. The pump moves water into the boiler to produce steam for the turbine. From the turbine, the steam goes to the condenser in which it condenses. From the condenser, the water is pumped back into the boiler, closing the cycle.

2.2.12 Reversibility of thermodynamic changes and entropy

The transition of a thermodynamic system – for example a volume of gas – from state i to state e is called reversible if it brings the system back from state e to the original state i by a reverse process, and the roundtrip from the initial state i to state e and back to i leaves no changes to the system itself and its surroundings (Potter and Somerton, 2014). This means that the same transformations of energy, work and heat sharing would take place in the opposite direction. Such reversible changes are an idealised abstraction. The conditions at which the reversibility definition holds are ones of quasi-equilibrium, at which there is no friction present, the heat transfer takes place at infinitesimal temperature differences and no unrestrained expansions occur.

These conditions are not met by real systems. Heat transfers always proceed only in one direction – from hotter to colder locations – over finite temperature differences. Another typical case is fluid flow in a tube. When flowing, due to the internal friction of the liquid, part of the kinetic energy is transformed into internal heat, where the kinetic energy for the motion comes at the expense of converting potential (pressure) energy. As a result, at the end of the pipe, the liquid has a lower pressure but a higher temperature (Balmer, 2011).

During the discovery of the various types of energy and the ways in which they are transformed, it became increasingly evident that all energy types can be spontaneously and completely transformed into heat. The opposite transformation that of heat into work and power is also possible, but this transformation is always partial, as a consequence of the *Second Law of Thermodynamics*. Furthermore, in the process of conversion of heat to work in the machines, reverse changes also take place in parallel, in which mechanical energy is transformed back into heat. For example, kinetic energy is transformed back into heat by friction inside the fluids and between the fluids and the machine parts. This heat is further lost to the ambient via the equipment casing.

In contrast to the heat exchanged by the working substance of the system with the environment, which can have both a positive and a negative sign, the heat generated by the reverse conversion of kinetic energy always has a positive sign, because it is associated with heat gain; it is equivalent to heat supply from outside the system.

Real thermodynamic changes are very complicated, but it is possible to evaluate quantitatively the irreversible energy transformation by means of the quantity of *entropy* defined and named in 1850 by Clausius (1822–1888) – for example Balmer (2011, pp. 218–219). Entropy is the amount of heat that is or can be transformed to work or mechanical energy per one degree of temperature difference, for the temperature level (measured from the absolute zero) at which the transformation takes place:

$$dS = \frac{dQ_{tot}}{T} \; ; \; \Delta S = \frac{\Delta Q_{tot}}{T} \tag{2.26}$$

$$Q_{tot} = Q_{transfer} + Q_z \tag{2.27}$$

Entropy is a measure of the degree of molecular disorder. The relation to the temperature provides information on how the internal thermal energy of a working substance changes – if its amount does not change, it means that all the shared heat has been transformed into mechanical energy. This is the case of a reversible engine – for example, the ideal Carnot engine model.

The definition of entropy in Eqs. (2.26)–(2.27) is given in differential and difference forms, where ΔS (also dS) [kJ/K] is the entropy change of the system during the thermodynamic state transition; Q_{tot} [kJ] is the overall heat transfer; $Q_{transfer}$ [kJ] is the amount of heat transferred to the system; Q_z [kJ] is the heat generated via friction, dissipation and other irreversibilities; T [K] mean temperature at which thermodynamic change occurred in the system.

Entropy can be also expressed per unit of substance – this is called specific entropy, usually denoted with a small letter s [kJ/kg/K]. Such a notation is useful for modelling complex systems involving multiple fluid streams – for instance, a chemical process or a site steam system.

Temperature–entropy diagrams (T–s diagrams) (Fig. 2.6) are used to illustrate the thermodynamic transitions, where the specific entropy is used on the X-axis. This provides the convenience that, in the T–s diagram, the specific heat transfer q and the specific heat generated by the irreversibilities in the considered processes q_z can be expressed as the areas enclosed under the integral of the T–s curve. The p–V diagram does not provide such an opportunity.

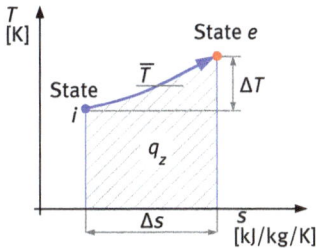

Fig. 2.6: An example of a T–s diagram.

The T–s diagram in Fig. 2.6 can be considered to describe the example thermodynamic transition resulting from the fluid flowing through a pipe, discussed in the beginning of the current section. In this case, the thermodynamically irreversible change occurring during fluid flow through the pipe can be modelled as follows. Assuming that there is no external heat supply to the system, the overall heat supply comes only from friction heat generation:

$$dq_{transfer} = 0 \Rightarrow dq_{tot} = dq_z \tag{2.28}$$

Assuming a non-compressible flow implies constant volume, which means that all the generated heat will contribute to the change of the fluid internal energy:

$$v = 0 \Rightarrow dq_{tot} = du \tag{2.29}$$

Consequently, the entropy change for this transition is given by the integral:

$$\Delta s = \int_i^e ds = \int_i^e \frac{1}{T} \times du \qquad (2.30)$$

The change in entropy is used to track the status of working substances depending on the supplied heat. For example, a T–s diagram of a fluid (gas) can be obtained by plotting temperature–entropy change curves at constant pressure (isobars) and at constant specific volume (isochores). Figure 2.7 shows an example of tracking the water state in a T–s diagram. The notation of the states is similar as the previous diagram: "i" denotes the initial state; e_P denotes the final state of a process that takes place at constant pressure; e_v denotes the final state of a process taking place at constant specific volume. The plot starts at the temperature and pressure corresponding to the initial liquid state of the water. The entropy change is calculated from the supplied heat based on the temperature recording against the initial state. For constructing such diagrams, accurate equations of state can be used – see Section 2.2.7.

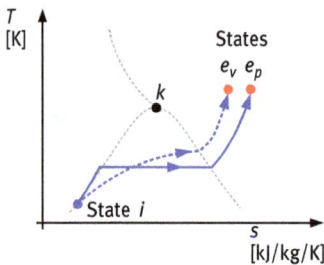

Fig. 2.7: A qualitative T–s diagram for the process of water heating.

The state of a substance can be tracked using other variables instead of temperature, which are related to the change in entropy. In engineering practice, enthalpy is often used for this, yielding the enthalpy–entropy diagram, also known as the Mollier diagram – named after its author Richard Mollier (Rajput, 2010, p.75). In this diagram, the Y-axis instead of the temperature tracks the enthalpy of the working fluid. The T–s and h–s diagrams have been constructed for various substances – most notably for water and steam, see Balmer (2011, p.266). Diagrams for other substances are also available – for example dichlorodifluoromethane (Poling et al., 2001, p. 1.7).

The T–s and h–s diagrams are used in engineering practice as an aid to visualise the status of the working substance and to determine the energy flows for the monitored thermodynamic changes or whole cycles of the systems. Particularly widespread is the use of the diagrams in the evaluation of thermodynamic changes of gases as working substances of power machinery.

For example, when gas flows through a pipe, pressure loss occurs, and the gas is heated, but unlike liquid, it expands by heating, so that its velocity (kinetic energy) increases. At the same time, according to the first law of thermodynamics, the total energy content of the gas does not change during the expansion. Such a change can be plotted in the corresponding T–s and h–s diagrams as shown in Fig. 2.8.

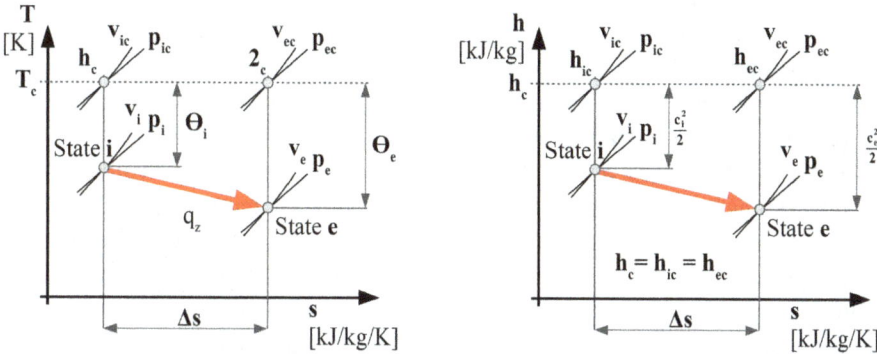

Fig. 2.8: Entropy-based diagrams for gas flow in a pipe: T–s (a) and h–s (b) diagrams.

The notation in Fig. 2.8 includes h_c [kJ/kg] total specific enthalpy which does not change with the flow; Θ [K] equivalent temperature rise of the gas velocity (gas temperature increase assuming complete conversion of the kinetic energy into enthalpy); T_c [K] – total gas temperature; Δp_Z [kPa] – pressure drop. During the flow, the internal thermal energy of the gas decreases. This drop in the internal energy is transformed into kinetic similarly as part of the lost (friction) heat, as the specific volume increases.

The thermodynamic change that occurs without changing the entropy is called *isentropic change*. From the definition of entropy [Eq. (2.26)] follows that any transformations in an isolated system will be either isentropic, for reversible processes, or will be accompanied by entropy increase – for real processes. This observation is valid for any transformations of energy between its forms – including for the transformation chain from chemical to thermal, mechanical, kinetic and the related friction transformation of the latter back to heat and its dissipation.

In order for entropy to decrease, the numerator of Eq. (2.26) should have a negative sign. This can only be achieved if heat is removed from the system under investigation, meaning that such a system would not be isolated – a direct consequence of the second law of thermodynamics.

The increase in entropy is related to the number of conditions that the working substance can achieve at the molecular level, which is related to the heat-sharing mechanism – heat is transferred from a warm body to a cold one and never in the opposite direction. In fact, the mechanism of heat dissipation and loss from a system involves a local increase in the temperature around a certain spot in the volume under investigation, followed by heat transfer to the immediate environment, which is cooler. This mechanism reduces the temperature at the point of heat loss and hence the ability of the working substance to perform work.

While the irreversibility of a process determines its direction, the difference between reversible and irreversible thermodynamic changes may not be significant by magnitude in some cases of technical practice. Irreversible changes are often neglected as a simplification at the first stages of process design, in order to calculate the process parameters faster, arriving at the order of magnitude for more detailed design parameters. For instance, in the case of pipe flow, such a simplification means that at the beginning and at the end of the pipe, the same liquid state would be initially assumed, as if the irreversible change would not occur.

The term entropy is used not only in thermodynamics but also in informatics, where it represents the uncertainty of decision making – known as Shannon entropy (Velázquez Martínez et al., 2019), in what state the members of the observed system may be. It has the measurement unit [bit] and is calculated using the expression

$$H = - \sum_{i=1}^{n} \{p_i \times \log_2 p_i\} \qquad (2.31)$$

where p denotes the probability of occurrence of one of the states and n is the number of states that the system can have (Gray, 2013). The logarithm has the base two because the state can either occur or not. The equation was derived based on what the uncertainty of decision making should mean. Maximum uncertainty is achieved by the system if all the possible conditions (states) of the system have the same occurrence probability. The Shannon entropy was developed as a concept by analogy with statistical thermodynamics and is used as the measure of uncertainty of the system states. Maximum Shannon entropy is obtained when the probability of the various states are equal. The amount of information in the bits is then the difference in decision uncertainty before the information is received and when the information is received. Most often, the decision uncertainty is related to a data entity that contains binary members, so each of them can take only the values 0 or 1. A data entity that contains one binary value can have a maximum of 1 bit. A data entity that contains two binary members can have a maximum of 2 bits, and so on.

2.2.13 Some well-known reversible thermodynamic processes

The gas state after some thermodynamic change can be calculated from the initial state and the type of change if the process can be sufficiently well described. For example, in Fig. 2.4, a general thermodynamic change between states 1 and 2 is shown, during which the pressure increases and then decreases again, accompanied by volume expansion and contraction. Such a complex state change is not straightforward to describe. The simple thermodynamic processes usually include the so-called adiabatic processes and some special cases of polytropic transition.

2.2.14 Adiabatic processes

Adiabatic transition is one in which the observed control volume, containing a working substance, is entirely isolated from external influence. In the case of a gas enclosed in a piston-cylinder (Fig. 2.4), the system is closed – it does not exchange matter with the surroundings. For an adiabatic change to take place, the cylinder has to be also perfectly insulated against heat transfer with the environment ($dq = 0$). The reversible adiabatic change can be described by the exponential equation called the adiabatic equation, so that the relationship between the pressure and the volume is exponential, as shown in Fig. 2.9a. This is an example of reversible adiabatic expansion, which is indicated by the pressure reduction. A reversible adiabatic compression would be the opposite of reversible adiabatic expansion. It would follow the opposite process trajectory. In the case of reversible adiabatic compression, following the transition from state e to state i, the same amount of work is consumed as it is performed during reversible expansion from state i to state e.

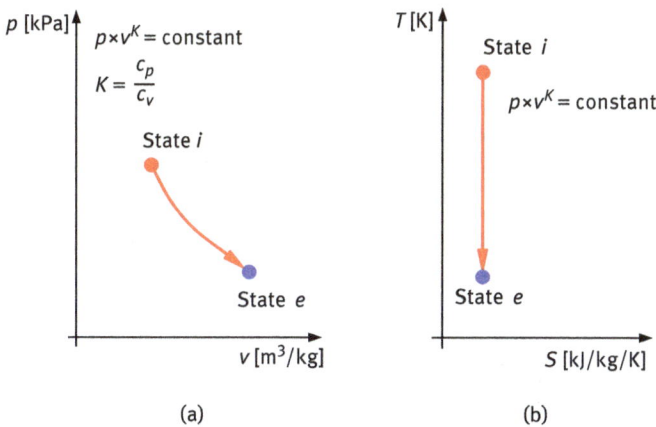

Fig. 2.9: Adiabatic expansion and the related equation: (a) p–v diagram and (b) T–s diagram.

In Fig. 2.9, κ [1] is the Poisson's constant also called an adiabatic exponent. It is a constant for a fixed gas composition. The equation is derived in Weinhold (2009). Reversible adiabatic process is at the same time isentropic (Fig. 2.9b) because it is not accompanied by heat transfer. Under these conditions, any changes take place at the expense of altering the internal energy of the working substance. For the case of expansion, this means temperature reduction.

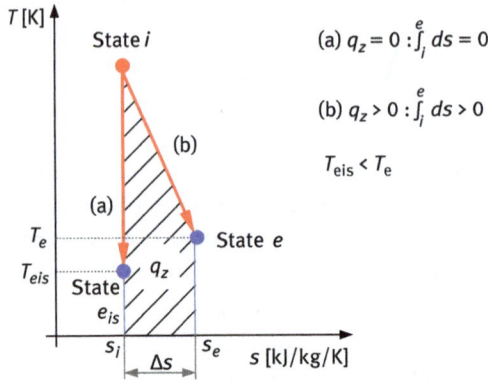

Fig. 2.10: Comparison of adiabatic expansion models: (a) reversible and (b) irreversible.

In the case of an irreversible adiabatic process, entropy is generated, which is illustrated in Fig. 2.10, where reversible adiabatic expansion is compared with irreversible adiabatic expansion. In the case of a reversible process, the expansion is isentropic and there are no dissipation losses ($q_z = 0$). In the case of an irreversible process, the expansion is accompanied by entropy generation and with dissipation losses ($q_z > 0$). The state denoted as e_{is} is the one after the isentropic expansion of the working gas in the cylinder. If the real expansion is modelled, heat transfer and dissipation heat losses take place and the working gas state at the end of expansion will correspond to point e. The state e is characterised by higher temperature (higher internal thermal energy of gas) and higher entropy than state e_{iz}.

In practical processes, entropy is always generated, following the pattern indicated in the transition $(i) \rightarrow (e)$ in Fig. 2.10. Examples of near-adiabatic processes with entropy generation include the expansion of the working fluid in a steam or gas turbine or the gas compression of in turbo compressors. Another practical process, which approaches the adiabatic assumption, is the steam expansion in a letdown valve. In many modelling cases, these valves are even assumed isenthalpic.

2.2.15 Polytropic processes

Polytropic processes are a more general class of thermodynamic transitions. The literal etymology of the term "polytropic" has the meaning of "multipath" (Balmer, 2011, p. 111). During such a process, the considered control volume can exchange heat with its surroundings. In the case of a gas enclosed in a piston-cylinder, a polytropic change will occur if the cylinder shares the heat with the environment ($dq \neq 0$). A reversible polytropic process (*not to be confused with a reversible cycle*) is described with an exponential equation, where the power factor is a given constant, similar to the reversible adiabatic process. This equation is known as the "polytropic equation":

$$p \times v^n = \text{const} \tag{2.32}$$

Such processes can be also illustrated on the $p-v$ and the $T-s$ diagrams, as shown in Fig. 2.11.

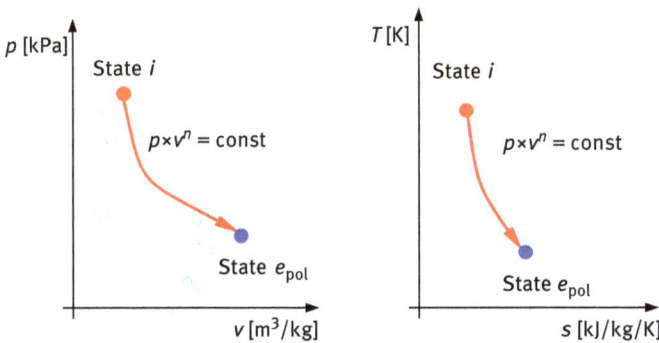

Fig. 2.11: Polytropic expansion.

In Fig. 2.11 and Eq. (2.32), n [1] is the polytropic index (Klein and Nellis, 2012, p. 480). In the simplest case, this is constant throughout the duration of the transition. In principle, nothing prevents the value of n from changing. The values of n for an ideal gas are bounded by two extreme cases. One is the isothermal process, for which $n = 1$. The other bounding case is the isentropic process, where $n = y = C_P/C_V$ (the ratio of the specific heat capacities at constant pressure and at constant volume). The expansion in Fig. 2.11 is associated with heat loss. The opposite of that reversible polythropic expansion is reversible polytropic compression with heat input. The reversible adiabatic compression from state e_{pol} to state i would take the same amount of work as the reversible expansion in the forward direction.

Special cases of reversible polytropic processes are those in which, in the course of heat exchange with the surroundings, some of the system properties remain constant. In the case of constant-temperature, the process is referred to as isothermal and the corresponding curve – an *isotherm* (Fig. 2.12). If the pressure is kept constant, it is termed isobaric (the curve is an *isobar*), and at constant volume, the process is called isochoric and the curve an *isochor*.

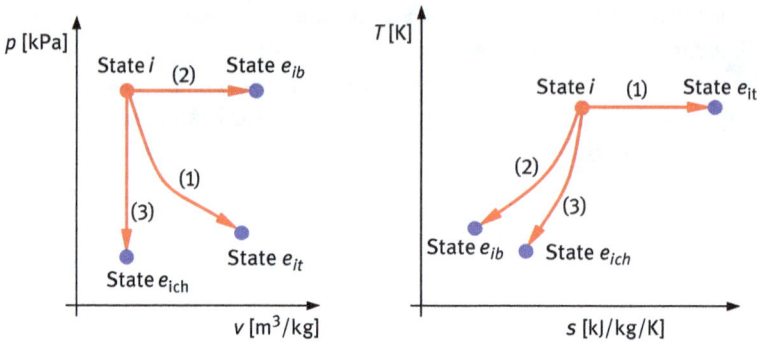

Fig. 2.12: Special cases of reversible polytropic processes. Referring to Eq. (2.32): (1) isothermal process (t = const, n = 1); (2) isobaric process (p = const, n = 0); (3) isochoric process (v = const, n = ∞).

Identification of the relevant reversible processes for the thermodynamic machines is of prime importance in their design. Such idealised representations of the real processes help the designers in obtaining estimates of the best possible process performance which is guaranteed that cannot be exceeded. By comparing such ideal processes of energy transformation within a machine with the actual one, the designer can obtain information on the type of the losses taking place in the machine and their magnitude. By studying the losses, the machine design can also be improved. One of the most common ideal cycles, used for estimation of the thermodynamic limits of a system, is the Carnot cycle.

2.2.16 Carnot cycle

The working substance of a Carnot cycle (Balmer, 2011) is ideal gas perfectly sealed in a cylinder with a movable piston, which performs a linear reciprocating motion and drives, for example, a crankshaft. Such an arrangement is shown in Fig. 2.13.

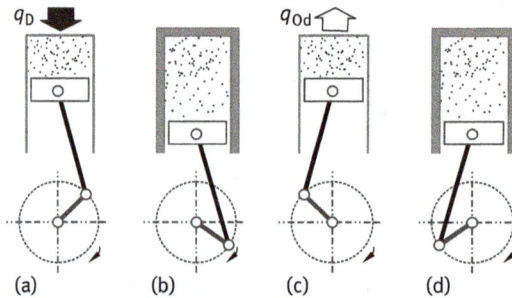

Fig. 2.13: One possible implementation of the Carnot cycle. (a) Isothermal expansion; (b) Adiabatic expansion; (c) Isothermal compression; (d) Adiabatic compression.

In Fig. 2.13, q_D [kJ/kg] is the delivered specific heat; q_{od} is the dissipated specific heat [kJ/kg]. One working cycle (one crankshaft revolution) consists of four reversible thermodynamic transitions of the working gas, which can be plotted on p–v and T–s diagrams.

Figure 2.14 shows those diagrams. The notation in the figure involves: 1–2 is isothermal expansion at temperature T_T (heat is supplied to the working gas); 2–3 adiabatic expansion (the working gas is thermally isolated from the environment); 3–4 isothermal compression at temperature T_S (heat is removed from the working gas); 4–1 adiabatic compression (the working gas is thermally insulated from the environment). These stages of the Carnot cycle are discussed next.

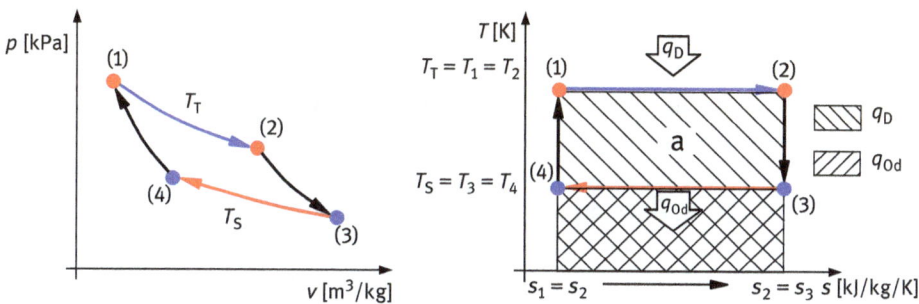

Fig. 2.14: p–v and T–s diagrams of the Carnot cycle.

Isothermal expansion (Fig. 2.13a, **1–2**): At the beginning of the cycle (*State 1*), the piston is at the top dead centre. In the process, the working gas is at temperature T_T, pressure p_1 and specific volume v_1. Heat is supplied to the working gas through the cylinder walls (external heating). The gas starts to expand and moves the piston, performing work. The heat is supplied to the working gas in such a way that its temperature does not change during the expansion, forming an isothermal process. The pressure drops and specific volume of the gas increases, until *State 2* (located between the top and bottom dead centre positions), when the heat supply is stopped. The heat supplied to the working gas for this step is q_D [kJ/kg]. The overall amount q_D can be expressed as a function of the specific volumes ratio and the temperature of the expansion:

$$q_D = \int_1^2 dq = \int_1^2 p \times dv = \int_1^2 \frac{R \times T_T}{v} = R \times T_T \times \ln\frac{v_2}{v_1} \qquad (2.33)$$

where R is the universal gas constant (see Section 2.2.7).

On the other hand, the same amount of heat can be related to the entropy change resulting from the transition:

$$q_D = \int_1^2 T \times ds = T_T \times (s_2 - s_1) \tag{2.34}$$

Adiabatic expansion (Fig. 2.13b, **2–3**): *State 2* is followed by an adiabatic process (expansion). During this transition, the cylinder is perfectly thermally insulated from the surroundings, preventing heat transfer. The gas is still expanding and acting on the piston by force and performing work. The pressure and the temperature of the working gas drop, as the gas is not supplied with heat. The transition ends in *State 3*, when the piston reaches the bottom dead centre. The temperature of the working gas at this point is T_S.

Isothermal compression (Fig. 2.13c, **3–4**): The piston moves from *State 3* at the bottom dead centre to the top dead centre while compressing the working gas in the cylinder isothermally. In this process, work is performed by the surroundings on the working gas. The piston gains energy from the rotating crankshaft. When compressing, the gas is heated. For the compression to be isothermal (T_S = const.), heat is removed from the gas. This transition takes place until the piston reaches *State 4*, which occurs between the lower and upper dead centre of the piston. For the entire course of this process, heat is generated through conversion of the work performed on the gas. The amount of heat removed from the gas to the surroundings is denoted as q_{Od} [kJ/kg] – having a negative value. The amount of heat is related to the work done by the piston on the gas:

$$q_{Od} = \int_3^4 P \times dv = \int_3^4 \frac{R \times T_S}{v} = R \times T_S \times \ln \frac{v_4}{v_3} \tag{2.35}$$

as well as to the entropy change in the course of the isothermal compression:

$$q_{Od} = T_S \times (s_4 - s_3) \tag{2.36}$$

Adiabatic compression (Fig. 2.13d, **4–1**): For the duration of this step, starting at *State 4*, the cylinder is perfectly thermally insulated from the surroundings, preventing heat transfer. The piston keeps performing work on the gas and compresses it. In the process, both the pressure and the temperature of the working gas increase, as the heat is not removed from the cylinder. This process is completed at State 1 (the initial state), at which the piston reaches the top dead centre. At this point, all state variables of the working gas assume the same values, as at the start of the cycle – including the temperature becoming equal to T_T.

In summary, the transitions of between states 2–3 and 4–1 are adiabatic, heat is added to the working gas during the isothermal expansion 1–2 and heat is removed during the isothermal compression 3–4. The heat flow of the isothermal expansion 1–2 balance is positive, which, besides the definition that heat supply is positive, can also be inferred from the natural logarithm in Eq. (2.33), where $v_1 < v_2$. Similarly, for the isothermal compression, heat is removed from the working gas and $v_4 > v_3$, resulting in the negative value of q_{Od}. The net work a [kJ/kg] of the cycle is equal to the closed integral of the heat flow, producing:

$$a = q_D + q_{Od} = T_T \times (s_2 - s_1) + T_S \times (s_4 - s_3) \tag{2.37}$$

Combining the assumptions of reversibility and adiabatic transition for steps 2–3 and 4–1, implies that

$$s_3 = s_2; \ s_4 = s_1 \tag{2.38}$$

which yields

$$a = T_T \times (s_2 - s_1) + T_S \times (s_1 - s_2) \tag{2.39}$$

$$a = (T_T - T_S) \times (s_2 - s_1) \tag{2.40}$$

Taking the sum of Eqs. (2.33) and (2.35) also follows the summation in Eq. (2.37) and produces:

$$a = R \times \left(T_T \times \ln \frac{v_2}{v_1} + T_S \times \ln \frac{v_4}{v_3} \right) \tag{2.41}$$

It has to be stressed that the Carnot cycle cannot be technically implemented. The main obstacle is the rapid change of heating of the cooling on the same heat transfer surface of the cylinder. However, this cycle is useful because it estimates the maximum work extractable from a combination of heat reservoirs at given temperatures (T_T and T_S), providing also an estimate of the maximum thermal efficiency. As such, it is often used as the benchmark for measuring the efficiency of real thermal cycles (Balmer, 2011).

2.2.17 Energy conversion efficiency

Heat can be transformed to work by the means of thermal cycles, implemented in energy conversion equipment. The thermal efficiency of such a transformation is the ratio of the performed work to the supplied heat, also termed the cycle efficiency:

$$\eta_t = \frac{A}{Q_D} = \frac{a}{q_D} \tag{2.42}$$

where η_t [1] denotes the cycle thermal efficiency, A [kJ] is the overall amount of work for the considered system, Q_D [kJ] is the overall amount of heat supplied to the system, both related to the specific amounts a and q_D.

The thermal efficiency of a cycle, as defined in Eq. (2.42), cannot reach the value of 1 even for the ideal Carnot cycle. This is due to the fact that part of the work performed (which was generated by the transformation of the supplied heat) is consumed internally, to return the working substance to its original state, so that the cycle can be repeated. This takes place regardless of the assumption for reversible processes.

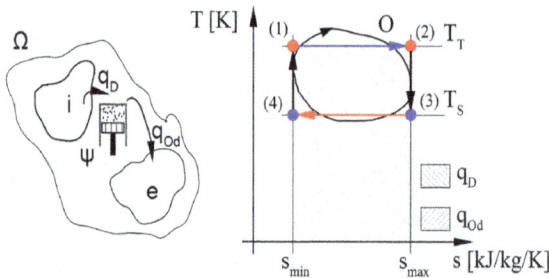

Fig. 2.15: Thermal efficiency derivation for the Carnot cycle.

The expression for the thermal efficiency of the Carnot cycle can be derived starting from the entropy balance (Fig. 2.15). The reversibility assumption also implies that the overall entropy change of the system for the cycle is zero. Out of the four stages, two are adiabatic (2–3 and 4–1). Taken together with the assumption of reversible processes, this means that there are no entropy changes during these stages. The other two stages – the isothermal expansion 1–2 and the isothermal compression 3–4 – are associated with heat absorption and heat rejection. The heat absorption during the isothermal expansion 1–2 is accompanied by entropy increase Δs_{1-2} and the isothermal compression 3–4 is accompanied by entropy decrease Δs_{3-4}:

$$\Delta s_{1-2} = \frac{q_D}{T_T}; \ \Delta s_{3-4} = \frac{q_{Od}}{T_S} \tag{2.43}$$

The cyclic integral of the entropy is equivalent to the summation of Δs_{1-2} and Δs_{3-4} and equals zero:

$$\oint_{\text{Carnot}} ds = \Delta s_{1-2} + \Delta s_{3-4} = \frac{q_D}{T_T} + \frac{q_{Od}}{T_S} = 0 \tag{2.44}$$

Re-arranging the terms in the rightmost equality of Eq. (2.44) yields:

$$q_{Od} = -q_D \times \frac{T_S}{T_T} \tag{2.45}$$

Recall that the specific work of the cycle is the algebraic sum of the heat exchanges, which can be further rearranged:

$$a = q_D + q_{Od} = q_D - q_D \times \frac{T_S}{T_T} = q_D \times \left(1 - \frac{T_S}{T_T}\right) \tag{2.46}$$

Taking the ratio of the specific work and the heat supply from Eq. (2.46) yields the thermal efficiency if the Carnot cycle:

$$\eta_t = \frac{a}{q_D} = 1 - \frac{T_S}{T_T} \tag{2.47}$$

In this derivation, the heat supply temperature T_T and the heat release temperature T_S are in [K], to adhere to the zero entropy definition at the absolute zero. This relationship was first derived by William Thomson (Lord Kelvin), who has first stated the assumption that there was an absolute zero and introduced the term thermodynamics. The ratio of the highest and the lowest temperatures of the Carnot cycle in Eq. (2.47) can be regarded as characteristic for the cycle and referred to as the cycle temperature ratio τ [1]:

$$\tau = \frac{T_S}{T_T} \tag{2.48}$$

The efficiency calculation in Eq. (2.47) is not applicable to thermodynamic cycles with real (irreversible) process featuring energy losses. One possible explanation can be given by the fact that the area enclosed by the curve in the $T–s$ diagram for a real cycle is not equivalent to the specific work of the cycle, and the entropy difference at the heat supply is different from that of the heat removal so that the equality in (2.44) used in the derivation of Eq. (2.47) does not hold.

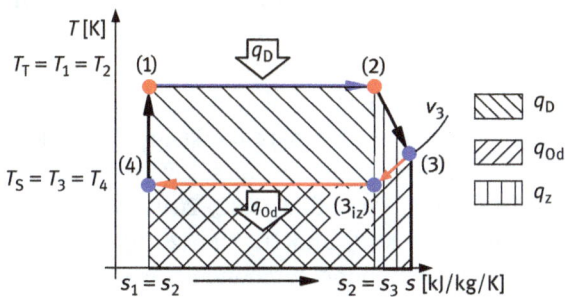

Fig. 2.16: Carnot cycle modification featuring energy losses.

Figure 2.16 shows a variation of the Carnot cycle with imperfect implementation. In contrast to the ideal cycle (Fig. 2.14), the expansion of the working gas in the transition 2–3 proceeds adiabatically but with losses. In order to return the working gas to the state with the entropy of *State 2* (i.e. to *State 3$_{iz}$*), equivalent to the isentropic expansion, part of this loss has to be discharged as heat outside the cycle at a constant volume corresponding to the area under the isochore in the transition section (*3–3$_{iz}$*). This becomes equivalent to an increase of the mean temperature of heat removal from the circulation compared to the ideal cycle, leading to a decrease of the work performed by the cycle.

For technology practice, the thermal efficiency of the cycle is very important because a higher efficiency of the thermal cycle translates to lower demand for primary energy resource – most often a fossil fuel, further incurring lower environmental footprints. Further implications of achieving higher conversion efficiency at the same power demand include smaller equipment sizes for boilers and other conversions (lower investment), as well as lower resource purchase and operating costs.

2.2.18 Thermodynamic properties of substances

It is clear from the above discussion that it is necessary to have accurate estimates of the substance properties and their relationships to accurately calculate the work and heat flows in the machines. Those include relationships of the enthalpy, entropy and other dependent variables as functions of the pressure, temperature and the operating conditions. Essentially, these are the data and relationships that allow constructing the *h*–*s* and *T*–*s* diagrams.

When it comes to water, freely available data are available over a wide range of conditions. These data are managed by the *IAPWS* – The International Association for the Properties of Water and Steam, which promotes the research, processing and disclosure of water properties. It is possible for instance to find the IAPWS'97 industrial formulation for the thermodynamic properties of water and steam in (Wagner and Kretzschmar, 2008a). These data are also known as "steam tables" (Wagner and Kretzschmar, 2008b). Thermodynamic data for air and other gases are also freely available – including gas mixtures (Lemmon et al., 2000).

Such data, in a consistent way, are part of built-in data banks and thermodynamic calculation packages in process simulators and similar software. Examples of software tools using such packages include Petro-SIM (KBC, 2019), Aspentech software tools ('AspenONE Product Portfolio', 2019) and DWSIM (2019). An Open Source library (CoolProp, 2019) for selected compound components, mainly refrigerants are available as well.

2.3 Summary

This chapter discusses the fundamental concepts necessary to understand and fruitfully use the material given in the main chapters on energy conversion equipment, Total Sites and utility networks. Since the main focus of the current book is on the overall utility systems modelling, covering the aspects of heat and power flows on the industrial sites, only the very basic thermodynamic concepts are introduced and references are provided to the specialised sources that threat those subjects in more detail.

The fundamental elements introduced include the means of utility heating and cooling for industrial processes. This is followed by laying out the basics of thermodynamics that are used later in the main chapters – the laws of thermodynamics, the definitions of the main thermodynamic concepts as temperature, heat, work, internal energy, enthalpy, entropy and the equations of state. Several simple thermodynamic cycles are introduced, starting from a simple piston configuration and a turbine expansion configuration. The pressure–volume, temperature–entropy and enthalpy–entropy diagrams are introduced and explained. Their significance to the analysis and calculation of energy conversion processes is also shown. These diagrams are used then to introduce and illustrate the concept of a polytropic process and then the idealised Carnot cycle.

Further detailed information on thermodynamic issues can be found in the multitude of monographs and textbooks dedicated to thermodynamics. Some of them are referred to in the current chapter. To summarise, the following sources have been found useful as sources of further information:

- The 8th edition of *Thermodynamics: An Engineering Approach* (Çengel and Boles, 2015), and the 9th edition is also expected soon.
- For readers interested in fundamentals of the power machines, the book by a similar team can be useful – *Fluid Mechanics: Fundamentals and Applications* (Çengel and Cimbala, 2014).
- *Modern Engineering Thermodynamics* (Balmer, 2011).
- *Schaum's Outline of Thermodynamics for Engineers* (Potter and Somerton, 2014)
- *Perry's Chemical Engineers' Handbook* (Green, 2018)
- *Thermodynamics* (Klein and Nellis, 2012)
- *Process Steam Systems* (Merritt, 2016) – although not strictly a thermodynamics book, it provides important pointers to fundamental knowledge about steam and how to use it in utility systems.
- *The Properties of Gases and Liquids* (Poling et al., 2001)
- *Engineering Thermodynamics* (Rajput, 2010)

Nomenclature

Symbol	Measurement unit	Description
0	–	Working gas position before displacement (Fig. 2.2)
A	m^2	Area
A	J, kJ	Work
A_{1-2}	kJ	The work performed by the gas during the movement of the piston from position 1 to position 2
A_{2-3}	kJ	The gas work during piston movement from position 2 to position 3. At position 3
a, a_i	J/kg, kJ/kg	Specific work
C_P	kJ/kg/K	Specific heat capacity at constant pressure
C_V	kJ/kg/K	Specific heat capacity at constant volume
c	m/s	Linear velocity of the falling sphere (Fig. 2.3)
E	J	Energy
E_k	J	Kinetic energy
E_p	J, kJ	Potential energy
F_g	N	Force exerted by the weight on the piston
F_p	N	Force of the gas under the piston on the piston by its pressure
g	m/s^2	Acceleration due to gravity
H	kJ	Enthalpy
H	bit	Shannon entropy
h	kJ/kmol	The specific enthalpy
h_c	kJ/kg	Total specific enthalpy (Fig. 2.8)
M_{gas}	kmol/kg	Molar weight of a gas
M_K	N × m	Torque on a rotor or a shaft
m	kg	Mass (of working substance, piston weight)
n	[1]	The polytropic index
n	[1]	The number of states that a system can have
p	N/m^2, (Pa)	Pressure
p	[1]	Probability of an event occurrence
p_0	N/m^2, (Pa)	Pressure exerted by the gas at position "0" – before displacement (Fig. 2.2)
p_x	N/m^2, (Pa)	Pressure exerted by the gas at position "x" – after displacement (Fig. 2.2)
Q	J, kJ	Heat transfer rate
Q_D	kJ	Overall amount of heat supplied to the system
Q_{tot}	kJ	Overall heat transferred
q_{tot}	kJ/kg	Specific, overall heat transferred
$Q_{transfer}$	kJ	The amount of heat transferred to a system
$q_{transfer}$	kJ/kg	Specific amount of heat transferred to a system
Q_z	kJ	Heat generated via friction, dissipation and other irreversibilities
q_z	kJ/kg	Specific heat generated via friction, dissipation and other irreversibilities; heat loss due to irreversibilities
q	kJ/kg	Specific heat transfer; specific heat dissipation
q_D	kJ/kg	Delivered specific heat
q_{Od}	kJ/kg	The amount of heat removed from the gas to the surroundings (specific)

(continued)

Symbol	Measurement unit	Description
R	J/mol/K	The universal gas constant
S	J/K, kJ/K	Entropy
s	kJ/kg/K	Specific entropy
r	kJ/kg/K	The individual gas constant
t	°C	The temperature in the Celsius scale
T	K	Temperature in the Kelvin scale
T_0	K	Temperature of the working gas at position "0" – before displacement (Fig. 2.2)
T_c	K	Total gas temperature
T_S	K	Heat release temperature (Fig. 2.14)
T_T	K	Heat supply temperature (Fig. 2.14)
TT	F	Temperature in Fahrenheit degrees
T_x	K	Temperature of the working gas at position "x" – after displacement (Fig. 2.2)
T_v	K	Temperature of the water (Fig. 2.2)
U	J, kJ	Internal energy of a system
U_v	J	Internal energy of the water volume in Fig. 2.2
U_{vz}	J	Internal energy of the air volume in Fig. 2.2
u	J/kg	Specific internal energy of the working substance
V	m^3	Volume
v	m^3/kg	Specific volume of a substance
V_0	m^3	Volume of the piston position "0" – before displacement (Fig. 2.2)
V_x	m^3	Volume of the gas at position "x" – after displacement (Fig. 2.2)
v_M	m^3/mol	Specific molar volume
X	–	Working gas position after displacement (Fig. 2.2)
z, z_1, z_2	m	Elevation of various positions
K	1	Poisson's constant also – as an adiabatic exponent
η_t	1	Thermal efficiency
Θ	K	Equivalent temperature rise of the gas velocity
ρ	kg/m^3	Density
τ	1	Carnot cycle temperature ratio
Δp_z	kPa	Pressure drop
ΔU	J, kJ	Change of the internal energy
ΔS (also dS)	kJ/K	Entropy change of a system during a thermodynamic state transition

Indices

Symbol	Description
e	Output (exit) of the fluid
ec	Stagnation equivalent state at the outlet
e_P	The final state of a process that takes place at constant pressure
e_V	The final state of a process taking place at constant specific volume
i	Inlet, initial state
ic	Stagnation equivalent state at the inlet
is	Outlet after isentropic process
v	Water
vz	Air

References

'AspenONE Product Portfolio'. 2019. 21 November 2019. https://www.aspentech.com/en/products/full-product-listing, accessed 13/08/2020.

Atkins, P.W. 2007. Four Laws That Drive the Universe. Oxford UK; New York, USA: Oxford University Press.

Balmer, R.T. 2011. Modern Engineering Thermodynamics. Amsterdam ; Boston: Academic Press.

Bird, J.O. and Ross, C.T.F. 2015. Mechanical Engineering Principles. Third edition. London, UK; New York USA: Routledge, Taylor & Francis Group.

Çengel, Y.A. and Boles, M.A. 2015. Thermodynamics: An Engineering Approach. Vol. 8. edition in SI units. New York, NY, USA: McGraw-Hill Education.

Çengel, Y.A. and Cimbala, J.M. 2014. Fluid Mechanics: Fundamentals and Applications. Third ed. New York, USA: McGraw Hill.

CoolProp. 2019. 'Welcome to CoolProp – CoolProp 6.3.0 Documentation. 28 November 2019. http://www.coolprop.org/#what-is-coolprop, accessed 13/08/2020.

De Sanctis, E., Monti, S., and Ripani, M. 2016. Energy from Nuclear Fission: An Introduction, Heidelberg, Germany: Springer. https://doi.org/10.1007/978-3-319-30651-3.

Dick, E. 2015. Fundamentals of Turbomachines. New York, USA: Springer.

DWSIM. 2019. DWSIM – Chemical Process Simulator. 11 August 2019. http://dwsim.inforside.com.br, accessed 18/08/2020.

Engineering ToolBox. 2003. Convective Heat Transfer. 2003. https://www.engineeringtoolbox.com/convective-heat-transfer-d_430.html, accessed 18/08/2020.

ESA. 2020. Gravity in Detail – Content – Earth Online – ESA. 22 January 2020. https://earth.esa.int/web/guest/-/gravity-in-detail-5728, accessed 22/01/2020.

Faramarzi, L., Kontogeorgis, G.M., Thomsen, K., and Stenby, E.H. 2009. Extended UNIQUAC Model for Thermodynamic Modeling of CO_2 Absorption in Aqueous Alkanolamine Solutions. Fluid Phase Equilibria 282(2): 121–32. https://doi.org/10.1016/j.fluid.2009.05.002.

Fredenslund, A., Jones, R.L., and Prausnitz, J.M. 1975. Group-Contribution Estimation of Activity Coefficients in Nonideal Liquid Mixtures. AIChE Journal 21(6): 1086–99. https://doi.org/10.1002/aic.690210607.

Gray, R.M. 2013. Entropy and Information Theory. New York, NY USA: Springer Science & Business Media.

Green, D.W. 2018. Perry's Chemical Engineers Handbook 9th ed. New York, NY USA: McGraw-Hill Education.

Hills, R.L. 1995. Power from Steam: A History of the Stationary Steam Engine. Repr. Cambridge UK: University Press.

KBC. 2019. Petro-SIM | Process Simulation Software | KBC. 19 August 2019. https://www.kbc.global/software/process-simulation-software, accessed 18/08/2020.

Kim, K.-Y., Samad, A., and Benini, E. 2019. Design Optimization of Fluid Machinery: Applying Computational Fluid Dynamics and Numerical Optimization. Hoboken, NJ, USA: Wiley.

Kittel, C. 2005. Introduction to Solid State Physics. 8th ed. Hoboken, NJ USA: Wiley.

Klein, S. and Nellis, G. 2012. Thermodynamics. New York, USA: Cambridge University Press.

Lemmon, E.W., Jacobsen, R.T., Penoncello, S.G., and Friend, D.G. 2000. Thermodynamic Properties of Air and Mixtures of Nitrogen, Argon, and Oxygen From 60 to 2000 K at Pressures to 2000 MPa. Journal of Physical and Chemical Reference Data 29(3): 331–85. https://doi.org/10.1063/1.1285884.

Liu, Z.-K., Shang, S.-L., and Wang, Y. 2017. Fundamentals of Thermal Expansion and Thermal Contraction. Materials 10(4): 410. https://doi.org/10.3390/ma10040410.

Lopez-Echeverry, J.S., Reif-Acherman, S., and Araujo-Lopez., E. 2017. Peng-Robinson Equation of State: 40 Years through Cubics. Fluid Phase Equilibria 447 (September): 39–71. https://doi.org/10.1016/j.fluid.2017.05.007.

Merritt, C. 2016. Process Steam Systems | Wiley Online Books. Hoboken, NJ, USA: John Wiley & Sons, Inc. https://doi.org/10.1002/9781119085454.

Organ, A.J. 2007. The Air Engine: Stirling Cycle Power for a Sustainable Future. Woodhead Publishing in Mechanical Engineering. Boca Raton, FL, USA: CRC Press.

Peng, D.-Y. and Robinson, D.B. 1976. A New Two-Constant Equation of State. Industrial & Engineering Chemistry Fundamentals 15(1): 59–64. https://doi.org/10.1021/i160057a011.

Poling, B.E., Prausnitz, J.M., and O'Connell, J.P. 2001. The Properties of Gases and Liquids. 5th ed. New York USA: McGraw-Hill.

Potter, M.C. and Somerton, C.W. 2014. Schaum's Outline of Thermodynamics for Engineers. Third ed. Schaum's Outline. New York USA: McGraw-Hill Education.

Rajput, R.K. 2010. Engineering Thermodynamics. Sudbury, Mass., USA: Jones and Bartlett Publishers.

Soave, G. 1972. Equilibrium Constants from a Modified Redlich-Kwong Equation of State. Chemical Engineering Science 27(6): 1197–1203. https://doi.org/10.1016/0009-2509(72)80096-4.

Velázquez Martínez, O., Van Den Boogaart, K.G., Lundström, M., Santasalo-Aarnio, A., Reuter, M., and Serna-Guerrero, R. 2019. Statistical Entropy Analysis as Tool for Circular Economy: Proof of Concept by Optimizing a Lithium-Ion Battery Waste Sieving System. Journal of Cleaner Production 212 (March): 1568–79. https://doi.org/10.1016/j.jclepro.2018.12.137.

Wagner, W. and Kretzschmar, H.-J., eds 2008a. IAPWS Industrial Formulation 1997 for the Thermodynamic Properties of Water and Steam. In: International Steam Tables: Properties of Water and Steam Based on the Industrial Formulation IAPWS-IF97, 7–150. Berlin, Heidelberg, Germany: Springer Berlin Heidelberg. https://doi.org/10.1007/978-3-540-74234-0_3.

Wagner, W. and Kretzschmar, H.-J., eds. 2008b. International Steam Tables. Berlin, Heidelberg: Springer Berlin Heidelberg. https://doi.org/10.1007/978-3-540-74234-0.

Weinhold, F. 2009. Classical and Geometrical Theory of Chemical and Phase Thermodynamics: A Non-Calculus Based Approach. Hoboken, USA: John Wiley & Sons.

3 Utility systems – an overview

Chemical and petrochemical plants operate on existing industrial sites, where a number of production processes are grouped together and are supplied with power and heat by a site utility system. Combined Heat and Power (CHP) production, also called cogeneration, is typical for modern utility systems, which usually comprise boilers, gas turbines and steam turbines.

3.1 The utility system from the process perspective

Very frequently there are several hot and several cold utilities available for providing the process heating and cooling requirements after internal energy recovery. The typical set of utilities available at an industrial site, from the perspective of production processes, is presented in Fig. 3.1 At the site level, the supplier of these utilities is the utility system and it is necessary to find and evaluate the cheapest and most effective combination for fulfilling its function.

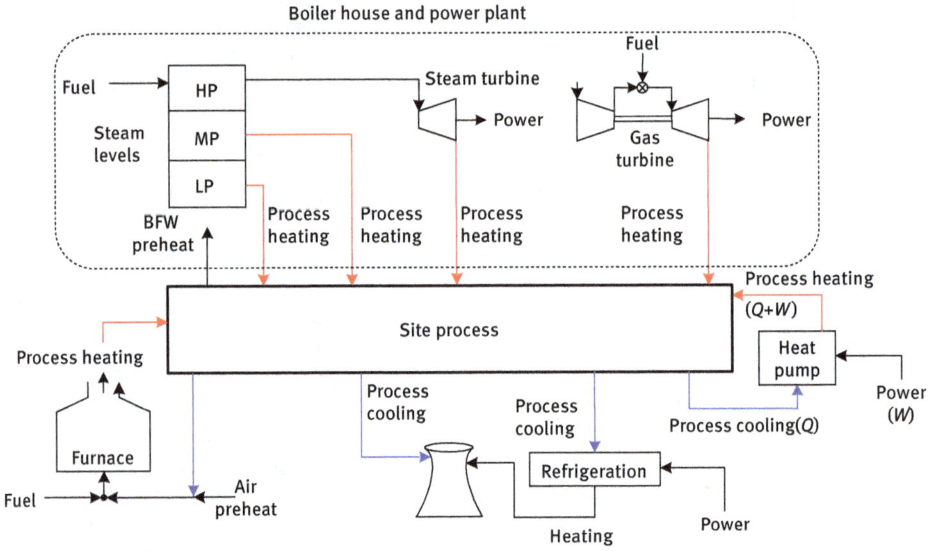

Fig. 3.1: The site utilities from the process viewpoint (amended after Klemeš et al., 2010).

As can be seen from Fig. 3.1, utility systems usually provide process heating and cooling services via various heat carriers. The heating options include:
- Steam at various levels, for serving heating requirements up to 250–350 °C (Kapil et al., 2010).

https://doi.org/10.1515/9783110630091-003

- Mineral oils and other specialised heat transfer fluids are known to be used for heating from about 50 °C to 400 °C – see Merritt (2016).
- Direct heating with flue gas, for higher temperature requirements (Broughton, 1994). The heaters of this type are referred to as "fired heaters" or "furnaces". Direct flue gas heating can also be provided by using gas turbine exhaust.
- It is also possible to use hot water (HW) for process heating if the required temperatures are below 100 °C (Klemeš, 2013). This is the case, for, instance in breweries (Eiholzer et al., 2017) and other food-processing plants.

The options for process cooling similarly feature several levels (Klemeš et al., 2018):
- Cooling water (CW): This can be used for efficient cooling due to the high specific heat capacity of water. The attainable temperatures are usually a few degree Celsius below the local current ambient temperature, due to the evaporation effect in the cooling water towers. The towers are part of the overall cooling water circuits, where the water lost with evaporation and blowdown is replaced by deionised water.
- Air cooling (using ambient air): This can be used for serving cooling demands down to about 10 °C above the ambient temperature at the current location and season. This option generally has lower operating costs than cooling with water but can have much higher capital costs and substantial space requirements, because of the lower heat transfer coefficients.
- Chilled water (ChW) (between the ambient temperature and the freezing point): This is a variation of the cooling water option, where the circulating water is cooled below the ambient temperature using refrigeration.
- Refrigeration at various sub-ambient temperatures, usually lower than the freezing point of water (0 °C. for P = 1 atm = 101.325 kPa). Linde Gas (Refrigeration: Processes & Temperatures | Linde Gas, 2019) classifies the temperature ranges as high-temperature refrigeration (above 0 °C), medium-temperature (0 to –25 °C), low-temperature (–25 °C to –50 °C) and very low-temperature (below –50 °C).

An important utility service, which is also shown in Fig. 3.1, is the use of heat pumps for providing process heating and cooling simultaneously, at the expense of using power. This option is becoming more and more popular with the increase of the share of renewables in power generation.

3.2 Utility networks overview

Within the scope of the current book, the utility systems are considered as providing heating in the form of steam and power in the form of electricity or direct drive. A typical configuration of such a system is shown in Fig. 3.2.

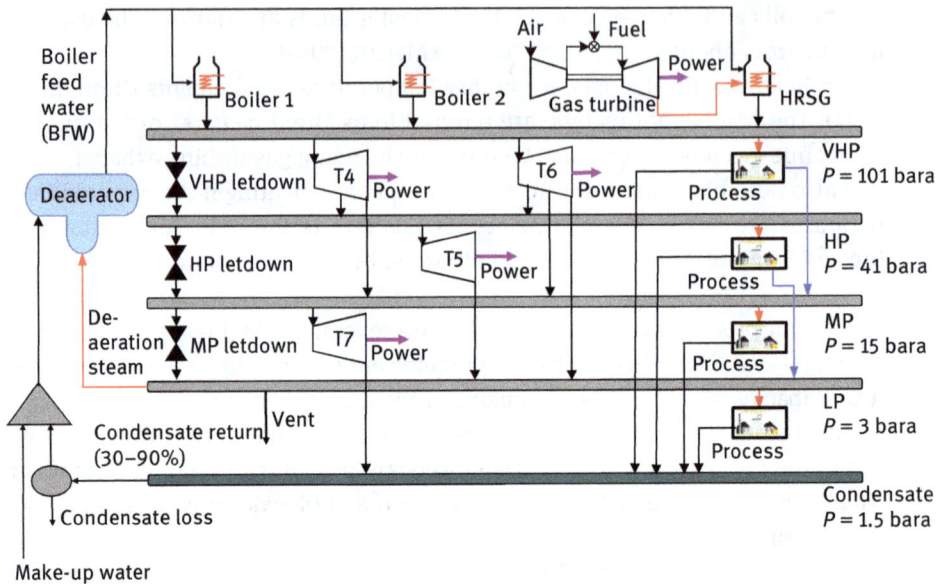

Fig. 3.2: A site utility system with all essential components.

Within the considered utility system functions, providing process heating and facilitating site-level heat recovery, based on the Total Site concept (Centre for Process Integration, 2004) is the main one, while power co-generation is an important secondary goal, stemming from the strive to maximise the fuel utilisation.

Referring back to Fig. 3.2, a utility system has several steam mains connected with each other by steam turbines and letdown stations. Each steam main stores, takes in, and provides steam with certain parameters, of which the pressure and the saturation temperature are the most important, followed by the degree of superheat. The latter is a certain excess temperature difference, above the saturation temperature of steam at the current pressure, implemented for steam generation and distribution. The main idea of having steam superheat is to minimise the possibilities for steam condensing inside the steam mains and the connecting pipes. After expansion to the various pressure levels, steam is also distributed to site processes. Some processes, releasing waste heat at sufficiently high temperature, can also supply steam, generated from that waste heat. The primary steam usually referred to as "Very-High Pressure" (VHP) steam is usually generated in steam generators (also referred to as boilers). Gas turbines (GT) are frequently installed as part of utility systems, for serving base loads or load tracking, depending on the cost of the available fuel and the dynamics of the process power demands. The gas turbine exhausts are most often routed to specialised steam generators – referred to as Heat Recovery Steam Generators (HRSG). In the HSRG, the heat of the GT exhaust can be used directly, or after adding more heat by so-called supplementary firing.

Site processes use the steam for process heating, which results in high-temperature condensate streams. A particular part of that condensate is typically lost in various ways, while the remainder, termed "condensate return", is recycled back to the utility system. The returned condensate is complemented with freshwater, usually treated to remove minerals and other impurities. For this reason, the freshwater stream is usually termed "demineralised water". The mixed stream is then fed to a device, named "de-aerator". That has the function to remove any air and other dissolved gases from the incoming water stream, for minimising the opportunities for corrosion and unwanted dynamic effects of steam systems. The de-aeration is achieved by using steam – usually from the lowest-pressure steam main. In Fig. 3.2, this is the LP steam main.

From Fig. 3.2, it can be noticed that the utility system interacts with site processes in several ways. The processes can be either users or generators of steam at different pressure levels. The heat and all or some of the power needed are commonly generated by firing fuel. Some of the VHP steam generated by the boilers is expanded through steam turbines to produce medium- and low-pressure steam, which is further provided to the site for heating proposes. Through this, the steam turbines generate power to satisfy some of the site requirements. Steam turbines by connectivity can be of two types. One type groups backpressure steam turbines, which exhaust to steam mains of lower pressure than the inlet. In Fig. 3.2, T4, T5 and T6 are backpressure turbines. On the other hand, if a steam turbine exhausts to a condenser, it is termed a condensing turbine. This is the case with T7 in Fig. 3.2.

A usually smaller share of the steam is expanded through the letdown stations for ensuring the steam balances and adjusting the steam condition for each steam main. Sometimes import of power from the grid is needed. In other cases, there can be an excess power generation on the site and it could be exported.

It is a known fact that some industrial processes require process cooling at high temperatures. For instance, Molten Carbonate Fuel Cells have cooling requirement above 650 °C (Varbanov et al., 2006) and at petroleum refineries, processes routinely feature hot streams (cooling demands) at temperatures above 250 °C (Čuček et al., 2015). This comes to indicate that the site utility systems can be used as utility exchange marketplaces. This is the case for all the spectrum of process industries – including petroleum, chemical and food. Site processes that have residual demands for utility cooling at temperatures above those of most heating demands of other processes are exploited for generating the relevant utilities – for instance HW or steam, and then further using these utilities for process heating as well as for power cogeneration. This approach has become known as Total Site Heat Integration. It is discussed in Chapter 11.

3.3 Combined cycles and Combined Heat and Power generation (cogeneration)

The terms Combined Cycle and cogeneration are often mistaken for reflecting the same practice. While they sound similar, they do differ. Combined Cycle refers to a combination of power-generating cycles, operating within different temperature ranges. This can be the case in Gas Turbine Combined Cycle (GTCC) power stations, where the GT exhaust is used to generate steam, which in turn is expanded through a condensing steam turbine. This arrangement is a pure Combined Cycle.

On the other hand, if both heat and power are generated from the same primary heat source, this is termed a CHP generation or cogeneration. It is important to note that cogeneration is the most efficient technology for utilising the combustion heat from fuel sources.

In the light of these definitions, a GTCC power station is not a cogeneration plant, but merely a simple generation one, because the waste heat from the gas turbine is used only for additional power generation. On the other hand, a GT-based cogeneration plant in a city district distributes the GT exhaust heat to heat users such as residential/commercial buildings and/or industrial processes. Combined Cycles and cogeneration are not mutually exclusive. In fact, many industrial utility systems, as the one shown in Fig. 3.2, have both features. This is because they do combine GT and steam turbines for power generation in shifted temperature ranges, but they also generate process heat.

3.4 Typical tasks

Utility systems are built, operated and decommissioned for serving industrial sites. Their typical life duration spans for decades. This is the case with the industry backbone sites as petrochemicals, metallurgy and speciality chemicals. Within their lifetime, utility systems operate, get maintained and retrofitted. This variety of conditions also defines the diversity of modelling tasks that may be appropriate. This section briefly overviews the most common tasks, without the pretence of being exhaustive.

The life cycle of a utility system follows the typical phases of an industrial plant. It is conceived conceptually, designed, built, commissioned, exploited and decommissioned (Fig. 3.3). Within most of these phases, there are essential tasks that require modelling and scoping. They are summarised below.

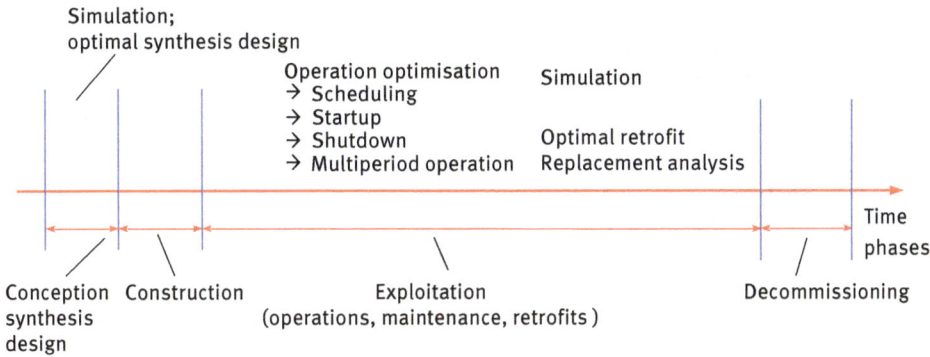

Fig. 3.3: Life-cycle phases of a site utility system.

3.4.1 Conceptual formulation and design phase

When the utility system is first conceived, estimates are derived for the main capacities of heating, cooling and power to be delivered. This activity involves a simulation of utility systems – in part or in whole, for proof of concept and obtaining estimates of the possible capacities and efficiencies. Certain optimisation models may also be run, mainly to obtain performance targets. Such performance targets are usually related to the possible heat recovery and power cogeneration (see Chapter 11). The objective function for these models is usually the minimisation of the fuel use, emissions or operating cost. Combinations of those are also possible.

During the system synthesis, the main tasks solved are system performance targeting (Liew et al., 2013), synthesis of optimal utility network (Varbanov et al., 2005) and optimal design. It has to be distinguished between the utility system synthesis and the utility system design (Westerberg, 2004). Synthesis is the activity of composing the system topology and selecting only the key design parameters, such as equipment sizes. Design is the follow-up activity of working out the further detailed design characteristics of the selected equipment, following the already determined system topology, including specific manufacturers, precise shapes and physical locations. Solving the design task also requires modelling by simulation and optimisation. The usual objective functions in this type of tasks include minimisation of total annualised costs or maximisation of net present value (Towler and Sinnott, 2013).

3.4.2 Exploitation phase

The most complicated from the viewpoint of possible tasks is the exploitation phase. Naturally, the overall network may be simulated for establishing the parameters of the existing system and its equipment items. Such an evaluation is usually part of an overall campaign for identification of a running system and implementation of an accurate model of that system in a computational model.

The identified model of the utility system can be further used for scheduling the operation in terms of loads of the various devices – steam/gas turbines, steam boilers and HRSG and letdown stations. The scheduling can involve a variation of the loads and throughputs as well as shutting down or starting up the devices. However, it should be noted that besides the gas turbines, which have a rapid start-up time of the order of 15–30 min, the other devices are much more sluggish. Steam turbines, depending on the size, may be started from a cold state for a few hours (Shirakawa and Nakamoto, 2003). Steam boilers are larger as a mass but also as dimensions, which results in slower start-up times – of the order of 10–11 h or longer (Taler et al., 2015). The main reasons for the slow start-up are the avoidance of thermal stresses in the facility.

The discussed characteristics of the dynamics of utility system components add complexity to the scheduling of utility system operation for optimal economic and/ or environmental performance. The complexity increases even further when one realises that steam boilers can have both a cold standby state and a warm standby state. In the latter, the boilers are kept warm, to enable fast ramp-up of the steam generation, but in this state, no steam is generated.

For cases when the site processes have relatively stable steam and power demands in time, one can apply operational optimisation to a multiperiod case, where the inactive states of the devices can be neglected, or the dynamic effects of the start-ups and shut downs can be neglected.

Finally, another significant activity, which requires modelling, is the retrofit of utility systems. Sometimes this task is as simple as replacing older equipment with a new one, after the expiration of the certified lifetime. In other cases, changes on the site and/or in the market conditions may require more extensive changes to the utility system, to the extent of a full system retrofit – adding, moving, replacing various components, maybe even steam mains.

3.5 Other utility system components and functions, beyond the scope of the book

Besides heating, cooling and power utilities, on an industrial site, there are other utility sub-systems, usually considered separately. One such utility is compressed air, generally used for driving some equipment and other specialised applications such as airlift.

Another example is process water. In fact, process water management is such an essential activity for industrial sites that there have been specialised research and engineering efforts for water management (Bandyopadhyay et al., 2006), water purification (Sahinkaya et al., 2017) and Water Integration (Wan Alwi and Manan 2013). The water networks integration approaches have evolved from the pioneering Water Pinch (Wang and Smith, 1994) to the more complex "total water systems" (Gunaratnam et al., 2005) and the holistic design of cost-optimal water networks (Sujak et al., 2017).

Further utilities can be hydrogen in refineries (Elsherif et al., 2015), the demineralised water itself – distributing any excess after supplying the steam and cooling water cycles, inert gases, effluent treatment and disposal.

3.6 Developments of utility system modelling

3.6.1 Supertargeting of utility systems: Accounting for investment costs and emissions

Industrial sites spend large amounts of energy emitting consequently considerable CO_2 emissions. Heat recovery at Total Sites provides one option for energy saving. A related option is the onsite power cogeneration, which is in a trade-off with the amount of fuel used and the related emissions. Varbanov et al. (2013) provided a model and concepts how to obtain the relevant targets and evaluate the trade-off of capital cost versus greenhouse gas footprint for a utility system with CHP generation, having a set of specified steam pressure levels.

The targeting model for a capital cost for power cogeneration at site level is based on the Total Site heat recovery targets. They estimate the overall needs for heat and power as well as energy losses, also incorporating the R-curves analysis initially developed by Kimura and Zhu (2000). After the heating/cooling requirements of the processes are identified, a simplified model of the steam network can be established, as shown in Fig. 3.4a. Another, more popular representation of the same information is the Utility Grand Composite Curve (UGCC) (Klemeš et al., 1997) shown in Fig. 3.4b.

The capital cost of power cogeneration is a function of the desired power production capacity. It depends on the inlet and outlet temperatures of steam in the turbines, the turbine efficiency, capacity and also the turbine design. The UGCC is constructed from the differences between utility generation and utility use at each utility level, usually considering only steam (Smith, 2016). The steam levels partition the overall temperature range for the Total Site into so-called expansion zones (Varbanov et al., 2005) – see Fig. 3.4.

The power generated by steam turbines is calculated using the model developed by Varbanov et al. (2004b), using the Willans Line approximation for the calculation. Power cogeneration potential is calculated for each expansion zone by assuming that all steam available in the zone is used for power generation and that a steam turbine matching the load is used for that. After that, a total sum over all

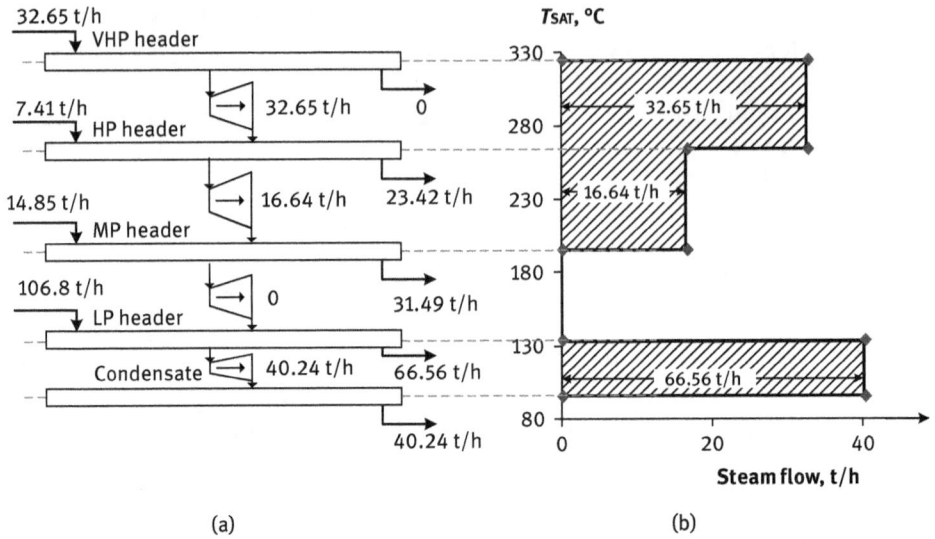

Fig. 3.4: A simple steam network: (a) steam network and (b) Utility Grand Composite Curve (UGCC) (after Varbanov et al., 2013).

expansion zones is produced. The capital cost corresponding to the targeted power cogeneration is calculated as:

$$CC = A \times N_{\text{MIN}} + B \times \sum_{i=1}^{n} W_i \qquad (3.1)$$

where A is the coefficient of turbine installation (USD); B is the coefficient of marginal capital cost per 1 MW power generation, (USD/MW); N_{min} is a minimum number of turbines based on the overall power target; n is the number of expansion zones; W_i is the power generated by expansion zone "i" (MW).

The minimum number of steam turbines can be derived from eqs. (3.2) to (3.4):

$$N_A = \left(\sum_{i=1}^{n} W_i \right) \text{div } W_{\text{UTHR}} \qquad (3.2)$$

$$N_B = \left(\sum_{i=1}^{n} W_i - N_A \times W_{\text{UTHR}} \right) \text{mod } W_{\text{LTHR}} \qquad (3.3)$$

$$N_{\text{MIN}} = N_A + N_B \qquad (3.4)$$

where *div* and *mod* are the integer division and remainder operators. The calculation is based on the total power generated. Combined with the lower (W_{LTHR}) and upper (W_{UTHR}) bounds on steam turbine capacities, N_A and N_B are intermediate variables meaning the number of largest turbines allowed on the site and number of smallest ones. Based on the logic of the equations, N_B can take only values of 0 and 1.

The carbon footprint (CFP, better greenhouse gas footprint including more GHGs) (Čuček et al., 2012) for the site has two major components – the CO_2 emissions from building the steam turbines and that from the fuel burned for steam generation. Both components can be joined to estimate the total annualised CFP of the site:

$$CFP \left[t_{CO_2}/y\right] = Q_{FUEL}[MW] \times EF_{FUEL}\left[t_{CO_2}/MWh\right] \times 8,760 \, [h/y] \qquad (3.5)$$

where Q_{FUEL} is the heat load to be covered by fuel, EF_{FUEL} is the CO_2 emission factor for the used fuel.

Using CFP [t/y], an intensive indicator can be defined – the CFP intensity of the site I_{CFP}:

$$I_{CFP} \left[t_{CO_2}/MW/y\right] = \frac{CFP \left[t_{CO_2}/y\right]}{W_{GEN}[MW]} \qquad (3.6)$$

The embedded CFP in the steam turbines is also possible to estimate. But because of spreading its value over many years of service life (usually a few decades), its intensity per unit of power generated is an order of magnitude smaller than the fuel-related component (Varbanov et al., 2018). Water and other relevant footprints can be calculated in a similar way, linking them to the fuel consumption and the equipment size.

The industrial site example from Fig. 3.4 was evaluated in the study by plotting the capital cost and the cogeneration efficiency against the power-to-heat ratio (*R*) for the site demands. The curves are shown in Fig. 3.5, which exhibits a linear trend of the capital cost with increasing onsite power generation and the value of the power-to-heat ratio, while the site cogeneration efficiency understandably decreases asymptotically to the value typical for the efficiency of condensing steam turbines.

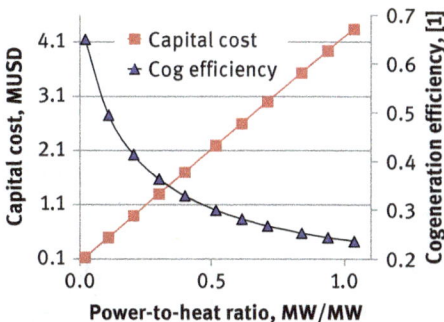

Fig. 3.5: Targets for capital cost and cogeneration efficiency (after Varbanov et al., 2013).

The CFP and its intensity were then investigated, as shown in Fig. 3.6. The trend of the CFP is a linear growth with increasing the power generation, which is due to the directly proportional relationship between the CFP and the fuel burnt in the boilers. More interesting is the trend for I_{CFP}. It represents an asymptotic decrease of the CFP intensity of the site with increasing the power generation. When compared with the trend of the cogeneration efficiency, this looks strange. However, the decreasing CFP intensity is a result of a trade-off between the increasing power generation efficiency and the decreasing overall cogeneration efficiency.

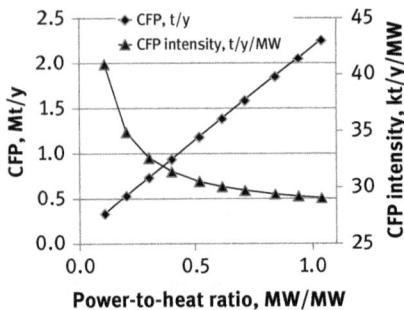

Fig. 3.6: Targets for the CFP and its intensity (after Varbanov et al., 2013).

3.6.2 Utility system cogeneration targeting with a process simulator

It is crucial to target the cogeneration potential of utility systems in the process of designing or retrofitting Total Site utility systems. Simulation software has been widely used in process design and analysis, but rarely for utility system specifically. The cogeneration potential of utility systems has been evaluated using a process simulator by Ren et al. (2018).

3.6.2.1 The principle and main advantages

The simulator is used to calculate the temperatures of steam mains, steam flow rates and shaft power rates. Shaft power generation by steam turbines in expansion zones is computed with the built-in simulation module in Aspen Plus (2019). The advantage of this method is that it needs neither establishing iterative calculation nor a programming procedure, and it requires fewer parameters. The simulator is sufficiently flexible to allow calculating also the capital cost targets and emission targets, building upon the model from the previous section. The calculation of the steam turbine capital cost and environmental footprint targets provides the decision-makers with a more complete picture of the designed sites. This should allow them to efficiently screen and compare the major design options on the basis of economic and environmental performance.

The simulation software provides a flexible and robust calculation framework that ensures convergence of material and energy calculations for process industries (Taufiq et al., 2015). The simulation software also has many built-in model blocks and a comprehensive database of the fluid properties – including for water and steam, which can be directly used in the simulation of utility systems. The simulators can guarantee the data accuracy in the calculation process when the input data are credibile. The mentioned features would make it much easier for the engineers to calculate the cogeneration potential of the site utility systems. The input data interface for the cogeneration estimation can all be collected using the same simulator. This makes it much easier for engineers to carry out the calculation inside the simulator too.

3.6.2.2 Steam turbine simulation and thermodynamic property method

The study, described in that paper, uses built-in modules of Aspen Plus for simulating single-stage steam turbines. This was performed based on an assumption of no steam condensation inside the turbine, which is also a standard requirement for real steam turbines. The model takes a specification for a given isentropic efficiency.

In the simulation, the thermodynamic model plays an important role because it directly affects the accuracy of the physical properties of the calculation and accuracy of the results. From the literature, it can be seen that linear or nonlinear thermodynamic models should be established for calculation of the cogeneration potential of the site utility systems. The simulator offers several thermodynamic property packages suitable for the purpose – STEAM-TA, STEAMNBS and IAPWS-95. The authors selected the STEAMNBS package, pointing out that it provides smooth correlations seamlessly handling the phase changes between water and steam.

The cogeneration targeting procedure involves a series of simulation runs, where the algorithm ensures that each steam main has the degree of superheat specified by the user initially, applying a convergence loop with a variation of the temperature of the boiler steam. Another convergence loop is used to adjust and minimise the discrepancy between the specified and the simulated flow of LP steam for process heating.

Figure 3.7 shows the process flow diagram (PFD) of a simplified utility system for the targeting and Fig. 3.8 shows its counterpart flowsheet representation in Aspen Plus. It can be seen that the approach includes modelling each of the steam mains as a sequence of a mixer and a splitter, as also discussed previously by Varbanov and Klemeš (2011). The capital cost and emission targets are also obtained, using the estimates for power cogeneration, steam turbine sizes and fuel use.

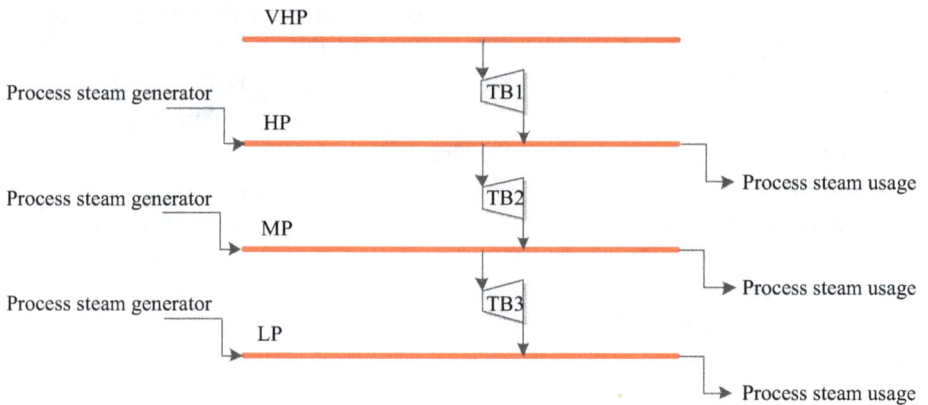

Fig. 3.7: A process flow diagram for the utility system under targeting (after Ren et al., 2018).

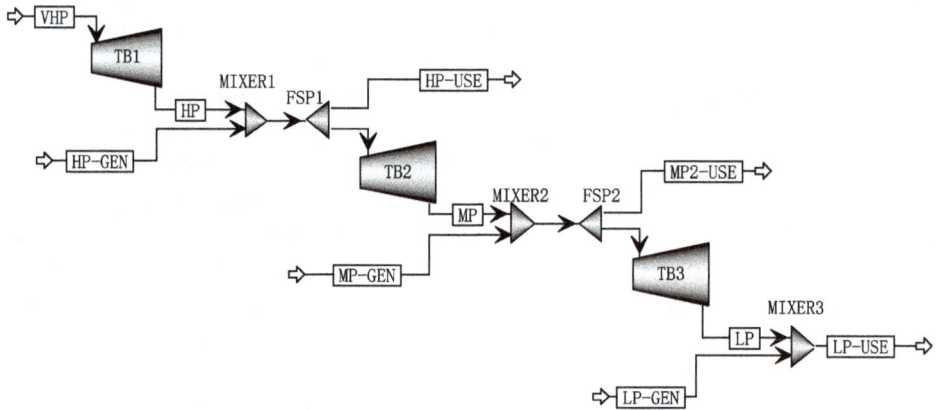

Fig. 3.8: Aspen Plus flowsheet corresponding to the PFD in Fig. 3.7 (after Ren et al., 2018).

3.7 Water minimisation and integration

Water is an essential utility on industrial sites, for both process use and thermal applications. Water supply together with the minimisation of freshwater use and wastewater generation define one of the fields where the Pinch Methodology has been widely implemented. Water Pinch Analysis (WPA) is the systematic technique for analysing water networks and reducing expenditures in different water-using processes

(Klemeš et al., 2018). Various WPA approaches have been developed, which are mainly categorised into two groups: insight-based Pinch Analysis and Mathematical Programming-based targeting approaches (Foo, 2012).

3.7.1 The initial Water Pinch method

Based on the mass exchange network theory initiated by El-Halwagi (1989), Wang and Smith (1994) developed the insight-based approach of Pinch Analysis and extended this approach to wastewater treatment network integration. The insight-based techniques mainly include graphical targeting techniques, that is Material Recovery Pinch Diagram (MRPD) developed by EI-Halwagi et al. (2003), and algebraic targeting techniques, that is Material Cascade Analysis (MCA). Widespread studies have also been conducted on the topic of mathematical programming targeting (Mughees and Al-Ahmad, 2015). They have been reported in review papers (Foo, 2009) as well as in books (Foo, 2012). Industrial applications have also been reported, for example (El-Halwagi, 2017). Several chapters in the *Handbook of Process Integration* (Klemeš, 2013) further interpreted the approaches for water management and minimisation based on the WPA and introduced the Total Site concept. The application of WPA and superstructure-based optimisation techniques are described, and a new process-based graphical approach is proposed for the simultaneous targeting and design of water networks.

3.7.2 Follow-up developments for Water Integration

A typical WPA solution has two steps of setting the water targets and followed by network design to achieve the targets (Manan et al., 2006). In the work of Wang and Smith (1994), the targets of using freshwater, maximise water reuse and water regeneration are discussed, and both single and multiple contaminants are addressed. Composite Curves are used to represent the water using process. The approach is able to identify the bottleneck for water minimisation in industrial processes, and the conceptual framework can help to specify the optimal type of regenerator and regenerator specification. That work is one of the earliest contributions of graphical methods for water minimisation studies using WPA.

In recent studies, the method has been further discussed and developed. For example, Tan et al. (2007) developed a systematic method for the retrofit of water network with regeneration based on WPA. The procedure consists of two parts: retrofit targeting and design for a water network with regeneration unit(s). At the targeting stage, retrofit targets (utility savings and capital investment) were determined for a range of process parameters (total flowrate and/or outlet concentration of the regeneration unit) to obtain savings versus investment curve. Next, the existing water network was re-designed to meet the chosen targets. A case study on a papermaking process was used to demonstrate the methodology.

Tan et al. (2009) started to combine water footprint and WPA. In that study, a graphical Pinch Approach for the analysis of water footprint constraints on biofuel production systems is presented. The method is based on the Composite Curve method, which was originally developed for carbon-constrained energy planning, which is extended in this paper based on the underlying similarities of source-sink allocation problems. The Pinch Analysis approach enables limiting water footprint conditions to be identified and provides insights that are useful for planning the large-scale cultivation of biofuel crops. An illustrative case study based on the bioethanol program of the Philippines is solved using the proposed approach.

Jia et al. (2015) extended the Water Footprint Pinch Analysis procedures, which are based on the decomposition of total water footprint into external and internal footprint components. Results show that water is mainly consumed in the utility processes, and it is possible to achieve a goal for water-saving of 16 %.

Several studies have considered the combination of different methods. For instance, Mughees and Al-Ahmad (2015) applied the WPA in water minimisation by combing graphical and mathematical methods using the LINGO optimisation environment (LINGO, 2019). Their results showed that this approach is easy to apply and can provide more precise results than other techniques. The authors noted that this approach is still a single-contaminant-oriented method, and the multiple-contaminant problems should be further investigated in future works. Liu et al. (2016) presented a "plant-based" model for solving water allocation problems within industrial parks. Superstructures are established and mathematically formulated, aiming to minimise freshwater consumption as well as the total annualised cost. At last, three integration cases are explored based on an example from literature for illustration.

3.7.3 Recent industrial implementations

Pinch Analysis has been implemented in industrial water saving, including industrial water using processes, evaporation systems and water recycling systems. Hu et al. (2011) used WPA to investigate the effect of different process decomposition, strategies on saving freshwater usage by applying a concentration–mass load diagram in the analysis. For multiple-contaminant water systems, the approach for the determination of interim concentrations for concentration decomposition is explored. Three sequential mathematical models and a related optimisation procedure to optimise regeneration reuse water networks are proposed.

Mohammadnejad et al. (2012) applied WPA to analyse the water network of a Tehran oil refinery and considered three key contaminants including suspended solids, hardness as well as COD. Results show that water minimisation through single contaminant approach was more considerable, while the method based on the double contaminant gives more precise results rather than a single contaminant.

Mughees and Al-Ahmad (2015) implemented the WPA of water minimisation at a refinery and set COD and hardness contaminants with single and double contaminants approach. The new technique was demonstrated to favourably compare to previous studies. The method resulted in a high percentage reduction of 43.8 % and 61.2 % of fresh water in three processes regarding COD and hardness.

Skouteris et al. (2018) proposed an algebraic technique and utilised the approach for the targeting of water regeneration, where an interception unit is used to partially purify the water sources for further reuse/recycle. The study applied Pinch Analysis for the water management and optimisation in a brick-manufacturing industry. They concluded that combined use of water footprint with Pinch Analysis can provide water-intensive manufacturing industries with a sound and a robust water management tool that can significantly improve their water consumption and consequently their long-term sustainability.

3.7.4 Possible directions of development

Graphical methods are practical to solve single-contaminant problems, but are complicated and sometimes impossible to apply to multiple-contaminant problems. Mathematical methods are more exact but sometimes complicated especially in the case of multiple contaminants (Ataei and Yoo, 2010). Modelling such problems can be performed using an algebraic language and environment – such as those provided by GAMS (2019). The development and implementation of WPA showed that regarding water minimisation, the consideration of water quality, especially the multiple contaminants, is still an issue that needs further effort. Integration of different methods is one of the possible directions for future developments, in order to tackle the increasingly complex water shortage and degradation issues.

3.7.5 Hydrogen Integration

Hydrogen networks play an important part in petroleum refineries by providing hydrogen for optimising the fuel production in a flexible way (Hallale et al., 2017). Besides providing the hydrogen to processes for the generation of lighter fuels, decreasing its content in fuel gas allows lower temperature combustion in fired heaters and, in turn – lower cost materials for the tubes. Having that in mind, the hydrogen management has been also one of the core domains of applying the Pinch Principle. Since the initial introduction of the Hydrogen Pinch (Alves, 1999), there have been a number of applications and developments, as reviewed in Elsherif et al. (2015). Analysing the spread of those works over the sub-topics, it has revealed a substantial share of Pinch Analysis papers. Most works use Mathematical Programming. This can be explained by the lower level of domain expertise required to start building the mathematical models as well as the need in some cases to model hydrogen networks with multiple contaminants. On the other hand, Hydrogen Pinch Analysis (HPA) provides a good conceptual basis and leads to a proper understanding of the refinery hydrogen flows and the system limitations. That is the driver to have also the substantial share of 15 papers based on HPA.

A recent study (Marques et al., 2017) focused on reviewing the Hydrogen Integration methods involving targeting. It found an extensive list of works based on Composite Curves and the Pinch Principle. The paper summarises the data extraction procedure for hydrogen network optimisation and reviews the insight-based (mainly Pinch) methods and the ones using mathematical programming. Among the Pinch-based methods, they start with the pioneering work of Alves (1999) and the follow-up developments accounting for pressure changes (Ding et al., 2011), introduction of a "gas cascade" (Foo and Manan, 2006) – an equivalent to the Problem Table (Linnhoff and Flower, 1978) in Heat Integration – as well as the extension to handling multiple impurities with the HPA (Zhao et al., 2007) – a traditionally difficult issue, considering the nature of intensive versus extensive indicator analysis to arrive at a Pinch Point. The reviewed fundamental tools include the Hydrogen Composite Curves, the Hdrogen Surplus Diagram (Alves, 1999), Average Pressure Profiles (Ding et al., 2011), Material Surplus Composite Curve (Saw et al., 2011) and Material Recycle Pinch Diagram (MRPD) (El-Halwagi et al., 2003) – differing from other representations that it uses the dimensions of impurity load versus stream flowrate. There has been even a spreadsheet implementation of HPA (Marques et al., 2017). The authors described the software features, including identification of bottlenecks in refinery hydrogen networks and generation of options for appropriate placement of purification units within the networks. This software tool is intended for analysis of the potential for hydrogen reuse in refineries.

Using a variation of the MRPD, Yang et al. (2016) have proposed a non-iterative method for hydrogen networks integration with consideration of purification units and reuse. The authors analysed also the method limitations, which include fixed concentrations of the feed streams and those of the purified and tail gas streams exiting the purification units.

A further improvement of the MRPD (El-Halwagi et al., 2003) has been the "relative concentration based Pinch Analysis" for targeting and design of resource networks (Zhang and Feng, 2012) where key the resources considered are hydrogen and water. They have replaced the contaminant load versus resource flowrate dimensions with contaminant concentration versus resource flowrate. The proposed Composite Curves diagram is used for deriving design targets and design rules, allowing to save more primary resource by applying appropriate streams purification and reuse.

A recent novel concept, transferred from water networks to hydrogen management, has been discussed by Wang et al. (2015). They introduced the concept of "concentration potential" to hydrogen networks, to enable their targeting and design in the presence of multiple contaminants. They have formulated a heuristic design procedure, which is reported to produce results comparable with previous methods based on Mathematical Programming.

The site and inter-plant level of Hydrogen Integration has been also investigated (Deng et al., 2014). The study uses a two-stage procedure, following the logic of Total Site Heat Integration, where the integration targets are first obtained at the process level and a hydrogen cascading tool named "Improved Problem Table" is used for targeting the overall inter-plant hydrogen network.

3.8 Water–Energy Integration

In many industrial systems, water, heat and power are intrinsically linked together. An example can be given from a recent case study on ethanol production (Pina et al., 2017), where just appropriate Heat Integration resulted in a simultaneous reduction of fuel and water consumption up to 20–30 %. A comprehensive review of Water Integration involving temperature change and heat transfer has been presented by Ahmetović et al. (2015), emphasising the network synthesis. The review covers both Pinch Analysis–based and Mathematical Programming methods, as well as their combined use. The authors analyse the main development and challenges in the area, identifying the relative lack of considering batch water networks.

Integrating the water and heat reuse/recovery has notable advantages for saving resources of both types, as discussed in Savulescu and Alva-Argaez (2013). That book chapter provided an introduction to the basic concepts of Water–Energy Integration, which deals with a water reuse networks while also accounting for the heating and cooling requirements of the involved streams. The concepts include the general problem setup, the fundamental assumptions and heuristics used, water-saving paths and the Two-Dimensional Grid Diagram (temperature vs. water quality). The further conceptual overview also mentions the use of the Heat Integration Composite Curves (Klemeš et al., 2018) within the context of water–energy networks (WENs), which adds a "Modified Cold Composite Curve" – lumping cold water streams for indirect heat exchange, and the definition of "separate systems" based on enthalpy blocks identified on the Composite Curves and non-isothermal stream mixing. These and other fundamental concepts of Water-Energy Integration are described in full details in the underlying Works – Water-Energy Integration without water reuse (Savulescu et al., 2005a) and the extension to systems with maximum water reuse (Savulescu et al., 2005b). These initial works have been then revisited by Leewongtanawit and Kim (2009), who have added the concept of "stream merging", which combined with the generation of "Separate Systems", besides simplifying the network design, also allows implementing vertical heat transfer in the surface heat exchangers, resulting in minimal heat transfer area. They have also defined a procedure for minimising the number of Separate Systems based on the concept of stream merging for maximising the temperature driving forces within each Separate System. The results showed a substantial decrease of both thermal utility demands and heat transfer area. An additional analysis of the strategic industrial implications of water–energy interactions is given in (Varbanov, 2014).

As non-isothermal mixing is used as one of the degrees of freedom in Water–Energy Integration, the analysis of its influence on the energy targets has been performed (Luo et al., 2014). Using the Composite Curves as a tool and thermodynamic reasoning, the authors revealed the possible cases and their implications on the utility targets. They have found that there are several configurations of streams. Mixing the same-type streams always causes energy penalty, while mixing a hot stream below the Pinch with a cold stream above the Pinch brings no energy penalty.

A natural extension of Water–Energy Integration within a single process is the integration between plants – at the site level. Liu et al. (2018) explored the issue, using a Mathematical Programming model. While the underlying model is discrete and non-linear (MINLP), the authors applied a model reformulation, obtaining a continuous (NLP) model. The positive trend from this is the development at the site level. However, that still left a conceptual gap for providing a site-level targeting model, rooted in Process Integration.

Another case, where energy and water are intrinsically related, is the production of biofuels based on dedicated crops (Tan et al., 2009). That study presents a model to evaluate the water footprint constraints on the production of biofuels within a given set of regions. The developed tool is a pair of Composite Curves – demand and supply, representing the energy demands with their maximum allowed water footprints on the one hand and the possible biofuel energy sources with their water requirements for production. The targeting with the tool allows establishing the minimum energy deficit to be covered by input and the potential energy export,

Considering the combined Water–Energy Integration as a special case of the Energy–Water Nexus implies the need to develop a systematic tool for modelling the systems incorporating it. Such a tool has been proposed by Tsolas et al. (2018). They formulated a pair of representations – a Water–Energy Nexus Graph and the Water–Energy Nexus Diagram – in turn consisting of an Energy Composite Curve and a Water Composite Curve. The study provides also a procedure for identifying redundant water–energy loops and rules on how to avoid them for new designs or eliminate them for existing systems. The developed tools have been illustrated in case studies from power generation, desalination and regional water management, demonstrating its scalability.

3.9 Trigeneration System Cascade Analysis (TriGenSCA)

CHP generation (cogeneration) is widely recognised as a superior to simple generation (Klemeš et al., 1997). This is the best possible technology when the served industrial site has mainly heat and power demands and only moderate cooling demands. For sites that also have sizeable demands for cooling below the ambient temperature, generating simultaneously heating, cooling and power from the same primary energy source becomes the preferred option, maximising the fuel efficiency. This is termed as trigeneration (Al Moussawi et al., 2016).

Trigeneration System Cascade Analysis (TriGenSCA) is a new numerical method (Jamaluddin et al., 2019) to minimise power, heating and cooling targeting for a site, as well as to optimise sizing of the turbines (gas and steam), absorption chiller, cooling tower and steam generator . It applies a method for Total Site cooling, heating and power integration, extended from Total Site Heat Integration. The focus is on integrating heating, cooling and power for multiple sites. The procedure follows eight steps which are: data extraction, Problem Table Algorithm (PTA) (Liew et al., 2012) for an individual process, Multiple Utility Problem Table Algorithm (MU-PTA) for an individual process, Total Site Problem Table Algorithm (TS-PTA) for all processes, estimation of energy source from the trigeneration system, TriGenSCA, Trigeneration Storage Cascade Table (TriGenSCT) and Total Site Utility Distribution (TSUD) to obtain the optimal sizes of trigeneration system parts.

The trigeneration system is implemented as a centralised energy system to supply power, heating and cooling services to the demands, as shown in Fig. 3.9. Following the figure, VHP steam is produced in a steam generator (boiler) and passes through a double extraction turbine simultaneously producing power, and lower pressure steams such as HP steam and LP steam. The HP steam produced from the double extraction turbine can be supplied to meet the demands directly or expanded to LP steam using a letdown station. Excess HP and LP steam can be cooled down by using CW or condensed by using the condensing turbine to generate more power. Condensing turbines can adjust their electrical output by varying the proportion of steam passing through the turbine. HW, on the other hand, is generated by using the condensation system. HW can then be used either directly for serving heating demands or converted to cooling utilities such as ChW by using an absorption chiller.

The cooling tower is generally used to cool the circulating CW via evaporation (Pitcher, 2015). Operation of the cooling tower starts with pumping the returning CW to enter at the top of the cooling tower through nozzles. The CW flowing through the nozzles is dispersed onto a large surface area which is also known as a fill. The fill is used to delay water from reaching the bottom of the tower and allow more time for contact between the falling water and the air.

The water then slowly makes its way down through the fill tanks via gravity until it reaches the bottom of the tower and a fan forces air up across the water path. The CW utility is then produced at the bottom of the tower and supplied to the

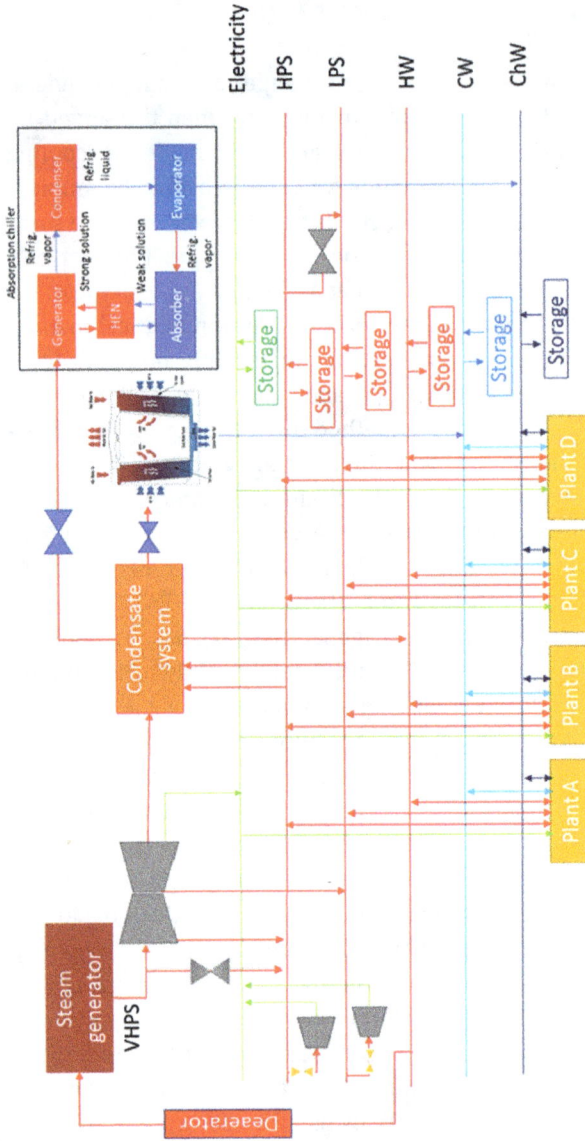

Fig. 3.9: A trigeneration configuration for a site utility system (after Jamaluddin et al., 2019).

process cooling demands. The absorption chiller, on the other hand, consists of four main components, which are generator, condenser, evaporator and absorber (ARANER, 2017). The process of producing ChW by using absorption chiller can be summarised as:

1) Generator – the HW produces refrigeration vapour from a strong refrigerate solution by transferring heat from the HW to cool the solution. The refrigeration vapour needs to pass through a rectifier for dehydration before it enters the condenser.

2) Condenser – dehydrated and high-pressure refrigerant enters the condenser where it is condensed. The refrigerant goes through an expansion valve after cooling. The expansion reduces the pressure and temperature of the refrigerant. The temperature of the refrigerant at this point must be lower than the stream posing the cooling demand, which passes through the evaporator.

3) Evaporator – the cooled refrigerant enters the evaporator, absorbs heat from the served stream and then leaves as saturated refrigeration vapour.

4) Absorber – the refrigeration vapour is exposed to a spray of the weak refrigerant-absorbent solution. As a result, the weak solution changes to a strong solution. The new solution passes through regenerator, which is also known as a heat exchanger. The solution that arrives at the generator has the same pressure as before. The process is then repeated.

3.10 Industrial implementation issues

While modelling industrial utility systems has its benefits, it is important to remember that all computational models are simplified representations of the existing systems or the ones being designed (Dym, 2004). The utility system models are no exception to this pattern. Many simplifications have been adopted in utility system models. Some of them include the assumption for fully isothermal profiles of steam generation and steam use in the early works – for example, Dhole and Linnhoff (1993). Another was the assumption of a uniform lower bound on the temperature driving forces for site-level heat recovery. For bringing the recommendations of computational optimisation studies to real solutions, implementation issues have to be accounted for.

Such issues have been identified by experts with considerable, multi-decade, industrial experience (Chew et al., 2013). These can be grouped into categories for design, operation, reliability/availability/maintenance, regulatory/policy and economics. Some of the key design issues include plant layout and pressure drop. Operation issues of importance include start-up and shutdown. Reliability, availability and maintenance are also important, to ensure uninterrupted and risk-free production. Government policies are also important and they have to be considered within the context of the process economic performance.

Within the context of design issues, layout is important for grassroots design and even more so for retrofit. This aspect of the process design affects multiple system properties such as piping, pressure drops, energy demand for pumping. Connected to this is the issue of appropriate selection of operating pressures. Such a selection influences not only the power demands and costs, but also the cost of equipment, as higher pressure equipment cost more than lower pressure analogues.

Various risks, such as contamination and explosion, have also to be considered. A common contamination problem on an industrial site relates to condensate return from used steam. Another risk from the domain of petro-chemical industries is the explosion risk of process units handling hydrogen-rich streams. Most often, such issues are handled during the stage of data extraction for Process Integration, where the engineers should clearly describe the potential risks associated with each candidate process and stream for integration and resource recovery. Further practical constraints that should be considered, include fluid flow (high/low viscosity), risks of sudden pressure fluctuations and even phase changes, materials of construction in relation to the desired temperatures,

Similarly, operability requirements for the utility systems have also to be considered. A major requirement is to ensure sufficient flexibility for accommodating planned or unexpected variations in the process throughput and external conditions – including operation scenarios, start-up and shutdown, emergency situations, external disturbances. The flexibility of the system has also to be supported by sufficient means of control.

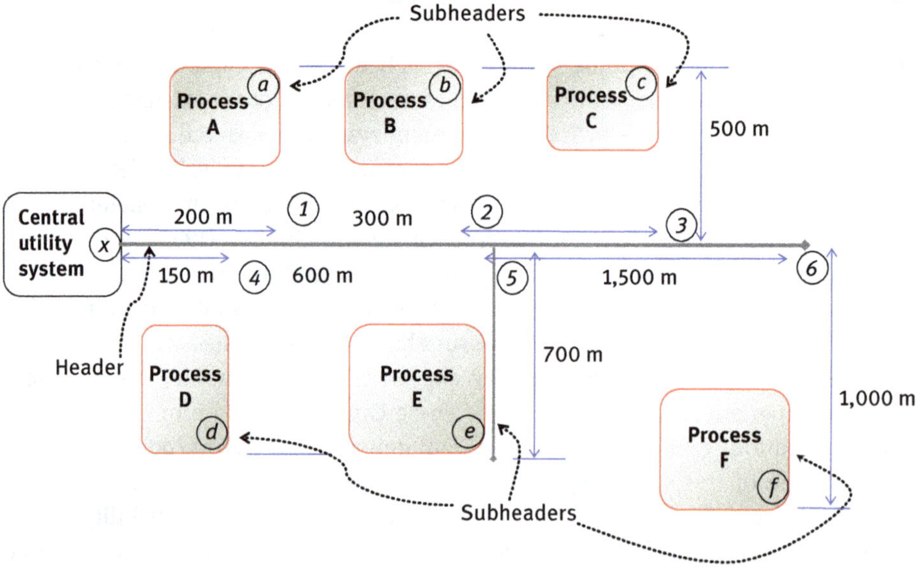

Fig. 3.10: An example of layout of processes that can supply low-pressure steam (amended from Chew et al., 2013).

The overview of practical issues in utility systems and Total Sites (Chew et al., 2013) illustrates the significance of such considerations on an example with process layout. They show a part of a utility system for generating low-pressure steam from process heat, which features horizontal distances and elevation variations (Fig. 3.10).

The various process units also have different pressure levels available – ranging from 3.0 to 5.25 barg. Without accounting for the distances and elevation differences of the layout in Fig. 3.10, an engineer may be misled to estimate the potential for low-pressure steam generation to the overall sum of capacities offered by all six process units (12.1 MW). Out of that potential, however, after accounting for the layout constraints, it becomes clear that only 2.1 MW of steam is feasible to be generated.

3.11 Summary

This chapter provided an overview of utility systems – including the main function, architecture, involved equipment types and typical arrangements – steam mains, steam and gas turbines, letdown stations, boilers, connected to the processes and arranged according to the operating pressure. The typical options for utility heating and cooling are covered. The life cycle of the utility system as part of the overall site is discussed and related to the optimisation tasks solved at each stage. The main recent developments in the modelling and evaluation methods concerning utility systems are discussed – including the targeting of power cogeneration, utility and investment costs and environmental footprints, Water/Water–Energy Integration as well as Hydrogen Integration in the context of site utility systems and trigeneration. Finally, the industrial implementation issues are overviewed, pointing the readers to the appropriate sources for further detailed information.

Nomenclature

Symbol	Measurement unit	Description
A	USD	Fixed-term coefficient (regression) for calculating steam turbine capital cost
B	USD/MW	Regression coefficient of marginal capital cost per 1 kW steam turbine power generation
BFW	–	Boiler Feed Water
CC (equation)	USD	Capital cost of steam turbines
CFP, GHGFP	t_{CO_2}/y	Carbon Footprint, GHG Footprint
CHP	–	Combined Heat and Power
ChW	–	Chilled water
COD	mg/L	Chemical Oxygen Demand

(continued)

Symbol	Measurement unit	Description
EF_{FUEL}	t_{CO_2}/MWh	CO_2 emission factor for the used fuel
GHG	–	Greenhouse gas
GT	–	Gas turbine
GTCC	–	Gas turbine combined cycle
HP(S) (steam)	–	High Pressure
HPA	–	Hydrogen Pinch Analysis
HP-GEN, MP-GEN, LP-GEN	t/h	Flowrate of steam generation at high, medium and low pressure
HP-USE, MP-USE, LP-USE	t/h	Flowrate of steam use at high, medium and low pressure
HRSG	–	Heat Recovery Steam Generator
HW	–	Hot water
I_{CFP}	$t_{CO_2}/MW/y$	CFP intensity
LP(S) (steam)	–	Low-Pressure steam
MCA	–	Material Cascade Analysis
MP(S) (steam)	–	Medium Pressure
MRPD	–	Material Recycle Pinch Diagram
MU-PTA	–	Multiple Utility Problem Table Algorithm
N	[1]	Number of expansion zones
N_A, N_B	[1]	Intermediate variables for calculating the minimum number of steam turbines
N_{min}	[1]	Minimum number of steam turbines
PFD	–	Process Flow Diagram
PTA	–	Problem Table Algorithm
Q	MW	Heat flow
Q_{FUEL}	MW	Heat load to be covered by fuel
R	[1]	Power-to-heat ratio
T4, T5, T6, T7	–	Steam turbines
TriGenSCA	–	Trigeneration System Cascade Analysis
TriGenSCT	–	Trigeneration Storage Cascade Table
TS-PTA	–	Total Site Problem Table Algorithm
TSUD	–	Total Site Utility Distribution
UGCC	–	Utility Grand Composite Curve
VHP(S) (steam)	–	Very High Pressure
W	MW	Power
WEN	–	Water–energy network
W_{GEN}	MW	Generated power
W_i	MW	Power generated by expansion zone
W_{LTHR}	MW	Lower bound on steam turbine capacities
WPA	–	Water Pinch Analysis
W_{UTHR}	MW	Upper bound on steam turbine capacities

Reerences

Ahmetović, E., Ibrić, N., Kravanja, Z., and Grossmann, I.E. 2015. Water and Energy Integration: A Comprehensive Literature Review of Non-Isothermal Water Network Synthesis. Computers & Chemical Engineering 82 (November): 144–171. https://doi.org/10.1016/j.compchemeng.2015.06.011.

Al Moussawi, H., Fardoun F., and Louahlia-Gualous, H. 2016. Review of Tri-Generation Technologies: Design Evaluation, Optimization, Decision-Making, and Selection Approach. Energy Conversion and Management 120 (July): 157–196. https://doi.org/10.1016/j.enconman.2016.04.085.

Alves, J. 1999. Analysis and Design of Refinery Hydrogen Distribution Systems. PhD Thesis, Manchester, UK.

ARANER. 2017. How Do Absorption Chillers Work? 10 September 2017. https://www.araner.com/blog/how-do-absorption-chillers-work/, accessed 18/08/2020.

'Aspen Plus'. 2019. 2019. https://www.aspentech.com/en/products/engineering/aspen-plus, accessed 18/08/2020.

Ataei, A. and Yoo, C.K. 2010. Simultaneous Energy and Water Optimization in Multiple-Contaminant Systems with Flowrate Changes Consideration. International Journal of Environmental Research 4(1): 11–26. https://doi.org/10.22059/IJER.2010.151.

Bandyopadhyay, S., Ghanekar, M.D., and Pillai, H.K. 2006. Process Water Management. Industrial & Engineering Chemistry Research 45(15): 5287–5297. https://doi.org/10.1021/ie060268k.

Broughton, J., ed. 1994. Process Utility Systems: Introduction to Design, Operation and Maintenance. Rugby, UK: Institution of Chemical Engineers.

Centre for Process Integration. 2004. Heat Integration and Energy Systems (MSc Course). School of Engineering and Analytical Science, Manchester, United Kingdom: The University of Manchester.

Chew, K.H., Klemeš, J.J., Alwi, S.R.W., and Zainuddin Abdul, M. 2013. Industrial Implementation Issues of Total Site Heat Integration. Applied Thermal Engineering 61(1): 17–25. https://doi.org/10.1016/j.applthermaleng.2013.03.014.

Čuček, L., Mantelli, V., Yong, J.Y., Varbanov, P.S., Klemeš, J.J., and Kravanja, Z. 2015. A Procedure for the Retrofitting of Large-Scale Heat Exchanger Networks for Fixed and Flexible Designs Applied to Existing Refinery Total Site. Chemical Engineering Transactions 45: 109–114. https://doi.org/10.3303/CET1545019.

Čuček, L., Klemeš, J.J., and Kravanja, Z. 2012. A Review of Footprint Analysis Tools for Monitoring Impacts on Sustainability. Journal of Cleaner Production 34 (October): 9–20. https://doi.org/10.1016/j.jclepro.2012.02.036.

Deng, C., Zhou, Y., Li, Y., and Feng, X. 2014. Flowrate Targeting for Interplant Hydrogen Networks. Chemical Engineering Transactions 39 (August): 19–24. https://doi.org/10.3303/CET1439004.

Dhole, V.R. and Linnhoff, B. 1993. Total Site Targets for Fuel, Co-Generation, Emissions, and Cooling. Computers & Chemical Engineering 17 (January): S101–9. https://doi.org/10.1016/0098-1354(93)80214-8.

Ding, Y., Feng, X., and Chu, K.H. 2011. Optimization of Hydrogen Distribution Systems with Pressure Constraints. Journal of Cleaner Production 19(2): 204–211. https://doi.org/10.1016/j.jclepro.2010.09.013.

Dym, C.L. 2004. Principles of Mathematical Modeling. Amsterdam, the Netherlands; Boston, USA: Elsevier Academic Press.

Eiholzer, T., Olsen, D., Hoffmann, S., Sturm, B., and Wellig, B. 2017. Integration of a Solar Thermal System in a Medium-Sized Brewery Using Pinch Analysis: Methodology and Case Study. Applied Thermal Engineering 113 (February): 1558–1568. https://doi.org/10.1016/j.applthermaleng.2016.09.124.

El-Halwagi, M.M. 2017. Sustainable Design Through Process Integration. 2nd ed. New York, USA: Elsevier.

El-Halwagi, M.M. and Manousiouthakis, V. 1989. Synthesis of Mass Exchange Networks. AIChE Journal 35(8): 1233–1244. https://doi.org/10.1002/aic.690350802.

El-Halwagi, M.M., Gabriel, F., and Harell, D. 2003. Rigorous Graphical Targeting for Resource Conservation via Material Recycle/Reuse Networks. Industrial & Engineering Chemistry Research 42(19): 4319–4328. https://doi.org/10.1021/ie030318a.

Elsherif, M., Manan, Z.A., and Kamsah, M.Z. 2015. State-of-the-Art of Hydrogen Management in Refinery and Industrial Process Plants. Journal of Natural Gas Science and Engineering 24 (May): 346–356. https://doi.org/10.1016/j.jngse.2015.03.046.

Foo, D. and Yee, C. 2009. State-of-the-Art Review of Pinch Analysis Techniques for Water Network Synthesis. Industrial & Engineering Chemistry Research 48(11): 5125–5159. https://doi.org/10.1021/ie801264c.

Foo, D.C.Y., and Manan, Z.A. 2006. Setting the Minimum Utility Gas Flowrate Targets Using Cascade Analysis Technique. Industrial & Engineering Chemistry Research 45(17): 5986–5995. https://doi.org/10.1021/ie051322k.

Foo, D.C.Y. 2012. Process Integration for Resource Conservation. Florida, USA: CRC Press.

GAMS. 2019. General Algebraic Modeling System. https://www.gams.com/, accessed 20/08/2019.

Gunaratnam, M., Alva-Argáez, A., Kokossis, A., Kim, J.-K., and Smith, R. 2005. Automated Design of Total Water Systems. Industrial & Engineering Chemistry Research 44(3): 588–599. https://doi.org/10.1021/ie040092r.

Hallale, N., Moore, I., Vauk, D., and Robinson, P.R. 2017. Hydrogen Network Optimization. In: Springer Handbook of Petroleum Technology, edited by, Hsu, C.S., Robinson, P.R., Hsu, C.S., and Robinson, P.R., 817–831. Cham, Switzerland: Springer.

Hu, N., Feng, X., and Deng, C. 2011. Optimal Design of Multiple-Contaminant Regeneration Reuse Water Networks with Process Decomposition. Chemical Engineering Journal 173(1): 80–91. https://doi.org/10.1016/j.cej.2011.07.040.

Jamaluddin, K., Wan Alwi, S.R., Manan, Z.A., Khaidzir, H., and Klemeš, J.J. 2019. A Process Integration Method for Total Site Cooling, Heating and Power Optimisation with Trigeneration Systems. Energies 12(6): 1030. https://doi.org/10.3390/en12061030.

Jia, X., Zhiwei, L., Fang Wang, D.C., Foo, Y., and Qian, Y. 2015. A New Graphical Representation of Water Footprint Pinch Analysis for Chemical Processes. Clean Technologies and Environmental Policy 17(7): 1987–1995. https://doi.org/10.1007/s10098-015-0921-1.

Kapil, A., Bulatov, I., and Kim, J.K. 2010. Exploitation of Low-Grade Heat in Site Utility Systems. Chemical Engineering Transactions 21 (September): 367–372. https://doi.org/10.3303/CET1021062.

Kimura, H. and Zhu, X.X. 2000. R – Curve Concept and Its Application for Industrial Energy Management. Industrial & Engineering Chemistry Research 39(7): 2315–2335. https://doi.org/10.1021/ie9905916.

Klemeš, J.J., Varbanov, P.S., Alwi, S.R.W., and Zainuddin Abdul, M. 2018. Sustainable Process Integration and Intensification. Saving Energy, Water and Resources. 2nd ed. Berlin, Germany: Walter de Gruyter GmbH.

Klemeš, J.J., ed. 2013. Handbook of Process Integration (PI): Minimisation of Energy and Water Use, Waste and Emissions. Woodhead / Elsevier: Cambridge, UK.

Klemeš, J.J., Dhole, V.R., Raissi, K., Perry, S.J., and Puigjaner, L. 1997. Targeting and Design Methodology for Reduction of Fuel, Power and CO_2 on Total Sites. Applied Thermal Engineering 17(8–10): 993–1003. https://doi.org/10.1016/S1359-4311(96)00087-7.

Klemeš, J.J., Friedler, F., Bulatov, I., and Varbanov, P.S. 2010. Sustainability in the Process Industry: Integration and Optimization. New York, USA: McGraw-Hill Education.

Leewongtanawit, B. and Kim, J.-K. 2009. Improving Energy Recovery for Water Minimisation. Energy 34(7): 880–893. https://doi.org/10.1016/j.energy.2009.03.004.

Liew, P.Y., Wan Alwi, S.R., Varbanov, P.S., Manan, Z.A., and Klemeš, J.J. 2012. A Numerical Technique for Total Site Sensitivity Analysis. Applied Thermal Engineering 40: 397–408. https://doi.org/10.1016/j.applthermaleng.2012.02.026.

Liew, P.Y., Wan Alwi, S.R., Varbanov, P.S., Manan, Z.A., and Klemeš, J.J. 2013. Centralised Utility System Planning for a Total Site Heat Integration Network. Computers and Chemical Engineering 57: 104–111. https://doi.org/10.1016/j.compchemeng.2013.02.007.

LINGO. 2019. LINGO and Optimization Modeling. https://www.lindo.com/index.php/products/lingo-and-optimization-modeling, accessed 20/08/2019.

Linnhoff, B. and Flower, J.R. 1978. Synthesis of Heat Exchanger Networks: I. Systematic Generation of Energy Optimal Networks. AIChE Journal 24(4): 633–642. https://doi.org/10.1002/aic.690240411.

Liu, L., Song, H., Zhang, L., and Jian, D. 2018. Heat-Integrated Water Allocation Network Synthesis for Industrial Parks with Sequential and Simultaneous Design. Computers & Chemical Engineering 108 (January): 408–424. https://doi.org/10.1016/j.compchemeng.2017.10.002.

Liu, L., Wang, J., Song, H., Jian, D., and Yang, F. 2016. Synthesis of Water Networks for Industrial Parks Considering Inter-Plant Allocation. Computers & Chemical Engineering 91 (August): 307–317. https://doi.org/10.1016/j.compchemeng.2016.03.013.

Luo, Y., Liu, Z., Luo, S., and Yuan, X. 2014. Thermodynamic Analysis of Non-Isothermal Mixing's Influence on the Energy Target of Water-Using Networks. Computers & Chemical Engineering 61 (February): 1–8. https://doi.org/10.1016/j.compchemeng.2013.10.008.

Manan, Z.A., Wan Alwi, S.R., and Ujang, Z. 2006. Water Pinch Analysis for an Urban System: A Case Study on the Sultan Ismail Mosque at the Universiti Teknologi Malaysia (UTM). Desalination 194(1–3): 52–68. https://doi.org/10.1016/j.desal.2005.11.003.

Marques, J.P., Matos, H.A., Oliveira, N.M.C., and Nunes, C.P. 2017. State-of-the-Art Review of Targeting and Design Methodologies for Hydrogen Network Synthesis. International Journal of Hydrogen Energy 42(1): 376–404. https://doi.org/10.1016/j.ijhydene.2016.09.179.

Merritt, C. 2016. Process Steam Systems | Wiley Online Books. Hoboken, NJ: USA: John Wiley & Sons, Inc. https://doi.org/10.1002/9781119085454.

Mohammadnejad, S., Ataei, A., Bidhendi, G.R.N., Mehrdadi, N., Ebadati, F., and Lotfi, F. 2012. Water Pinch Analysis for Water and Wastewater Minimization in Tehran Oil Refinery Considering Three Contaminants. Environmental Monitoring and Assessment 184(5): 2709–2728. https://doi.org/10.1007/s10661-011-2146-z.

Mughees, W. and Al-Ahmad, M. 2015. Application of Water Pinch Technology in Minimization of Water Consumption at a Refinery. Computers & Chemical Engineering 73 (February): 34–42. https://doi.org/10.1016/j.compchemeng.2014.11.004.

Pina, E.A., Reynaldo Palacios-Bereche, M.F., Chavez-Rodriguez, A.V., Ensinas, M.M., and Nebra, S.A. 2017. Reduction of Process Steam Demand and Water-Usage through Heat Integration in Sugar and Ethanol Production from Sugarcane – Evaluation of Different Plant Configurations. Energy 138 (November): 1263–1280. https://doi.org/10.1016/j.energy.2015.06.054.

Pitcher, J. 2015. Part I: How Heat Loads Affect Evaporative Cooling Tower Efficiency – Flow Control Network. https://www.flowcontrolnetwork.com/part-i-how-heat-loads-affect-evaporative-cooling-tower-efficiency/, accessed 18/04/2020.

'Refrigeration: Processes & Temperatures | Linde Gas'. 2019. https://www.linde-gas.com/en/processes/refrigeration_and_air_conditioning/refrigeration_processes_and_temperatures/index.html, accessed 14/08/2019.

Ren, X.-Y., Jia, -X.-X., Varbanov, P.S., Klemeš, J.J., and Liu, Z.-Y. 2018. Targeting the Cogeneration Potential for Total Site Utility Systems. Journal of Cleaner Production 170: 625–635. https://doi.org/10.1016/j.jclepro.2017.09.170.

Sahinkaya, E., Yurtsever, A., and Çınar, Ö. 2017. Treatment of Textile Industry Wastewater Using Dynamic Membrane Bioreactor: Impact of Intermittent Aeration on Process Performance. Separation and Purification Technology 174 (March): 445–454. https://doi.org/10.1016/j.seppur.2016.10.049.

Savulescu, L., Kim, J.-K., and Smith, R. 2005a. Studies on Simultaneous Energy and Water Minimisation – Part I: Systems with No Water Re-Use. Chemical Engineering Science 60(12): 3279–3290. https://doi.org/10.1016/j.ces.2004.12.037.

Savulescu, L., Kim, J.-K., and Smith, R. 2005b. Studies on Simultaneous Energy and Water Minimisation – Part II: Systems with Maximum Re-Use of Water. Chemical Engineering Science 60(12): 3291–3308. https://doi.org/10.1016/j.ces.2004.12.036.

Savulescu, L.E. and Alva-Argaez, A. 2013. Process Integration Concepts for Combined Energy and Water Integration. In: Handbook of Process Integration (PI): Minimisation of Energy and Water Use, Waste and Emissions, edited by, Klemeš, J.J. and Klemeš, J.J., 461–483. Cambridge, UK: Woodhead / Elsevier.

Saw, S.Y., Liangming Lee, M., Lim, H., Foo, D.C.Y., Chew, I.M.L., Tan, R.R., and Klemeš, J.J. 2011. An Extended Graphical Targeting Technique for Direct Reuse/Recycle in Concentration and Property-Based Resource Conservation Networks. Clean Technologies and Environmental Policy 13(2): 347–357. https://doi.org/10.1007/s10098-010-0305-5.

Shirakawa, M. and Nakamoto, M. 2003. Start-Up Schedule Optimizing System of a Combined Cycle Power Plant. IFAC Proceedings Volumes 36(20): 261–266. https://doi.org/10.1016/S1474-6670(17)34477-4.

Skouteris, G., Ouki, S., Foo, D., Saroj, D., Altini, M., Melidis, P., Cowley, B., Ells, G., Palmer, S., and Sean, O. 2018. Water Footprint and Water Pinch Analysis Techniques for Sustainable Water Management in the Brick-Manufacturing Industry. Journal of Cleaner Production 172 (January): 786–794. https://doi.org/10.1016/j.jclepro.2017.10.213.

Smith, R. 2016. Chemical Process Design and Integration. 2nd ed. Chichester, West Sussex, United Kingdom: Wiley.

Sujak, S., Handani, Z.B., Sharifah Rafidah, W.A., Manan, Z.A., Hashim, H., and Lim Jeng, S. 2017. A Holistic Approach for Design of Cost-Optimal Water Networks. Journal of Cleaner Production, 146 (March): 194–207. https://doi.org/10.1016/j.jclepro.2016.06.182.

Taler, J., Węglowski, B., Taler, D., Sobota, T., Dzierwa, P., Trojan, M., Madejski, P., and Pilarczyk, M. 2015. Determination of Start-up Curves for a Boiler with Natural Circulation Based on the Analysis of Stress Distribution in Critical Pressure Components. Energy, Special Issue devoted to The 12th International Conference on Boiler Technology (ICBT 2014) – Current Issues of Construction and Operation of Boilers, 92 (December): 153–159. https://doi.org/10.1016/j.energy.2015.03.086.

Tan, R.R., Dominic Chwan Yee, F., Aviso, K.B., and Ng, D.K.S. 2009. The Use of Graphical Pinch Analysis for Visualizing Water Footprint Constraints in Biofuel Production. Applied Energy 86(5): 605–609. https://doi.org/10.1016/j.apenergy.2008.10.004.

Tan, Y.L., Manan, Z.A., and Foo, D.C.Y. 2007. Retrofit of Water Network with Regeneration Using Water Pinch Analysis . Process Safety and Environmental Protection 85(4 B): 305–317. https://doi.org/10.1205/psep06040.

Taufiq, B.N., Kikuchi, Y., Ishimoto, T., Honda, K., and Koyama, M. 2015. Conceptual Design of Light Integrated Gasification Fuel Cell Based on Thermodynamic Process Simulation. Applied Energy 147 (June): 486–499. https://doi.org/10.1016/j.apenergy.2015.03.012.

Towler, G.P. and Sinnott, R.K. 2013. Chemical Engineering Design: Principles, Practice, and Economics of Plant and Process Design. 2nd ed. Boston, MA, USA: Butterworth-Heinemann.

Tsolas, S.D., Karim, M.N., and Hasan, M.M.F. 2018. Optimization of Water-Energy Nexus: A Network Representation-Based Graphical Approach. Applied Energy 224 (August): 230–250. https://doi.org/10.1016/j.apenergy.2018.04.094.

Varbanov, P., Perry, S., Klemeš, J., and Smith, R. 2005. Synthesis of Industrial Utility Systems: Cost-Effective de-Carbonisation. Applied Thermal Engineering 25 (7 SPEC. ISS.): 985–1001. https://doi.org/10.1016/j.applthermaleng.2004.06.023.

Varbanov, P., Perry, S., Makwana, Y., Zhu, X.X., and Smith, R. 2004a. Top-Level Analysis of Site Utility Systems. Chemical Engineering Research and Design 82(6): 784–795. https://doi.org/10.1205/026387604774196064.

Varbanov, P.S., Boldyryev, S., Nemet, A., Klemeš, J.J., and Kapustenko, P.O. 2013. Targeting of the Trade-Off of Capital Cost and Carbon Footprint for CHP. In, 227–232. Kuala Lumpur, Malaysia. https://doi.org/10.13140/rg.2.1.2361.2247.

Varbanov, P.S., Doyle, S., and Smith, R. 2004b. Modelling and Optimization of Utility Systems. Chemical Engineering Research and Design 82(5): 561–578. https://doi.org/10.1205/026387604323142603.

Varbanov, P.S. and Klemeš, J.J. 2011. Integration and Management of Renewables into Total Sites with Variable Supply and Demand. Computers and Chemical Engineering 35(9): 1815–1826. https://doi.org/10.1016/j.compchemeng.2011.02.009.

Varbanov, P.S. 2014. Energy and Water Interactions: Implications for Industry. Current Opinion in Chemical Engineering 5: 15–21. https://doi.org/10.1016/j.coche.2014.03.005.

Varbanov, P.S., Klemeš, J., Shah, R.K., and Shihn, H. 2006. Power Cycle Integration and Efficiency Increase of Molten Carbonate Fuel Cell Systems. Journal of Fuel Cell Science and Technology 3(4): 375–383. https://doi.org/10.1115/1.2349515.

Varbanov, P.S., Walmsley, T.G., Klemeš, J.J., Wang, Y., and Jia, X.-X. 2018. Footprint Reduction Strategy for Industrial Site Operation. Chemical Engineering Transactions 67 (September): 607–612. https://doi.org/10.3303/CET1867102.

Wan Alwi, S.R., and Manan, Z.A. 2013. Water Pinch Analysis for Water Management and Minimisation: An Introduction. In: Handbook of Process Integration (PI), edited by, Klemeš, J.J., 353–382. Cambridge, UK: Woodhead Publishing/Elsevier. https://doi.org/10.1533/9780857097255.3.353.

Wang, Y.P. and Smith, R. 1994. Wastewater Minimisation. Chemical Engineering Science 49(7): 981–1006. https://doi.org/10.1016/0009-2509(94)80006-5.

Wang, Y., Sidong, W., Feng, X., and Deng, C. 2015. An Exergy-Based Approach for Hydrogen Network Integration. Energy 86 (June): 514–524. https://doi.org/10.1016/j.energy.2015.04.051.

Westerberg, A.W. 2004. A Retrospective on Design and Process Synthesis. Computers & Chemical Engineering 28(4): 447–458. https://doi.org/10.1016/j.compchemeng.2003.09.029.

Yang, M., Feng, X., and Liu, G. 2016. A Unified Graphical Method for Integration of Hydrogen Networks with Purification Reuse. Chinese Journal of Chemical Engineering 24(7): 891–896. https://doi.org/10.1016/j.cjche.2016.04.018.

Zhang, Q. and Feng, X. 2012. Hydrogen Network Integration with Both Pressure and Impurity Constraints. In: Computer Aided Chemical Engineering, Edited by, Karimi, I.A., Srinivasan, R., Karimi, I.A., and Srinivasan, R., Vol. 31: 630–634. https://doi.org/10.1016/B978-0-444-59507-2.50118-9.

Zhao, Z., Liu, G., and Feng, X. 2007. The Integration of the Hydrogen Distribution System with Multiple Impurities. Chemical Engineering Research and Design 85(9): 1295–1304. https://doi.org/10.1205/cherd07014.

4 Primary resources for energy supply

The recent edition of Merriam-Webster dictionary defines energy (Merriam-Webster, 2019) in the physics sense as
- An entity transferred between parts of a system, which results in the production of physical change.
- Usable heat or electricity.

A similar definition can also be found in the recent Cambridge English Dictionary (2019). Thermodynamics elaborates on these definitions (Wu, 2007), referring to energy as the property of the considered system, having components related to
- Temperature, giving rise to internal energy;
- Velocity, associated with kinetic energy;
- Position/potential in a field (e.g. gravitational, magnetic, electrical) – defining several types of potential energy.

There are various forms of energy – stored or being transferred between systems. Some examples can be given as follows.
- **Radiation:** This is associated with a flow of photons of various wavelengths. The solar radiation contains energy in photons over a wide range of the wavelengths (Garner, 2015), arriving to the Earth mainly in the infrared, visible and ultraviolet ranges. The radiation from a fire or a candle is within the visible and infrared ranges.
- **Chemical energy:** Many materials, mainly carbon-based – for example, wood and oil – contain energy stored in a chemical form. These are combustible materials. Other materials, capable of an electrochemical reaction of releasing electrical energy, are used for constructing batteries.
- **Potential energy:** This is energy stored in material within a field; there exists a potential for changing the position within the field. An example can be given with the energy of a water reservoir at a certain height. The water has the potential to fall. The fall from the height to a lower point would generate energy. The amount of potential energy is proportional to the amount of water, and the height difference between the reservoir and the location of the final point of the fall.
- **Kinetic energy:** This is the energy of moving bodies. Examples include a football flying after a kick, wind (moving air mass) and a water stream. The faster the speed and the more mass involved, the more energy is contained.
- **Thermal energy (heat):** The indicator for the level of thermal energy is temperature. Higher temperature indicates faster movement or vibrations of the

https://doi.org/10.1515/9783110630091-004

molecules of a body. As a result, this is translated to larger energy quantity per unit mass. The overall amount of heat is proportional to body mass.

– **Electrical energy**: This is an energy flow, associated with the movement of electrons within an electrical field – from a location with a negative charge, in the direction of a positive charge. The difference between the electrical charges is termed voltage. The flow of electrons is characterised by its voltage and current. Both of them are proportional to the delivered amount of electrical energy.

Tracking the overall energy demands for an economy is essential, as demand levels and trends determine to a large degree the emissions and market conditions, and especially the purchase prices for procuring energy resources. From the publicly available data, the datasets provided by the US Energy Information Administration are regularly updated and kept consistent, allowing analysis for regional, national and international levels. The data for the year 2018 containing historical statistics and projections have been shown in Fig. 1.2a and reproduced in Fig. 4.1, following the data published by the US Energy Information Administration (AEO, 2018). They show that, for the case of the United States, the most significant energy demands come from the industrial sector, exceeding those of the transport sector by 10–15 % and almost twice the demands of the residential and commercial sectors.

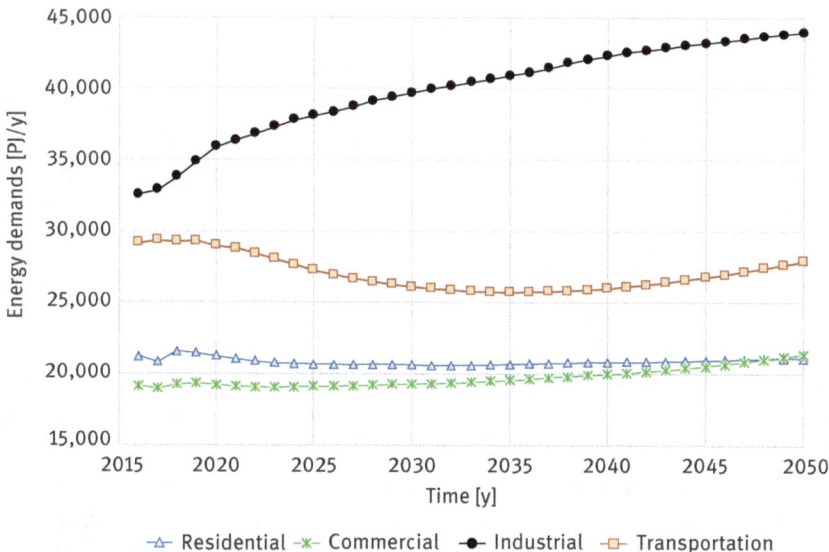

Fig. 4.1: US energy-consumption projections in PJ/y, based on data from AEO (2018).

From the viewpoint of the planet Earth and the humans living on its surface, there are three main energy supply flows (Fig. 4.2). They are the solar irradiation, geothermal heat and gravitational attraction forces of the earth, the moon and the other bodies in the solar system (Trenberth et al., 2009).

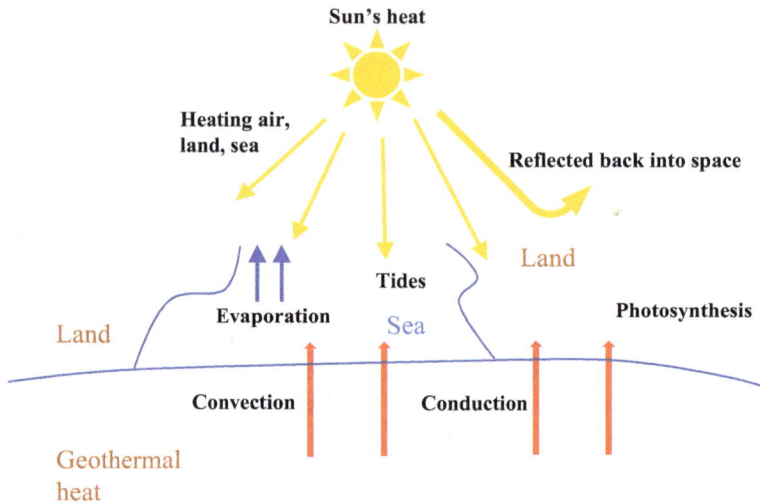

Fig. 4.2: Global energy flows at the planetary level.

All other sources for generating useable energy are derived from those flows in some way – solar energy can power various conversion technologies for obtaining heating, cooling and power. Geothermal heat can also be used directly for heat and power generation (Akrami et al., 2017). On the other hand, the other widely known renewable energy sources – for example, hydropower, wind, wave, derived from the global inflows of solar radiation, geothermal heat and gravity interactions.

A particular form of chemical energy is manifested in fuels – fossil and biomass. These are all various forms of energy storage. Biomass is material for short- to medium-term energy storage, while fossil fuels – crude oil, natural gas, coal – are derived from fossilised biomass and stored for millions of years in the Earth's crust. They are all renewable. However, the fossil fuel cycle is enormously long and not feasible for practical use. It is also widely practised to generate power using nuclear reactions. That raises some questions concerning safety and creates the need to treat the resulting radioactive waste. This is a separate, strategic branch of energy engineering, under the very stringent control of international bodies – for instance, the International Atomic Energy Agency, mainly for safety reasons (IAEA, 2019).

4.1 Solar radiation as a source of energy

The history of technologies for the direct use of solar energy is relatively short (Škorpík, 2019b). However, its recent development is quite vigorous (Ramos et al., 2017). This trend is boosted by the fact that the use of solar energy is available over most of the Earth surface, however, in various intensity and time span intensity.

Solar energy is a flow of electromagnetic radiation, termed solar radiation (Myers, 2017). The sun may be regarded as a black body (Phillips, 1992), which according to Planck's radiation law radiates in the wavelength range corresponding to the surface temperature of the Sun – around 5,527 °C (5,800 K) (Nobel, 1999). At the same time, most energy is emitted in the wavelength range of visible light, see the spectral luminosity of the black body and the Sun (Stewart and Johnson, 2017).

The Earth does not absorb the entire radiant influx coming from the Sun. The absorbed part of the radiation, after various transformations, increases the Earth's internal thermal energy and a part of the heat is radiated back into space.

4.1.1 The intensity of solar radiation on the surface of the Earth

The intensity of solar radiation falling on the Earth's surface is equal to the amount of solar energy per unit surface area, usually measured in W/m^2. The intensity depends on the latitude and the weather – more precisely the amount of cloud cover. The intensity of solar radiation can be expressed (Myers, 2017) as the sum of the intensity of direct radiation and the intensity of diffuse radiation:

$$I = I_P + I_D \qquad (4.1)$$

where I (W/m^2) is the overall intensity of solar radiation at the Earth's surface; I_P (W/m^2) is the intensity of direct solar radiation, and I_D (W/m^2) is the intensity of diffuse solar radiation. These quantities can be defined as:
- **Direct radiation**: Sunlight that is not reflected or absorbed and re-emitted as it passes through the atmosphere.
- **Diffuse radiation**: Sunlight reflected from particles in the atmosphere (water droplets, dust) and has changed its direction. The wavelength of this radiation remains the same as before the reflection. The amount of diffuse radiation depends on cloudiness and air pollution. These phenomena, in turn, reduce the amount of direct radiation. For example, when the sky is cloudy, only diffuse radiation falls on the Earth's surface.

Solar irradiation is measured and the measurement is integrated over annual periods by research and development institutions. For instance, the maps for 2016 for the European Union and Associated Countries can be found in the EU Photovoltaic Geographical Information System (European Commission, 2019).

As an example, the country map for the Czech Republic indicates the range of solar radiation intensity from 1,100 to 1,300 kWh/(m² × y), referring to measurements at the surface of the photovoltaic modules.

4.1.2 Use of solar energy

The heat supplied by solar radiation to the atmosphere and the Earth's surface is a significant climate factor. It enables a climate suitable for life. In addition, sunlight enables photosynthesis and vision. Solar energy is used to generate energy services, mainly in the form of heating and power. Solar radiation is directly used for generating useful heat with solar thermal collectors (Martínez-Rodríguez et al., 2019) and useful electrical power converted by photovoltaic panels (PHOTON SOLAR Energy GmbH, 2019). Hybrid devices combining photovoltaic power and heat generation from the sunlight are also known as Ramos et al. (2017). Another possible principle for solar radiation use for power generation is by concentrating solar rays to heat the working medium at high temperatures, capable of driving power cycles. There are known commercial applications of steam cycle-based concentrated solar power plants (Siemens, 2019), but developments on powering gas turbines (Mazzoni, Cerri, and Chennaoui, 2018) and solar combined cycles (Elmohlawy et al., 2019) are also underway.

4.2 Biomass as a source of energy

Biomass is an essential local energy source (Škorpík, 2019a). It has been the traditional fuel powering human activities for centuries. However, it should be used close to its production sites. This is due to the high transport cost of biomass and due to its low specific energy content (Čuček et al., 2012a). The performance of biomass-fired power plants depends on the size of the catchment area on which biomass can be produced, in addition to the equipment construction and quality. This is because of the economy of scale (Smith, 2016) – larger plants tend to be more energy- and economically efficient than smaller ones and the size of the catchment area defines the size of the power plants. This results in a trade-off (Lam et al., 2011). Broader catchment area is associated with potentially more significant power plant but this also means longer transportation routes of the biomass from the harvesting locations to the power plant.

Wood and wood waste is the usual biomass fuel for water/steam boilers. However, that may be replaced by the straw or other crop residues. Also in animal production, waste is produced in the form of slurry and manure, which can be decomposed into biogas in bioreactors. Biogas can be burned directly in gas boilers (Zhang et al., 2016) and even in internal combustion engines, gas (micro-)turbines (Capstone, 2019) and reciprocating engines (Santos et al., 2016). Today, the term biodiesel – hydrocarbon-containing naphtha derived from the treatment of biomass (e.g. pressing the fruits of oilseed rape) – is topical. These and similar uses of biomass are called energy use of biomass.

4.2.1 Plant biomass characterisation

Plant biomass consists of organic matter, water and low content of non-combustible minerals, remaining as ash after biomass combustion. The organic matter composition includes elements with affinity to oxidation: C, H, O as well as N. The latter can also be oxidised at very high temperatures, which is an undesirable side effect producing oxides of N, collectively referred to as NO_x (Klemeš, 2015). In biomass, it is possible to find also flammable elements, typical for inorganic compounds – most often S. Detailed analyses of biomass are performed for each case. An example of such a study can be found in Vassilev et al. (2010).

Biomass should not contain sulphur (S) and chlorine (Cl) in large amounts to be useable for combustion. These elements get into the biomass from the air, where they get partly due to human activity (burning fossil fuels) and partly due to natural disasters (e.g. volcano eruption). The ash content from the biomass boiler is relatively low, usually limited to below 10 % (mass, dry basis), with only several of biomass fuel types having higher ash content (Vassilev et al., 2010). However, for the large capacity combustion, they can accumulate substantially and have to be dealt with. The ashes can contain soil-derived minerals and sometimes metals (Vassilev et al., 2010). In addition, the ash may contain non-combustible residues of coarse impurities, which were brought to the boiler together with the fuel (e.g. clay or dust).

4.2.2 Types of biomass conversion for energy purposes

Biomass decomposition processes are used in various ways to obtain fuel, energy and other products. Decomposition of biomass in the presence of air and/or oxygen is called aerobic and is usually implemented as combustion. If the decomposition takes place in the absence of oxygen, then it is termed anaerobic. An example of the anaerobic decomposition process is pyrolysis. The main processes are outlined below.

Combustion: This involves almost complete oxidation of the biomass. During combustion, the carbon and hydrogen contained in the fuel burn, releasing gaseous substances and leaving a small amount of ash. Biomass is burned to generate steam or hot water usually. There are various arrangements for the air supply to the combustion chamber. More details on biomass boilers can be found in Vakkilainen (2017).

Gasification: In gasification (incomplete combustion), the carbon contained in the fuel is partially oxidised (Basu, 2010). Some of the carbon content oxidises to CO and another part to CO_2. Some of the hydrogen content reacts to form liquid water (H_2O), and another part forms hydrogen gas (H_2). The resulting gaseous product is referred to as synthesis gas or syngas for short. This is on its own a useful fuel, which can be used in boilers and even gas turbines and internal combustion engines, after cleaning of tars, particulates and other impurities. Syngas, however, has also another potential use – for synthesising hydrocarbons – including liquid fuels. The process is known as Fischer-Tropsch Synthesis (Davis, Occelli, and American Chemical Society, 2016). The advantage of gasification is the high efficiency of energy use in the fuel and lower harmful emissions compared to conventional combustion. The disadvantage is the more complicated process, leading to the complexity of the equipment items and their arrangement.

Pyrolysis: This is the thermal decomposition of biomass, in the absence of oxygen, into solid carbon (charcoal) and pyrolysis gas (Wang and Luo, 2017). The pyrolysis chamber is heated by burning part of the pyrolysis gases taken from the furnace. Hot inert gas (oxygen-free) can also be used to heat the biomass in the furnace. The pyrolysis process can be divided into three parts, according to the temperature reached. In the temperature range up to 200 °C, water vapour is formed and released from the material. This stage is strongly endothermic. Between 200 and 500 °C is the range of the so-called dry distillation – at this stage, gaseous products are released from the biomass. This is related to the thermal destruction of organic compounds with high molecular weight, leaving gases and stable carbon. After that follows the gas formation phase, in the temperature range of 500–1,200 °C. During that, the dry distillation products are further released and transformed. Stable gases such as H_2, CO, CO_2 and CH_4 are produced from both solid carbon and liquid organic substances. Charcoal from pyrolysis combustion is used, for example, as a fertiliser or as a fuel for grilling. Charcoal is pure carbon, and the product of its combustion is the only CO_2, which is colourless and odourless, however the primary greenhouse gas (GHG).

The further options of biomass processing for energy purposes include alcohol fermentation (aerobic process) and anaerobic digestion. The first type of processes produces bioethanol (Wyman, 1996) and the second type produces biogas (Wellinger et al., 2013).

4.3 Other renewable energy sources

4.3.1 Wind

Wind energy is mainly used for the generation of electricity using wind turbines. In wind turbines, some of the kinetic energy of the wind is transformed first into the work of a wind turbine (Fig. 4.3), which is then transformed into electrical energy in an electric generator. Wind power output ranges from very small around 200 W

Fig. 4.3: An off-shore wind turbine (Hollman, 2019). Photo credit: Phil Hollman from London, UK, Vestas V90-3 MW wind turbine of the Kentish Flats Offshore Wind Farm, Thames Estuary. This file is licensed under the Creative Commons Attribution 2.0 Generic license. No editing was applied, and the image has been scaled down, to be suitable for pagination.

(up to about 4 kW it is considered as home-scale) to extensive installations with capacity exceeding 10 MW (Haymarket, 2020). A comprehensive overview of wind turbine technologies can be found in the handbook (Letcher, 2017). It provides discussions and information on several significant areas concerning wind turbines – wind resources and their properties worldwide, technologies used for wind power generation, environmental impacts, the economics of the installed facilities including operation and investments, as well as projections for the future outlook. The discussions also include energy storage, which is a key aspect of wind power generation, since the wind energy source is highly intermittent.

Figures 4.4 and 4.5 show the wind resource availability for the Czech Republic (Technical University of Denmark – DTU 2019) and for Germany (Technical University of Denmark – DTU 2020), at 100 m above the surface, in terms of power generation potential (W/m^2). Such maps are an essential factor in choosing the sites for installation of wind turbines.

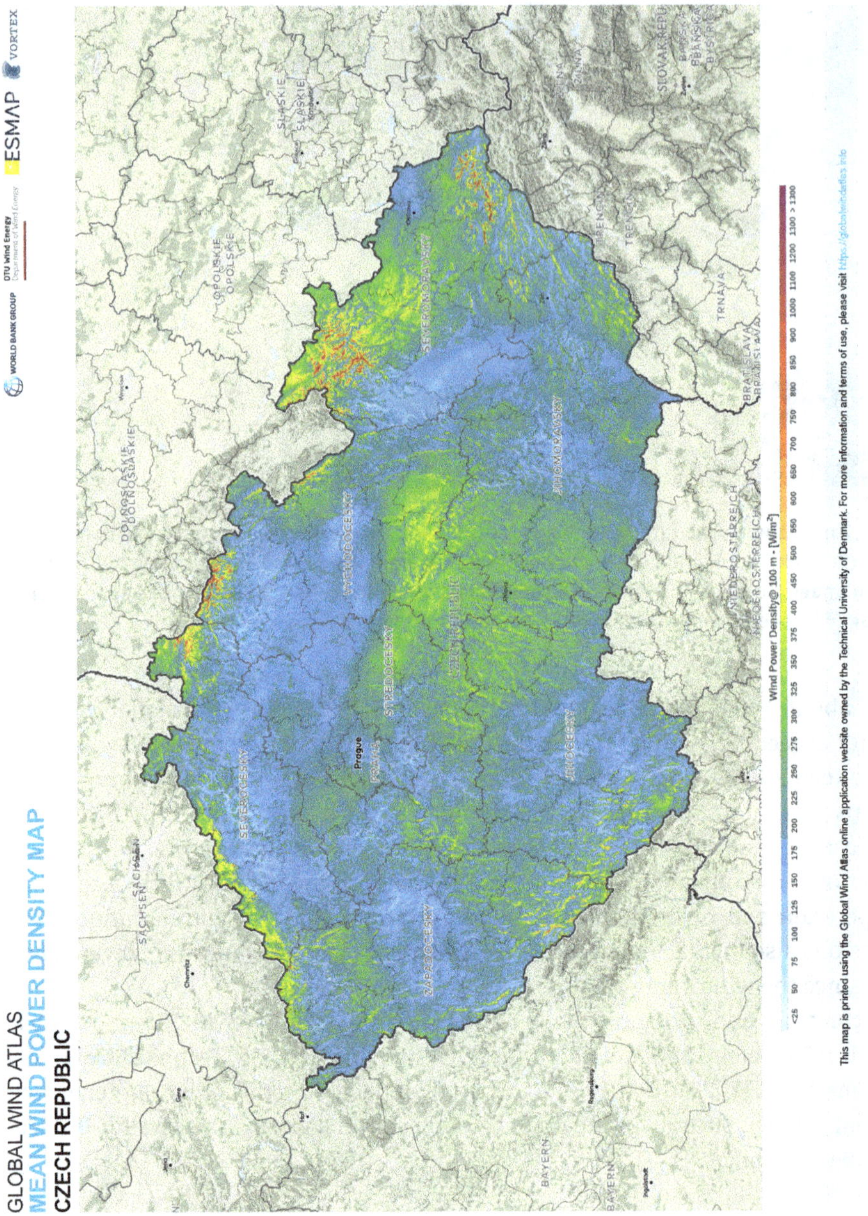

Fig. 4.4: Wind Atlas of the Czech Republic (Technical University of Denmark – DTU 2019). This image is licensed under the Creative Commons License – Attribution 4.0 International (CC BY 4.0). No editing was applied, and the image has been downloaded as a PDF, extracted as a JPEG and scaled down, to be suitable for pagination.

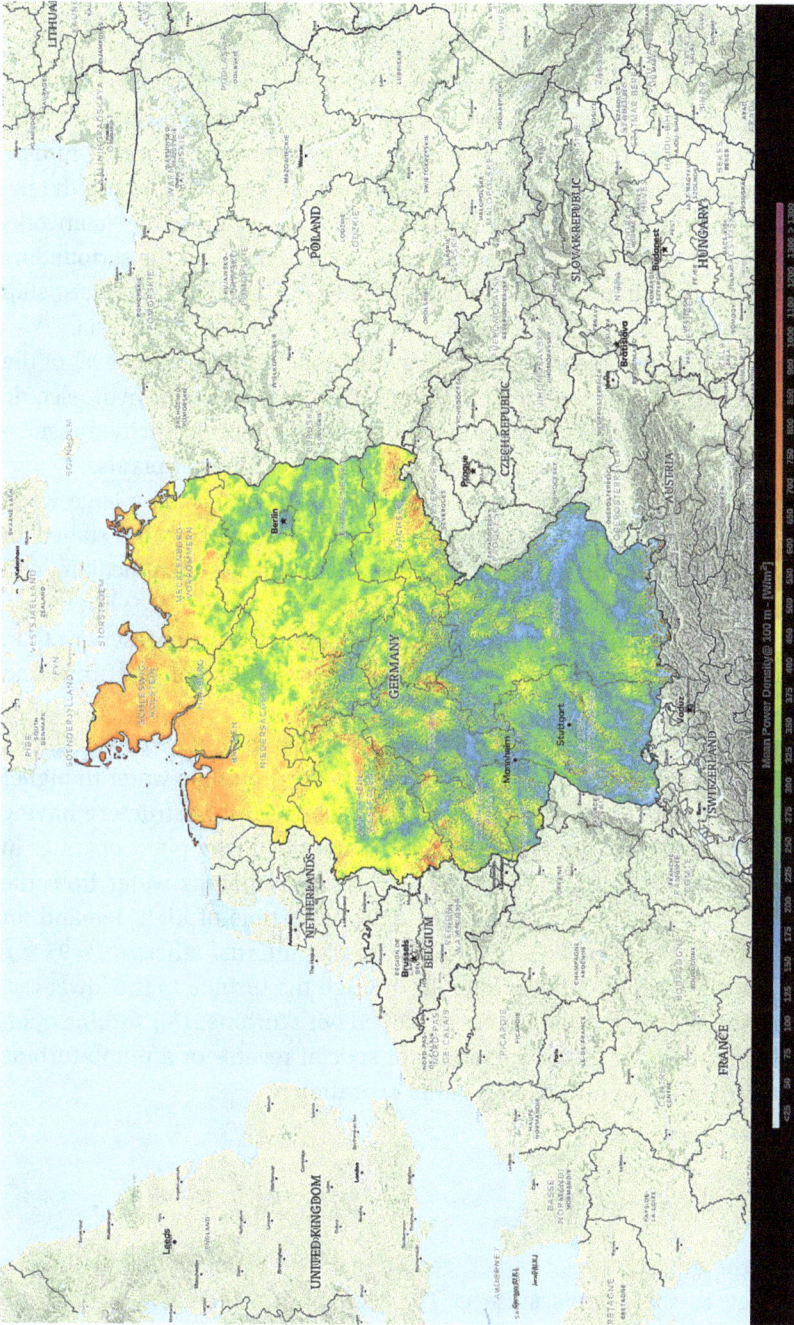

Fig. 4.5: Wind Atlas of Germany (Technical University of Denmark – DTU 2020). This image is licensed under the Creative Commons License – Attribution 4.0 International (CC BY 4.0). No editing was applied, and the image has been downloaded as a PDF, extracted as a JPEG and scaled down, to be suitable for pagination.

4.3.2 Hydropower

The history using the energy of falling water goes back to the past, thanks to the water wheels (Reynolds, 2003), but today's technologies in the hydropower industry are dominated by hydraulic turbines (Blair, 2017). It is necessary to combine a hydraulic turbine set, a generator and a dam reservoir, sometimes two or more reservoirs, to use the hydraulic head. All these facilities and structures are part of the waterworks. Waterworks almost always perform other tasks than energy and have an impact on the surrounding landscape. They are also often used to regulate the flow on the rivers and assist ship navigation. However, the lifespan of the reservoirs can be curtailed by the alluvia.

Hydroelectric plants use the potential energy difference between the level of the upper reservoir and the lower reservoir. There are three basic types of hydroelectric power stations that differ from each other in their function in the electricity transmission system. These are storage, flow and pumped storage hydropower plants.

The hydroelectric works with the accumulation power plant include a large water reservoir. This type of power plant is triggered in cases of power shortage. It is started for a certain part of the day, during peak hours, at a flow rate greater than medium, and during the rest of the day, the reservoir is gradually refilled.

Waterworks with a through-flow power plant has a much smaller reservoir than storage plants. These are power plants with a small slope but a constant flow and in constant operation.

Pumped-storage power stations have a similar function to storage power plants, supplemented by the possibility of "storing electricity" by pumping water to higher altitude storage for reuse. These power plants have a special infrastructure having at least two reservoirs. At the time of excess electricity, the power plant operates in a pumping mode (internal efficiency ≈ 90 %), in which it pumps water from the lower to the upper reservoir, consuming electricity. At a time of high demand for electricity, the power plant operates in a turbine mode (internal efficiency ≈ 95 %), by discharging water from the upper reservoir through the turbine to the lower reservoir. Pumped-storage power stations are equipped with turbines (for turbine operation) as well as pumps (for pump operation) or special reverse or a pump-turbine capable of operating in both turbine and pump operation.

4.3.3 Geothermal energy sources

As discussed at the outset of this chapter, geothermal heat is one of the continuous inflows of energy to the Earth's surface. The temperature increases with depth under the surface (Chiozzi et al., 2017). Geothermal energy is considered to originate from the heat of the Earth's core (Glassley, 2015), which has been cooling for several billion years. It comprises the energy released by nuclear reactions within the Earth and the energy of the Earth's plates.

Direct use of geothermal energy is linked to sites with natural occurrence of hot sites near the Earth's surface and rising hot springs. In the Czech Republic, the occurrence of such sites is minimal and known springs are used only for spa purposes. For instance, the typical temperature is from 30 to 80 °C – in the resort of Karlovy Vary is 73 °C. However, some European countries have plentiful sources of geothermal water, such as Hungary. In Iceland, there are such abundant sources of hot springs that they are even used for powering combined heat and power generation plants (Karlsdóttir et al., 2015). An Iceland-produced survey (Enex and Geysir Green Energy, 2008) analysed the capacities for geothermal energy harvesting – looking into installed and potential capacities for power generation (Fig. 4.6) and heat services (Fig. 4.7). From the surveyed countries, the most use of geothermal energy for power generation is practised by the United States, Indonesia, Mexico, Italy, New Zealand and Iceland. Similarly, for heating, Iceland is leading by installed capacity, followed by Turkey, Germany, Hungary and Switzerland.

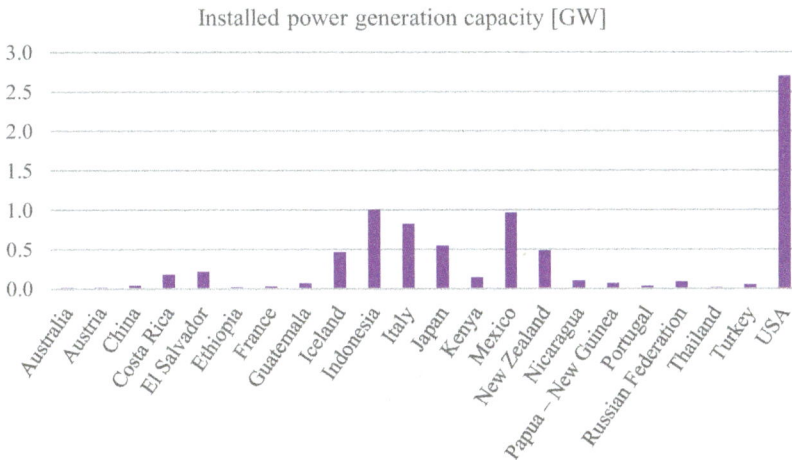

Fig. 4.6: Installed power-generation capacities from geothermal sources, as of 2005, based on data from Enex and Geysir Green Energy (2008).

A more recent source, focusing on power generation from geothermal sources, shows the installed power generation capacities, as of the end of 2019 (ThinkGeoEnergy, 2020). The chart shows a moderate increase in the spread of the technology for the 15 years since 2005 (Fig. 4.8).

In terms of industrial relevance, geothermal energy is most important for the so-called "High-temperature geothermal countries" – those with the availability of geothermal heat from 150 °C and higher. As such, in the survey, are identified Iceland, Italy, Portugal and Turkey.

The places most suitable for geothermal-based power generation are at locations where geothermal energy is concentrated at relatively low depth – close to

Installed thermal capacity [GW]

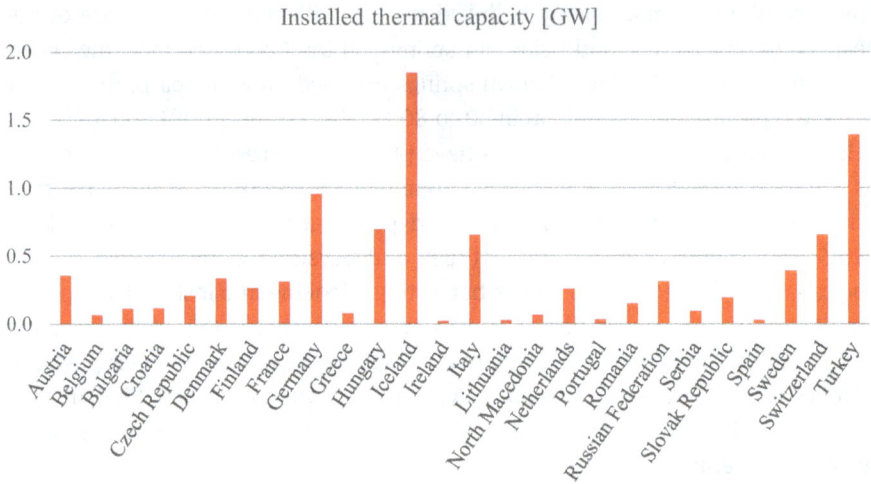

Fig. 4.7: Installed thermal service capacities from geothermal sources, as of 2005, based on data from Enex and Geysir Green Energy (2008).

Installed power generation capacity [GW]

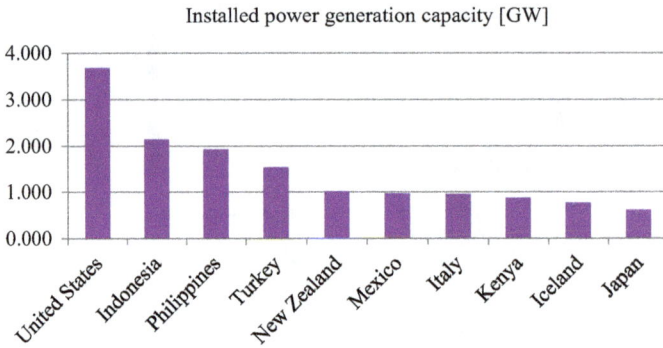

Fig. 4.8: Installed power generation capacities from geothermal sources, as of 31 December 2019, based on data from ThinkGeoEnergy (2020).

volcanic activity or near tectonic faults. Locations with high rock temperatures, which can be used for heating of working substances in thermodynamic cycles, can be used for commercial power generation (Karlsdóttir et al., 2015). For implementing a steam cycle (Rankine cycle), the temperature of geological layers should reach at least 200 °C (Desideri and Bidini, 1997). At lower temperatures, it is possible to use the Kalina cycle (Prananto et al., 2018) or organic Rankine cycle (Wang et al., 2017).

Another, more frequent use of geothermal energy is the application of heat pumps, whose evaporator (cold reservoir) is linked to a circulating fluid passing underground (Chiasson, 2016). However, there are frequent problems caused by scaling and fouling caused by geothermal water. There have been even recent research works on scaling this up to district heating scope (Arat and Arslan, 2017).

4.4 Fossil fuels

Fossil fuels are deposits formed from prehistoric organic matter. They occur in solid, liquid and gaseous forms. Fossil fuels contain mainly carbon and hydrogen, which are not chemically bonded to other elements and therefore have a relatively high heating value. They are mainly used as a source of energy released during combustion in internal combustion engines and boilers (prior to combustion, the treatment of raw fossil fuel). They are also important raw materials for the production of synthetic materials and chemicals such as plastics, asphalt and even medicines.

Without fossil fuels, energy services would be more expensive and less available, and today's world would certainly look different from what is known – especially in terms of cost and the effort spent on generating energy to meet human needs. The availability of fossil fuels worldwide is still sufficient to supply the world economy. However, the distribution of these resources among the various countries and regions is uneven. A good source of information on that is the BP Statistical Review of World Energy (BP p.l.c. 2019). The latest document provides the reserves-to-production ratio (R/P) (also known as the years of reserves left) as shown in Tab. 4.1.

Tab. 4.1: Overall world reserves of fossil fuels (end of 2018), based on data from BP (2019).

Fuel type	Reserves	R/P [y]	Most significant reserves
Crude oil	244.1×10^9 t	50	Venezuela, Saudi Arabia, Iran, Iraq, Russian Federation
Natural gas	196.9×10^{12} m^3	50.9	Russian Federation, Iran, Qatar, Turkmenistan, United States
Coal	1.054×10^{12} t	132	United States, Russian Federation, Australia, China

4.4.1 Coal

Coal is the charred remains of predominantly terrestrial plants. Most of the coal comes from the Carboniferous period – the fifth interval of the Paleozoic Era (Encyclopedia Britannica, 2019) when the conditions for the formation of coal were very favourable – characterised by major changes of the landmasses and continents and mountain-forming activities.

The essence of coal formation (Osborne, 2013) is to cut dead biomass from the air. This can happen, for example, if it is covered with a layer of mud with approximately 50 cm thickness. At first, the biodegradation of biomass was carried out by bacteria. In the next phase of coal formation, thermal decomposition at high pressure took place, which could be caused by the sinking of the deposit to a greater depth. During the carbonation of biomass, gases (CH_4, CO_2) and water were formed, which partly remain in the deposit.

The formation of coal happens over millions of years. The process develops via intermediates, which can be classified into the sequence: peat → lignite → brown coal → hard coal → anthracite.

Peat is formed through the decomposition of biomass by aerobic bacteria, and after dropping to a depth of several meters of anaerobic bacteria layer. At this stage, most of the volatile substances contained in the biomass escape into the environment. The structure of the peat still resembles the shapes of the original biomass material.

Lignite and brown coal: Lignite is formed from peat deposits. For the formation of brown coal, a significant decrease in the volume of the deposit occurs by piling up and accumulating further layers of biomass on the top of the original layer, and the subsequent sinking to a greater depth. Under these conditions of increased pressure, the temperature of the current layer raises to 150–200 °C. The transition phase between peat and brown coal is lignite, which has higher moisture content than brown coal. The deposit of lignite and brown coal has the properties of soft rock, but it is possible to recognise imprints of the shape of the original material.

Hard coal and anthracite. When brown coal deposits shrink further and sink to greater depths, the temperature of the deposit reaches 300–500 °C, and the pressure becomes sufficient to make coal more compact, and hard coal is formed. If the share of carbon in the deposit is greater than 92 %, then the coal is called anthracite. In extreme cases of high pressures over a long period of time in some anthracite deposits, a regional metamorphosis occurred, where the organic matter was converted to graphite. Hard coal deposits resemble rock and it is not possible to discern traces of shapes and imprints of the original material.

The composition of coal and peat varies with the sequence "peat → lignite → brown coal → hard coal → anthracite". In this order, the carbon share increases and the share of moisture decreases. With this, the heating value changes, too (Tab. 4.2).

Tab. 4.2: Approximate ranges of heating values for peat and coal (after International Energy Agency, 2019).

Material	Lower heating value [MJ/kg]
Peat	8.5–12.5
Lignite	5–20
Bituminous coal	22–27
Hard coal	28–31
Anthracite	25–32

4.4.2 Petroleum

Petroleum refers to the overall category of liquid materials related to hydrocarbons – crude oil, as well as the intermediates and the products of its processing (US EIA, 2019). Liquid or semi-liquid bituminous deposits have been used as building materials, light sources or pharmaceutical products since the Middle Ages, especially in Southeast Asia. On a larger scale, oil and its products began to be used in the United States during the nineteenth century, due to the commencement of industrial oil extraction in 1859 (Cleveland, 2009, p.204). After extraction, the crude was further processed (by distillation) into various fractions. At first, medium-weight fractions (kerosene) were used for combustion and lighting, later lighter as diesel and petrol were used to drive internal combustion engines. During the twentieth century, oil has become the dominant energy source, especially in transport, where adequate large-scale alternatives cannot be found yet.

Crude oil is a mixture of liquid and gaseous hydrocarbons and other organic compounds. The exact composition varies depending on the crude oil origin. It contains bituminous substances, which may also include natural gas, asphalts and ozokerite (natural wax). During mining, water and mineral impurities also get into the oil. Some crude oils contain mainly alkanes (paraffinic crude oil), while in others dominate cycloalkanes or aromatic hydrocarbons. Crude oil formation is not as well-studied and described as that of coal. It is thought as originating from biomass decomposition near coastlines, which later dropped to greater depths. Some theories also assume the non-biological formation of oil-based primarily on geological processes in which organic compounds contained in an inanimate rock are separated. A comprehensive source on the formation and properties of crude oil is (Tissot and Welte, 1984).

The net heating value (NHV) of crude oil is within the range of 40–50 MJ/kg, according to available sources. Since crude oil is a general term for a variety of hydrocarbon mixtures, there is no standardised value. From the publicly available sources, there is a published energy statistics guide by the OECD (International Energy Agency, 2004) listing the NHV of petroleum products, and the range of the values, quoted there, is within 44–52 MJ/kg. Finding open information on the heating value of crude oil itself is more difficult. An energy information handbook for New Zealand (Eng et al., 2008) quotes a range of 40–44 MJ/kg.

In this context, it is much easier to obtain precise measurements for crude oil derivatives – such as pure hydrocarbons. For instance, the values for light hyrocarbons are given in Lackner et al. (2013) as
- Ethane: 47.652 MJ/kg (64,616 MJ/m^3)
- n-Hexane: 45.240 MJ/kg (173,720 MJ/m^3)

For standardised hydrocarbon fuel products, it is possible to find (Demirel, 2012) the following values:
- Petrol (gasoline): 44 MJ/kg
- Light diesel (780–840 kg/m³): 43.2 MJ/kg
- Heavy diesel (820–880 kg/m³): 42.8 MJ/kg

Because of the disparity of reserves and demands among countries, crude oil is traded on a large scale internationally. The review by BP (BP p.l.c. 2019) provides statistical information on the global oil trade movements for the year 2018. As expected, there are significant exports of oil from the Middle East to China (223.3 Mt), Europe (157.2 Mt) and Japan (146.4 Mt). Exports from the Russian Federation and Central Asia go mostly to Europe (308.7) and China (73.9 Mt). There are also trade flows within the Americas and from South America (mainly Brazil) to China and India.

4.4.3 Natural gas

Natural gas is a mixture, which consists mainly of methane (CH_4) – within 88–99 % (Tissot and Welte, 1984) It often accompanies deposits of crude oil and sometimes coal. Natural gas accumulates in coal and oil deposits during biomass decomposition if it is prevented from escaping to the surface. The natural gas pressure in the deposit can reach up to 100 MPa (Ibler, 2002). A distinguishing characteristic of natural gas, related to its composition, is the relatively low share of carbon emissions from its combustion – the lowest among all fossil fuels. This makes it the preferred fuel for substituting coal and even petroleum-based fuels for heat and power generation, delivering CO_2 emission reduction because of the fuel composition and the effect of increased energy utilisation efficiency.

As can be seen from the overview of proven reserves (Tab. 4.1), the availability of natural gas is not uniform over the World. Similar to crude oil, natural gas is subject to an intensive international trade. The trade takes place over various pipelines and by ships (Liquefied Natural Gas, LNG). The natural gas global trade is also well developed, but the overall trading pattern is simpler than that for crude oil. The main trade flows are from the Russian Federation to Europe and China, followed by the imports to China from the Middle East, South-East Asia and Australia. Significant natural gas exchanges take place among the United States and Canada, as well as some export from the United States to Mexico.

4.5 Combustion of fuels and environmental impacts

After combustion, the exhaust gas has a different composition than the combustion air. Flue gases contain all elements and compounds from the air, including oxygen remaining after the combustion, because thermodynamically is not possible to construct a technical device that consumes all the air oxygen during combustion. In addition, there are other substances in the flue gas that come from unburned fuel and the combustion products. Some of the components in the flue gas are harmful to human health and nature. If they are not removed or transformed before flue gas is released into the atmosphere, they could be harmful to humans and nature, not only in the immediate vicinity of the source but in extended regions. This effect is further amplified by the multitude of combustion installations all over the World.

Some compounds emitted into the atmosphere have high water solubility and are deposited back with the water droplets on the Earth's surface. Other compounds escape to the stratosphere, where, due to solar radiation, they can decompose into other particles, which react better with the environment and again form other compounds.

Environmental impacts are resulting from the combustion of both fossil and biomass-derived fuels (Klemeš, 2015). While fossil fuels are bound mainly to the GHG Footprint (GHGFP), there are specifics for each fuel type:
- **Coal**, besides GHGFP, also releases significant emissions of SO_x and NO_x, as well as fine particles (particulate matter, PM).
- **Crude oil** processing and the use of its derivative fuels are responsible also for significant emissions of GHG, SO_x, NO_x and PM.
- **Natural gas**, although releasing the smallest amount of GHG per unit of use, still results in significant overall emissions due to the large-scale use.
- **Biomass** is considered a renewable fuel. It is indeed renewable in terms of constant availability, under the condition that harvesting does not exceed the natural replenishment (for natural) or production (for crops) rates. However, the renewability does not imply complete GHG neutrality of biomass as fuel (Klemeš, 2015). There are net GHG emissions from biomass use as fuel on a life cycle basis, resulting from various activities – from seeding, to crop care, harvesting, transportation and processing to fuel form.

The environmental and health effects from industrial activities – including fuel use and industrial production – are quantified using the unified indicators, accepted for measuring all types of activities. These include the families of impact potentials and footprints (Klemeš, 2015). This section provides a review of the main emission types and the indicators that quantify their impacts.

4.5.1 SO_x and NO_x

These are greenhouse effect gases, but also have pollution effects on their own. An extended treatment of the effects and the associated footprints is given in Klemeš (2015).

If the fuel contains sulphur, SO_2 can be produced during combustion, and some of the SO_2 (2 % to 3 %) reacts to SO_3 in the combustion plant. SO_2 in the atmosphere is also oxidised to SO_3, and further reacting to sulphuric acid is (H_2SO_4) upon contact with moisture, which gets into the soil during rainfall (acid rain). The removal of sulphur compounds from the flue gas is called desulfurisation. In the case of fluidised bed combustion, the sulphur contained in the fuel is partially removed in the fluidised bed by adding lime.

In combustion, nitrogen is present in the fuel and in the combustion air. Both nitrogen components react with oxygen in the combustion air producing NO (about 95 % of the total amount of nitrogen oxides) and NO_2 (about 5 % of the total amount of nitrogen oxides), collectively called NO_x. The formation depends primarily on the oxygen concentration, residence time and temperature in the combustion chamber where oxygen is present. NO_x causes ozone reduction and thus contributes to the greenhouse effect and contributes to climate change. Under certain conditions, nitric acid can also be formed in the atmosphere with the aid of nitrogen compounds. The main contributors to SO_x and NO_x emissions are processes burning coal and crude (petroleum) oil deriatives.

4.5.2 Greenhouse effect and global warming

Sunlight falling on the surface of the Earth has a wavelength that most often corresponds to visible light. Part of this radiation is reflected by the Earth's atmosphere or surface into outer space without changing the wavelength. Most of the radiation is absorbed by the surface of the Earth and converted into its internal thermal energy. The heated surface emits radiation with a wavelength corresponding to its temperature (thermal radiation with a wavelength of 5 to 40 mm). For thermal radiation, it is much harder to pass through the atmosphere than sunlight and it is partially absorbed by the GHGs in the atmosphere.

GHG heat up and emit thermal radiation either to space or back to earth. In this way, some of the radiation returns to the earth's surface and heats it even more until the wavelength of radiation is such that it passes through the atmosphere in sufficient volume. In order to balance the radiated and the received heat of the Earth's surface, as the concentration of GHGs increases, the surface temperature must increase.

The radiant spectrum of the Earth does not change dramatically even with global warming, but there are several kinds of GHGs and, by their absorption, cover virtually the entire spectrum except the spectrum of visible light. Active GHGs in the atmosphere include (NWE, 2017) water vapour (H_2O), CH_4, CO_2, N_2O, O_3, chlorofluorocarbons and hydrofluorocarbons. In particular, increasing the concentration in the atmosphere of such gases is attributed to a large share of climate change due to global warming (Letcher, 2019).

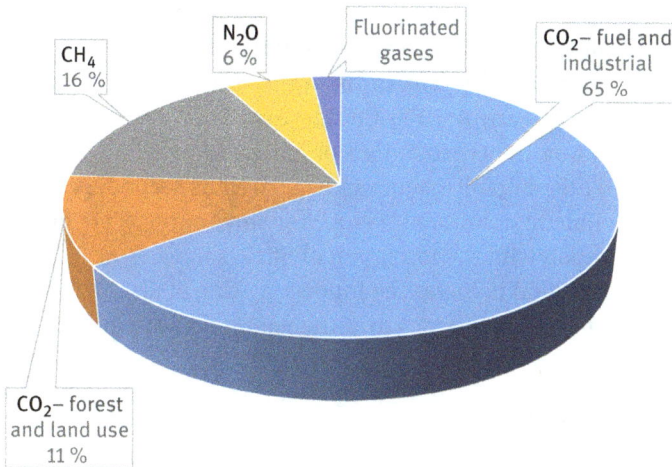

Fig. 4.9: Global GHG emissions structure, based on data from US EPA (2016a).

The GHG inventory is continuously monitored and is presented elsewhere. For instance, a summary can be found in US EPA (2016a) - see (Fig. 4.9), and the full IPCC report is in (Edenhofer, 2014). The key GHGs emitted by human activities include:

- **Carbon dioxide (CO_2):** These emissions come mainly from fossil fuel use. There are also other sources related to forestry and other land use. The activities causing such effects include deforestation and land clearing for agriculture. Increasing forests (reforestation) and other activities can also remove CO_2 from the atmosphere.
- **Methane (CH_4):** This is mainly released from agricultural activities, waste management and other decay processes.
- **Nitrous oxide (N_2O)** comes from agricultural activities, as well as from fossil fuel combustion.
- **Fluorinated gases** (hydrofluorocarbons (HFCs), perfluorocarbons (PFCs), sulphur hexafluoride (SF_6), and nitrogen trifluoride (NF_3)) are related to industrial processes – mainly refrigeration.

Greenhouse effect can also be caused by other media – such as black carbon in aerosol form. In this regard, it is important to note that less than two-thirds the effect can be attributed to CO_2, while several gasses originating from the combustion processes and boilers contribute also a considerable share.

4.5.3 Particulate matter and haze

Haze is the cause of severe health issues, affecting both the physiological and psychological wellbeing of the urban population. While this problem is monitored and tackled, – for example, costing to China nearly 4 % of its gross domestic product (Ouyang et al., 2019) – it is still far from a satisfactory solution. The main constituents of haze are particulate matter and aerosols, also including metal-containing fractions (with components such as Zn, As, Pb, Cd, Mo and Cu) (Li et al., 2016). Haze is formed (To et al., 2017) by the combination of emissions of $PM_{2.5}$ (IRCEL-CELINE 2020, 10) NO_x. SO_2, volatile organic compounds, as well as secondary aerosol formation, and specific climatic conditions favouring slow surface air flows. Fuel use shows a strong correlation with the formation of haze (smog).

The mechanism of haze formation involves the following (Xiao et al., 2011):
–   An active link of fuel use to GHG emissions and the generation of particulates (Particulate Matters, PM)
–   In addition to the general use of fuels in industry and transport sectors, transport activities emit PMs also from tyre friction with the road.
–   PM formation is linked to the NO_x emissions and those of O_3.
–   PMs ($PM_{2.5}$) are related to the haze formation.
–   GHG emissions and energy use are intrinsically and strongly linked.

The important sources of PM and haze pollution inland are industrial processes burning goal, as well as the transport sector combining fossil fuel combustion and PM generation from tyre friction. Additionally, substantial quantities of PMs and aerosols are generated by ships using low-quality fuels in seas and ports (Wan et al., 2016).

4.5.4 Environmental impact assessment concepts – "impact potentials"

At the dawn of Life Cycle Assessment (LCA) development, the attention of scientists and decision-makers was focused on the so-called impact potentials (Saur, 1997). These are defined as indicators of environmental impact dealing with the potential effects and impacts on humans, the environment and resources. They come from the Life Cycle Inventories (LCI). The most prominent such indicators, stemming from those initial studies, are:
–   Global Warming Potential (GWP)
–   Acidification Potential (AP)
–   Eutrophication Potential (EP)
–   Human Toxicity Potential (HTP)
–   Ozone Depletion Potential (ODP)

The most prominent of these indicators is the GWP, which has been conceived as a term by Lashof and Ahuja (1990). The footprint encyclopaedia kind of book (Klemeš, 2015) stresses several critical points concerning the indicator:
–   It quantifies the potential change in climate conditions due to increased concentrations of CO_2, CH_4, and other GHGs in the atmosphere, which entrap heat.

- Some of the effects are increased droughts, floods, loss of polar ice caps, sea-level rise, soil moisture loss, forest loss, change in wind and ocean patterns, and changes in agricultural production.

The GWP is expressed in CO_2 equivalents usually for a time horizon 100 y. According to the US EPA (2016c), it is a measure of how much radiative energy the emissions of 1 t of gas would absorb over a given period of time, relative to the emissions of 1 t of CO_2.

Acidification potential assesses the acid-forming potential of various gases (Clark and Macquarrie, 2008), taking the effect of acid deposition of SO_2 as a unit, so the measurement unit is kg of SO_2 equivalent per functional unit. The assessed gases include SO_2, NO_x, HCl, NH_3. The further indicators (Clark and Macquarrie, 2008), enumerated above, measure the effects of oxygen depletion and biomass over-growth (EP), the depletion of the Earth's ozone layer (ODP), the toxic effect on humans via air, water and soil (HTP).

A very illustrative website is provided by a non-profit organisation of scientists, presenting in the real-time Air Quality Index (AQI; Tab. 4.3) (WAQI, 2020b). It shows in several colour-coded air quality categories, based on the air quality standard by the US-EPA (2016b). The AQI values, measured over the World, are displayed on a map (Fig. 4.10), updated hourly. The air pollution levels categorising the AQI values are summarised in Tab. 4.3. The AQI is based on the measurement of particulate matter (PM2.5 and PM10), ozone (O_3), nitrogen dioxide (NO_2), sulphur dioxide (SO_2) and carbon monoxide (CO) emissions. Most of the stations on the map are monitoring both PM2.5 and PM10 data, but there are few exceptions where only PM10 is available. For more information about the pollutants included in the AQI scale and the method for its calculation, please refer to WAQI (2020a).

Tab. 4.3: Air Quality Index values mapped to air pollution levels (after WAQI, 2020b).

AQI	Air pollution level	Health issues
0–50	Good	Satisfactory air quality, little or no risk to health
51–100	Moderate	Acceptable air quality, but there may be a moderate health concern for people most sensitive to air pollution.
101–150	Unhealthy for sensitive groups	Health problems may occur for sensitive people, while the majority are not likely to be affected.
151–200	Unhealthy	Everyone may begin to experience health problems, while sensitive people may experience serious health problems
201–300	Very unhealthy	Emergency conditions. The effects concern the entire population.
300 +	Hazardous	Pervasive very serious health issues

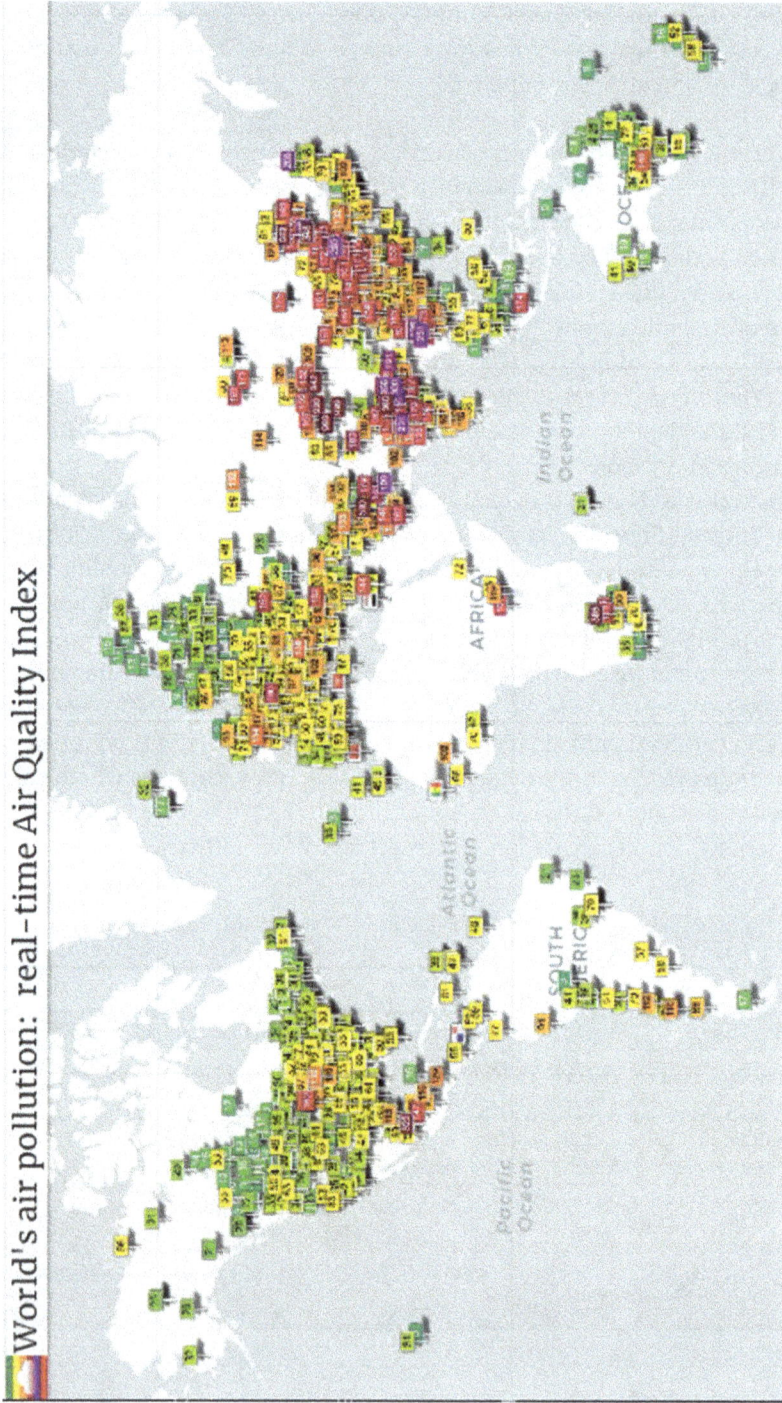

Fig. 4.10: World Air Quality Map, with permission from the World Air Quality Index Project (WAQI, 2020b).

4.5.5 Environmental footprints

Footprints are indicators able to track human pressures on the planet from different angles (Klemeš, 2015). They represent the major categories of impacts – both in terms of resource extraction and emissions. Overall, all these impacts converge to the state of the environment and availability of specific resources – mineral, water, energy and human workforce. Many footprints have been developed, including GHGFP, ecological footprint, water footprint and energy footprint. They are related to climate, food, water and energy security.

Besides these footprints, nitrogen, phosphorus, biodiversity and land footprints are recognised as vital environmental footprints, as they are amongst those essential for health security, food and water security, and land and species security.

The GHGFP is also known as the carbon footprint. The most quoted definition is that it stands for the amount of CO_2 and other GHGs, emitted over the full life cycle of a process or a product. It is expressed in mass of CO_2eq (Klemeš, 2015). Policies aimed at reducing GHGFP include identifying and attaining targets for emissions reduction, increased use of renewable energy and increased energy efficiency. In the context of industrial utility systems, the primary source of GHGFP is the use of fossil fuels.

The structure of the released GHG emissions can be monitored from statistical data. For the United States, the Energy Information Administration provides such statistics and projections, alongside the data on energy demands. The 2018 data, initially given in Fig. 1.2b, are given again in Fig. 4.11 for convenience. It can be seen that the sectors responsible for the most emissions are the industrial, transportation and power generation.

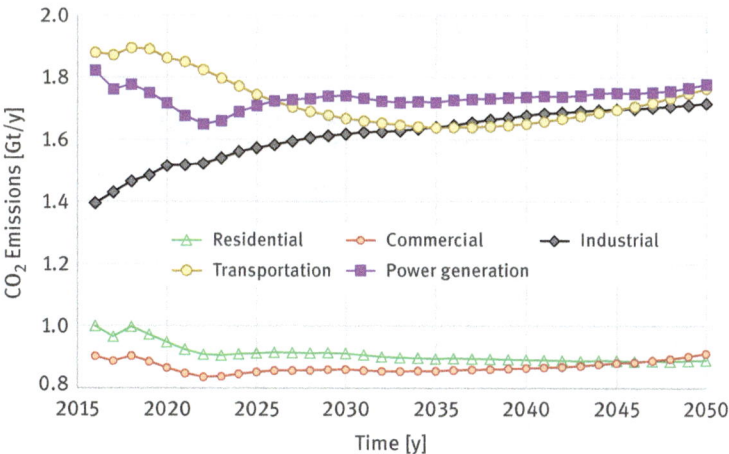

Fig. 4.11: US energy-related CO_2 emissions – statistics and projections, based on data from AEO (2018).

The need for measuring the water footprint has come about due to freshwater availability and security issues, also related to pollution. The quality and quantity of surface and groundwater resources are affected by population growth, rising resource consumption and climate change. A significant contributor to this pattern is industrial activity.

WF is an indicator of direct and indirect water use by an individual, a community, a business or a nation (Hoekstra, 2008). It is closely linked to the concept of virtual water (Hoekstra and Chapagain, 2007) and represents the total volume of direct and indirect freshwater used, consumed and/or polluted. Within the context of industrial utility systems, the main contributors to WF are the cooling water circuits with their cooling tower evaporative losses, the boiler and cooling tower blowdown flow and the WF caused by the production and delivery of the used fossil fuels.

Nitrogen is essential for life, and it is a critical limiting element for food production (Smil, 1997). Nitrogen footprint is the total amount of reactive nitrogen released into the environment as a result of an entity's resource consumption, expressed in mass units of reactive nitrogen (Leach et al., 2012). It represents the disruption of the regional to global nitrogen cycle and its consequences due to human activities (Čuček et al., 2012b).

It is important to stress that the use of renewable energy sources, while offers the potential to reduce the GHGFP , may not result in real or tangible GHGFP reductions if the logistics and delivery distances are not adequately evaluated. An additional aspect of the potential trade-off in fuel substitution scenarios is the use of biomass or biomass-derived fuels, which exhibit a substantial nitrogen footprint, as can be seen from Tab. 4.4.

Tab. 4.4: GHG and nitrogen footprints for fuel options, based on the evaluation in Čuček et al. (2012b).

	GHG Footprint [kg C/GJ]	Nitrogen Footprint [kg N/GJ]
Rape methyl ester	15.52	0.66
Ethanol	15.88	0.88
Diesel	26.26	0.11
Petrol	27.03	0.09

4.6 Summary

This chapter provides an overview of the primary sources used for supplying the energy demands, extending the discussion also to the environmental impacts associated with the use of the various options. Of those, the use of coal appears to be the most harmful to the environment, followed by other fossil fuels. From the renewable energy sources, biomass fuels have a potentially controversial pattern of emissions, where the replacement of some fossil fuels exhibits an adversarial relationship between the GHG and the nitrogen footprints.

In summary, using energy sources more efficiently also raises the question of whether this really reduces emissions on a global scale. There has been the argument that increased energy efficiency is bound to lead to decrease in the prices for the energy services, in turn inducing increased demand and finally – compensating or even overshooting the prior consumption of primary energy resources, which is referred to as the "rebound effect". An empirical study (Greening et al., 2000) based on United States data sources corroborates this argument to a moderate extent. A more recent study for China (Li and Lin, 2017) also supports the existence of such an effect, varying between 50 % and a two- to threefold increase in energy demands as a result of energy efficiency improvement.

While the reasons for these trends are under investigation, one has also to pay attention to a subtle difference in the argument to reduce energy waste. While Varbanov et al. (2018) put forward the argument that energy waste has to be reduced, the official statistics detect only the energy waste within the supply chain of delivering energy services. This implicitly excludes the waste of the energy-based services themselves. If one analyses the complete life cycle, also including the user demands, the picture becomes much clearer. The use of food, lighting and direct energy use all involve waste. Further emission reduction attempts should consider all opportunities for wasting energy and minimise them.

Nomenclature

Symbol	Measurement unit	Description
AP	–	Acidification potential
AQI	[1]	Air Quality Index
EP	kg $PO_4^{3-}{}_{eq}$/kg product, kg N_{eq}/kg product	Eutrophication potential (Acero et al., 2017)
GHG	–	Greenhouse Gas
GHGFP	kg CO_{2-eq}/kg product, kg C/GJ (Tab. 4.4)	Greenhouse Gas Footprint. The exact measurement units vary. The internationally accepted one is [kg CO_{2-eq}/kg product]. (Acero et al., 2017)
GWP	[1], kg CO_{2-eq}/kg product	Global Warming Potential. The exact measurement unit depends on the used definition (Acero et al., 2017)
HTP	$kg_{1,4-DB-eq}$	Human Toxicity Potential (Acero et al., 2017)
I	W/m^2	Overall intensity of solar radiation at the Earth's surface

(continued)

Symbol	Measurement unit	Description
I_D	W/m^2	Intensity of diffuse solar radiation
I_P	W/m^2	Intensity of direct solar radiation
IPCC	–	Intergovernmental Panel on Climate Change
LCA	–	Life Cycle Assessment
LCI	–	Life Cycle Inventory
LNG	–	Liquefied Natural Gas
ODP	–	Ozone Depletion Potential
OECD	–	Organisation for Economic Cooperation and Development
PM	$\mu g/m^3$	Particulate Matter
PM10	$\mu g/m^3$	Fraction of Particulate Matter with particle size 10 µm
$PM_{2.5}$	$\mu g/m^3$	Fraction of Particulate Matter with particle size 2.5 µm

References

Acero, A.P., Rodríguez, C., and Ciroth, A. 2017. LCIA Methods. Impact Assessment Methods in Life Cycle Assessment and Their Impact Categories. https://www.openlca.org/wp-content/uploads/2015/11/openLCA_LCIA_METHODS-v.1.5.6.pdf, accessed 19/08/2020.

AEO. 2018. AEO. Annual Energy Outlook 2018 with Projections to 2050. https://www.eia.gov/outlooks/aeo/pdf/AEO2018.pdf, accessed 29/07/2018.

Akrami, E., Chitsaz, A., Nami, H., and Mahmoudi, S.M.S. 2017. Energetic and Exergoeconomic Assessment of a Multi-Generation Energy System Based on Indirect Use of Geothermal Energy. Energy 124(April): 625–39. https://doi.org/10.1016/j.energy.2017.02.006.

Arat, H. and Arslan, O. 2017. Exergoeconomic Analysis of District Heating System Boosted by the Geothermal Heat Pump. Energy 119 (January): 1159–70. https://doi.org/10.1016/j.energy.2016.11.073.

Basu, P. 2010. Biomass Gasification and Pyrolysis: Practical Design and Theory. Burlington, MA, USA: Academic Press.

Blair, T.H. 2017. Energy Production Systems Engineering. Hoboken, NJ, USA: John Wiley & Sons, Inc.

BP. 2019. BP Statistical Review of World Energy 2019, 68th ed. June 2019. https://www.bp.com/content/dam/bp/business-sites/en/global/corporate/pdfs/energy-economics/statistical-review/bp-stats-review-2019-full-report.pdf, accessed 19/08/2020.

BP p.l.c. 2019. Statistical Review of World Energy | Energy Economics. BP Global 2019. https://www.bp.com/content/dam/bp/business-sites/en/global/corporate/xlsx/energy-economics/statistical-review/bp-stats-review-2019-all-data.xlsx, accessed 17/10/2019.

Cambridge English Dictionary. 2019. ENERGY | Meaning in the Cambridge English Dictionary. https://dictionary.cambridge.org/dictionary/english/energy, accessed 21/08/2019.

Capstone. 2019. Capstone Microturbines to Convert Biogas Into Heat and Power for Irvine Water Recycling Plant: Capstone Turbine Corporation (CPST). https://www.capstoneturbine.com/news/press-releases/detail/3406/, accessed 22/08/2019.

Chiasson, A. 2016. Geothermal Heat Pump and Heat Engine Systems: Theory and Practice. Hoboken, N.J, USA: Wiley.

Chiozzi, P., Barkaoui, A.-E., Rimi, A., Verdoya, M., and Zarhloule, Y. 2017. A Review of Surface Heat-Flow Data of the Northern Middle Atlas (Morocco). Journal of Geodynamics 112 (December): 58–71. https://doi.org/10.1016/j.jog.2017.10.003.

Clark, J.H. and Macquarrie, D.J. 2008. Handbook of Green Chemistry and Technology. New York, NY, USA: John Wiley & Sons. http://nbn-resolving.de/urn:nbn:de:101:1-20141023424.

Cleveland, C.J. 2009. Concise Encyclopedia of the History of Energy. San Diego, CA, USA: Elsevier Science & Technology Books. http://international.scholarvox.com/book/88812164.

Čuček, L., Varbanov, P.S., Klemeš, J.J., and Kravanja, Z. 2012a. Total Footprints-Based Multi-Criteria Optimisation of Regional Biomass Energy Supply Chains. Energy 44(1): 135–45. https://doi.org/10.1016/j.energy.2012.01.040.

Čuček, L., Klemeš, J.J., and Kravanja, Z. 2012b. Carbon and Nitrogen Trade-Offs in Biomass Energy Production. Clean Technologies and Environmental Policy 14(3): 389–97. https://doi.org/10.1007/s10098-012-0468-3.

Davis, B.H., Occelli, M.L., and American Chemical Society, eds.2016. Fischer-Tropsch Synthesis, Catalysts and Catalysis: Advances and Applications. Chemical Industries 142. Boca Raton, USA: CRC Press, Taylor & Francis Group, CRC Press is an imprint of the Taylor & Francis Group, an informa business.

Demirel, Y. 2012. Energy. Green Energy and Technology. London, UK: Springer London. https://doi.org/10.1007/978-1-4471-2372-9.

Desideri, U. and Bidini, G. 1997. Study of Possible Optimisation Criteria for Geothermal Power Plants. Energy Conversion and Management, Efficiency, Cost, Optimization, Simulation and Environmental Aspects of Energy Systems, 38(15). 1681–91. https://doi.org/10.1016/S0196-8904(96)00209-9.

Edenhofer, O., ed. 2014. Climate Change 2014: Mitigation of Climate Change: Working Group III Contribution to the Fifth Assessment Report of the Intergovernmental Panel on Climate Change. New York, NY, USA: Cambridge University Press.

Elmohlawy, A.E., Ochkov, V.F., and Kazandzhan, B.I. 2019. Thermal Performance Analysis of a Concentrated Solar Power System (CSP) Integrated with Natural Gas Combined Cycle (NGCC) Power Plant. Case Studies in Thermal Engineering 14 (September): 100458. https://doi.org/10.1016/j.csite.2019.100458.

Encyclopedia Britannica. 2019. Carboniferous Period | Geochronology. Encyclopedia Britannica. https://www.britannica.com/science/Carboniferous-Period, accessed 22/08/2019.

Enex, and Geysir Green Energy. 2008. Geothermal Utilization in Europe. January 2008. https://www.stjornarradid.is/media/atvinnuvegaraduneyti-media/media/frettir/080119_geothermal_europe_memo_for_ossur.pdf, accessed 19/08/2020.

Eng, G., Bywater, I., and Hendtlass, C.A. 2008. New Zealand Energy Information Handbook. Christchurch. NZ: New Zealand Centre for Advanced Engineering.

European Commission. 2019. JRC Photovoltaic Geographical Information System (PVGIS). https://re.jrc.ec.europa.eu/pvg_download/map_index.html, accessed 22/08/2019.

Garner, R. 2015. Solar Irradiance. Text. NASA. 3 April 2015. http://www.nasa.gov/mission_pages/sdo/science/solar-irradiance.html, accessed 19/08/2020.

Glassley, W.E. 2015. Geothermal Energy: Renewable Energy and the Environment, Boca Raton, FL, USA: CRC Press.

Greening, L.A., Greene, D.L., and Difiglio, C. 2000. Energy Efficiency and Consumption – the Rebound Effect – a Survey. Energy Policy 28(6): 389–401. https://doi.org/10.1016/S0301-4215(00)00021-5.

Haymarket. 2020. Wind Turbine Manufacturers | Windpower Monthly. https://www.windpowermonthly.com/turbine-manufacturers, accessed 09/04/2020.

Hoekstra, A.Y. 2008. Water Neutral: Reducing and Offsetting Water Footprints. https://research.utwente.nl/en/publications/water-neutral-reducing-and-ofsetting-water-footprints, accessed 19/09/2020.

Hoekstra, A.Y. and Chapagain, A.K. 2007. Water Footprints of Nations: Water Use by People as a Function of Their Consumption Pattern. Water Resources Management 21(1): 35–48. https://doi.org/10.1007/s11269-006-9039-x.

Hollman, P. 2019. File:Off-ShoreWind Farm Turbine.Jpg – Wikimedia Commons, Creative Commons Attribution 2.0 Generic License, <https://Creativecommons.Org/Licenses/by/2.0/Deed.En>. https://commons.wikimedia.org/wiki/File:Off-shore_Wind_Farm_Turbine.jpg, accessed 22/08/2019.

IAEA. 2019. Safety Standards. Text. https://www.iaea.org/resources/safety-standards, accessed 18/02/2019.

Ibler, Z. 2002. Technical Guide for Energetics. Prague, Czech Republic: BEN – technická literatura (in Czech).

International Energy Agency. 2004. Energy Statistics Manual. Paris, France: OECD/IEA. https://doi.org/10.1787/9789264033986-en.

International Energy Agency. 2019. Coal Information – 2019 Edition. Database Documentation. http://wds.iea.org/wds/pdf/Coal_documentation.pdf, accessed 07/08/2019.

IRCEL-CELINE. 2020. What Is PM10 and PM2.5? – English. FAQ entry. 2020. https://www.irceline.be/en/documentation/faq/what-is-pm10-and-pm2.5, accessed 19/08/2020.

Karlsdóttir, M.R., Pálsson, Ó.P., Pálsson, H., and Maya-Drysdale, L. 2015. Life Cycle Inventory of a Flash Geothermal Combined Heat and Power Plant Located in Iceland. The International Journal of Life Cycle Assessment 20(4): 503–19. https://doi.org/10.1007/s11367-014-0842-y.

Klemeš, J.J. 2015. Assessing and Measuring Environmental Impact and Sustainability. Oxford, UK; Waltham, MA, USA: Butterworth-Heinemann/Elsevier.

Lackner, M., Palotás, Á.B., and Winter, F. 2013. Combustion: From Basics to Applications. Weinheim, Germany: Wiley-Vch.

Lam, H.L., Varbanov, P.S., and Klemeš, J.J. 2011. Regional Renewable Energy and Resource Planning. Applied Energy 88(2): 545–50. https://doi.org/10.1016/j.apenergy.2010.05.019.

Lashof, D.A. and Ahuja, D.R. 1990. Relative Contributions of Greenhouse Gas Emissions to Global Warming. Nature 344(6266): 529–31. https://doi.org/10.1038/344529a0.

Leach, A.M., Galloway, J.N., Bleeker, A., Erisman, J.W., Kohn, R., and Kitzes, J. 2012. A Nitrogen Footprint Model to Help Consumers Understand Their Role in Nitrogen Losses to the Environment. Environmental Development 1(1): 40–66. https://doi.org/10.1016/j.envdev.2011.12.005.

Letcher, T.M. 2019. Managing Global Warming: An Interface of Technology and Human Issues. London, United Kingdom: Academic Press.

Letcher, T.M., ed. 2017. Wind Energy Engineering: A Handbook for Onshore and Offshore Wind Turbines. London San Diego, CA Cambridge, MA, USA: Academic Press, an imprint of Elsevier.

Li, H., Qin'geng Wang, M.S., Wang, J., Wang, C., Sun, Y., Qian, X., Hongfei, W., Yang, M., and Fengying, L. 2016. Fractionation of Airborne Particulate-Bound Elements in Haze-Fog Episode and Associated Health Risks in a Megacity of Southeast China. Environmental Pollution 208 (January): 655–62. https://doi.org/10.1016/j.envpol.2015.10.042.

Li, J. and Lin, B. 2017. Rebound Effect by Incorporating Endogenous Energy Efficiency: A Comparison between Heavy Industry and Light Industry. Applied Energy 200 (August): 347–57. https://doi.org/10.1016/j.apenergy.2017.05.087.

Martínez-Rodríguez, G., Fuentes-Silva, A.L., Lizárraga-Morazán, J.R., and Martín, P.-N. 2019. Incorporating the Concept of Flexible Operation in the Design of Solar Collector Fields for Industrial Applications. Energies 12(3): 570. https://doi.org/10.3390/en12030570.

Mazzoni, S., Cerri, G., and Chennaoui, L. 2018. A Simulation Tool for Concentrated Solar Power Based on Micro Gas Turbine Engines. Energy Conversion and Management 174 (October): 844–54. https://doi.org/10.1016/j.enconman.2018.08.059.

Merriam-Webster. 2019. Definition of ENERGY. https://www.merriam-webster.com/dictionary/energy, accessed 21/08/2019.

Myers, D. 2017. Solar Radiation: Practical Modeling for Renewable Energy Applications, Boca Raton, FL, USA: CRC Press.

Nobel, P.S. 1999. Physicochemical & Environmental Plant Physiology. 2nd ed. San Diego, CA, USA: Academic Press.

NWE. 2017. Greenhouse Gas – New World Encyclopedia. https://www.newworldencyclopedia.org/entry/Greenhouse_gas, accessed 14/07/2017.

Osborne, D. 2013. The Coal Handbook: Towards Cleaner Production. Woodhead Publishing Series in Energy, no. 50, Cambridge, UK: Woodhead Publishing Limited.

Ouyang, X., Zhuang, W., and Sun, C. 2019. Haze, Health, and Income: An Integrated Model for Willingness to Pay for Haze Mitigation in Shanghai, China. Energy Economics 84 (October): 104535. https://doi.org/10.1016/j.eneco.2019.104535.

Phillips, K.J.H. 1992. Guide to the Sun. Cambridge, UK; New York, USA: Cambridge University Press.

Photon Solar Energy GmbH. 2019. Photon-Solar.Eu - Photovoltaic Trade Germany - Home. http://www.photon-solar.eu/, accessed 21/08/2019.

Prananto, L.A., Zaini, I.N., Mahendranata, B.I., Juangsa, F.B., Aziz, M., and Soelaiman, T.A.F. 2018. Use of the Kalina Cycle as a Bottoming Cycle in a Geothermal Power Plant: Case Study of the Wayang Windu Geothermal Power Plant. Applied Thermal Engineering 132 (March): 686–96. https://doi.org/10.1016/j.applthermaleng.2018.01.003.

Ramos, A., Chatzopoulou, M.A., Guarracino, I., Freeman, J., and Markides, C.N. 2017. Hybrid Photovoltaic-Thermal Solar Systems for Combined Heating, Cooling and Power Provision in the Urban Environment. Energy Conversion and Management 150 (October): 838–50. https://doi.org/10.1016/j.enconman.2017.03.024.

Reynolds, T.S. 2003. Stronger than a Hundred Men: A History of the Vertical Water Wheel, Baltimore, MD, USA: Johns Hopkins University Press.

Santos, I.F.S.D., Barros, R.M., and Filho, G.L.T. 2016. Electricity Generation from Biogas of Anaerobic Wastewater Treatment Plants in Brazil: An Assessment of Feasibility and Potential. Journal of Cleaner Production 126 (July): 504–14. https://doi.org/10.1016/j.jclepro.2016.03.072.

Saur, K. 1997. Life Cycle Impact Assessment. The International Journal of Life Cycle Assessment 2 (2): 66–70. https://doi.org/10.1007/BF02978760.

Siemens. 2019. Solar Power Night and Day. https://assets.new.siemens.com/siemens/assets/api/uuid:a9f15e27-3998-4a5f-a560-2a0c65401d56/version:1560516622/csp-brochure-2019.pdf, accessed 19/08/2020.

Škorpík, J. 2019a. Biomass as a source of Energy. https://www.transformacni-technologie.cz/03.html#menu, accessed 22/08/2019 (in Czech).

Škorpík, J. 2019b. Sunshine as a source of energy. https://www.transformacni-technologie.cz/02.html#menu, accessed 22/08/2019 (in Czech).

Smil, V. 1997. Global Population and the Nitrogen Cycle. Scientific American 277(1): 76–81. https://doi.org/10.1038/scientificamerican0797-76.

Smith, R. 2016. Chemical Process Design and Integration. 2nd ed. Chichester, West Sussex, United Kingdom: Wiley.

Stewart, S.M. and Johnson, R.B. 2017. Blackbody Radiation: A History of Thermal Radiation Computational Aids and Numerical Methods. Boca Raton, FL, USA: CRC Press.

Technical University of Denmark – DTU. 2019. Global Wind Atlas – Czech Republic, Creative Commons License – Attribution 4.0 International (CC BY 4.0), <https://Creativecommons.Org/Licenses/by/4.0/>. https://globalwindatlas.info/en/area/Czech%20Republic, accessed 22/08/2019.

Technical University of Denmark – DTU. 2020. Global Wind Atlas – Germany. Global Wind Atlas. https://globalwindatlas.info/en/area/Germany?print=true, accessed 09/04/2019.

ThinkGeoEnergy. 2020. The Top 10 Geothermal Countries 2019 – Based on Installed Generation Capacity (MWe). Think GeoEnergy – Geothermal Energy News. https://www.thinkgeoenergy.com/the-top-10-geothermal-countries-2019-based-on-installed-generation-capacity-mwe/, accessed 09/04/2020.

Tissot, B.P. and Welte, D.H. 1984. Petroleum Formation and Occurrence, Berlin, Heildelberg. Germany: Springer Verlag. https://doi.org/10.1007/978-3-642-87813-8.

Wai-Ming, T., Lee, P.K.C., and Ng, C.T. 2017. Factors Contributing to Haze Pollution: Evidence from Macao, China. Energies 10(9): 1352. https://doi.org/10.3390/en10091352.

Trenberth, K.E., Fasullo, J.T., and Kiehl, J. 2009. Earth's Global Energy Budget. Bulletin of the American Meteorological Society 90(3): 311–24. https://doi.org/10.1175/2008BAMS2634.1.

US EIA. 2019. What Is the Difference between Crude Oil, Petroleum Products, and Petroleum? – FAQ – U.S. Energy Information Administration (EIA). https://www.eia.gov/tools/faqs/faq.php?id=40&t=6, accessed 23/08/2019.

US EPA. 2016a. Global Greenhouse Gas Emissions Data. Overviews and Factsheets. US EPA. https://www.epa.gov/ghgemissions/global-greenhouse-gas-emissions-data, accessed 12/01/2016.

US EPA. 2016b. Air Topics. Collections and Lists. US EPA. https://www.epa.gov/environmental-topics/air-topics, accessed 16/11/2016.

US EPA, OAR. 2016c. Understanding Global Warming Potentials. Overviews and Factsheets. US EPA. https://www.epa.gov/ghgemissions/understanding-global-warming-potentials, accessed 12/01/2016.

Vakkilainen, E.K. 2017. Steam Generation from Biomass: Construction and Design of Large Boilers. Oxford, United Kingdom: Butterworth-Heinemann: Elsevier.

Varbanov, P.S., Sikdar, S., and Lee, C.T. 2018. Contributing to Sustainability: Addressing the Core Problems. Clean Technologies and Environmental Policy 20(6): 1121–22. https://doi.org/10.1007/s10098-018-1581-8.

Vassilev, S.V., David Baxter, L.K.A., and Vassileva, C.G. 2010. An Overview of the Chemical Composition of Biomass. Fuel 89(5): 913–33. https://doi.org/10.1016/j.fuel.2009.10.022.

Wan, Z., Zhu, M., Chen, S., and Sperling, D. 2016. Pollution: Three Steps to a Green Shipping Industry. Nature News 530(7590): 275. https://doi.org/10.1038/530275a.

Wang, S. and Luo, Z. 2017. Pyrolysis of Biomass. GREEN Alternative Energy Resources, volume 1, Berlin, Germany: De Gruyter; Science Press.

Wang, Y.Z., Zhao, J., Wang, Y., and An, Q.S. 2017. Multi-Objective Optimization and Grey Relational Analysis on Configurations of Organic Rankine Cycle. Applied Thermal Engineering 114 (March): 1355–63. https://doi.org/10.1016/j.applthermaleng.2016.10.075.

WAQI. 2020a. A Beginner's Guide to Air Quality Instant-Cast and Now-Cast. Aqicn.Org. 2020. https://aqicn.org/search/vn/, accessed 11/04/2020.

WAQI. 2020b. World's Air Pollution: Real-Time Air Quality Index. Waqi.Info. https://waqi.info/, accessed 10/04/2020.

Wellinger, A., Murphy, J., and Baxter, D., eds 2013. The Biogas Handbook: Science, Production and Applications. Woodhead Publishing Series in Energy, Number 52. Oxford, UK: Woodhead Publishing.

Wu, C. 2007. Thermodynamics and Heat Powered Cycles: A Cognitive Engineering Approach. New York, USA: Nova Science Publishers.

Wyman, C.E., ed 1996. Handbook on Bioethanol: Production and Utilization. Applied Energy Technology Series. Washington, DC, United States: Taylor & Francis.

Xiao, Z.-M., Zhang, Y.-F., Hong, S.-M., Xiao-hui, B., Jiao, L., Feng, Y.-C., and Wang, Y.-Q. 2011. Estimation of the Main Factors Influencing Haze, Based on a Long-Term Monitoring Campaign in Hangzhou, China. Aerosol and Air Quality Research 11(7): 873–82. https://doi.org/10.4209/aaqr.2011.04.0052.

Zhang, G., Li, Y., Dai, Y.J., and Wang, R.Z. 2016. Design and Analysis of a Biogas Production System Utilizing Residual Energy for a Hybrid CSP and Biogas Power Plant. Applied Thermal Engineering 109 (October): 423–31. https://doi.org/10.1016/j.applthermaleng.2016.08.092.

Part 2: **Components of utility networks**

5 Steam generators

This chapter provides an overview of the main concepts in steam generation and generators, providing links to specialised sources for obtaining more details. That is followed by an overview of Boiler Feed Water (BFW) – the need for treatment, stages of treatment, then by a discussion of boiler blowdown. The chapter is completed by a description of the steam generator models for utility network optimisation.

Steam generators are mostly referred to as "steam boilers" in the industry (Ganapathy, 2017). The primary function of the boiler is to generate steam. In industry, the main use of the steam is for process heating and any steam generated at a pressure higher than the required use can be expanded via steam turbines, to cogenerate power. A boiler consists of two principal parts:

- The furnace, where heat is generated, usually by burning a fuel
- The boiler chamber/drum in which the heat is supplied to the water stream to become steam.

Steam boilers can be of various sizes, which depend on the desired capacity and the type. Steam generators can be implemented as single devices, where one possible design is, as described previously, a furnace with high-temperature-resistant tubes built into the facility. Sometimes, a heat exchanger used for process cooling can play the role of a steam generator – provided that the process cooling demand is at sufficiently high temperature (Klemeš et al., 2018). A steam boiler converts a particular heat input – from a hot stream or from combusting fuel, into steam at the specified pressure and temperature. In some cases, for lower temperature applications such as in the food industry, hot water is used instead of steam. The commonly used fuels in the industry are fossil fuels as coal, fuel oil and natural gas, as well as biomass – mostly wood. This makes it very important to maximise the efficiency of steam boilers, to improve the fuel utilisation rate and minimise the contribution of steam generation to GHG, nitrogen and water footprints. The potential utilisation of waste heat from site processes is even better, as it does not generate additional emissions. In the US pulp-and-paper industry, the share of wood increased in the two decades (1990–2010), as GHG reduction legislation and regulations progress (Bhander and Jozewicz, 2017). However, not only carbon emissions should be considered, but all GHG and also the other emissions contributing to haze and smog. There is also a general trend to recommend natural gas for fuel switching as the fossil fuel containing the lowest fraction of carbon (C2ES, 2019).

https://doi.org/10.1515/9783110630091-005

5.1 Fuel-based steam generators

Steam boiler designs can be classified as belonging to two main types:
- **Fire-tube boilers**: For these devices, the hot flue gas, after the burner, passes through bundles of tubes, which are in turn immersed within a water drum. The flue gas passing through the tubes releases its heat to the water in the drum. The water boils and releases steam, which is subsequently delivered to the other parts of the utility network – steam mains, steam turbines and letdown valves.
- **Water-tube boilers**: In these, the water passes through the tubes and the flue gas from the burners passes through the shell around the tube bundles, heating the water and giving rise to steam.

One variation of water-tube boilers is the drum boiler, shown schematically in Fig. 5.1. The BFW enters the drum. The BFW consists of varying parts of recovered condensate (returned from process heaters and other steam users) and freshwater – referred to as "make-up water". The make-up water is drawn from natural sources in the vicinity of the particular site and is then treated to eliminate, as much as possible, any suspended and also dissolved solids and any other impurities. In engineering practice, it is also referred to as "demineralised water". It is important to note that, although the content of minerals and dissolved solids in the make-up water is significantly reduced after the treatment, some small concentrations of impurities remain. BFW composition, therefore, depends on the quality of the make-up water and the amount of condensate returned from the processes. The water in the drum gradually

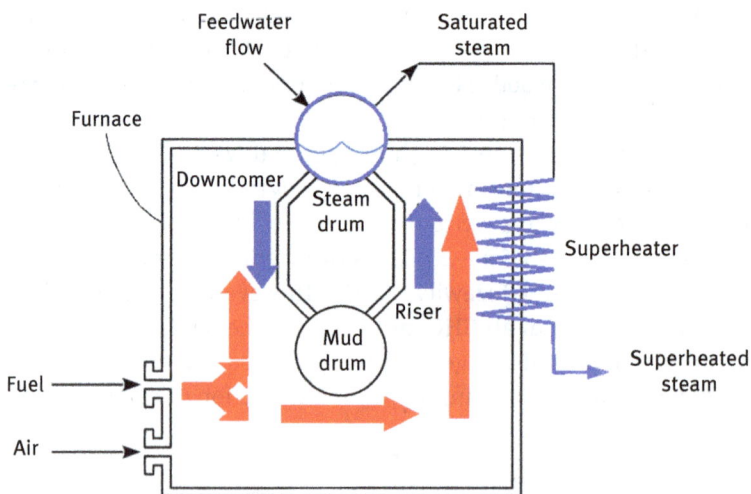

Fig. 5.1: A schematic diagram of a drum boiler (water-tube) (adapted from Yapur and Adam, 2018), under a Creative Commons Attribution-ShareAlike 4.0 International License (Creative Commons, 2019).

accumulates dissolved solids (minerals) by virtue of the constant inflow of make-up water and evaporation of pure water into the steam stream. If this process is not counteracted in some fashion, the result would be a too high concentration of the impurities in the drum, increased scaling and corrosion, then frequent stoppages as a consequence. For this reason, in engineering practice, it is practised to drain out partly the content of the drum as a kind of a purge, then to allow the drum level to be replenished by some of the incoming BFW. In this way, the impurities are purged (blown down), which has given rise to the term for this operation as "boiler blowdown".

While cleaning the drum and maintaining a desired low level of impurities in the drum is essential, blowdown operations bear cost in terms of lost value of treated water and lost value of the heat contained in the blowdown stream. To minimise such losses, two types of measures are practised:

- Minimisation of the blowdown amounts. This makes it necessary to operate the boilers on the edge of the dissolved solids concentration – just below the level that would result in scaling and other detrimental effects.
- The blowdown stream is flashed – an operation that reduces suddenly the stream pressure, allowing it to enter a vessel with a large volume. The effect of the sudden pressure drop and availability of additional volume cause the water stream to become overheated relative to the saturation temperature for the new pressure, and part of the water content evaporates. That is called "flash steam" and can be further reused with its heat and water contents.
- The boiler efficiency manual by the US Department of Energy (US DOE, 2012) goes further, to also recommend recovering heat in a heat exchanger after the steam flashing.

More detailed information on the operation of blow-down and how to calculate the necessary flows can be found in (Ganapathy, 2017). Other types of treatment of the BFW are also applied before entering the boilers, most notably, deaeration. This is an operation where the BFW passes through the device called deaerator, coming in contact with steam (usually the lowest pressure steam available in the utility system), which drives out any dissolved gases.

Another classification of the boilers refers to their compactness and the way of building them. By the principle of construction, boilers are grouped into package and field-erected (Ganapathy, 2003). A package boiler is entirely produced and assembled at the factory of production. The assembly also includes the burner, control equipment, mechanical draft and the accessories. A distinguishing characteristic of these boilers is the compactness. Field-erected boilers are more massive, not so compact and they, as the category name implies, are constructed and assembled at the place of their service. In addition, steam can be generated at one or several pressure levels. In industry, single-pressure level boilers are more common (Smith, 2016).

A well-known supplier of package boilers is Babcock and Wilcox (2019). For field-erected boilers, the picture is diverse. Building them requires local expertise. For example, in the United States, in Pennsylvania, field-erected boilers are available from Indeck Keystone Energy (2019a).

The temperature levels on the tube side reach 300–400 °C (Rajput, 2015), at supercritical conditions – pressure and temperature are higher than the critical point of water: 374 °C and 22.064 MPa = 217.75 bar (Wagner et al., 2000). Typical pressure is 100 bar and higher (Roberts et al., 2017). On the fireside, the temperature is determined by the flame and the inherent heat losses from the burner and the casing and reaches 1,300–1,600 °C (Rajput, 2015). This makes it necessary to use stainless steel and special materials.

Many different fuels are used in industrial boilers – virtually any of the fuels discussed in Chapter 4 are suitable. For instance, in oil refineries, standard hydrocarbon fuels are burned – natural gas and fuel oil. In some chemical plants, especially close to deposits, coal is used. It is possible also to burn waste materials. Such a case is the Brno Waste-to-Energy plant (SAKO, 2019). Some boilers can burn more than one fuel – for example, gas or oil. This flexibility is useful for assuring uninterrupted steam generation under conditions of variable or intermittent supply of one of the fuels, however mostly reducing the efficiency.

A steam boiler can generate steam at a single level, which is widely practised. However, multiple-level steam generators are also known.

The boiler efficiency is usually expressed as the percentage of the energy delivered with the steam within the fuel input energy. Efficiency can also be expressed as a dimensionless energy fraction. It is usually within the range 80–85 % (Rajput, 2015), depending on the boiler technology, ambient temperature and the water quality (determining the blowdown fraction). This efficiency range is a result of the losses due to various reasons. From the heat released from fuel combustion, there are mainly heat losses through the stack, the device casing and with the boiler blowdown. Other losses are also possible, depending on the case. Of these, the stack losses are the most significant.

For a given steam generation flowrate, the heat loss with the flue gas is influenced by several factors. The main factors, acting in combination, are the steam temperature and pressure, the flue gas temperature after the burners, and the minimum stack temperature. The role of these factors is discussed in detail later in this chapter. The minimum stack temperature is a constraint for preventing condensation of corrosive substances from the flue gas and depends on the used fuel, usually ranging between 100 and 200 °C – see Tab. 5.1. The data are from the *Handbook of Energy Engineering* (Thumann and Mehta, 2008). The selection of the specific value of the minimum stack temperature depends on several factors and their trade-off:

- The acid dew point (Ganapathy, 2017) of the flue gas depends on the fuel mixture burned. For burning coal, the main acid components in the flue gas are related to the CO_2 and SO_x content, and the typical range of the acid dew point is 120–150 °C (Vandagriff, 2001). More generally, the range is w (80–180 °C), and it depends on the content of both SO_3 and water in the flue gas (Green, 2018). As a general trend, higher content of each of the two components increases the temperature at which the dew point takes place.

- Materials of construction used for the boiler stack. Regular steel is more susceptible to corrosion than stainless steel and other alloys. However, the special materials come at a higher cost than normal steel (Rayaprolu, 2009).
- It is also possible to use special cover materials like borosilicate glass or Teflon (Ganapathy, 1989). A case study on condensing boilers (Chen et al., 2012) has indicated that stainless steel costs more than twice the price of the carbon steel boiler.

Tab. 5.1: Minimum stack temperature limits for steam boilers (after Thumann and Mehta, 2008).

Fuel	°F	°C
Fuel oil with $S > 2.5$ %	390	198.9
Fuel oil with $S < 1.0$ %	330	165.6
Black (bituminous) coal with $S > 3.5$ %	290	143.3
Black (bituminous) coal with $S < 1.5$ %	230	110.0
Pulverised anthracite	220	104.4
Natural gas, $S = 0$ %	220	104.4

During boiler operation, the actual flue gas temperature in the stack, compared with the design specification, is an indicator of how well the combustion is performed and how efficient is the heat transfer from the flue gas to the water/steam side.

A second significant source of losses is the boiler blowdown – a purge of some of the water from the boiler drum practised to prevent a build-up of dissolved solids in the water, circulating through the steam system. The heat is lost with the purged water flow, although some of the heat can be recovered by applying flash operations (Ganapathy, 2003). It is indicated in (Thermodyne, 2019) that the typical losses of this type range between 1 % and 3 % of the burned fuel. The US Department of Energy (US DOE, 2012) recommends that boilers, exceeding 5 % blowdown rate, as good candidates for flash steam recovery.

Another non-negligible part is the heat loss through the boiler casing. Its rate depends on the presence and the quality of the thermal insulation around the boiler shell, as well as on the ambient temperature. According to Smith (2016), these losses range between 0.25 % and 1.5 % of the burned fuel – depending on the boiler size, design and implementation. Boiler performance and losses are discussed in more detail later in this chapter.

5.2 Heat recovery steam generator (HRSG)

Heat recovery steam generator (HRSG) is a device or a facility for raising steam from high-temperature flue gases and other heat flows. As such, they perform the same functions as the dedicated steam boilers, but their primary source of heat is from waste streams and effluents, or other waste heat (Ganapathy, 2003). The most frequent field of application for HRSG is the steam generation from gas turbine exhausts, as part of the gas turbine combined cycles. In industry, HRSG is used in cogeneration mode, to supply steam to the steam mains for combined power generation by steam turbines and process heating (Varbanov et al., 2005). In addition to the heat contained in the flue gas source, additional heat can be added by burning extra fuel in a burner, using the flue gas residual oxygen content. This mode of operation is called "supplementary firing".

While the most frequent heat source for HRSGs is gas turbine exhausts for forming combined cycles, other heat sources are well-known (Ahmed et al., 2018):
- Furnace exhausts
- Jacket cooling from reactors
- Compressor cooling
- Cooling of finished products

As in dedicated boilers, HRSG can also deliver steam at one or several pressure levels, aiming to maximise the energy recovery from the heat source. The temperature-enthalpy *(T-H)* profiles inside an HRSG follow the typical shapes. On the flue gas side, the profile is a straight line (assuming constant specific heat capacity). On the waterside, there are three distinctive segments: water preheat – with a slope, a flat segment for the evaporation at the saturation temperature of the set pressure, and steam superheat segment with a slope. The *T–H* profiles are discussed further in the model description section.

HRSG can be offered as field-erected units (Indeck Keystone Energy, 2019b) to complement gas turbine installations, as well as packaged units (Cleaver-Brooks, 2019). A dedicated discussion on HRSG can be found in (Eriksen, 2017).

HRSGs can be of single-, dual- and triple-pressure types, having one drum per steam level. The simplest configuration is the one with a single pressure level. Dual-pressure HRSGs, however, are very popular, because they offer higher efficiencies than single-pressure systems and also tend to have a lower specific investment cost (per unit of energy delivered) than the other types.

5.3 Boiler feed water and boiler blowdown

BFW is an essential part of steam systems. It is formed (Fig. 5.2) by mixing condensate return from process steam use and fresh make-up water and then deaeration of the mixture. The make-up water is obtained from local sources and is added to the returning condensate after a thorough treatment.

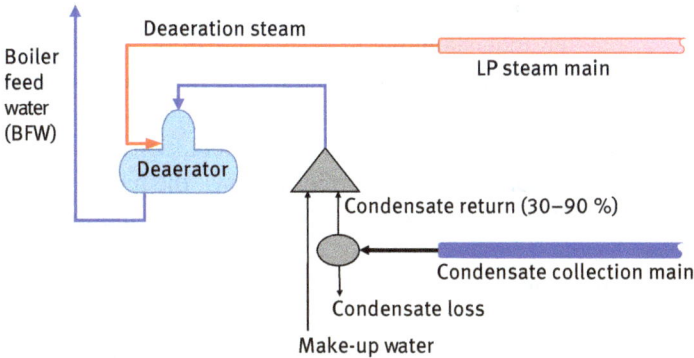

Fig. 5.2: Formation of the boiler feed water.

The make-up water is treated (Ganapathy, 2003) for removal of harmful impurities. Mechanical impurities, including also suspended solids, are removed by settling and filtering. The dissolved solids are removed by water softening, also referred to as deionisation. The main goal of such treatment is to prevent scaling and corrosion in the steam system and especially in the boilers. The impurities include dissolved solids, suspended solids and organic material, including various metal ions (iron, calcium, magnesium, calcium), as well as carbonate and similar ions. There are also dissolved gases, the removal of which is the task for the deaerator (Fig. 5.2).

The make-up water treatment removes the impurities to a very high degree but not completely. As a result, BFW contains some impurities (both suspended and dissolved solids). They accumulate inside the water-steam circuits and leave scaling on the heat exchange surfaces – mainly in the boilers. The dissolved solids may also cause carryover of water droplets into the steam with follow-up damage to downstream equipment – most notably steam turbines. On the other hand, the build-up of suspended solids can form sludge, causing significant mechanical problems affecting the boiler efficiency.

Water has to be periodically or continuously released (purged) – referred to as blowdown – to avoid those effects on the steam system. The most convenient location for the blowdown are the steam boilers. It is important to stress that the blowdown operation has to be optimised, as it is involved in a trade-off. On the one hand, insufficient blowdown would lead to an accelerated build-up of impurities. On the other hand, excessive blowdown would lead to increased fuel consumption, additional chemical treatment spending and heat losses.

5.4 Main factors driving boiler efficiency

Boiler efficiency, as calculated using a rigorous heat balance method, includes stack losses and casing losses (also referred to as radiation and convection losses) (Cleaver-Brooks, 2010). For a better understanding of steam generators and composing adequate models, it is important to understand which factors have the most significant effect on boiler efficiency. First of all, boiler efficiency, as understood in this book, is the fuel-to-steam efficiency. All other partial efficiency definitions are essential to detailed boiler designers, but for the scope of utility systems, the fuel-to-steam efficiency is the relevant one. Following are the major factors, influencing the efficiency of a boiler of a given design:
(1) Stack temperature of the flue gas
(2) Fuel type and composition
(3) Oxygen content in the combustion mixture – essentially the air-to-fuel ratio
(4) The temperature of the ambient air
(5) Casing losses

These factors are discussed for revealing the trends. Firstly, the stack temperature is the one, at which the flue gas is released to the environment. This defines, at a given flue gas flow rate, the amount of energy that will be lost with the outflow of the flue gas. This heat loss is termed as "stack loss". The proportion of heat starting from the combustion temperature (flame temperature), down to the stack temperature, is retrievable to the boiler and the remaining part is lost directly with the flue gas at its exit from the stack – see Fig. 5.3.

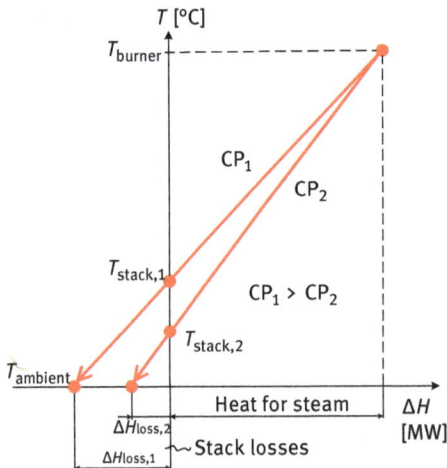

Fig. 5.3: Formation of the stack losses in a boiler (amended after Klemeš et al., 2018).

The boiler efficiency usually drops with raising the stack temperature, in a linear way (Fig. 5.4). It has to be noted that the efficiency line would be at a different level for different fuels. In the case of hydrocarbons, which are related to natural gas and petroleum products, the efficiency trend is opposite to that of the hydrogen proportion in the fuel chemical composition. This is due to the water formation and evaporation as a result of the fuel combustion, which absorbs part of the heat of combustion.

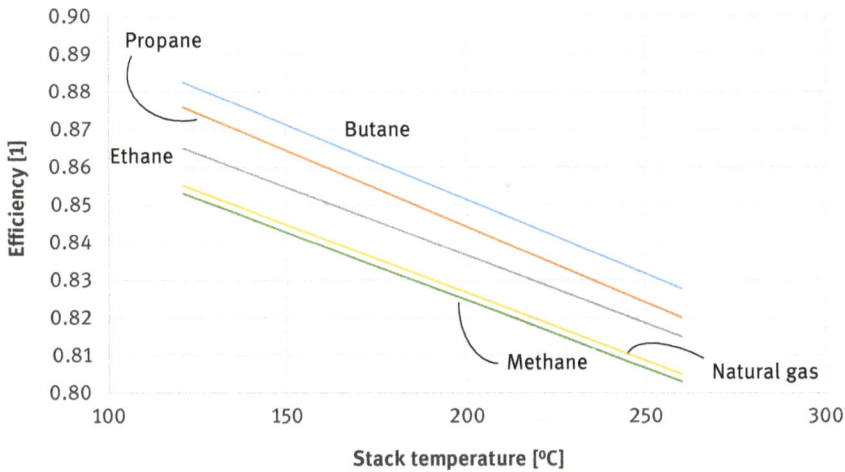

Fig. 5.4: Boiler efficiency variation with stack temperature (amended after Cleaver-Brooks, 2010).

Excess air is the additional proportion of air, mixed with the fuel, above the strictly stoichiometric ratio corresponding to the chemical reaction. This controls the fraction of oxygen in the mixture. Within this aspect, a fuel-air mixture which contains less air than the stoichiometric ratio is referred to as "fuel-rich" and the opposite – as "fuel-lean". Usually, fuel-rich combustion causes fuel to burn incompletely, which results in both low combustion efficiency and contamination of the environment through the exhaust flue gas. On the other hand, higher excess air tends to increase the combustion temperature. This is the cause of both higher efficiency and increased probability of NO_x formation. Additionally, higher excess air also means increased carryover of heat with the increased flue gas flow rate – refer to Fig. 5.3, where $CP_1 > CP_2$ (CP: heat-capacity flow rate, MW/°C). The Cleaver-Brooks efficiency manual (Cleaver-Brooks, 2010) cites values of 15–25 % excess air with tight combustion temperature control as the means of achieving high efficiency, at low-emission (CO and NO_x) formation.

The temperature of the ambient air is also an important factor for boiler efficiency. Lower temperature means more heat spent on pre-heating up the air to the combustion temperature, which translates to a direct heat loss. The trend of the boiler efficiency with increasing the ambient temperature is a line with a positive slope – higher ambient temperature means higher boiler efficiency.

The boiler efficiency also depends on the casing heat losses. The latter is related to the surface temperature of the casing. Besides the casing heat losses, boiler efficiency is also related to the rate of part load. The reason is that, at a given firing temperature, higher load rate distributes the same casing heat losses among larger steam flow, reducing the relative losses. This results in the following trends:

– Higher-pressure boilers have a higher share of casing losses than lower pressure boilers
– At higher firing (steam generation) rate, closer to its rating, the boiler has lower casing losses than at lower part loads.

5.5 Steam generator models for utility network optimisation

This section, building upon the knowledge and the information of the technology overview, gives the description of the steam generator models that are suitable for optimising the complete utility system networks. The simplest optimisation case can be that of the day-to-day operation. Scheduling and longer term planning can also be optimised, alongside the system design. The various modelling use cases are discussed in Chapter 2.

5.5.1 Fuel-based steam generators (boilers)

5.5.1.1 Performance model
The steam boilers can be modelled in two alternative ways – with a constant efficiency model or with a variable efficiency model. The constant efficiency model is linear:

$$Q_{BF} = \frac{m_{stm} \times \Delta h_{gen}}{\eta_{blr}} \qquad (5.1)$$

This merely takes the net heat required for steam generation and divides it by the boiler efficiency. For quick calculations, and for cases when the boilers are expected to operate at or close to the rated loads, this model may be adequate.

In reality, the boiler efficiency varies significantly with the load. Industrial operations rarely stick to the design conditions. More often, reacting to market demands, it becomes necessary to operate at part-load or even overload may be required, leading to the necessity to expand production capacity. It is necessary to understand the nature of the heat losses from boilers to formulate a model that can cope with load variations. Figure 5.5 illustrates typical temperature – load profiles for a fired steam boiler and the associated heat losses. The main loss mechanisms are the stack losses, the boiler blowdown and the casing losses through the boiler walls.

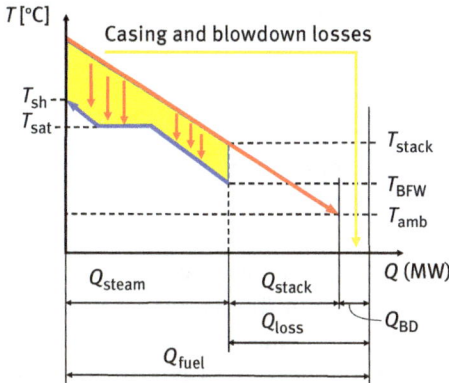

Fig. 5.5: Temperature profiles and heat losses in a boiler (adapted from Varbanov, 2004).

Boiler blowdown can be represented in two ways – as a fraction of the steam generation (blowdown ratio, R_{BD}) and as a fraction of the BFW (blowdown fraction, F_{BD}):

$$R_{BD} = \frac{m_{BD}}{m_{stm}} \qquad (5.2)$$

$$F_{BD} = \frac{m_{BD}}{m_{BFW}} \qquad (5.3)$$

Both these ways are equivalent, and one of the ratios can be calculated from the other. By taking the definitions and the mass balance of steam and water for a boiler (neglecting any water leakage losses), the relationship can be derived as

$$\frac{1}{F_{BD}} = \frac{1}{R_{BD}} + 1 \Leftrightarrow F_{BD} = \frac{R_{BD}}{R_{BD} + 1} \qquad (5.4)$$

If the model is to include the variation in boiler efficiency with load, avoiding non-linearity, the form from Eq. (5.1) is not suitable. Following the formulation in Shang and Kokossis (2004), the fuel heat is related linearly to the steam generation flow-rate and an additional term is added, accounting for the heat losses from boiler blowdown:

$$Q_{BF} = \Delta h_{gen} \times [(b_{blr} + 1) \times m_{stm} + a_{blr} \times m_{stm, max}] + R_{BD} \times m_{stm} \times \Delta h_{pre} \qquad (5.5)$$

The relationship in Eq. (5.5) is formulated as one relating the required fuel heat (Q_{BF}), versus the current steam generation rate (m_{stm}), the maximum steam generation rate ($m_{stm,max}$) and the boiler blowdown ratio (R_{BD}). The other attributes of the equation are the enthalpy difference for the steam generation (Δh_{gen}), the regression parameters (a_{blr} and b_{blr}) and the enthalpy difference for BFW preheat (Δh_{pre}). Note that the latter is part of Δh_{gen}. The enthalpy difference for steam generation is the sum of the enthalpy differences for the preheat, the evaporation and the superheat segments:

$$\Delta h_{gen} = \Delta h_{pre} + \Delta h_{evap} + \Delta h_{sh} \qquad (5.6)$$

The first term of Eq. (5.5) represents the combination of the heat demand for raising the steam and that for covering the casing heat losses. The second term accounts for the heat losses with the boiler blowdown. In this way, the model allows the boiler blowdown losses to be accounted for separately from the other losses, adding accuracy to the calculation, while keeping the linear form. The particular values of the regression parameters for this model, as given in Shang and Kokossis (2004), are $a_{blr} = 0.0126$ [1] and $b_{blr} = 0.2156$ [1], based on a regression of published boiler data. In practice, the values of the parameters depend on the design, operation and maintenance of each boiler.

The performance, according to Eq. (5.5), features interesting trends. The boiler efficiency is defined as the fraction of the fuel heat used to generate the useful steam:

$$\eta_{blr} = \frac{Q_{steam}}{Q_{BF}} \qquad (5.7)$$

The following additional dimensionless parameters can be defined:
-   Relative steam load m_{RL} (1) – mass fraction, $\in$ [0. .1], Eq. (5.8).
-   Preheat ratio R_{PH} (1) – energy fraction, $\in$ [0. .1], Eq. (5.9). For fixed BFW conditions, this represents the various boiler pressures:

$$m_{RL} = \frac{m_{stm}}{m_{stm,\,max}} \qquad (5.8)$$

$$R_{PH} = \frac{\Delta h_{pre}}{\Delta h_{gen}} \qquad (5.9)$$

Based on the given definitions, the following dimensionless equation for the boiler efficiency can be obtained:

$$\eta_{blr} = \frac{m_{RL}}{a_{blr} + m_{RL} \times (b_{blr} + 1 + R_{PH} \times R_{BD})} \qquad (5.10)$$

Using the relationship in Eq. (5.10) and the coefficients from Shang and Kokossis (2004), curves of boiler efficiency as a function of the relative boiler load have been plotted. The value of the preheat ratio was varied within [0.1 and 0.5], with a step of 0.1, each of these values yielding one curve. One set of curves has been plotted for $R_{BD} = 0.01$ [1] and another – for $R_{BD} = 0.10$ [1]. The curves are shown in Fig. 5.6.

It can be seen that for lower blowdown ratio values, the boiler pressure influence on the efficiency is less significant. In this case, for 1 % boiler blowdown, it is insignificant. However, at higher blowdown ratios, the pressure causes significant

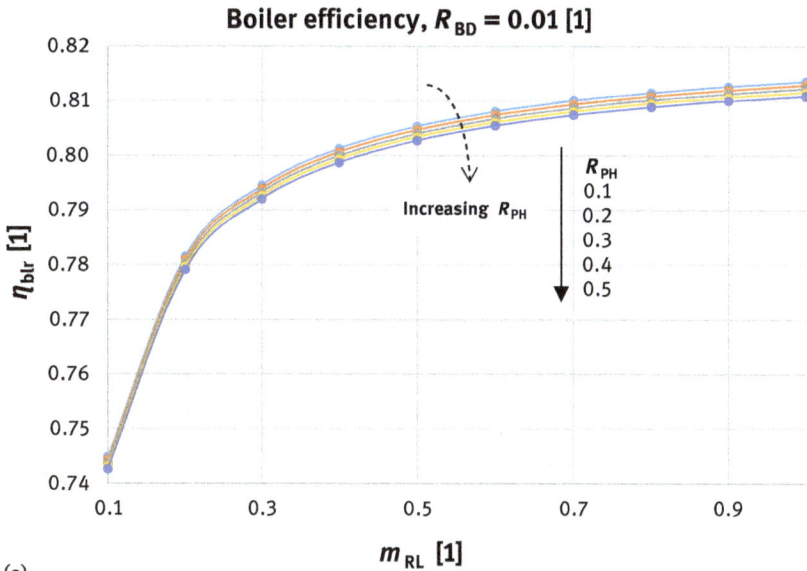

(a)

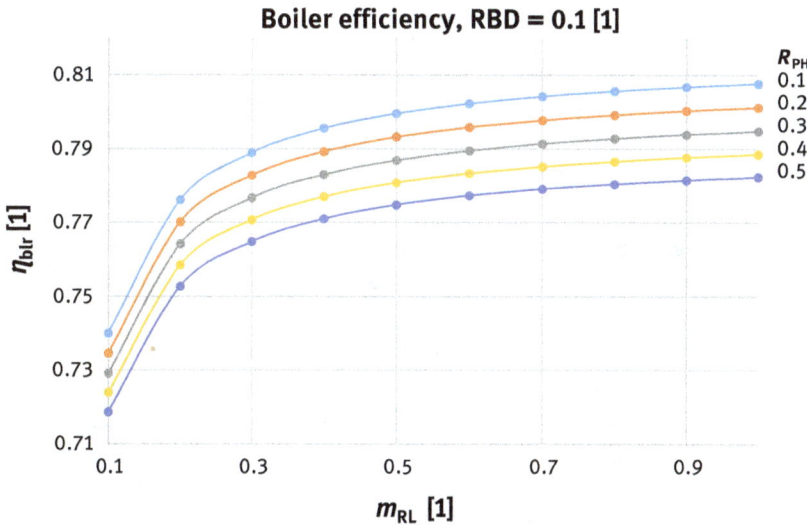

(b)

Fig. 5.6: Boiler efficiency curves for different values of the blowdown and preheat ratios.

efficiency variations. This can be explained with the fact that with increasing the boiler pressure, the relative amount of the required preheat also increases, together with the blowdown losses.

To summarise, the boiler performance model demonstrates that the boiler efficiency depends substantially on the boiler size, pressure and the current load.

Of course, different boiler types and designs would feature different efficiency levels, and the differences should be captured by the regression parameters a_{blr} and b_{blr}.

5.5.1.2 Estimation of the greenhouse gas emissions

The emissions of GHGs from fired boilers are estimated using fixed emission factors for the emissions formed per unit of fuel burnt. From the LCA perspective, it is also possible to add equivalent emissions/footprints for boiler manufacturing, delivery, erection, maintenance, and so on. However, in this model, only the boiler operation is concerned, so only the emissions concerned with the fuel use are modelled. The calculation procedure for obtaining values of the NO_x emission factors for boilers is slightly different from that for gas turbines. The difference originates from the measurement units of the published typical NO_x emission factors.

Tab. 5.2: NO_x emission factors for boilers (Nasruddin et al., 2015).

Fuel type	Combustion emission rate range [kg/GJ]
Natural gas	1.37×10^{-2}–8.28×10^{-2}
Liquid petroleum gas (LPG)	2.30–2.52
Residual fuel oil	3.83–6.59
Diesel	2.27–2.73

Table 5.2 lists average emission rates of NO_x emissions from industrial steam boilers for Malaysia. These emission factors can be converted to mass fractions of t NO_x emitted per t of burned fuel, knowing the fuel heating values.

For estimation and the correct assignment of NO_x emissions, it is necessary to understand that they are formed in different mechanisms, mainly in parallel:

- Thermal NO_x: The nitrogen oxides are formed by oxidising the nitrogen contained in the combustion air, as a result of very high combustion temperature. The reason for this is the high activation energy of the nitrogen oxidation reaction, which prevents it from happening at lower temperatures. The oxidation of N to NO below 760 °C is insignificant and accelerates to exhaust the available oxygen at temperatures above 1,300 °C (US EPA, 1999). The same guide also provides an observation that the NO_x formation is stronger for fuel-lean conditions.
- Fuel NO_x: This is formed by the oxidation of bound nitrogen, contained in the fuel.
- Prompt NO_x: This is a more complex mechanism, including stages of air nitrogen combining with the fuel compounds, then being oxidised in the process of the fuel combustion.

5.5.2 Heat recovery steam generators (HRSGs)

5.5.2.1 General considerations

An HRSG is typically connected to a gas turbine exhaust, and can come with different designs, depending on the desired steam parameters. This chapter discusses the key relationships for a HRSG with a single steam level. A typical pair of temperature profiles of a single-level HRSG is shown in Fig. 5.7. The HRSG of a gas turbine is constrained by heat and mass balances. In order to ensure the thermodynamic feasibility of the heat transfer for steam generation, it is necessary to account for the temperature differences in the HRSG sections. They have to be larger than the minimum allowed temperature difference (ΔT_{min}), specified for the HRSG. In addition, the temperature of the flue gases in the stack (T_{stack}) has to be higher than the minimum stack temperature ($T_{stack,min}$). The latter is derived from the acid dew point of the flue gas ($T_{DEW,FG}$) and depends mainly on the fuel composition.

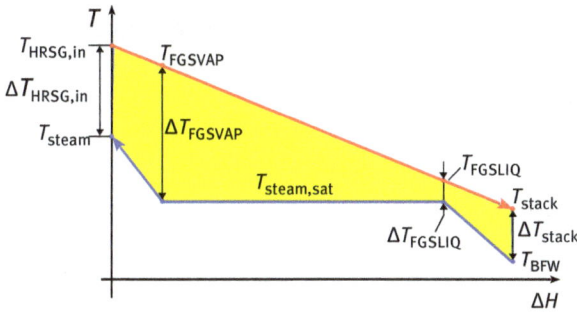

Fig. 5.7: Temperature profiles in a heat recovery steam generator.

5.5.2.2 Supplementary firing

HRSGs may apply supplementary firing, if necessary. As a result, the temperature of the gas turbine exhaust may rise up to 800–900 °C. The mass flow of the supplementary fuel is related to the additional heat needed and the fuel heating value:

$$Q_{sf} = \sum_{fuels} [m_{sf,fuel} \times NHV_{fuel}] \qquad (5.11)$$

The total heat flow entering the HRSG with the flue gas can be expressed as:

$$Q_{in} = Q_{ex} + Q_{sf} \qquad (5.12)$$

In general, the mass flow of the flue gas increases as a result of the supplementary firing. However, taking this into account may complicate the model without a significant gain in precision. This issue should be evaluated and decided upon, depending on the specific cases solved, together with the optimisation formulation. An example can be taken from a Gas Turbine Combined Cycle (González Díaz et al., 2016), where the fuel mass flow rate is given as 22.2 kg/s and the flue gas mass flow rate as 696 kg/s, meaning that the supplementary firing fuel is about 6 %, which can be significant for operating optimisation and significant or neglected for design calculations. Another case has been presented in Ameri et al. (2008), where the flue gas flow rate in unfired HRSG mode is 500 kg/s and in supplementary-fired mode is 500.8 kg/s – a difference of 0.16 %.

In the equations of this chapter, starting from Eq. (5.13), this mass flow increase is neglected for the calculation of the steam generation and flue gas cooling in the HRSG. By using this assumption, the flue gas inlet temperature at the hot end of the HRSG can be derived as

$$T_{\text{HRSG, in}} = T_{\text{ex}} + \frac{Q_{\text{sf}}}{m_{\text{ex}} \times C_{p,\text{ex}}} \tag{5.13}$$

5.5.2.3 Steam generation

The heat requirements for steam generation in each HRSG section are given in the following equations. Note that Eq. (5.14) also accounts for the heat losses with the blow-down.

$$Q_{\text{HRSG, pre}} = m_{\text{HRSG}} \times [R_{\text{BD, HRSG}} + 1] \times \Delta h_{\text{HRSG, pre}} \tag{5.14}$$

$$Q_{\text{HRSG, evap}} = m_{\text{HRSG}} \times \Delta h_{\text{HRSG, evap}} \tag{5.15}$$

$$Q_{\text{HRSG, sh}} = m_{\text{HRSG}} \times \Delta h_{\text{HRSG, sh}} \tag{5.16}$$

The total heat requirement for steam generation is a sum of those for the different sections:

$$Q_{\text{HRSG}} = Q_{\text{HRSG, pre}} + Q_{\text{HRSG, evap}} + Q_{\text{HRSG, sh}} \tag{5.17}$$

5.5.2.4 Flue gas calculations

The flue gas in the HRSG is consecutively cooled down in the different sections. The corresponding temperature calculations are given in the following equations:

$$T_{\text{FGSVAP}} = T_{\text{HRSG, in}} - \frac{Q_{\text{HRSG, sh}}}{m_{\text{ex}} \times C_{p,\text{ex}}} \tag{5.18}$$

$$T_{\text{FGSLIQ}} = T_{\text{FGSVAP}} - \frac{Q_{\text{HRSG, evap}}}{m_{\text{ex}} \times C_{p,\text{ex}}} \tag{5.19}$$

$$T_{\text{stack}} = T_{\text{FGSLIQ}} - \frac{Q_{\text{HRSG, pre}}}{m_{\text{ex}} \times C_{p,\text{ex}}} \tag{5.20}$$

The feasible temperature differences between the flue gas and the water/steam at the ends of the different sections are defined as

$$\Delta T_{HRSG,in} = T_{HRSG,in} - T_{steam} - \Delta T_{MIN,HRSG} \qquad (5.21)$$

$$\Delta T_{FGSVAP} = T_{FGSVAP} - T_{steam,sat} - \Delta T_{MIN,HRSG} \qquad (5.22)$$

$$\Delta T_{FGSLIQ} = T_{FGSLIQ} - T_{steam,sat} - \Delta T_{MIN,HRSG} \qquad (5.23)$$

$$\Delta T_{stack} = T_{stack} - T_{BFW} - \Delta T_{MIN,HRSG} \qquad (5.24)$$

As can be seen from Fig. 5.7, from the temperature constraints in Eqs. (5.21) to (5.24), that action can be only Eqs. (5.21) and (5.23), because of the slope ratios of the profiles. An additional constraint that may activate is the one posed by the acid dew point temperature:

$$T_{stack} \geq T_{stack,min} > T_{DEW,FG} \qquad (5.25)$$

The temperature feasibility conditions form the set of temperature constraints to be included in the HRSG model for any operational or design model. In order to apply them directly, the flue gas temperatures have to be calculated for each HRSG section from the heat balances. This introduces non-linearity in the model, which cannot be used in the MILP formulation. This problem can be overcome by replacing the temperature inequalities with equivalent heat flow constraints.

The derivation starts by expressing the critical HRSG temperatures as sums of the ambient temperature (a fixed parameter) and the corresponding temperature differences as follows:

$$\Delta T_{HRSG,in} = \underbrace{T_{amb} + \frac{Q_{in}}{m_{ex} \times C_{p,ex}}}_{T_{HRSG,in}} - T_{steam} - \Delta T_{MIN,HRSG} \geq 0 \qquad (5.26)$$

$$\Delta T_{FGSLIQ} = \underbrace{T_{amb} + \frac{Q_{in} - Q_{HRSG,sh} - Q_{HRSG,evap}}{m_{ex} \times C_{p,ex}}}_{T_{FGSLIQ}} - T_{steam,sat} - \Delta T_{MIN,HRSG} \geq 0 \qquad (5.27)$$

$$\underbrace{T_{amb} + \frac{Q_{in} - Q_{HRSG}}{m_{ex} \times C_{p,ex}}}_{T_{stack}} - T_{stack,min} \geq 0 \qquad (5.28)$$

The expressions are multiplied by the term $(m_{ex} \times C_{p,ex})$, and the following forms are produced:

$$Q_{in} - m_{ex} \times C_{p,ex} \times [T_{steam} + \Delta T_{MIN,HRSG} - T_{amb}] \geq 0 \qquad (5.29)$$

$$Q_{in} - Q_{HRSG,sh} - Q_{HRSG,evap} - m_{ex} \times C_{p,ex} \times [T_{steam,sat} + \Delta T_{MIN,HRSG} - T_{amb}] \geq 0 \qquad (5.30)$$

$$Q_{in} - Q_{HRSG} - m_{ex} \times C_{p,ex} \times [T_{stack,min} - T_{amb}] \geq 0 \qquad (5.31)$$

The final step of the linearisation is to replace the mass flow of the gas turbine exhaust at current load m_{ex} with the one at full load $m_{ex,max}$, since m_{ex} is calculated by non-linear relationships. This operation yields

$$Q_{in} - m_{ex,max} \times C_{p,ex} \times [T_{steam} + \Delta T_{MIN,HRSG} - T_{amb}] \geq 0 \qquad (5.32)$$

$$Q_{in} - Q_{HRSG,sh} - Q_{HRSG,evap} - m_{ex,max} \times C_{p,ex} \times [T_{steam,sat} + \Delta T_{MIN,HRSG} - T_{amb}] \geq 0 \quad (5.33)$$

$$Q_{in} - Q_{HRSG} - m_{ex,max} \times C_{p,ex} \times [T_{stack,min} - T_{amb}] \geq 0 \qquad (5.34)$$

There are two reasons supporting such a transformation:
(a) The gas turbine exhaust flow is approximately the same as the HRSG flue gas flow. This is because the most substantial supplementary firing fuel flow is often within a margin of 2 % of the gas turbine exhaust flow.
(b) The value of the gas turbine exhaust mass flow at any partial load is smaller than the one at full load, making the inequalities in Equations more constraining than those in Eqs. (5.29)–(5.31).

As a result, if the modified inequalities in Eqs. (5.32)–(5.34) hold, this also automatically satisfies the constraints in Eqs. (5.29)–(5.31).

The heat losses from HRSGs include losses from the stack, the heat loss by the boiler blowdown and the radiative (casing) heat losses. The calculation of the casing losses Q_{casing} depends on the exact design of the HRSG – shape, levels and sections. The boiler blowdown could reach significant levels – up to 5–6 %. These can be expressed as

$$Q_{HRSG,LOSS} = -m_{ex} \times C_{p,ex} \times [T_{stack} - T_{amb}] + m_{HRSG} \times R_{BD,HRSG} \times \Delta h_{HRSG,pre} + Q_{casing} \qquad (5.35)$$

The HRSG efficiency can be expressed as

$$\eta_{HRSG} = \frac{Q_{HRSG}}{Q_{in}} \qquad (5.36)$$

5.5.2.5 Modelling the greenhouse gas (GHG) emissions from heat recovery steam generators

Similar to gas turbines, the emissions of carbon dioxide, as well as sulphur and nitrogen oxides, are modelled as proportional to the amounts of the supplementary fuels burnt. The coefficients are the same as for gas turbines since in most cases, the supplementary firing uses the same fuels.

5.6 Summary and further information

This chapter discusses the basics of boiler design and efficiency trends. It proceeds to also present a model of boiler performance, sufficiently accurate to represent the performance, but also sufficiently simple for incorporating into large-scale optimisation models for utility networks.

In addition to the discussed issues and the cited literature, the readers may refer to recent developments concerning boiler start-up and dynamic operation – for example, the work (Taler et al., 2019) is a representative. It shows that steam boilers are very inert and starting them can expend a lot of fuel for heating up and reaching the required steady state. The appropriate planning of the energy generation of a utility system has to a large degree to account for this sluggishness of steam boilers, to minimise the losses from incorrect or inaccurate forecasting of the process steam demands.

Nomenclature

Symbol	Measurement unit	Description
Δh_{evap}	MWh/t	Enthalpy difference for evaporating water during steam generation
$\Delta h_{HRSG,evap}$	MWh/t	Enthalpy difference of water evaporation for steam generation in the HRSG
$\Delta h_{HRSG,pre}$	MWh/t	Enthalpy difference of preheating the BFW in the HRSG
$\Delta h_{HRSG,sh}$	MWh/t	Enthalpy difference of superheating for steam generation in the HRSG
Δh_{sh}	MWh/t	Enthalpy difference of steam superheating
$\Delta T_{MIN,HRSG}$	°C	Minimum allowed temperature difference in the HRSG
a_{blr}	[1]	Regression parameter
b_{blr}	[1]	Regression parameter
BFW	–	Boiler Feed Water
$c_{p,ex}$	MWh/(t×°C)	Specific heat capacity of the gas turbine exhaust
$CP_{<index>}$	MW/°C	Heat-capacity flow rate of a process stream
GHG	–	Greenhouse Gas
HRSG	–	Heat Recovery Steam Generator
LP (steam)	–	Low-pressure steam
LPG	–	Liquid petroleum gas
m_{ex}	t/h	Mass flow rate of the gas turbine exhaust at current load
$m_{ex,max}$	t/h	Mass flow rate of the gas turbine exhaust at full load
m_{HRSG}	t/h	Mass flow rate of the steam generated in the HRSG
m_{RL}	[1]	Relative steam load of a boiler

(continued)

Symbol	Measurement unit	Description
$m_{sf,fuel}$	t/h	Mass flow rate for supplementary firing of each fuel type
m_{stm}	t/h	Steam generation rate
$m_{stm,max}$	t/h	Maximum steam generation rate for a boiler
NHV_{fuel}	MWh/t	Net (lower) Heating Value of a fuel
Q	MW	Heat flow
Q_{BD}	MW	Heat lost with the boiler blowdown
Q_{BF}, Q_{fuel}	MW	Heat supplied by fuel combustion
Q_{casing}	MW	Casing heat losses in a boiler or HRSG
Q_{ex}	MW	Heat flow carried with the gas turbine exhaust at the entrance of the HRSG
Q_{HRSG}	MW	Total heat flow for steam generation in the HRSG
$Q_{HRSG,evap}$	MW	Heat of evaporation for steam generation in the HRSG
$Q_{HRSG,LOSS}$	MW	Total heat losses from the HRSG
$Q_{HRSG,pre}$	MW	Preheat for steam generation in the HRSG
$Q_{HRSG,sh}$	MW	Superheat for steam generation in the HRSG
Q_{in}	MW	Total heat flow entering the heat recovery steam generator with the flue gas
Q_{sf}	MW	Heat supplied to the HRSG by supplementary firing
Q_{stack}	MW	Heat lost in the stack
Q_{steam}	MW	Heat flow absorbed by steam generation
R_{BD}	[1]	Boiler blowdown ratio
$R_{BD,HRSG}$	[1]	Blowdown ratio for the HRSG
R_{PH}	[1]	Preheat ratio
T, $T_{<index>}$	°C	Temperature of a stream or at a given point in equipment
T_{amb}	°C	Ambient temperature
T_{BFW}	°C	Temperature of the boiler feed water
$T_{DEW,FG}$	°C	Acid dew point of the flue gas
T_{ex}	°C	Temperature of the gas turbine exhaust at the entrance of the HRSG and before eventual supplementary firing
T_{FGSLIQ}	°C	Temperature of the flue gas at the point of saturated liquid in the HRSG
T_{FGSVAP}	°C	Temperature of the flue gas at the point of saturated steam in the HRSG
$T_{HRSG,in}$	°C	Flue gas inlet temperature at the hot end of the HRSG
T_{sat}	°C	Saturation temperature of steam generation
T_{sh}	°C	Superheated temperature of generated steam
T_{stack}	°C	Temperature of the flue gas in the stack
$T_{stack,min}$	°C	Minimum allowed stack temperature
T_{steam}	°C	Temperature of steam generation after superheat, in the HRSG
T_{steam}	°C	Saturation temperature of steam generation, in the HRSG
ΔH, $\Delta H_{<index>}$	MW	Enthalpy difference flow
Δh_{gen}	MWh/t	Enthalpy difference for the steam generation
Δh_{pre}	MWh/t	Enthalpy difference for boiler feed water preheat in a boiler
$\Delta h_{HRSG,pre}$	MWh/t	Enthalpy difference for boiler feed water preheat in the HRSG

(continued)

Symbol	Measurement unit	Description
ΔT_{FGSLIQ}	°C	Temperature difference between the flue gas and the saturated liquid in the HRSG
ΔT_{FGSVAP}	°C	Temperature difference between the flue gas and the saturated steam in the HRSG
$\Delta T_{HRSG,in}$	°C	Temperature difference between the flue gas and the steam outlet for the HRSG
$\Delta T_{MIN,HRSG}$	°C	Minimum allowed temperature difference in the HRSG
ΔT_{stack}	°C	Temperature difference between the flue gas at the point of entering the stack and the entering boiler feed water
η_{blr}	[1]	Boiler efficiency
η_{HRSG}	[1]	HRSG efficiency

References

Ahmed, A., Esmaeil, K.K., Irfan, M.A., and Al-Mufadi, F.A. 2018. Design Methodology of Heat Recovery Steam Generator in Electric Utility for Waste Heat Recovery. International Journal of Low-Carbon Technologies 13(4): 369–79. https://doi.org/10.1093/ijlct/cty045.

Ameri, M., Ahmadi, P., and Khanmohammadi, S. 2008. Exergy Analysis of a 420 MW Combined Cycle Power Plant. International Journal of Energy Research 32(2): 175–83. https://doi.org/10.1002/er.1351.

Babcock and Wilcox. 2019. Package Boilers. https://www.babcock.com/products/package-boilers, accessed 26/08/2019.

Bhander, G. and Jozewicz, W. 2017. Analysis of Emission Reduction Strategies for Power Boilers in the US Pulp and Paper Industry. Energy and Emission Control Technologies; Macclesfield 5: 27–37. http://dx.doi.org/10.2147/EECT.S139648.

C2ES. 2019. Leveraging Natural Gas to Reduce Greenhouse Gas Emissions | Center for Climate and Energy Solutions. Center for Climate and Energy Solutions. https://www.c2es.org/document/leveraging-natural-gas-to-reduce-greenhouse-gas-emissions/, accessed 25/08/2019.

Chen, Q., Finney, K., Hanning, L., Zhang, X., Zhou, J., Sharifi, V., and Swithenbank, J. 2012. Condensing Boiler Applications in the Process Industry. Applied Energy 89(1): 30–36. https://doi.org/10.1016/j.apenergy.2010.11.020.

Cleaver-Brooks. 2010. Boiler Efficiency Guide. http://cleaverbrooks.com/reference-center/insights/Boiler%20Efficiency%20Guide.pdf, accessed 01/05/2019.

Cleaver-Brooks. 2019. Heat Recovery Steam Generators – HRSG. http://cleaverbrooks.com/products-and-solutions/boilers/hrsg/index.html, accessed 26/08/2019.

Creative Commons. 2019. Creative Commons – Attribution-ShareAlike 4.0 International – CC BY-SA 4.0. https://creativecommons.org/licenses/by-sa/4.0/, accessed 26/08/2019.

Eriksen, V.L., ed 2017. Heat Recovery Steam Generator Technology. Woodhead Publishing Series in Energy. Duxford, United Kingdom: Woodhead Publishing, an imprint of Elsevier.

Ganapathy, V. 1989. Cold End Corrosion: Causes and Cures. Hydrocarbon Processing 68 (January): 57–59.

Ganapathy, V. 2003. Industrial Boilers and Heat Recovery Steam Generators : Design, Applications, and Calculations. New York, USA: Marcel Dekker.

Ganapathy, V. 2017. Steam Generators and Waste Heat Boilers: For Process and Plant Engineers. Boca Raton, FL, USA: CRC Press.

González Díaz, A., Fernández, E.S., Gibbins, J., and Lucquiaud, M. 2016. Sequential Supplementary Firing in Natural Gas Combined Cycle with Carbon Capture: A Technology Option for Mexico for Low-Carbon Electricity Generation and CO_2 Enhanced Oil Recovery. International Journal of Greenhouse Gas Control 51 (August): 330–45. https://doi.org/10.1016/j.ijggc.2016.06.007.

Green, D.W. 2018. Perry's Chemical Engineers' Handbook 9th Edition. New York, NY, USA: McGraw-Hill Education.

Indeck Keystone Energy. 2019a. Field Erected Boilers. http://www.indeck-keystone.com/boilers/field-erected-boilers/, accessed 26/08/2019.

Indeck Keystone Energy. 2019b. Indeck Heat Recovery Steam Generators (HRSG). http://www.indeck-keystone.com/boilers/heat-recovery-steam-generators/, accessed 26/08/2019.

Klemeš, J.J., Varbanov, P.S., Alwi, S.R.W., and Manan, Z.A. 2018. Sustainable Process Integration and Intensification. Saving Energy, Water and Resources. 2nd ed. Berlin, Germany: Walter de Gruyter GmbH.

Nasruddin, H.N., Azid, A., Juahir, H., Abdullah, A.M., Amran, M.A., Mustafa, A.D., and Azaman, F. 2015. NOx Emission Modelling from Industrial Steam Boilers. Jurnal Teknologi 76: 1. https://doi.org/10.11113/jt.v76.4152.

Rajput, R.K. 2015. Thermal Engineering: Including : Thermodynamics, Heat Engines and Non-Conventional Power Generation : For Engineering Students Preparing for B.E./B. Tech., AMIE-Section B (India), GATE and UPSE (Engg. Services) Examinations : SI Units. Bengaluru, India: Laxmi Publications (P) Ltd.

Rayaprolu, K. 2009. Boilers for Power and Process. Boca Raton, FL, USA: CRC Press.

Roberts, I., Stoor, P., Carr, M., Höcker, R., and Seifert, O. 2017. Steam Handbook. 1st ed. Reinach, Switzerland: Endress+Hauser. https://portal.endress.com/wa001/dla/5001084/9862/000/00/CP01195DEN_0117.pdf, accessed 25/08/2019.

SAKO. 2019. Waste to Energy | SAKO – Svoz a Zpracování Odpadu Brno, Czech Republic. https://www.sako.cz/page/en/607/waste-to-energy/, accessed 23/08/2019.

Shang, Z. and Kokossis, A. 2004. A Transhipment Model for the Optimisation of Steam Levels of Total Site Utility System for Multiperiod Operation. Computers & Chemical Engineering 28(9): 1673–88. https://doi.org/10.1016/j.compchemeng.2004.01.010.

Smith, R. 2016. Chemical Process Design and Integration. 2nd ed. Chichester, West Sussex, United Kingdom: Wiley.

Taler, J., Zima, W., Ocłoń, P., Grądziel, S., Taler, D., Cebula, A., Jaremkiewicz, M., et al. 2019. Mathematical Model of a Supercritical Power Boiler for Simulating Rapid Changes in Boiler Thermal Loading. Energy 175 (May): 580–92. https://doi.org/10.1016/j.energy.2019.03.085.

Thermodyne, 2019. Boiler Efficiency Improvement & Heat Loss Explained in Boiler|Thermodyne Thermodyne Boilers (blog). http://www.thermodyneboilers.com/boiler-efficiency/, accessed 21/08/2019.

Thumann, A. and Mehta, D.P. 2008. Handbook of Energy Engineering. 6th ed. Lilburn, GA : Boca Raton, FL, USA: Fairmont Press; CRC Press.

US DOE. 2012. Recover Heat from Boiler Blowdown (DOE/GO-102012-3408). https://www.energy.gov/sites/prod/files/2014/05/f16/steam10_boiler_blowdown.pdf, accessed 26/08/2019.

US EPA. 1999. Nitrogen Oxides (NOx), Why and How They Are Controlled. https://www3.epa.gov/ttncatc1/dir1/fnoxdoc.pdf, accessed 16/02/2020.

Vandagriff, R. 2001. Practical Guide to Industrial Boiler Systems. Basel, Switzerland: Marcel Dekker/CRC Press.

Varbanov, P., Perry, S., Klemeš, J., and Smith, R. 2005. Synthesis of Industrial Utility Systems: Cost-Effective de-Carbonisation. Applied Thermal Engineering 25(7): 985–1001. https://doi.org/10.1016/j.applthermaleng.2004.06.023.

Varbanov, P.S. 2004. Optimisation and Synthesis of Process Utility Systems. PhD Thesis, Manchester, UK: University of Manchester Institute of Science and Technology.

Wagner, W., Cooper, J.R., Dittmann, A., Kijima, J., Kretzschmar, H.-J., Kruse, A., Mareš, R., et al. 2000. The IAPWS Industrial Formulation 1997 for the Thermodynamic Properties of Water and Steam. Journal of Engineering for Gas Turbines and Power 122(1): 150. https://doi.org/10.1115/1.483186.

Yapur, S.F. and Adam, E.J. 2018. A Comparison of MIMO Tuning Controller Techniques Applied to Steam Generator. Advances in Science, Technology and Engineering Systems Journal 3(3): 07–14. https://doi.org/10.25046/aj030302.

6 Steam turbines

6.1 Overview of steam turbines

Steam turbines are the most important class of machines in the energy industry, used to convert heat into work. Steam turbines need many auxiliary equipment units and use steam as a working fluid. Steam is generated separately and supplied to the turbines. This allows the design and operation of the steam turbine-based systems using diverse primary resources to generate the steam – starting from fuels as well as solar energy (see Chapter 4). When using fuels, steam can be generated in steam boilers or heat recovery steam generators (Chapter 5). Additionally, when using solar energy, steam is generated in specialised boilers, within concentrating solar power plants (Lovegrove and Stein, 2012). Other potential heat sources for generating steam are geothermal heat (Akrami et al., 2017) and nuclear reactions (Riznic, 2017).

The investment cost of the steam turbine units is very high, and the size of this unit is larger than other types of energy conversion equipment especially compared to internal combustion engines. According to the US Environmental Protection Agency (EPA) technology characterisation guide (US EPA, 2015), the equipment cost ranges from nearly 700 US$/kW for the smaller units of 500 kW capacity down to a level lower than 400 US$/kW for large units of 15,000 kW (15 MW) capacity. For this reason, the most common use of steam turbines is as a drive of an electric generator for stationary power generation, a direct drive for large process equipment, for instance a compressor, or for driving large ships.

The most common working fluid of Rankine cycle turbines is water steam, but other fluids are also used – such as organic mixtures in organic rankine cycles (Yu et al., 2018). The temperature range for admission of steam to steam turbines is wide, especially for industrial utility systems. It is always synchronised with the pressure of the steam flow, to ensure that as a result of the steam expansion in the turbine the outlet flow contains steam with quality (dryness fraction) no less than 0.95, preferably dry steam. For the typical equipment parameters, quoted in the US-EPA guide (US EPA, 2015), the following values are given:
- for 500 kW steam turbines – inlet pressure of 34.47 bar(g) equal to 35.49 bar(a); inlet temperature of 287.8 °C; outlet pressure of 3.45 bar(g) (4.46 bar(a)) and outlet temperature of 147.8 °C
- for 3 MW steam turbines – inlet pressure of 41.37 bar(g) (42.38 bar(a)) and inlet temperature of 301.7 °C; outlet pressure of 10.34 bar(g) (11.36 bar(a)) and outlet temperature of 189.4 °C
- for 15 MW steam turbines – inlet pressure of 48.26 bar(g) (49.28 bar(a)) and inlet temperature of 343.3; outlet pressure of 10.34 bar(g) (11.36 bar(a)) and outlet temperature of 193.2 °C

https://doi.org/10.1515/9783110630091-006

Steam is fed from the steam source to the turbine, where it expands and performs the work that delivered to other devices by a rotating shaft. Outlet steam is often passed to a condenser where the condensed water is pumped to the steam boiler and the cycle is repeated. Such a cycle is called the steam cycle or Rankine–Clausius cycle.

6.2 Rankine–Clausius cycle (R–C cycle, steam cycle)

The steam cycle is the oldest heat cycle which is used in technical practise. For transformation of heat to work through the steam cycle, initially steam piston engines were used (Schobert, 2014). The current state of the art in power generation uses steam turbines (US EPA, 2015). The use of a particular mechanism or machine does not change principle of the cycle, even though the steam turbine is a turbomachine and the piston steam engine is a reciprocating engine. Unlike the internal combustion engine cycle, the steam cycle connects several devices.

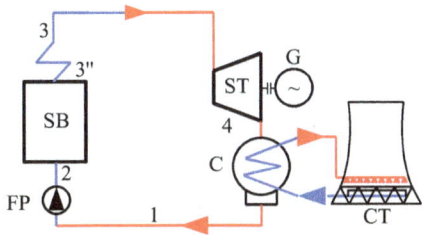

Fig. 6.1: Flow chart of Rankine–Clausius cycle. Notation: SB, steam boiler; ST, steam turbine or another type of steam engine; G, electric generator; C, condenser (for condensing the steam and creating low outlet pressure); CT, cooling tower; FP, feed pump.

Figure 6.1 shows the steam cycle for power generation only. The description of each part of the flow chart follows:

State 1–state 2: feed water pressure increase. This is performed by using a boiler feed water pump. The power input to the pump is lower compared with the work of the steam expansion in the steam turbine, because water is almost an incompressible fluid.

State 2–state 3: steam generation. Raising steam (steam generation) in the boiler (Chapter 5) can be divided into three steps. In the first step 2–3′, the water is heated up to the liquid saturation point. This is referred to as boiler feed water pre-heating. The second part of the process is the evaporation of water to the state of steam saturation (3′–3″). The steam is then superheated to state 3 in a super-heater section of the boiler. Obtaining superheated steam is essential for power generation using this cycle, as steam turbines use exactly the superheated part of the steam energy content and steam at the exit of the turbine has to be essentially dry (Varbanov, 2004). Note, states 3′ and 3″ are internal to the steam boiler.

State 3–state 4: steam expansion in the turbine. Inside the turbine, the steam expands from the higher pressure of state 3 to the lower pressure of state 4. In the process, the expanding fluid performs work on the turbine blades and rotates the shaft. The shaftwork is then passed to either a direct-drive device or converted to electrical power in an electric generator. In Fig. 6.1, the latter option is depicted.

State 4–state 1: steam condensation. The condensation of steam is performed inside the condenser after the expansion stage. The steam is condensed from the state 4 (steam) to state 1 (water). With this step the cycle is closed, because water is again in state 1. The condensation takes place by the inlet steam flow transferring heat to cooling water. The usual arrangement inside the condenser is to flow the cooling water inside the tubes and the steam to condense on the shell side. The reason for this is to accommodate the larger specific volume of the steam flow. In turn, the cooling water is cooled after the condenser by external means. The most frequent facility for performing this operation is a cooling tower. The cooling effect in the cooling tower is achieved at the expense of evaporating part of the circulating cooling water into the ambient air. This has several significant effects (Kim and Smith, 2004):

- Part of the cooling water is lost via the evaporation.
- As a consequence of the evaporation, the concentration of dissolved solids in the remaining cooling water flow inevitably increases. This causes the need for purging some of the cooling water, which is also known as **cooling water blowdown.**
- Both previous effects lead to the need to replenish the cooling water that circulates in the cooling circuit.
- The use of the evaporation to the ambient air constrains the achievable temperature for the cooling water to 8–9 °C below that of the ambient air.

A simple example of such a cycle is considered for illustration. Figure 6.2 illustrates a typical steam cycle representation on a temperature–entropy diagram (a) and on an enthalpy–entropy diagram (b). For conceptual understanding, a reversible Rankine cycle is considered first. Within that definition, the heat losses through the cycle are assumed negligible and the state transitions are assumed reversible. Since the individual thermodynamic processes are performed in several equipment units, the energy balances are written for each of the units using the balances for open systems, because all the equipment units form open systems – having their input and output streams. An additional assumption is that the change of potential energy of the working fluid is negligible.

Under those assumptions, the most significant interactions become more clearly identifiable (see Fig. 6.2). The heat supply flow to the working fluid is provided only in the boiler and heat is rejected only in the condenser:

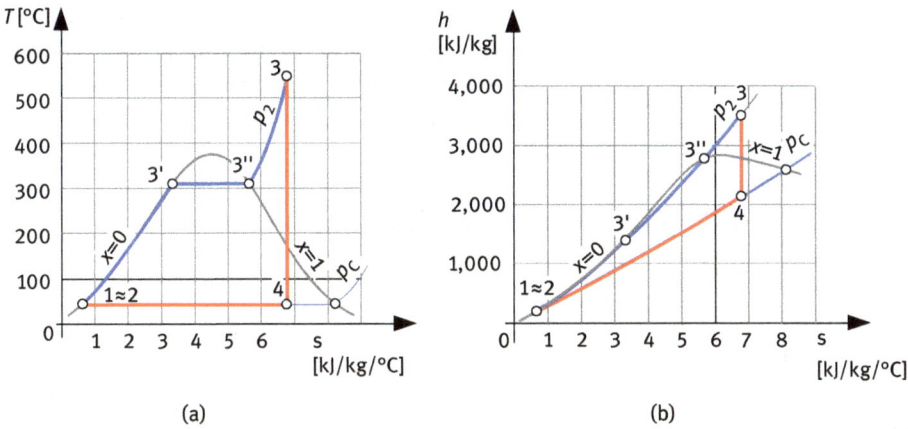

Fig. 6.2: The steam cycle represented on (a) T–s and (b) h–s diagrams. The state of steam in point 3 is 9.4 MPa and 550 °C; the condensation pressure is 9 kPa. Notation: p [Pa], pressure (p_2, steam pressure; p_c, condenser pressure); h [kJ/ kg], specific enthalpy; T [°C], temperature; x [1], steam quality (mass fraction of the vapour within the overall flow); s [kJ/kg/°C], specific entropy; w [kJ/kg], specific work of the cycle.

$$q_{2-3} = h_{3c} - h_{2c}; \; q_{4-1} = h_{1c} - h_{4c} \tag{6.1}$$

The turbine generates power (performs work and positive amount):

$$w_T = h_{3c} - h_{4c} \tag{6.2}$$

The boiler feed water pump consumes power (negative amount):

$$w_{BFWP} = h_{1c} - h_{2c} \tag{6.3}$$

The net power generation is equal to the algebraic sum of the turbine power generation and the pump power consumption:

$$w = w_T + w_{BFWP} = q_{2-3} + q_{4-1} = (h_{3c} - h_{2c}) + (h_{1c} - h_{4c}) \tag{6.4}$$

Therefore, the first-law efficiency of the reversible Rankine cycle becomes:

$$\eta_{RRC} = 1 + \frac{h_{1c} - h_{4c}}{h_{3c} - h_{2c}} \tag{6.5}$$

The notation of Eqs. (6.1) to (6.5) includes q [kJ/kg] – specific heat flows, $h_{<1\text{-}4>c}$ [kJ/kg] – specific stagnation enthalpies, η_{RRC} [1] – efficiency of the cycle, w_T [kJ/kg] – specific internal work of the turbine and w_{BFWP} [kJ/kg] – specific internal work of feed pump.

The actual Rankine cycle, as opposed to the reversible idealisation earlier, incurs heat, friction and dissipation losses of energy. For this reason, its representation on the temperature–entropy (T–s) and enthalpy–entropy (h-s, Mollier) diagrams is not as simple. At the very least, they involve entropy increase during each of the cycle stages. Also, the practical implementations of the cycle, for achieving higher power output and efficiency, perform heat regeneration, as well as practice steam extraction from some turbine stages for supplying process heating.

6.3 Principle of operation of steam turbines and main relationships

The simplest type of steam turbine is the Laval turbine, which describes the general principle of steam turbines as shown in Fig. 6.3. Steam expands from state 0 to the state 1 through the Laval nozzle (stator). In this Laval nozzle, enthalpy drop of the steam is transformed to kinetic energy. This stream enters the blade-to-blade passages of the rotor, where kinetic energy of the steam is transformed to work and pushes the blades, causing the rotor to rotate the shaft. The kinetic energy of the steam stream is lower behind the rotor than in front of the rotor. The difference of kinetic energy on the two sides of the rotor constitutes the amount of work done by the expanding steam.

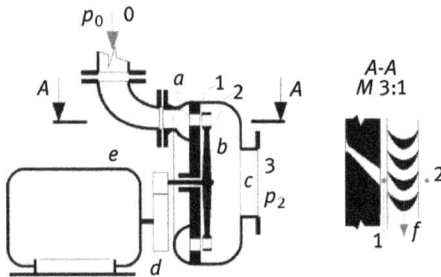

Fig. 6.3: A section view of a Laval turbine. Notation: a, nozzle (guide vanes disk usually has several nozzles for higher mass flow rate and power); b, rotor, c, exit flange, d, gearbox; e, generator; f, direction of rotation; 0, inlet of the steam; 1, gap between the guide vanes and the rotor; 2, exit of steam from rotor; 3, exhaust.

For higher power output at larger capacity, multi-stage steam turbines are built (Fig. 6.4). Each stage of the turbine contains one row of the stator blades fixed to the casing, which forms a nozzle row and one row of the rotor blades. The nozzle need not be only one, but the blades of the stator may form a few nozzles distributed around the periphery of the rotor. Steam turbine facilities with high power output can be also composed of several smaller turbines, which are arranged on a shared shaft.

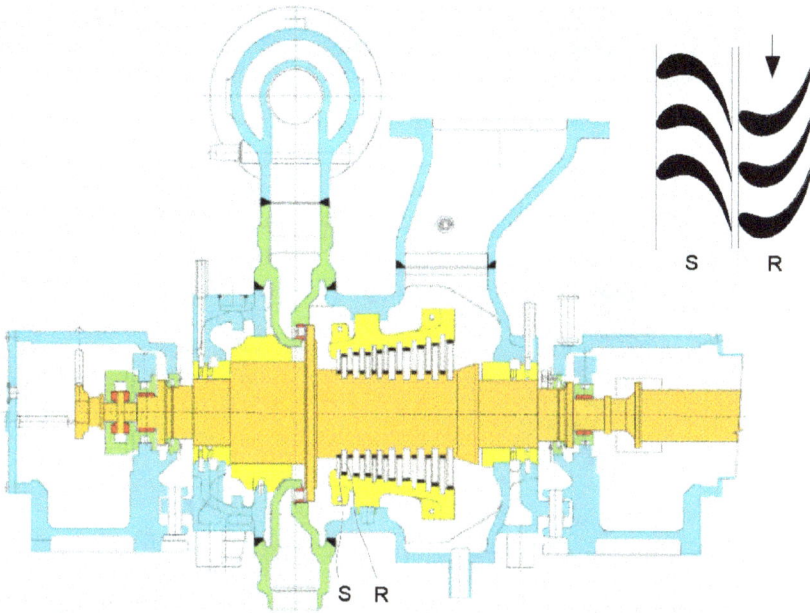

Fig. 6.4: A simplified longitudinal section view through a multi-stage steam turbine. S: stator blade row and R: rotor blade row. The steam turbine: 6 MW, 9.980 min^{-1}, the admission steam: 36.6 bar, 437 °C and the exhaust steam: 6.2 bar. Made in Alstom (factory PBS – CZ) (amended after Alstom Power, 2002).

6.3.1 Internal power output of steam turbine

An important property of the steam turbine is its internal power output. This is the power available at the turbine rotor:

$$WP_i = w_i \times m_{ST} \tag{6.6}$$

where WP_i [kW] is the overall power output transferred between the steam and the rotor inside the turbine, w_i [kJ/kg] is specific internal work (transferred energy between steam and rotor) and m_{ST} [kg/s] is mass flow rate of steam flowing through turbine. Note, this property is not the indicated shaft power, given by the manufacturers. The latter is a smaller energy flow, which results after the losses in the mechanical transmissions are accounted for.

Another internal parameter of the turbine is its internal efficiency η_i [1], which indicates the quality of transformation of energy inside turbine and is defined as the ratio of the real specific internal work with ideal specific internal work of the turbine. The difference between the real specific internal work and ideal specific internal work is called *specific internal losses* of turbine. The internal power and efficiency are not influenced only by the internal losses. There are other factors, which also include leakage through the machine parts.

6.3.2 Specific internal work

The specific internal work of a steam turbine can be derived by applying to the turbine the first law of thermodynamics for open systems. The derivation usually neglects the effect of Earth's gravity on the steam expansion and flow.

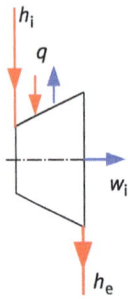

Fig. 6.5: A steam turbine stage for deriving the expression for the specific internal work.

Referring to the configuration in Fig. 6.5,

$$w_i = \left(h_i + \frac{c_i^2}{2} \right) - \left(h_e + \frac{c_e^2}{2} \right) + q = h_i - h_e + \frac{c_i^2 - c_e^2}{2} + q \qquad (6.7)$$

where q [kJ/kg] is specific heat transfer with surroundings per unit amount of steam. The subscript i denotes inlet and e denotes exit. The term containing the steam flow velocities is the required change of kinetic energy between the inlet and the outlet of the machine:

$$\Delta e_K = \frac{c_i^2 - c_e^2}{2} \ [kJ/kg] \qquad (6.8)$$

Usually it is required $\Delta e_K \approx 0$. The expansion of the steam can be adiabatic inside the turbine, meaning $q = 0$ (or $q \approx 0$). There are cases where heat transfer with surroundings has influence on the expansion (so called polytropic expansion $q \neq 0$). Inside the steam turbines, the temperature is higher than that of the surroundings. However, for machines with a good thermal insulation such heat transfer (loss) to the surroundings is negligible in practice. Polytropic expansion is considered for poorly insulated machines and especially for machines with chilled blades.

6.3.3 Adiabatic expansion inside heat turbine

The working gas (steam) inside the turbine expands from pressure p_i on pressure p_e. If the turbine is an isolated system, then entropy of working gas must be constant or increasing. The increase of entropy is caused by heat loss and local temperature differences, which arise during turbulence and friction of the steam flow, internal and with the machine parts, as well as leakages and mixing. For illustrating the heat flows inside the steam turbine, h–s (Mollier) and T–s diagrams (Fig. 6.6) are used.

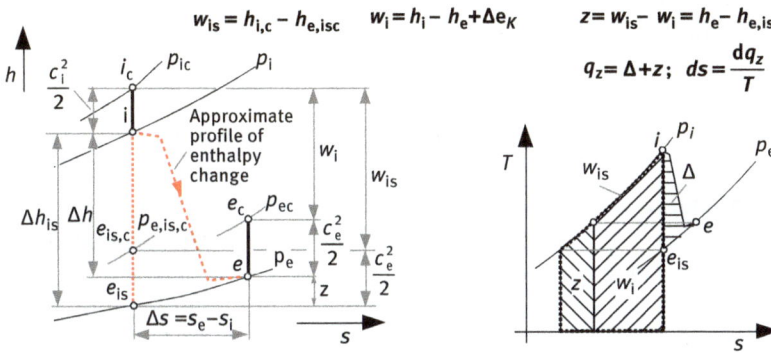

Fig. 6.6: The specific internal work of the heat turbine at adiabatic expansion on the h–s and T–s diagrams. Notation: T [K], absolute temperature; w_{is} [kJ/kg], specific internal work at isentropic expansion (adiabatic expansion without losses); v [m³/kg], specific volume; Δh_{is} [kJ/kg], specific enthalpy difference at isentropic expansion; Δh [kJ/kg], used enthalpy difference; q_z [kJ/kg], specific internal heat loss inside the turbine; z [kJ/kg], specific internal mechanical losses inside the turbine; Δ [kJ/kg], specific reusable heat (part of q_z, which can be transformed to work in other parts of the turbine).

For the difference of specific enthalpy across the turbine, assuming a constant specific heat capacity at constant pressure, the following equation can be written as

$$h_i - h_e = C_P \times (T_i - T_e) = \frac{k}{k-1} \times r \times T_i \times \left[1 - \left(\frac{P_e}{P_i}\right)^{\frac{n-1}{n}}\right] \qquad (6.9)$$

where $k = \frac{C_P}{C_V}$ [1] is the Poisson index (Potter and Somerton 2014), n [1] is polytropic index (for the case of a flow without losses $n = k$), $r = \frac{R}{MW_{Water}}$ [kJ/kg /°C] is the characteristic gas constant for water, the ratio of the universal gas constant and the molecular weight of water and C_P and C_V [kJ/kg /°C] – specific heat capacity of working gas at constant pressure and at constant volume.

From the previous equations it is evident that maximum specific work at adiabatic expansion is achieved for the case of an isentropic process. Therefore, the isentropic expansion is used as a benchmark for assessment of internal efficiency of steam turbines. Equation (6.10) shows the definition of the internal turbine efficiency of the heat turbine in relation to isentropic process, which is in turn referred to as *isentropic efficiency*:

$$\eta_{is} = \frac{w_i}{w_{is}} \ [1]$$

(6.10)

The isentropic efficiency of steam turbines varies in a wide range – from about 20 % (National Renewable Energy Laboratory (NREL), Golden, CO, 2006) to higher than 90 % (Luo et al., 2011).

6.4 Main steam turbine configurations and classifications

The steam turbines can be classified by several principles. One option is to distinguish by the service type – direct drive and power generation. Another classification criterion is their connection with the rest of the utility system (plant) units – they can be condensing steam turbines, backpressure steam turbines and turbines with a steam extraction.

6.4.1 Steam turbines by service type: direct drive and power generation

In industrial practice, steam turbines can be used in two types of applications. The most obvious application is to connect the steam turbine to an electrical power generator. This provides much flexibility and allows distribution of the generated power over the site among smaller power demands. This also allows to balance on-site power generation against import or export of power to/from a utility grid.

In some cases, the given fixed mechanical drive power demands by process units, which are large enough, to justify a dedicated steam turbine connected to them. In such cases, the steam turbine applications are referred to as **"direct drive"** or **"mechanical drive"**. Usually, in direct drive application mode, given fixed power demands have to be satisfied.

Finally, a steam turbine can be even connected to drive two potential users (Petchers, 2003), by attaching them to both ends of a common shaft, driven by the turbine. The two applications can be a direct drive and a generator. An important reason to select a steam turbine for directly driving a piece of equipment can also be safety. As commented in a technology assessment document by the US Department of Energy (US DOE, 2012), steam turbine drives are suitable for potentially explosive environments, due to the possibility to provide non-sparking operation.

From the viewpoint of utility systems and their flexible operation, the generator turbines are very flexible and allow variation of their steam throughput (and steam generation), giving the opportunity to optimise on-site power generation and following of process steam demand in a flexible way. On the other hand, any connection to direct drive applications, fixes the steam throughput through a turbine, or at least constrains it within very narrow intervals.

Steam turbine manufacturers optimise their fleets of steam turbines, usually offering equipment units suitable for both types of application. This is the case of General Electric (GE) (Scoretz and Williams, 2008) and Siemens (Siemens AG 2019).

In the simplest configuration, steam turbines are used for electrical power generation in an arrangement illustrated in Fig. 6.7. That arrangement features the steam turbine, where power is generated from the energy flow of the expanding steam. After incurring certain losses (friction of the parts, the fluid with the parts and internal fluid friction), the turbine passes a shaftwork (power) W_i to the mechanical transmission – consisting of the coupling and the gearbox. The coupling passes a further reduced power flow W_{co} to the gearbox and, after losses, passes to the power generator of power flow W_{gr}. Finally, the generator provides electrical power W at the electrical contacts.

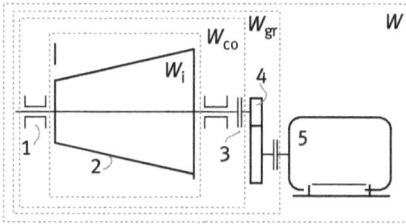

Fig. 6.7: Efficiency and power output of a steam turbine coupled with a power generator.

The power output of a set of a steam turbine and a generator is evaluated using several efficiency ratings, corresponding to the described contours. The overall efficiency of the set can be expressed as the product of the internal efficiency of the turbine, the mechanical efficiency of the turbine, the efficiency of the gearbox (if included in the set) and the efficiency of the generator (Fig. 6.7):

$$W = W_i \times \eta_{total} = W_i \times \eta_{turbine} \times \eta_m \times \eta_{generator} \qquad (6.11)$$

In Eq. (6.11), $\eta_{turbine}$ [1] represents the steam turbine efficiency, η_m [1] represents the combined efficiency of the transmissions – coupling and gearbox and $\eta_{generator}$ [1] is the efficiency of the electrical generator. All these efficiencies are under 1 (fraction).

Usually, the parameters of the steam turbine-generator set are given on a label of the generator. On this label, nominal power W_n [MW] and optimal power W_{opt} [MW] (power under maximal efficiency) that can be delivered by the equipment is shown.

6.4.2 Steam turbine arrangements in a site utility system

Another principle of classification, very frequently used in industrial utility systems, is by their arrangements within the utility system and the potential use of the exhaust steam flow. Within this classification, it is possible to distinguish condensing steam turbines and backpressure steam turbines . For both these types, it is possible to draw steam from an intermediate expansion stage, before the final exhaust outlet. This intermediate flow is referred to as **extraction.**

6.4.2.1 Condensing steam turbines
A condensing steam turbine is an arrangement of a steam turbine that is connected to a condenser at its exhaust (Fig. 6.8).

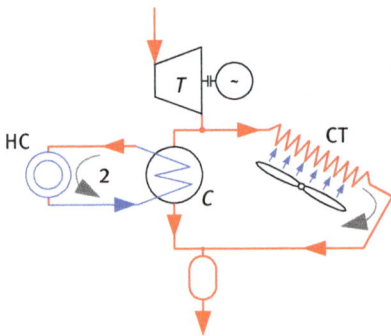

Fig. 6.8: Flowsheet of a condensing steam turbine. Notation: 1, cooling loop of a condenser at minimum condensing pressure; 2, cooling loop of condenser for the case of suppressed condensation; T, steam turbine; C, condenser for the suppressed condensation case; CT, condenser for the vacuum condensation case, often an air cooler or a mixing water condenser; HC, heat consumer (e.g. a site process).

The flowsheet in Fig. 6.8 is typical for combined heat and power (CHP) plants. Loop (2) is activated in the case of a requirement for heat supply in the temperature interval possible for the turbine exhaust (e.g. during a heating-winter season). If there is no process heating demand to be served, then Loop 1 is activated. In this case, the lower condensation temperature is possible, which allows achieving maximum power output. This must not be mixed with the maximum thermal efficiency of the steam cycle, as in the case of such condensation, a large part of the initial heat is rejected to the environment.

A surface condenser, if used, is usually located next to the turbine. In the case of an air cooled condenser, the distance between the turbine and the condenser is longer, as the air condensers are usually located on the roof of the boiler room.

6.4.2.2 Backpressure steam turbines and steam extraction
Additional to the condensing steam turbines, discussed in the previous section, there are also other configurations in industry. They pertain to steam turbines connecting various steam mains, whose steam pressure levels vary from the boiler

pressure level down to the lowest pressure in the utility system. The typical steam mains pressure level intervals are

- Very High Pressure (VHP) main: 97–164 bar and 480–565 °C or 593 °C (Dryden, 1982)
- High Pressure (HP) main: 35–55 bar
- Medium Pressure (MP) main: 10–15 bar
- Low Pressure (LP) main: 3–5 bar

Similar ranges are also quoted in a book on biomass-based energy use in industry (Sims, 2002). In this context, backpressure steam turbines draw steam mostly from VHP, HP and sometimes from the MP mains and exhaust to steam mains of lower pressure, as illustrated in Fig. 6.9.

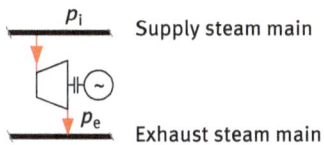

Fig. 6.9: Flowsheet of a backpressure steam turbine. Notation: p_i [bar] the admission pressure; p_e [bar] backpressure.

Some steam turbines take steam from higher-pressure mains and expand it down over several mains – including down to LP main or condensation level. In the course of this expansion, steam is extracted from the turbine at intermediate pressure (IP) levels. Two variants of a steam turbine with one extraction are shown in Fig. 6.10. Figure 6.10a shows a controlled extraction arrangement and Fig. 6.10b shows the variation with an uncontrolled extraction.

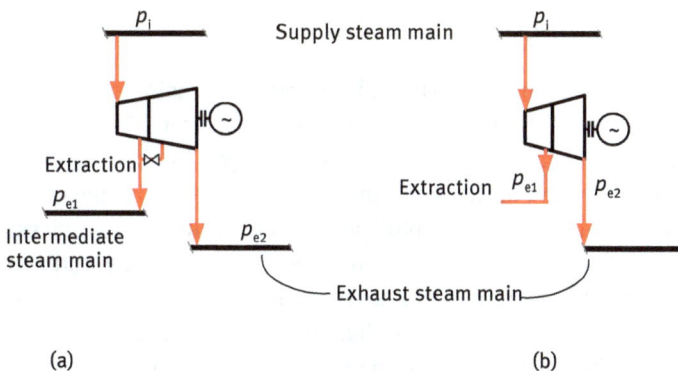

Fig. 6.10: Types of the extractions from steam turbines: (a) controlled extraction and (b) uncontrolled extraction.

The extracted steam is expanded from the admission pressure to the IP of the point of the extraction. The controlled extraction arrangement is used for ensuring a specified constant pressure of the drawn steam, to be supplied to a steam user directly or to a steam main. The amount of the steam extraction is controlled by the valve and manipulating the steam flow through the turbine stages downstream from the extraction point (Woodruff et al., 2012). The usual case when uncontrolled extraction is practiced is for steam turbines in power plants for regeneration of heat.

A CHP plant flowsheet is shown in Fig. 6.11. It contains two pressure letdown units that allow bypassing of the turbine. This allows to continue the heat supply to users even in the case of a failure of the turbine and/or the power generator.

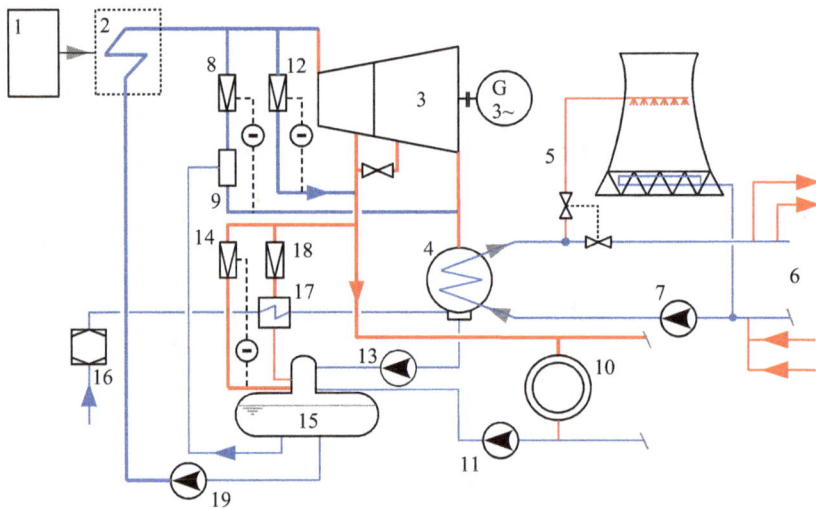

Fig. 6.11: CHP plant flowsheet featuring an extraction steam turbine.

The CHP plant from the flowsheet in Fig. 6.11 contains one condensing steam turbine with a controlled steam extraction for supplying industrial users. Such a system is used for steam turbines in CHP plants about power output from 10 up to 70 MWe. This is designed to supply in parallel a certain amount of heat via the steam to users. The typical power-to-heat ratios (useful power and useful heat services) achieved range from 0.2 to 1, depending on the efficiency of the turbine and the efficient management in the minimisation of the heat losses.

The full notation of the flowsheet in Fig. 6.11 includes 1, fuel storage; 2, steam boiler; 3, condensing steam turbine with a controlled extraction; 4, condenser; 5, parallel cooling loop, in this case with cooling tower; 6, supply and return of cooling water; 7, circulation pump of the cooling loop; 8, letdown valve for bypassing steam to the condenser (for start-up/shut-down and emergency operations); 9, cooling unit; 10, steam users; 11, pump for the condensate return; 12, steam letdown valve linking the boiler to the steam users; 13, condensate pump; 14, steam letdown valve connecting the turbine extraction and the deaerator; 15, feed water storage tank with deaerator; 16, make-up water supply from the chemical treatment of fresh water; 17, recovery heat exchanger transferring heat from extracted steam to the make-up water; 18, reducing pressure unit of steam for heating of refilling water; 19, boiler feed water pump.

CHP plants are usually built near the heat and steam users, with the most frequent applications being district heating CHP plants and industrial site utility systems. Among the other criteria for the selection of location for such energy conversion facilities are the proximity of the primary energy source (e.g. fuel or solar) proximity of water sources (rivers, dams, lakes) and electrical grid interface points.

In dedicated thermal power plants, condensing steam turbines are installed, having uncontrolled steam extractions for heat regeneration. The exact parameters, the flowsheet arrangement and the connections of this type of turbine usually depend on type of the steam generator.

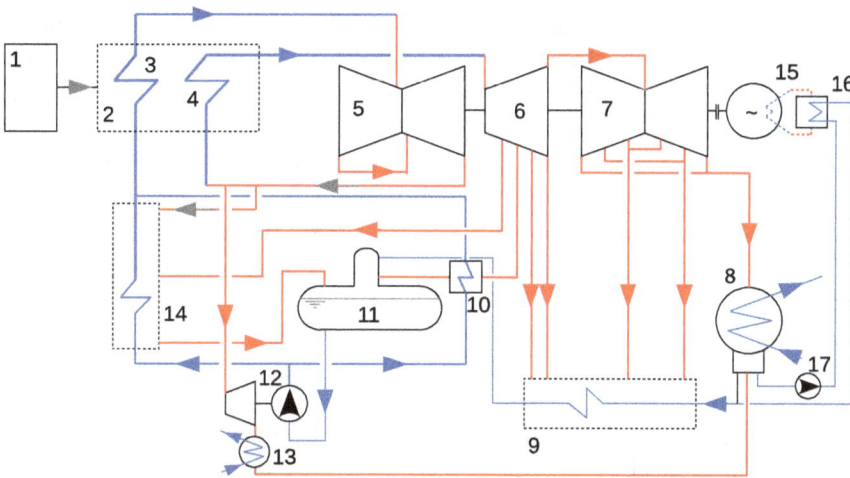

Fig. 6.12: The flowsheet for 200 MW Škoda thermal power plant (Škorpík, 2020).

Figure 6.12 shows the flowsheet of a thermal power plant, featuring a single shaft linked to several steam turbines. The flowsheet includes the following units: **1** –fuel storage; **2** – steam boiler; **3** – super-heater; **4** – re-heater; **5** – HP turbine stage; **6** – IP turbine stage; **7** – LP turbine stage; **8** – condenser; **9** – system of LP feed water heaters (FWH); **10** – cooler of steam for spray deaerator; **11** – feed water tank; **12** – feed pump driven by a direct-drive steam turbine; **13** – condenser of the feed pump driver turbine; **14** – system of HP FWH; **15** – cooling system of the generator and exciter (generator is cooled by hydrogen); **16** –heat recovery from the cooling system of the generator and exciter; **17** – condensate pump.

The reheating of steam is one of the approaches to increase the efficiency of a Rankine cycle implementation. The steam at the exit of the HP turbine stage is transferred again to the boiler (the re-heater section), where it is heated almost on the admission temperature, but at a lower pressure.

As a drive of the feed pump, an electric motor is typically used for smaller installations, while for larger plants a small steam turbine is installed. The power input of the feed pump is usually relatively small. Therefore, the steam turbine for driving the pump tends to have lower efficiency than the main steam turbine. If the efficiency of the direct-drive steam turbine for driving the pump is lower than the overall efficiency of electricity generation by the main turbine combined with the electric motor, then the electric motor is used.

6.4.2.3 Connecting a steam turbine

A steam turbine can be connected to other process units in a variety of ways. The exact connections depend on the type of those units. The steam turbine interfaces usually have a set of connection points as illustrated in Fig. 6.13. The picture is simplified, not including the interfaces necessary for start-up and shut-down operations. In that figure, the following items are marked: 1, steam supply to the turbine; 2, drainage of the inlet pipeline; 3, shut-off/emergency valve; 4, valve actuators (most often mechanical-hydraulic); 5, control valve(s); 6, drainage of control valve chambers; 7, controlled steam extraction; 8, steam control valve; 9, pipes of uncontrolled steam extraction; 10, steam outlet to the condenser; 11, drainage of the last stage of the turbine (several can be used); 12, HP labyrinth seals; 13, LP labyrinth seals; 14, oil circuit of the bearings; 15, turbine tracking quantities measured; 16–17, bearing stands.

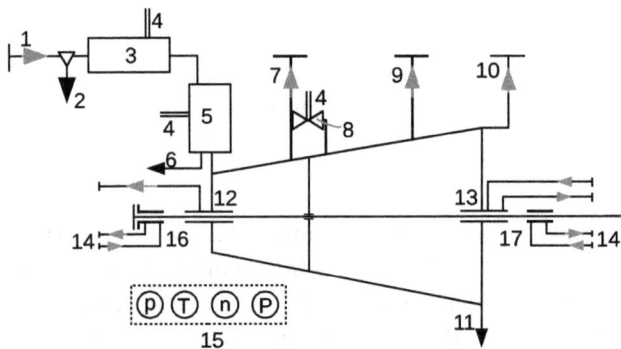

Fig. 6.13: An example of interface points for a steam turbine with extractions.

Steam turbine seals. The seals at key points in the turbine are intended to block and minimise the steam leakage and maximise the flow through the turbine stages and the power generation. A certain amount of steam still escapes through the seals. Although relatively small, up to 2 % for small turbines and up to 1 % for large turbines, depending on the seal wear, such amounts are not negligible, especially for larger-scale facilities.

Of course, the steam leaking through the seals can be collected and used. Steam leaking from the seals can be drained to the so-called seal condensers, where the heat from the condensation is used to heat the Boiler Feed Water (Chapter 5 on steam generators). In HP seals, the pressure gradually decreases from the duct to the outlet pressure, so that at appropriate pressure levels of the seal, it is possible to guide the escaping steam into extraction points. However, it is necessary to account for the potentially higher steam temperature at the seal that can be significantly higher than that at the extraction point, which would lead to an increase of the temperature of the extraction outlet after mixing. This may be dangerous for the potential users of the extracted steam.

The main controlled variables for a steam turbine are the steam pressures and temperatures at the inlet and outlet of the turbine and offtakes; lubricating oil pressure and temperature; speed and generator power parameters (voltage and current); control valve positions; position of main shut-off/emergency fitting (open-closed).

The typical arrangements for steam turbines with generators include placement over cast iron supports, on a concrete foundation. The units, installed on a common frame with accessories, are located directly on the machine floor of the machine room. The condenser is positioned behind the turbine, in the axial direction of the large turbines. Very small turbines of up to 1 MW are usually not placed directly on the floor of the engine room, but on a concrete base height from 40 to 60 cm for easier access of the personnel for service and maintenance.

6.5 Thermal efficiency of a steam turbine-based power plant

The **thermal efficiency** of a power generation facility is defined as the ratio of the amount of electrical energy provided at the output to the thermal energy supplied at the inlet to the facility. This is also referred to as the **First-Law Efficiency**. The input energy is usually in the form of fuel or an energy flow imported from elsewhere, for example, heat flow from a concentrating solar capture installation or from a geothermal source. A combination of fuel supplies and other heat sources is also possible. The definition for thermal efficiency accounts for the total sum of all energy inputs to the facility.

Figure 6.14 presents a diagram with the notation to illustrate the definition of the thermal efficiency. The notation includes **SG** –steam generator; Q_{fuel} [MW] – heat supplied to the boiler with the primary source (e.g. fuel); Q_S [MW] – heat output of the boiler carried by the steam flow; Q_{SD} [MW] – heat carried with the steam

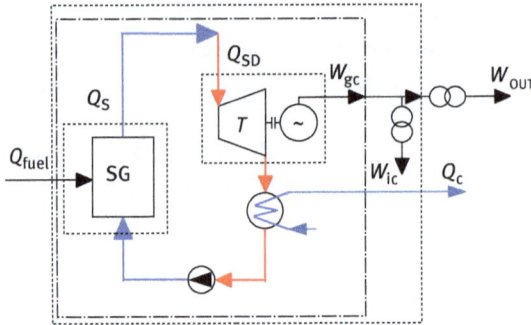

Fig. 6.14: Notation for the definition of thermal efficiency of a steam turbine-based power plant.

at the turbine inlet; W_{gc} [MW] – power output of the engines at the generator contacts; W_{ic} [MW] – internal power consumption by the auxiliary machines of the block; W_{OUT} [MW] – power plant output; Q_C [MW] – heat rejected to cooling water in the turbine condenser.

It should be noted that each of the steps in this energy system involves energy losses. Some of them are explicitly marked – as Q_C [MW] – the heat rejected to the condenser, as well as the internal (parasitic) power consumption in the system W_{ic} [MW]. Others include heat lost during steam generation and steam distribution.

The thermal efficiency of the power plant, as denoted in Fig. 6.14, is

$$\eta_{PP} = \frac{W_{OUT}}{Q_{fuel}} \tag{6.12}$$

Other efficiency characteristics, used by utility engineers include the thermal efficiency of steam generation:

$$\eta_{SG} = \frac{Q_S}{Q_{fuel}} \tag{6.13}$$

and the thermal efficiency of steam distribution:

$$\eta_{SD} = \frac{Q_{SD}}{Q_S} \tag{6.14}$$

Currently there are advanced real-time software applications for calculating the entire block (technological) including the computational fluid dynamics (CFD) model of the flow part of the turbine. This software is composed of individual modules for each block device that cooperates with each other, while the interface of two devices must result in the same marginal conditions for both devices. For quick optimisation, however, the fastest and most reliable analytical methods, which are also electronically processed and are usually part of the earlier software for determining initial calculation conditions and control calculations, are still optimized. Such

software solutions are as a rule proprietary and linked to the companies offering
the relevant hardware – an example of this line is ABB's "Ability" (ABB, 2018).
There is also an offering for power plant performance, linked to consultancy services for power plant optimisation (MapEx Software, Inc, 2020).

6.6 Steam turbine operation

The operation of the steam turbines affects their efficiency. Moreover, the selection
steam turbine operation strategies intrinsically affect the turbine design and equipment with the relevant controls. This section discusses the main steam turbine governing strategies and their implication to the turbine designs. This is followed by a
discussion of the steam turbine operation modes and steam turbine maps.

6.6.1 Steam turbine governing

The rate of power generation by a steam turbine depends on
- **The inlet and outlet pressures of the expanding steam.** While the exact values of these properties have their importance, the difference between the inlet
  and the outlet pressures is strongly correlated to the power generation capacity
  of the turbine, pointing to this difference as the main constituent of the power
  generation driving force. All remaining interactions and effects are secondary
  to the specific power generation (per unit of steam flow).
- **The flow rate of the expanding steam.** Generally, the trend is that larger flow
  rate results in more power generation. The trend lines are mostly linear, but
  with segmentation, as a result of discrete regimes changing each other.

While it is possible to manipulate these degrees of freedom simultaneously, the
usual practice in industry is to set the parameters of the boiler steam fixed for a
given utility system design and to use the steam flow rate as the degree of freedom
manipulated for each steam turbine. As will become apparent in the following explanation, the regulation of the steam flow also affects the pressure difference of
the steam expansion through the turbine. The basic way of regulating the flow
through the steam turbine is through throttle governing at the inlet (Sarkar, 2015).
In throttle governing, steam enters the turbine through one or more parallel but simultaneously operating control valves (illustrated in Sarkar, 2015, p.213). For throttle governing, single-acting valves, two-sided valves or diffuser valves are used.

Figure 6.15 shows a simple governing system, using one valve at the steam turbine inlet. The notation includes **1** – path for the valve status with fully open control
valve; **2** – path for the valve status with partially open control valve; **SV** – shut-off
valve; **GV** – governing control valve; state indicator "i" denoting inlet condition;

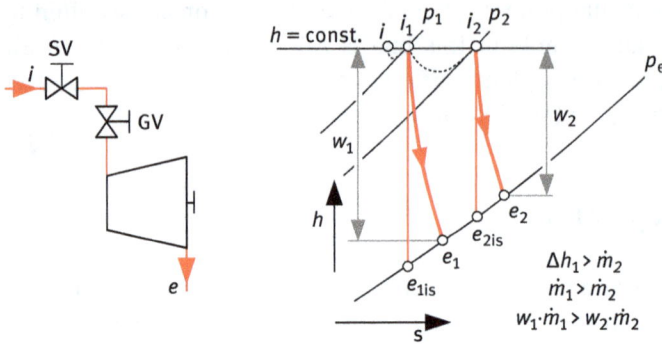

Fig. 6.15: The principle of steam turbine power control by throttle governing.

state indicator "**e**" denoting outlet condition; index suffix "**is**" meaning "isentropic". Within this context, the inlet pressure for governing variant 1 is p_1 and that for governing variant 2 is p_2. The outlet pressure is denoted as p_e.

When the control valve **GV** is fully open, the steam pressure at the inlet to the turbine is equal to the pressure just before the valve (p_1) and the steam flow through the turbine is maximal, corresponding to point i_1 on the Mollier diagram in Fig. 6.15. If the control valve GV is changed to a partially open state, there is a throttling, which results in isenthalpic steam expansion before the steam enters the turbine. This causes the pressure at the turbine inlet to reduce to p_2.

Since the pressure at the end of the turbine (p_e) is the same for both cases, the specific internal work of the turbine decreases when valve GV is changed from fully to partially open state. Therefore, this type of regulation can be inefficient, as reducing the specific work of the turbine also reduces the turbine efficiency. For this reason, this type of governing is used either for very small loads or when at high loads; the part-load rates are not typical, for example, in nuclear power plants or other baseload-type power plants.

A somewhat higher efficiency can be achieved by the means of group control, in the case of nozzle governing. In this case, the stator row of the blades (nozzles) of the first turbine stage is divided into several groups. Each nozzle group is assigned a dedicated control valve, for which reason this turbine stage is also referred to as a "control stage". The flow control is accomplished by manipulating the individual control valves, so that throttling occurs in a maximum of one valve, as shown in Fig. 6.16.

For steam turbines requiring occasional temporary surges of the turbine power generation, bypass governing (Sarkar, 2015) may be used. In such a control method, the last few stages of the turbine are sized to be able to handle larger than the nominal steam flow rate. Then, when necessary, an increase of the power generation is achieved by feeding the oversized stages with additional steam through the bypass from the first stage.

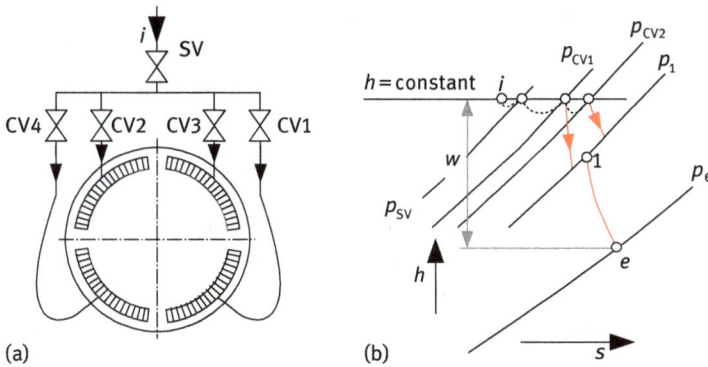

Fig. 6.16: The principle of nozzle governing of steam turbines. (a) nozzle governing with four control valves – CV{#}; (b) Mollier diagram (h–s) of steam expansion in the turbine with fully open CV1 and partially open CV2. Notation: p_1 [bar], the pressure at the inlet to the rotor blades of the first turbine stage; p_{SV}, p_{CV1}, P_{CV2} [bar], steam pressure after the shut-off valve SV and valves CV1 and CV2, respectively; w [kJ/kg], specific work by the steam expansion; 1, state of steam at the inlet of the first stage turbine rotor blade after mixing the steam exiting from the groups of blade passages controlled by CV1 and CV2; i and e, initial and final states of the steam in the turbine.

6.6.2 Steam turbine operating modes

The main operating modes of steam turbines are
- start-up – where the start-up has two sub-modes: starting and loading
- normal operation
- shut-down

These modes are illustrated in Fig. 6.17. This is typically referred to as the operation diagram. The times given correspond to a steam turbine of about 30 MW (Škorpík, 2020). Each turbine has its operation diagram supplied with the turbine by the manufacturer.

Usually, starting a steam turbine proceeds as follows. Before starting up from cold, (as the turbine is at ambient temperature) it is necessary to warm-up the turbine by letting it through a small amount of steam, while turning the rotor and draining out the formed condensate. At these low steam flow rates, the steam turbine is rotated by an externally powered rotating device, to achieve uniform and rapid heating of the steam turbine parts. This is necessary, to avoid deformation and mechanical damage to the turbine due to the thermal expansion of the parts.

When sufficient steam flow is achieved when the turbine is able to overcome losses in the mechanisms, the externally powered rotation of the shaft is stopped and the rotation becomes supported by the steam expansion alone. The amount of steam fed to the turbine is increased gradually with a uniform warm-up of the turbine. If the turbine is equipped with a hydrostatic bearing, then it is under pressure and active during the start-up. In this way, the speed is increased up to the rated value. At that point, the generator is connected to the power distribution network (local or to the grid). After the connection to the network, the turbine-generator set can be further loaded and when increasing the temperature, pressure and steam flow there is no increase in the rotation speed. Only the internal power generation of the turbine increases. This corresponds to the second stage of the turbine start (stage b in Fig. 6.17 – loading). During this stage, the steam parameters and the turbine load are gradually increased, while the turbine is still warming up and the internal power is increased to the required state. After reaching the required performance, the turbine-generator set goes into the normal operation mode (sector c in Fig. 6.17). The main operation mode is discussed further in the following sections.

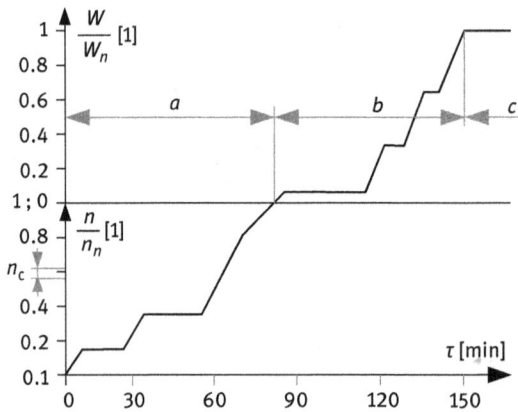

Fig. 6.17: An example of the distribution of the main steam turbine operating modes (amended after Škorpík, 2020). Notation: a, start-up; b, loading; c, normal operation. τ [min], elapsed time; W, W_n [MW], current power output and nominal power output of the turbine-generation set; n, nn [s^{-1}], current turbine speed and nominal turbine speed; n_k [s^{-1}] critical rotation speed of the turbine.

The shutdown of a steam turbine can be of two types:
- unloading shutdown – planned
- emergency shutdown

Unloading shutdown is planned. It is also referred to as "normal shut-down" (Sarkar, 2015). During such an operation, the regulating valve is closed gradually until the power generation delivered by the machine becomes zero. At that point, the machine is disconnected from the electrical network. The further closure of the control valve reduces the turbine speed until the rotation is completely stopped. When the steam supply is stopped, the turbine starts to cool and the turbine material begins to deform due to thermal contraction. The thermal deformation during the shutdown is prevented by the occasional rotating of the shaft (according to the manufacturer's instructions) by an externally powered roll device. When rotating, the hydrostatic bearing system is active.

During emergency shutdown, the emergency valve of the turbine, or another shut-off valve in the system, suddenly closes and the steam is directed away from the turbine inlet. The turbine rotor keeps rotating further until it gradually stops. At this state, the oil pump of the turbine is driven mechanically through the shaft. If the cause of the emergency stoppage is not a breakdown of the turbine, oil loss, vibration, sudden power drop, generator crash, for example, overheating and turbine destruction and the turbine is used for electricity generation, then this is followed by controlled cooling of the turbine in the same way as for normal shutdown. If the cause of the crash is in the turbine-generator, it is necessary to inspect and service the hardware. After emergency shutdown, the turbine has to be inspected visually and diagnostically and monitored closely afterwards.

6.6.3 Steam turbine maps and operating points

Steam turbine performance is commonly presented on diagrams, involving the power generation and the flow rates of the involved inlet and outlet streams. These diagrams are referred to as
- steam turbine performance map (McGowan et al., 2011)
- steam consumption diagram (Ohji and Haraguchi, 2017)
- extraction map (Scoretz and Williams, 2008)

The most commonly used consumption map is shown in Fig. 6.18. The diagram features typical trends of the efficiency and the steam consumption with the part-load operation. The denoted "nominal" operation point is at an overload condition, which should not be kept for extended periods of time.

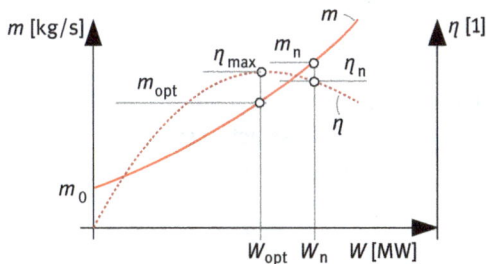

Fig. 6.18: Consumption map of a steam turbine. Notation: η [1], the steam turbine efficiency; m_0 [kg/s], steam mass flow at zero power output; W [MW], power output of the steam turbine. Subscript n indicates nominal operation and subscript opt indicates optimal performance.

The operating point of the steam turbine is determined by matching the properties (pressure and mass flow) of the steam source and of the steam consumption of the turbine. Visually, this is most conveniently illustrated on a pressure–steam consumption chart (Fig. 6.19). The figure shows the trends of the characteristic lines of the steam turbine and the steam line for a given temperature. The point of intersection of the two curves determines the operating point of the turbine at the fully open position of the control valve. If the parameters of the steam source or the opening position of the control valve change (to partially open), then a new flow through the steam turbine is set, featuring a lower pressure at the turbine inlet after the valve and a lower steam flow rate, indicated as OP, P_{CV} ΔP_{CV} and P_i in Fig. 6.19.

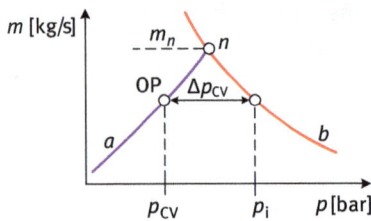

Fig. 6.19: Steam turbine operation point estimation. Notation: a, steam turbine characteristics (for t_i = constant, see the notation of Fig. 6.18); b, steam source characteristics (for t_i = const, see the notation of Fig. 6.18); p_i [bar], steam pressure at the turbine inlet (upstream of the control valve); OP, operating point; p_{CV} [bar], steam pressure after the valve; Δp_{CV} [bar], pressure drop in the control valve; n, nominal operation at fully open control valve.

In dedicated (standalone) power plants, the extracted steam is used exclusively for water preheaters to maximise the energy efficiency (Moran, 2011). However, in the context of industrial utility systems, the extractions deliver steam to the steam mains (Strouvalis et al., 1998). The flow rates of the uncontrolled extractions vary approximately proportionate (McKetta, 1996) to the flows through the turbine. This is illustrated in Fig. 6.20.

Using the diagram and the notation of Fig. 6.20, the following equations (Škorpík, 2020) express the approximate proportionality of the flow rates:

$$m_{R1} = \varphi_{R1} \times (m_e + m_{WH}) \tag{6.15}$$

$$m_{R2} = \varphi_{R2} \times (m_e + m_{WH}) \tag{6.16}$$

$$m_{R3} = \varphi_{R3} \times m_e \tag{6.17}$$

where φ_{R1}–φ_{R3} [1] – proportionality coefficients (can be determined experimentally); m_{R1}–m_{R3} [kg/s] – the flow rates through the uncontrolled extractions; m_e [kg/s] – the steam flow rate at the outlet towards the condenser; m_{WH} [kg/s] – the steam flow rate to the water heater.

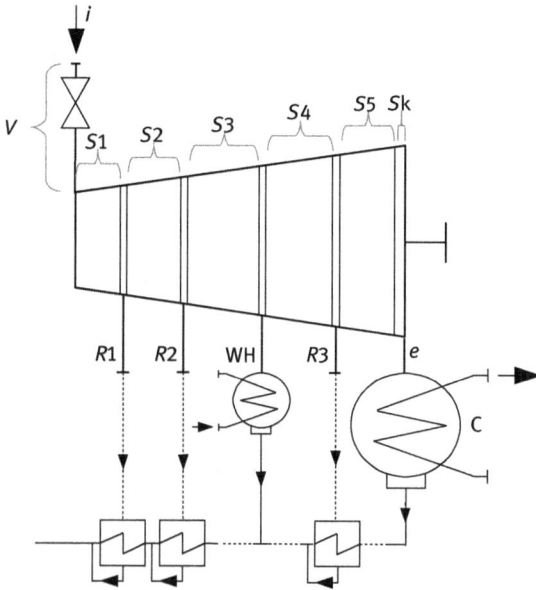

Fig. 6.20: Steam flow through uncontrolled extractions for regenerative heating of the feed water. Notation: i, steam turbine inlet; e, steam turbine outlet; R1–R3, uncontrolled extractions for the regenerative feed water heat exchangers; WH, water heater; C, condenser.

6.6.4 Simplified consumption maps of steam turbines

Steam turbines are designed for specific rated parameters that include the values of the steam inlet flow rate and the generated power. Any departure from the rated conditions, especially the change in the pressure, temperature or flow rate of the steam input and the flow rates of the outputs (in case of extractions) causes changes in the power output, the pressures and the steam flow rates at the other connections. Generally, to determine the values related to the steam turbine properties is possible only from measurements and correlations. Without measurement, they can be determined with acceptable accuracy from a virtual turbine model using CFD, which implies that it can only be developed by the turbine manufacturer. Usually, turbine manufacturers provide steam turbine performance software tools to prospective customers, for evaluation of the performance of potential installations. It is possible to use such tools to derive performance correlation of reasonable accuracy (Sun and Smith, 2015). There are also analytical procedures based on the similarity of steam turbine maps and the representation of their flows (Strouvalis et al., 1998). However, the accuracy of such calculations is usually sufficient only for limited intervals of the parameter variations. This is why they are referred to as simplified characteristics and used mainly for benchmarking.

The main advantage of the simplified steam turbine characteristics is that they are derived on the basis of similarity, and so in their construction, the exact design of the steam turbine is not necessary to be known. They are used for fast calculations, for example in initial designs of technological units, where the exact geometry of the turbine is not yet known. Another example of using such simple equations is when the turbine-related equipment is not available to the designer of the equipment connected to the turbine (for example, for older turbines, for business reasons, the turbine was supplied by a competing company).

A typical example of a change in the characteristics of a steam turbine is a turbine with a controlled extraction. This is caused by the wide range of variation of the inlet and outlet steam flows. An example of a performance map (extraction map) for a steam turbine with a controlled extraction is shown in Fig. 6.21. The turbine has one inlet and two outlet connections. From those, one control valve is placed at the inlet and the other at the extraction point. For each specific position of the main control valve (at the inlet), the extraction control valve can be adjusted to any position between fully open and fully closed. As a result, there are many characteristic curves (theoretically infinite) for the various combinations of control valve opening positions.

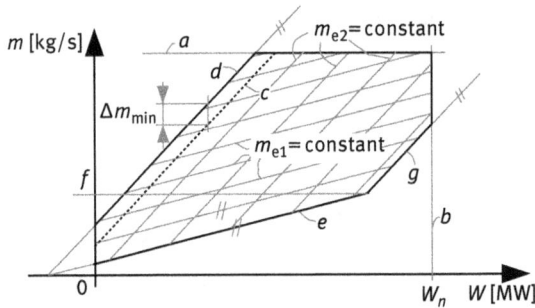

Fig. 6.21: A simplified steam turbine map with a controlled steam extraction.

The notation of Fig. 6.21 includes the following
- m [kg/s] – flow through the inlet control valve of the turbine
- m_{e1} [kg/s] – flow through the (controlled) steam extraction
- m_{e2} [kg/s] – steam flow at the exit of the turbine
- a – the line of maximum possible flow through the turbine (at the inlet)
- b – the line for maximum power output from the turbine
- c – theoretical map line of the turbine when the turbine extraction control valve of extraction is fully open and the turbine acts as partly a backpressure turbine;
- d – the line showing the minimum amount of steam that the turbine can handle and the minimal excess Δm_{min} which must be ensured in order for the

turbine to rotate. For operational reasons, for example, heat losses in the LP part of the turbine, the LP section cannot be completely closed.

- *e* – the characteristic line for fully closed extraction control valve, where $m_{e1} = 0$ and the turbine operates as a fully condensing type;
- *f* – a line representing the maximum flow through the LP part of the turbine;
- *g* – increase in turbine output due to increased flow of HP turbine section and through steam extraction.

It is possible to use the Stodola equations of flow rate through the turbines (Wilda and Salter, 2003) to obtain simplified regression equations whose characteristics are very close to those of straight lines, parabola-shaped or quadratic polynomial curves. The usual simplified characteristics used have linear or quadratic expressions relating the mass flow rate of steam and the generated power:

$$\text{Linear correlation: } m = la_0 + la_1 \times W \qquad (6.18)$$

$$\text{Quadratic correlation: } m = qa_0 + qa_1 \times W + qa_2 \times W^2 \qquad (6.19)$$

where la_0 [kg/s], la_1 [kg/MJ], qa_0 [kg/s], qa_1 [kg/MJ], qa_2 [kg × s/MJ²] are regression parameters, m [kg/s] is the steam mass flow rate and W [MW] is the generated power. For the same turbine, the regression parameters would have different values and, if a good precision is required, the regression may need to be performed for narrower intervals of the steam part-load performance. Generally, the linear shape of the equations can be used with sufficient accuracy over the range from 20 % of the turbine nominal output, up to the full-load condition, with maximum two regression intervals (Sun and Smith, 2015).

It has to be noted that the fixed-term coefficient in the approximations in Eqs. (6.18) and (6.19) should not be confused with the idle steam flow in the turbine, as these are approximate trend equations and the regression coefficients do not always have a direct physical meaning.

Examples of linearised steam turbine maps following Škorpík (2020) are shown in Fig. 6.22. The figure shows two cases – throttle governing and nozzle governing. In the case of a steam turbine, the nozzle governing divides the map into two parts, where in the second segment, after the optimal load point, the slope becomes steeper, indicating that the marginal steam rate increases as per unit of power generation increases and is higher (implying lower efficiency). The approximation modelling of steam turbines, in the context of utility networks, is discussed later in this chapter, in detail.

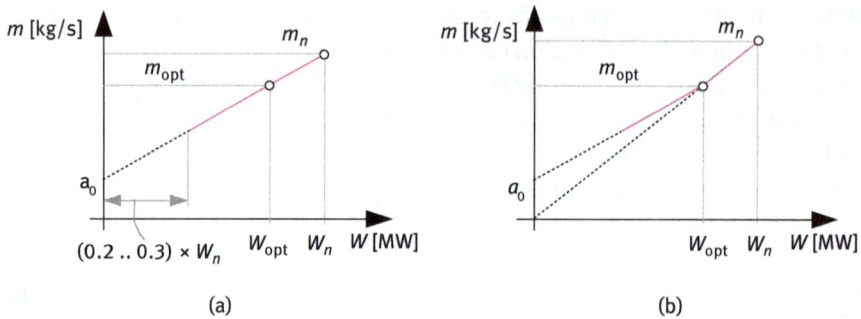

Fig. 6.22: Examples of linearised steam turbine maps, after (Škorpík, 2020): (a) case of throttle governing and (b) case of nozzle governing. Notation: n, nominal operation; opt, economically optimal operation of the steam turbine.

6.7 Global market and main suppliers of steam turbines

The global market of steam turbines has a stable demand, mainly driven by installations in Asia. According to a recent report (GlobalData Energy, 2019), the market size was estimated to 9,250 M$ (US) for the year 2018 and it is projected to the value of 36,700 M$ (US) for the period 2019–2023. The report indicates that the main contributors to the market demand are China (38.46 %), India (7.11 %), followed by the United States, Egypt and Vietnam. According to the market research web site "Market Watch" (MarketWatch Inc, 2019), the major players on the global steam turbine market are GE, Siemens, Elliott, Toshiba, Mitsubishi Hitachi Power Systems and MAN. This is supported by a statistical analysis of Defense Security Monitor (Slade, 2017).

The publicly available information from General Electric (2020a) shows a portfolio of steam turbines on offer, ranging from 100 MW to 1.9 GW capacity. The portfolio includes steam turbines for standalone fossil-based power plants (100–1,200 MW), offered in varieties with and without steam reheat. There are also turbines that specialise in combined cycle applications, for example, the STF-A650 line (85–300 MW) and the STF-D600 line (180–700 MW), as well as nuclear applications.

A material published by Siemens AG (2019) shows a similarly wide range of steam turbine offers – from 2 MW and up to 1.9 GW. Within that range, the industrial steam turbines take the niche 2–250 MW and the utility steam turbines are from 100 to 1,900 MW. The Dresser-Rand company, part of the Siemens group, offers also turbines within the smaller range of 10 kW to 25 MW (Siemens AG 2020a).

There are also further steam turbine suppliers, some with strong traditions and achievements. An example can be given with Škoda, currently a part of the Doosan Heavy Industries group (Doosan Skoda Power, 2020). The company has been a known supplier of high-quality steam turbines since 1904. It currently offers turbines for combined cycle and for district heating applications, incineration plants for municipal waste and biomass. They also serve the nuclear power plant industry with large-scale turbines of up to 1.2 GW capacity.

Another manufacturer, with world-wide applications, is G-Team a.s (2020). They offer steam turbines in the lower range of capacities, starting from micro steam turbines of 10 kW, filling up the range with turbines rated at 150 kW, 700 kW, 1.2 MW, 3 MW and 5 MW. The company positions itself as serving small to medium industry sector. Its portfolio of installed equipment includes a solar power plant in Australia, a distillery in the Czech Republic, thermal power plants from the Czech Republic, Poland and Germany, and further industrial customers from Cuba, Egypt, Pakistan, the Russian Federation, Slovakia and the United Kingdom.

In addition to the world's largest steam turbine manufacturers, which cover the market with unified steam turbine classes, there are also regional manufacturers focusing on single-piece production of steam turbines with parameters and properties outside the large manufacturers' unified classes, see Tab. 6.1. The classification is made according to the delivered turbine output, working substance and whether there are also steam turbines for ships in the production portfolio. Note, since November 2015, Alstom's power and grid businesses are part of GE's portfolio (General Electric, 2020b).

6.8 Steam turbine model for utility network optimisation

Modelling steam turbines is closely related to the modelling of all shaft-power devices, for the purpose of rating their efficiency. Various parameters have been rated, among which, for steam-based engines, the steam rate for unit power generation has been the most popular metric (Moyer, 1917). This section discusses the models available in the literature. It starts with the Willans Line concept, which is common for most models of steam turbines and is used for modelling other engines – including diesel engines (Vukovic et al., 2017).

6.8.1 The Willans Line origin

The concept is related to the power machinery company Willans & Robinson, originating from Thames Ditton, Surrey, the United Kingdom (Coulls, 2013). The heart of the company was the engineer and inventor Peter William Willans, known for the invention of a central-valve steam engine, which he experimented with and published an extensive study (Willans, 1888). The Willans & Robinson company has been known to also produce steam turbines. It was further acquired by other companies, finishing with Alstom (Coulls, 2013).

Tab. 6.1: Steam turbine manufacturers, including regional.

Manufacturer	≤ 100 kW	100 kW–1 MW	1–10 MW	10–100 MW	100–1,000 MW	≥1,000 MW	ORC	Marine applications
Siemens (2020b)	x	x	x	x	x	x	x	
General Electric (2020c)			x	x	x	x	x	x
Mitsubishi Hitachi Power Systems Ltd (2020)		x	x	x	x	x		x
Doosan Skoda Power (2020)			x	x	x	x		
MAN (2020)			x	x	x			
PBS Group, a. s, (2020)		x	x	x				
EKOL, spol. s r.o, (2020)		x	x	x				
G-Team a.s, (2020)	x	x	x					

The Willans Line for steam turbines is defined in the *Dictionary of Mechanical Engineering* (Atkins and Escudier, 2013) as a plot of steam consumption versus output power. The first well-documented mention of the Willans Line is in the book on *Steam Turbines* by Moyer (1917), where the plots of "shaft output" versus number of nozzles open are discussed as equivalent to the Willans Line. It is worth mentioning that, in general, these plots are not necessarily given as straight lines, but the term "line" is used with the meaning of a general curve (Church, 1928).

6.8.2 Overview of steam turbine models for network optimisation

A number of models have been evaluated for the performance of steam turbines within the context of utility systems. Mavromatis and Kokossis (1998) proposed a simple model of backpressure steam turbine performance. In this model, the performance of a steam turbine is related to its size (in terms of maximum shaft power) and to part-load performance. The shaft power is modelled as a linear function of the steam mass flow, known as the Willans Line. This model was later extended to condensing steam turbines by Shang (2000). All these works follow the same model structure and employ the same equations; however, they use different values for the turbine regression coefficients.

The intercept of the Willans Line was mapped by Mavromatis and Kokossis (1998), as well as by Shang (2000), as identical to the turbine energy losses, also assuming a fixed loss rate. Varbanov et al. (2004b) introduced improvements to those models by: (1) recognising that the Willans Line intercept has no direct physical meaning and is simply the intercept of a linearisation (see also the discussion related to Fig. 6.22) and (2) accounting for both inlet and outlet pressures of the steam turbines. The improved steam turbine models have been incorporated into methodologies for simulating and optimising steam networks and also have been used to target the cogeneration potential by assuming a single large steam turbine for each expansion zone between two consecutive steam headers (Varbanov et al., 2004a). A summary of the steam turbine models developed for steam network optimisation can be seen in Tab. 6.2.

Tab. 6.2: Summary of steam turbine models suitable for network simulation and optimisation.

Sources	Features	Intended uses
Raissi (1994) and later elaborated in Klemeš et al. (1997)	Based on heat flows, intended specifically for cogeneration targeting	Cogeneration targeting only
Bruno et al. (1998)	Non-linear model, relating efficiency to steam flow	Utility system synthesis, MINLP model

Tab. 6.2 (continued)

Sources	Features	Intended uses
Mavromatis and Kokossis (1998)	Backpressure steam turbine model based on the Willans Line. Performance related to size and part load. Willans Line intercept assumed identical with energy losses	Steam network modelling, cogeneration targeting
Shang and Kokossis (2004)	Extension of Mavromatis and Kokossis (1998) to condensing steam turbines	Steam network modelling, cogeneration targeting
Varbanov et al. (2004b)	Complete overhaul of Mavromatis and Kokossis (1998) and Shang and Kokossis (2004) models, accounting for the pressure at both inlets and outlets	Steam network modelling, cogeneration targeting
Aguilar et al. (2007)	Customisation of the model in (Varbanov et al., 2004b) accounting for flexibility	Steam network modelling, cogeneration targeting
Sun and Smith (2015)	Refinement of the Varbanov et al. (2004b) model accounting for more turbine types, producing a unified model	Steam network modelling, cogeneration targeting

A flexible steam turbine model, balancing the model complexity and precision, is the one in Varbanov et al. (2004b), termed here "the base model". It is suitable both for steam network modelling and targeting based on the Site Utility Grand Composite Curve information. Analysing the other models in Tab. 6.2, the base model features a linear relationship between the power generation and steam flow-rate, making it simpler than the non-linear model of Bruno et al. (1998). It also accounts for both steam pressures, at the inlet and the outlet for correlating the design-point performance to the model equations, clearly relating the performance to the steam pressure drop. In terms of the Willans Line, this model also recognises that the fixed term has no direct physical meaning but is instead part of the linearisation approximation of the real steam turbine performance. These features make it more adequate for modelling steam turbines than the initial turbine hardware model of Mavromatis and Kokossis (1998) for backpressure steam turbines and the extension to condensing turbines by Shang and Kokossis (2004).

A potentially useful extension of the base model is presented in Aguilar et al. (2007). This work describes a procedure for modelling the Willans Line intercept. In turn, Sun and Smith (2015) elaborates on the factors governing steam turbine performance. As a result, for the case of utility system design, a unified correlation has been obtained, which does not partition steam turbines by size. Here, the description of the model in Varbanov et al. (2004b) is provided, which should be appropriate for a majority of potential applications. For more specialised studies, the other models can be also adapted. The full discussion of the presented model can be found in Varbanov et al. (2004b).

6.8.3 A balanced model for steam network optimisation

This section describes the reasoning and the formulation of the model for steam turbines, balancing the needs for accuracy and simplicity for the needs of steam network optimisation.

6.8.3.1 Thermodynamics, factors and performance trends of steam turbines
Steam turbine performance is affected by a number of factors. The most significant among them are
- thermodynamic limitations and efficiency
- turbine size in terms of maximum power load
- pressure drop across the turbine
- current load

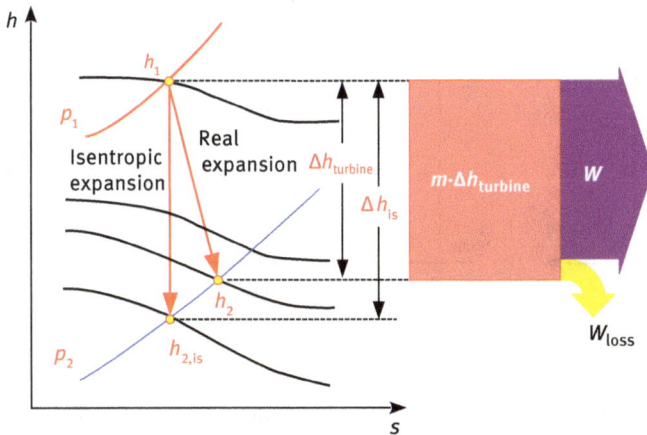

Fig. 6.23: A Mollier diagram for the expansion of steam in a turbine.

Thermodynamic limitations, pressure drop and efficiency. The turbine thermodynamics and design as factors are best understood from the representation of the expansion of steam in a turbine on a Mollier Diagram (Fig. 6.23). The isentropic efficiency for a given turbine load is defined as

$$\eta_{is} = \frac{\Delta h_{turbine}}{\Delta h_{is}} = \frac{h_1 - h_2}{h_1 - h_{2,\,is}} \tag{6.20}$$

The expansion process transforms part of the energy of the inlet steam into power. The magnitude if this part depends on the turbine design. Different ways of expanding steam would result in different isentropic efficiency. The total power from the expansion is further split (Fig. 6.23) into useful power, delivered to the shaft and energy losses. Losses occurring in steam turbines are mechanical friction losses casing heat losses and kinetic energy losses with the turbine exhaust – see the discussion in Fig. 6.7.

Current load. For a particular steam turbine and a specified maximum steam flow, it is possible to calculate its maximum shaft power. During operation, however, it is likely that many turbines will work at part-load. Therefore, it is necessary to estimate the actual enthalpy change, isentropic efficiency and the energy losses with changing turbine load. From operating practice, it is known that steam turbine efficiency varies with part-load (Madejski, 2018, p.24). At lower loads it is lower and gradually increases with the turbine reaching its maximum rated load Fig. 6.24a, which can be maintained for sizeable periods of time. Note: this excludes the temporary overload conditions.

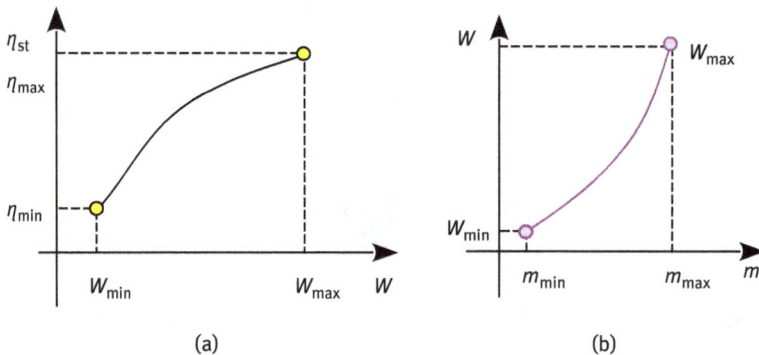

Fig. 6.24: Typical steam turbine performance trends. (a) Efficiency variation at part-load and (b) power generation as a function of steam flow rate.

The change in overall efficiency can be also represented in a different way – by evaluating the relationship of the shaft power and the mass flow rate of steam. From the behaviour illustrated in Fig. 6.24a, it follows that at smaller loads, a unit increase in the steam flow will result in smaller power increase than at higher loads. This is illustrated in Fig. 6.24b.

The overall turbine efficiency can be represented as consisting of two components: the isentropic efficiency and the machine efficiency. The machine efficiency is generally higher than the isentropic efficiency (Bhatt and Rajkumar, 1999) and changes in a relatively narrow range. In contrast, the isentropic efficiency changes substantially with load:

$$\eta_{st} = \eta_{is} \times \eta_{m}; \; \eta_{m} \gg \eta_{is} \tag{6.21}$$

The variation of the overall efficiency is generally non-linear (Fig. 6.24a). It can also be interpreted as the equivalent relationship of the shaft power versus the mass flow of steam (Fig. 6.24b). The latter tends to be less non-linear and when approximated with piecewise-linear segments, usually leads to better accuracy. Typically, for given values of the inlet and outlet steam specifications, the overall turbine load interval is sufficiently well represented by one or two linear segments.

To model the turbine in a way suitable for efficient optimisation applications requires deriving a linear relationship and characterising the line coefficients. The line slopes are related to the actual enthalpy change and pressure drop across the turbine. For any load of a steam turbine with fixed inlet and back pressures and fixed inlet temperature, the isentropic enthalpy change remains constant. Because of changing isentropic efficiency, the actual enthalpy drop across the turbine varies with the load, remaining smaller than the isentropic enthalpy drop. The actual non-linear power, steam flow relationship in Fig. 6.24b can be approximated by linear segments (Fig. 6.25) usually termed Willans Lines (Church 1950), linearised pieces of the original plots (Atkins and Escudier, 2013). The actual steam turbine performance curve is replaced with one or more straight lines with different but fixed coefficients.

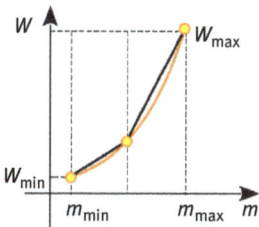

Fig. 6.25: Linear approximation segments (Willans Lines) superimposed on the performance curve.

6.8.3.2 Derivation and regression coefficients

The performance of a real steam turbine is given by

$$W = \Delta h_{turbine} \times m_{ST} - W_{Loss} \tag{6.22}$$

where $\Delta h_{turbine}$ (see also Fig. 6.23) is the actual enthalpy change of steam across the turbine and W_{Loss} designates the energy losses. The expression in Eq. (6.22) has a typical linear form. However, it should be noted that both the actual enthalpy change and the loss term, are functions of the current load, which makes the resulting plot a curve.

A useful simplification of the direct model from Eq. (6.22) is the "isentropic efficiency" model. It assumes a fixed isentropic efficiency, neglects the loss term and represents the power generation rate as proportional to the product of the steam mass flow rate, the isentropic enthalpy change and the isentropic efficiency:

$$W_{is} = \Delta h_{is} \times m_{ST} \tag{6.23}$$

The performance approximation in each modelling interval, such as those in Fig. 6.25, has the form of the Willans Line, as follows:

$$W = n_{ST} \times m_{ST} - W_{INT} \tag{6.24}$$

For existing utility systems, often the values of the steam parameters, pressures and temperatures, are fixed. To model steam turbines in such cases, it is sufficient to apply the Willans Line method directly.

Turbine size. For designing new systems, including Total Site targeting, evaluation of potential configuration changes and buying new turbines, it is necessary to also model the turbine performance more broadly, including operation at design and off-design conditions, across many steam turbines of varying size.

Assuming that newly installed steam turbines will operate at their maximum (design) load or in the vicinity of that, the operation can be modelled with a single linear segment for any specified set of steam pressures and temperatures. For the maximum load, the general Willans Line expression from Eq. (6.24) takes the following form:

$$W_{max} = n_{ST} \times m_{max} - W_{INT} \tag{6.25}$$

Within that, the Willans Line slope n_{ST} is estimated based on the isentropic enthalpy drop across the turbine and the maximum steam flow rate:

$$n_{ST} = \frac{L + 1}{B_{ST}} \times \left(\Delta h_{is} - \frac{A_{ST}}{m_{max}} \right) \tag{6.26}$$

In eq. (6.26), L is referred to as the intercept ratio, as defined in Eq. (6.27), A_{ST} and B_{ST} are regression parameters:

$$W_{INT} = L \times W_{max} \tag{6.27}$$

The Willans Line intercept is calculated from

$$W_{INT} = \frac{L}{B_{ST}} \times (\Delta h_{is} \times m_{max} - A_{ST}) \tag{6.28}$$

The steam turbine parameters A_{ST} and B_{ST} used in Eqs. (6.26) and (6.28) are calculated from the regression relationships:

$$A_{ST} = b_0 + b_1 \times \Delta T_{sat} \tag{6.29}$$

$$B_{ST} = b_2 + b_3 \times \Delta T_{sat} \tag{6.30}$$

The intercept ratio L from Eqs. (6.26) and (6.28) has also been found to depend on the saturation temperature drop across steam turbines, following the linear expression

$$L = a_L + b_L \times \Delta T_{sat} \tag{6.31}$$

The coefficients in Eqs. (6.29) to (6.31) are b_0 [MW], b_1 [MW/°C], b_2 [1], b_3 [1/°C], a_L [1] and b_L [1/°C]. They are regressed for particular coherent selections of steam turbines. The coefficient values are related to the pressure drops across the turbines. However, in the presented model, the pressure drop is replaced by its equivalent saturation temperature difference. The use of temperature difference provides an easier interface to utility calculations with process heating and cooling demands.

For appropriate estimation of the steam condition in the headers downstream of steam turbines, as well as for the correct estimation of the condenser duties, it is necessary to evaluate the enthalpy of the steam at the turbine exhaust. This evaluation is based on the total energy extracted from the expanding steam. It can be estimated using a fixed specification for the combined mechanical efficiency of the steam turbine and generator, as follows:

$$W_{total} = \frac{W}{\eta_{mech}} \tag{6.32}$$

The difference between W and W_{total} constitutes the energy losses – a part of the consumed energy is lost to the environment in different forms – mainly friction losses, kinetic energy losses and casing heat losses. The mechanical efficiency η_{mech} from Eq. (6.32) is a collective representation of the efficiency after accounting for all losses. The overall mechanical efficiency of steam turbines is relatively high and does not change as much as the isentropic efficiency. As an illustration can be taken from published measured data for large power-station steam turbines, with capacities in the range 30–50 MW (Bhatt and Rajkumar, 1999) – see Tab. 6.3.

If a specification for η_{mech} is available, that would be the simplest way to model the turbine. This is usually the case when the steam turbines being modelled are already installed and a specific site utility system is being evaluated. Alternatively, for new system designs, the evaluation can be done from a linear regression model if regression data are available. For the specific steam turbine catalogue data, analysed in Varbanov et al. (2004b), the linear expression in Eq. (6.33) has been fitted:

$$W_{total} = a_{total} + b_{total} \times W \tag{6.33}$$

Tab. 6.3: Published efficiency data for large steam turbines at full load, extracted from Bhatt and Rajkumar (1999).

No.	W_{max} [MW]	$\eta_{generator}$ [1]	$\eta_{transmission}$ [1]	$\eta_{adiabatic}$ [1]	η_{is} [1]
1	500.0	0.9862	0.9941	0.9999	0.8907
2	210.0	0.9855	0.9953	0.9988	0.8875
3	210.0	0.9855	0.9953	0.9978	0.8852
4	110.0	0.9845	0.9945	0.9983	0.8740
5	62.5	0.9830	0.9941	0.9980	0.8580
6	30.0	0.9810	0.9939	0.9979	0.8320

The latter allows estimation of the enthalpy of the turbine exhaust from the enthalpy balance:

$$h_{ST, EX} = h_{ST, IN} - \frac{W_{total}}{m_{ST}} \tag{6.34}$$

Utility system design problems are characterised by lack of initial information about any particular steam turbine choice. In this case, it is necessary to obtain consistent and complete performance estimates for the steam turbines of the preferred steam turbine product lines. An example data set from a manufacturer catalogues has been analysed in Varbanov et al. (2004b). The values of the performance regression parameters from that study are listed in Tab. 6.4. The validity intervals of the capacity for these coefficients are
- backpressure steam turbines: from 1.165 to 34.707 MW
- condensing steam turbines: from 8.232 to 59.298 MW

Additionally, further simplifications of this model may be used, where some of the calculated coefficients are fixed based on additional sensitivity analysis. Other steam turbine models are also possible to apply for simplified evaluations. These include
- Assuming certain isentropic efficiency of the turbine and neglecting the mechanical losses, the power generation can be estimated as the product of the isentropic enthalpy drop Δh_{is}, the steam mass flow rate m_{STEAM} and the isentropic efficiency. This also includes the case of assuming isentropic expansion ($\eta_{is} = 1$).
- Accounting for the mechanical losses in the isentropic model can be performed by adding the mechanical efficiency to the product.

It is necessary to emphasise that the coefficients given in Tab. 6.4 have been obtained for a specific line of condensing steam turbines based on manufacturer

Tab. 6.4: Regression coefficients for steam turbine performance estimation based on manufacturer data (after Varbanov et al., 2004b).

		Back pressure turbines		Condensing turbines
		$W_{max} \leq 8$ MW	$W_{max} > 8$ MW	W_{max} : single range
b_0	MW	0.0250	0.0900	-2.080×10^{-08}
b_1	MW/°C	0.00463	0.0130	0.000297
b_2	[1]	1.39	1.22	1.602
b_3	1/°C	-0.000940	-0.000570	-0.00160
a_L	[1]	0.248	0.190	-0.0100
b_L	1/°C	-0.00126	-0.000790	0.000326
a_{total}	MW	0.0741	0.1376	0.1422
b_{total}	MW/°C	0.00101	0.001015	0.001017

catalogues. This data set may not necessarily be appropriate to predict the performance of all steam turbines. Therefore, it is recommended that an independent regression be performed, for particular steam turbine designs, if data from equipment manufacturers are available, or better still, operating data. The data required to determine the coefficients are

- W [MW] – the shaft power
- $T_{sat,in}$ [°C] – the inlet steam saturation temperature
- T_{in} [°C] – the *inlet* steam actual temperature
- $T_{sat,out}$, [°C] – the outlet steam saturation temperature
- m_{ST} [t/h] – the mass flow rate of the expanding steam

6.9 Summary and further information

This chapter is dedicated to steam turbines. It starts with an introduction and an overview of the Rankine cycle of steam use for power generation. It proceeds to the description of the ideal steam cycle assuming isentropic steam expansion, the main steam turbine types, and configurations within steam-based power plants. The operation and governance principles are discussed, followed by an overview of the steam turbine market. The last part of the chapter presents the steam turbine model for utility system optimisation.

Further information on steam turbines can be found in supplier catalogues, for example, those of General Electric (2020a), Siemens AG (2019), Škoda (Doosan Skoda

Power, 2020) or G-Team (2020). There are also dedicated books on construction (Wilda and Salter, 2003), operation (Sarkar, 2015), including performance maps in general (McGowan et al., 2011) or emphasising the steam extraction aspects of the turbine operation (Scoretz and Williams, 2008). Particularly useful books are *Steam Turbines for Modern Fossil-Fuel Power Plants* (Leĭzerovich, 2008) and *Blade Design and Analysis for Steam Turbines* (Singh and Lucas, 2011).

Nomenclature

Symbol	Measurement unit	Description
$A–A$	–	A drawing plane (Fig. 6.3)
A_{ST}	MW	Intermediate regression parameter in the steam turbine model (Eqs. (6.26), (6.28) and (6.29))
B_{ST}	[1]	Intermediate regression parameter in the steam turbine model (Eqs. (6.26), (6.28) and (6.30))
a, b, c, d, e, f, g (Fig. 6.21)	–	Lines identifying operation policies in a steam turbine map
a_L	[1]	Regression parameter for the intercept ratio in a steam turbine, Eq. (6.31)
b_L	1/°C	Regression parameter for the intercept ratio in a steam turbine, Eq. (6.31)
a_{total}	MW	Regression parameter for the total energy flow in a steam turbine, Eq. (6.33)
b_{total}	[1]	Regression parameter for the total energy flow in a steam turbine, Eq. (6.33)
b_0	MW	Regression parameter in the steam turbine model Eq. (6.29)
b_1	MW/°C	Regression parameter in the steam turbine model Eq. (6.29)
b_2	[1]	Regression parameter in the steam turbine model Eq. (6.30)
b_3	1/°C	Regression parameter in the steam turbine model Eq. (6.30)
C	–	Condenser (Fig. 6.1, Fig. 6.8)
CHP	–	Combined heat and power (generation)
c_i, c_e	m/s	Steam flow velocities in states **i** and **e**
C_p	kJ/kg/°C	The specific heat capacity of working gas at constant pressure
CFD	–	Computational fluid dynamics
C_v	kJ/kg/°C	The specific heat capacity of working gas at constant volume
$CV_{\{\#\}}$	–	Control valve
CT	–	Cooling tower (Fig. 6.1) and steam vacuum condenser (Fig. 6.8)
E	–	Outlet steam state assuming real expansion
e_c	–	Outlet steam state assuming real expansion and no kinetic energy loss
$e_{is,c}$	–	Outlet steam state assuming isentropic expansion and no kinetic energy loss

(continued)

Symbol	Measurement unit	Description
e_{is}	–	Outlet steam state assuming isentropic expansion
e_K	kJ/kg	The required change of kinetic energy between the inlet and the outlet of the machine
FP	–	Feed pump
FWH	–	Feed water heater
G	–	Power generator
GV	–	Governing control valve (Fig. 6.15)
h	kJ/kg	Specific enthalpy
HC	–	Heat consumer (Fig. 6.8)
HP	–	High pressure (steam)
$h_1, h_2, h_{2,is}$ (Fig. 6.23)	kJ/kg	Specific steam enthalpy in states 1 and 2, and at pressure p_2, assuming isentropic expansion
$h_{<1-4>c}$	kJ/kg	Specific stagnation enthalpies in the Rankine cycle diagram (Fig. 6.2)
h_i, h_e,	kJ/kg	Specific enthalpy of steam at the inlet and at the outlet of a turbine
$h_{e,isc}$	kJ/kg	Specific enthalpy of steam at the outlet of a turbine assuming isentropic expansion
Δh_{is}	kJ/kg	Isentropic change in the specific enthalpy
Δh	kJ/kg	Specific enthalpy difference
$\Delta h_{turbine}$ (Fig. 6.23)	kJ/kg	The actual enthalpy change of steam across the turbine
$h_{ST,EX}$	kJ/kg	Specific enthalpy of the steam turbine exhaust
$h_{ST,IN}$	kJ/kg	Specific enthalpy of the steam turbine inlet
k	[1]	The Poisson index
		Approximate profile of enthalpy change during steam expansion in a turbine
i	–	Inlet steam state
i_c	–	Inlet steam state assuming no kinetic energy loss
IP	–	Intermediate-pressure (steam)
L	[1]	Intercept ratio
la_0	kg/s	Fixed term coefficient in a Stodola-type linear regression equation
la_1	kg/MJ	Proportional term coefficient in a Stodola-type linear regression equation
qa_0	kg/s	Fixed term coefficient in a Stodola-type quadratic regression equation
qa_1	kg/MJ	Proportional term coefficient in a Stodola-type quadratic regression equation
qa_2	kg × s × MJ^{-2}	Quadratic term coefficient in a Stodola-type quadratic regression equation
LP	–	Low Pressure (steam)

(continued)

Symbol	Measurement unit	Description
m_{ST}	kg/s	Mass flow rate of steam flowing (expanding) through a turbine
m_0	kg/s	The steam mass flow at zero power output
m_{R1}	kg/s	The flow rates through the uncontrolled extractions
m_{R2}	kg/s	The flow rates through the uncontrolled extractions
m_{R3}	kg/s	The flow rates through the uncontrolled extractions
m_{min}, m_{max} (Fig. 6.24)	kg/s	Lower and the upper bounds for mass flow rates inlet to a turbine
m_e	kg/s	Steam flow rate at the outlet towards the condenser
m_{WH}	kg/s	Steam flow rate to the water heater
m	kg/s	Steam mass flow through a turbine
m_{opt}	kg/s	Mass flow rate of steam for economically optimal operation of the turbine
m_n	kg/s	Mass flow rate of steam for the nominal operation point of the turbine
MP	–	Medium Pressure (steam)
m_{e1}	kg/s	The flow through the (controlled) steam extraction
m_{e2}	kg/s	The steam flow rate at the outlet towards the condenser
Δm_{min}	kg/s	Minimal excess steam flow for ensuring that the turbine rotates
m_{ST}	kg/s, t/h	Mass flow rate of steam expanding in a turbine
n	[1]	The polytropic index
n_{ST}	kJ/kg, MWh/t	The slope coefficient of the Willans Line
n_k	s^{-1}	Rotation speed of a steam turbine
n	–	Nominal operation point/state of a steam turbine
OP	–	Operating point
p	Pa, bar	Pressure
p_1	bar	The pressure at the inlet to the rotor blades of the first turbine stage
p_1, p_2 (Fig. 6.23)	bar	Steam pressure in states 1 and 2
p_e	bar	Pressure at the steam turbine outlet – state **e**
p_{ec}	bar	Pressure at the steam turbine outlet – state $\mathbf{e_c}$
$p_{e,is,c}$	bar	Steam pressure at the inlet state $e_{is,c}$
p_i	bar	Steam pressure at the inlet state i
p_{ic}	bar	Steam pressure at the inlet state i_c
p_{SV}	bar	The steam pressure after the shut-off valve SV
p_{CV1}	bar	The steam pressure after the shut-off valve CV1
p_{CV2}	bar	The steam pressure after the shut-off valve CV2
Δp_{CV}	bar	The pressure drop in the control valve
q	kJ/kg	The specific heat transfer with surroundings per unit amount of steam
q_{2-3}	kJ/kg	Heat supply flow to the Rankine cycle
q_{4-1}	kJ/kg	Heat rejection flow from the Rankine cycle
q_z	kJ/kg	Specific internal heat loss inside the turbine

(continued)

Symbol	Measurement unit	Description
Q_{fuel}	MW	Heat supplied to a boiler with the primary source – typically fuel
Q_S	MW	Heat output of the boiler carried by the steam flow
Q_{SD}	MW	Heat carried with the steam at the turbine inlet
Q_C	MW	Heat rejected to cooling water in the turbine condenser
R	kJ×kmol× °C^{-1}	The universal gas constant
$R1–R3$	–	Uncontrolled extractions for the regenerative feed water heat exchangers
r	kJ/kg/°C	The characteristic gas constant – the ratio of the universal gas constant and the molecular weight
s	kJ/kg/°C	Specific entropy
s_e	kJ/kg/°C	Specific entropy of steam in state **e**
s_i	kJ/kg/°C	Specific entropy of steam in state **i**
Δs	kJ/kg/°C	Specific entropy change
$S_{\{\#\}}$	–	Steam turbine stages (Fig. 6.**20**)
SB	–	Steam boiler
SG	–	Steam generator
ST	–	Steam turbine
SV	–	Shut-off valve (Fig. 6.15, Fig. 6.16)
SUGCC	–	Site Utility Grand Composite Curve
T	°C	Temperature
T	–	Turbine (steam) – Fig. 6.8
ΔT_{sat}	°C	Saturation temperature difference between the Inlet and the outlet of a steam turbine
$T_{sat,in}$	°C	The inlet steam saturation temperature
T_{in}	°C	The inlet steam actual temperature
T_i, T_e	°C	Temperatures of steam in states *i* and *e*
$T_{sat,out}$	°C	The outlet steam saturation temperature
τ	min	Elapsed time
v	m^3/kg	Specific volume
VHP	–	Very High Pressure (steam)
W	MW	Generated power, delivered to the users
W_{co}	MW	Power left after the coupling and before the gearbox (Fig. 6.7)
w	kJ/kg	specific work by the steam expansion
w_T	J/kg	Specific internal work generation of the steam turbine (Fig. 6.2)
w_{BFWP}	J/kg	The specific internal work used by the feed pump (Fig. 6.2)
W_{gr}	MW	Power left after the gearbox (Fig. 6.7)
W_i	MW	Power passed from the steam turbine to the mechanical transmission
W_{is}	MW	Power generation assuming isentropic expansion
WP_i	kW	The overall power output transferred between the steam and the rotor inside the turbine
w_i	kJ/kg	The specific internal work of a steam turbine

(continued)

Symbol	Measurement unit	Description
w_{is}	kJ/kg	The specific internal work at isentropic expansion
W_n	MW	The nominal power of the steam turbine or the turbine-generator assembly
W_{opt}	MW	Optimal power generation – under maximal efficiency
W_{gc}	MW	Power output of the engines at the generator contacts
W_{ic}	MW	Internal power consumption by the auxiliary machines of the block
W_{OUT}	MW	Power plant output
W_{loss}	MW	Losses in power generation by a steam turbine
W_{min}, W_{max} (Fig. 6.24)	MW	Steam turbine power generation values for the lower and the upper bounds for inlet steam flow
W_{INT}	MW	Intercept coefficient of the Willans Line equation
W_{total}	MW	Total energy flow consumed in a steam turbine
X	[1]	Steam quality (dryness fraction)
z	kJ/kg	Specific internal mechanical losses inside a steam turbine
Δ	kJ/kg	Specific reusable heat
η	[1]	Steam turbine efficiency
η_{RRC}	[1]	Efficiency of the reversible Rankine cycle [Fig. 6.2, eq. (6.5)]
η_i	[1]	Internal efficiency of a steam turbine
η_{is}	[1]	Isentropic efficiency of a steam turbine
$\eta_{adiabatic}$	[1]	Adiabatic efficiency of a steam turbine
η_{total}	[1]	Overall efficiency of the assembly of steam turbine, coupling, gearbox and generator [Eq. (6.11)]
$t\eta_{urbine}$	[1]	The efficiency of the steam turbine
η_m	[1]	The combined efficiency of the transmissions
$\eta_{generator}$	[1]	The efficiency of the electrical generator
η_{PP}	[1]	Thermal efficiency of a power plant
η_{SG}	[1]	Thermal efficiency of steam generation
η_{SD}	[1]	Thermal efficiency of steam distribution
η_{st} (Fig. 6.24)	[1]	Steam turbine efficiency
η_{min}, η_{max} (Fig. 6.24)	[1]	Steam turbine efficiency values for the lower and the upper bounds for inlet steam flow
$\eta_{mech}/\eta_{transmission}$	[1]	Mechanical efficiency of a steam turbine
$\varphi_{R1}-\varphi_{R3}$	[1]	Proportionality coefficients in modelling uncontrolled extractions

References

ABB. 2018. ABB Ability™. Performance Monitoring for Power Generation. https://library.e.abb.com/public/ba3e8f9490964a0cb4804b6f446372a1/PerformanceMonitoring_Brochure_20190114.pdf, accessed 27/02/2020.

Aguilar, O., Perry, S.J., Kim, J.-K., and Smith, R. 2007. Design and Optimization of Flexible Utility Systems Subject to Variable Conditions: Part 1: Modelling Framework. Chemical Engineering Research and Design 85(8): 1136–48. https://doi.org/10.1205/cherd06062.

Akrami, E., Chitsaz, A., Nami, H., and Mahmoudi, S.M.S. 2017. Energetic and Exergoeconomic Assessment of a Multi-Generation Energy System Based on Indirect Use of Geothermal Energy. Energy 124 (April): 625–39. https://doi.org/10.1016/j.energy.2017.02.006.

Alstom Power. 2002. 100 let: historie a současnost vývoje a výroby parních turbín v Brně (Hundred Years: history and current development, and production of steam turbines in Brno). Prague, Czech Republic: For company ALSTOM Power Brno published by studio Trilabit in collaboration with Association PCC Pro společnost ALSTOM Power Brno vydalo studio Trilabit ve spolupráci s Asociací PCC.

Atkins, A.G. and Escudier, M. 2013. A Dictionary of Mechanical Engineering: Over 8,500 Definitions for Students and Professionals. Oxford Paperback Reference. Oxford, UK: Oxford Univ. Press.

Bhatt, M.S. and Rajkumar, N. 1999. Performance Enhancement in Coal Fired Thermal Power Plants. Part II: Steam Turbines. International Journal of Energy Research 23(6): 489–515. https://doi.org/10.1002/(SICI)1099-114X(199905)23:6<489::AID-ER494>3.0.CO;2-T.

Bruno, J.C., Fernandez, F., Castells, F., and Grossmann, I.E. 1998. A Rigorous MINLP Model for the Optimal Synthesis and Operation of Utility Plants. Chemical Engineering Research and Design. Techno-Economic Analysis, 76(3): 246–58. https://doi.org/10.1205/026387698524901.

Church, E. F. 1928. Steam Turbines. New York, USA: McGraw-Hill.

Church, E.F. 1950. Steam Turbines. 3rd ed. New York, USA: McGraw-Hill.

Coulls, P. 2013. Willans Archive. Industrial Archaeology News 165, 20–21, https://industrial-archae ology.org/wp-content/uploads/2016/04/ian165.pdf, accessed 20/08/2020.

Doosan Skoda Power. 2020. Steam Turbines : Doosan Škoda Power, the Czech Republic. http://www.doosanskodapower.com/en/steam/, accessed 08/03/2020.

Dryden, I.G.C., ed 1982. The Efficient Use of Energy. 2nd ed. London, UK: Butterworth Scientific in collaboration with the Institute of Energy acting on behalf of the UK Dept. of Energy.

EKOL, spol. s r.o. 2020. Comprehensive Turbines Solutions | Steam Turbines | Products. https:// www.ekolbrno.cz/en/s1405/Products/Steam-Turbines/c2643-Comprehensive-Turbines-Solutions, accessed 09/04/2020.

General Electric. 2020a. Steam Turbine Technology | GE Steam Power. https://www.ge.com/power/steam/steam-turbines, accessed 08/03/2020.

General Electric. 2020b. Alstom Energy Acquisition | GE Power. https://www.ge.com/power/about/alstom-acquisition, accessed 09/04/2020.

General Electric. 2020c. GE Power | General Electric. https://www.ge.com/power, accessed 09/04/2020.

GlobalData Energy. 2019. GlobalData: Global Steam Turbine Market Valued at $36.7bn in 2019–2023. Power Technology | Energy News and Market Analysis (blog). https://www.power-technology.com/comment/globaldata-global-steam-turbine-market/, accessed 22/08/2019.

G-Team a.s. 2020. Steam Turbines. Steam Turbo. https://www.steamturbo.com/steam-turbines.html, accessed 08/03/2020.

Kim, J.-K. and Smith, R. 2004. Cooling System Design for Water and Wastewater Minimization. Industrial & Engineering Chemistry Research 43(2): 608–13. https://doi.org/10.1021/ie020890m.

Klemeš, J.J., Dhole, V.R., Raissi, K., Perry, S.J., and Puigjaner, L. 1997. Targeting and Design Methodology for Reduction of Fuel, Power and CO_2 on Total Sites. Applied Thermal Engineering 17(8–10): 993–1003. https://doi.org/10.1016/S1359-4311(96)00087-7.

Leĭzerovich, A.S. 2008. Steam Turbines for Modern Fossil-Fuel Power Plants. Lilburn, GA: Boca Raton, FL, USA: Fairmont Press; Distributed by Taylor & Francis/CRC Press.

Lovegrove, K. and Stein, W., eds 2012. Concentrating Solar Power Technology: Principles, Developments and Applications. Woodhead Publishing Series in Energy 21. Cambridge, UK: Woodhead Publishing.

Luo, X., Zhang, B., Chen, Y., and Songping, M. 2011. Modeling and Optimization of a Utility System Containing Multiple Extractions Steam Turbines. Energy 36(5): 3501–12. https://doi.org/10.1016/j.energy.2011.03.056.

Madejski, P. 2018. Thermal Power Plants: New Trends and Recent Developments. London, UK: IntechOpen, https://www.intechopen.com/books/thermal-power-plants-new-trends-and-recent-developments, accessed 20/08/2020.

MAN. 2020. MAN Energy Solutions Steam Turbines. MAN Energy Solutions. https://www.man-es.com/process-industry/products/steam-turbines, accessed 09/04/2020.

MapEx Software, Inc. 2020. Power Plant Performance Monitoring. MapEx Software. https://powerplantperformance.com/about-us, accessed 27/02/2020.

MarketWatch Inc. 2019. Steam Turbine Market 2019 – Business Revenue, Future Growth, Trends Plans, Top Key Players, Business Opportunities, Industry Share, Global Size Analysis by Forecast to 2024 × 360 Research Reports. MarketWatch. https://www.marketwatch.com/press-release/steam-turbine-market-2019-business-revenue-future-growth-trends-plans-top-key-players-business-opportunities-industry-share-global-size-analysis-by-forecast-to-2024-360-research-reports-2019-09-25, accessed 25/09/2019.

Mavromatis, S.P. and Kokossis, A.C. 1998. Conceptual Optimisation of Utility Networks for Operational Variations – I. Targets and Level Optimisation. Chemical Engineering Science 53 (8): 1585–1608. https://doi.org/10.1016/S0009-2509(97)00431-4.

McGowan, T.F., Brown, M.L., Bulpitt, W.S., and Walsh, J.L. 2011. Biomass and Alternate Fuel Systems: An Engineering and Economic Guide. New York, USA: Wiley.

McKetta, J.J., ed 1996. Steam Reforming, Operating Experience to Storage Tank Explosion Safeguards. Encyclopedia of Chemical Processing and Design, executive ed.: John J. McKetta; 54. New York, USA: Marcel Dekker.

Mitsubishi Hitachi Power Systems Ltd. 2020. Product Lineup | Gas Turbines | MITSUBISHIHITACHI POWER SYSTEMS, LTD. Mitsubishi Hitachi Power Systems Official Website. https://www.mhps.com/products/gasturbines/lineup/, accessed 13/03/2020.

Moran, M.J., ed 2011. Fundamentals of Engineering Thermodynamics. 7th ed. Hoboken: N.J., USA Wiley.

Moyer, J.A. 1917. Steam Turbines; a Practical and Theoretical Treatise for Engineers and Students, Including a Discussion of the Gas Turbine. 3rd ed. New York, USA: John Wiley & Sons, Inc.

National Renewable Energy Laboratory (NREL), Golden, CO, USA. 2006. Consider Installing High-Pressure Boilers with Backpressure Turbine-Generators. DOE/GO-102006-2267, 875766. https://doi.org/10.2172/875766.

Ohji, A. and Haraguchi, M. 2017. Steam Turbine Cycles and Cycle Design Optimization. In: Advances in Steam Turbines for Modern Power Plants. 11–40. Elsevier. https://doi.org/10.1016/B978-0-08-100314-5.00002-6.

PBS Group, a. s. 2020. Turbines. https://pbs.cz/en/our-business/powergineering/turbines, accessed 09/04/2020.

Petchers, N. 2003. Combined Heating, Cooling & Power Handbook: Technologies & Applications: An Integrated Approach to Energy Resource Optimization. Lilburn, GA, USA: Fairmont Press; Distributed by Marcel Dekker.

Potter, M.C. and Somerton, C.W. 2014. Schaum's Outline of Thermodynamics for Engineers. Third ed. Schaum's Outline. New York, USA: McGraw-Hill Education.

Raissi, K. 1994. Total Site Integration. PhD Thesis, Manchester, UK: University of Manchester Institute of Science and Technology.

Riznic, J.R., ed 2017. Steam Generators for Nuclear Power Plants. In: Woodhead Publishing Series in Energy. Oxford, UK: Woodhead Publishing.

Sarkar, D.K. 2015. Thermal Power Plant: Design and Operation. Amsterdam, The Netherlands: Elsevier.

Schobert, H.H. 2014. Energy and Society: An Introduction. Boca Raton, FL, USA: CRC Press.

Scoretz, M. and Williams, R. 2008. Industrial Steam Turbine Value Packages. GE Energy, Atlanta, GA, United States. https://www.ge.com/content/dam/gepower-pgdp/global/en_US/docu ments/technical/ger/ger-4191b-industrial-steam-turbine-value-packages-2008.pdf, accessed 21/02/2020.

Shang, Z. 2000. Analysis and Optimisation of Total Site Utility Systems. PhD Thesis, Manchester, UK: University of Manchester Institute of Science and Technology.

Shang, Z. and Kokossis, A. 2004. A Transhipment Model for the Optimisation of Steam Levels of Total Site Utility System for Multiperiod Operation. Computers & Chemical Engineering 28(9): 1673–88. https://doi.org/10.1016/j.compchemeng.2004.01.010.

Siemens AG. 2019. Efficiency: More Value to Your Facility. https://assets.new.siemens.com/sie mens/assets/public.1560517188.c3192f5e-0979-4c71-9028-45f1913a80f2.steam-turbine-overview-2019.pdf, accessed 21/02/2020.

Siemens AG. 2020a. Dresser-Rand Steam Turbines - a Siemens Business. Newton_ps-detail. Siemens.Com Global Website. https://new.siemens.com/global/en/products/energy/power-generation/steam-turbines/d-r-steam-turbines.html, accessed 08/03/2020.

Siemens AG. 2020b. Energy Products & Services. Newton_ps-access. Siemens.Com Global Website. https://new.siemens.com/global/en/products/energy.html, accessed 09/04/2020.

Sims, R.E.H. 2002. The Brilliance of Bioenergy: In Business and in Practice. London, UK: James & James.

Singh, M.P. and Lucas, G.M. 2011. Blade Design and Analysis for Steam Turbines. New York, USA: McGraw-Hill.

Škorpík, J. 2020. Steam turbines as part of overall plants (Parní turbína v technologickém celku). Energy Conversion Technologies (Transformační technologie), January. https://www.transfor macni-technologie.cz/25.html, accessed 26/02/2020 (in Czech).

Slade, S. 2017. General Electric to Dominate Industrial Power Market. Defense Security Monitor. https://dsm.forecastinternational.com/wordpress/2017/06/14/general-electric-to-dominate-industrial-power-market/, accessed 06/03/2020.

Strouvalis, A.M., Mavromatis, S.P., and Kokossis, A.C. 1998. Conceptual Optimisation of Utility Networks Using Hardware and Comprehensive Hardware Composites. Computers & Chemical Engineering 22 (March): 175–82. https://doi.org/10.1016/S0098-1354(98)00052-0.

Sun, L. and Smith, R. 2015. Performance Modeling of New and Existing Steam Turbines. Industrial & Engineering Chemistry Research 54(6): 1908–15. https://doi.org/10.1021/ie5032309.

US DOE. 2012. Consider Steam Turbine Drives for Rotating Equipment, DOE/GO-102012-3396. https://www.energy.gov/sites/prod/files/2014/05/f16/steam21_rotating_equip.pdf, accessed 21/02/2020.

US EPA. 2015. Catalog of CHP Technologies. Section 4. Technology Characterization - Steam Turbines. https://www.epa.gov/sites/production/files/2015-07/documents/catalog_of_chp_technologies_section_4._technology_characterization_-_steam_turbines.pdf, accessed 03/06/2019.

Varbanov, P., Perry, S., Makwana, Y., Zhu, X.X., and Smith, R. 2004a. Top-Level Analysis of Site Utility Systems. Chemical Engineering Research and Design 82(6): 784–95. https://doi.org/10.1205/026387604774196064.

Varbanov, P.S. 2004. Optimisation and Synthesis of Process Utility Systems. PhD Thesis, Manchester, UK: University of Manchester Institute of Science and Technology.

Varbanov, P.S., Doyle, S., and Smith, R. 2004b. Modelling and Optimization of Utility Systems. Chemical Engineering Research and Design 82(5): 561–78. https://doi.org/10.1205/026387604323142603.

Vukovic, M., Leifeld, R., and Murrenhoff, H. 2017. Reducing Fuel Consumption in Hydraulic Excavators – A Comprehensive Analysis. Energies 10(5): 687. https://doi.org/10.3390/en10050687.

Wilda, H. and Salter, C. 2003. Steam Turbines: Their Theory and Construction. Palm Springs, CA, USA: Wexford College Press.

Willans, P W. 1888. 'ECONOMY TRIALS OF A NON-CONDENSING STEAM-ENGINE: SIMPLE, COMPOUND AND TRIPLE. (INCLUDING TABLES AND PLATE AT BACK OF VOLUME)'. Minutes of the Proceedings of the Institution of Civil Engineers 93 (1888): 128–88. https://doi.org/10.1680/imotp.1888.21059.

Woodruff, E.B., Lammers, H.B., and Lammers, T.F. 2012. Steam Plant Operation. 9th ed. New York, USA: McGraw-Hill.

Yu, H., Gundersen, T., and Feng, X. 2018. Process Integration of Organic Rankine Cycle (ORC) and Heat Pump for Low Temperature Waste Heat Recovery. Energy 160 (October): 330–40. https://doi.org/10.1016/j.energy.2018.07.028.

7 Gas turbines

The working fluid of gas turbines is heated gas or airflow mixed with products of combustion. The gas turbines are most often used with combustion chambers, which are based on the so-called open cycle (Razak, 2007). Therefore, they are also often referred to as combustion turbines. An open-cycle gas turbine typically contains a turbocompressor section, a combustion chamber and a turbine section. For most gas turbines, simple designs are developed, since the technology has been tightly related to the jet engines for propulsion. In fact, many gas turbine designs for industrial use are derived from jet engines, which have given rise to the classification "aeroderivative" gas turbines. Figure 7.1 shows a cross-sectional visualisation of an industrial gas turbine.

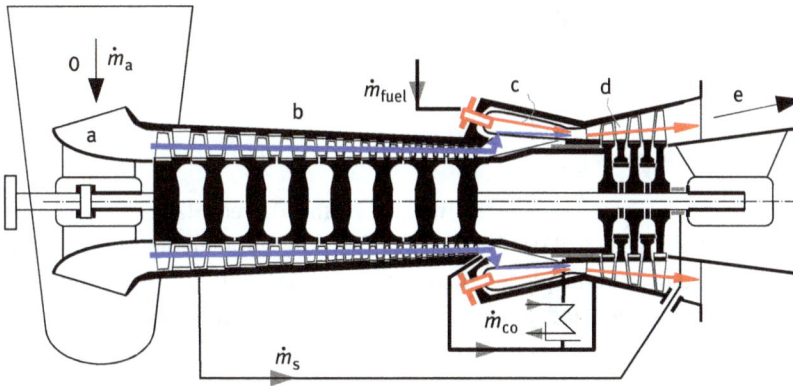

Fig. 7.1: A principle diagram of a gas turbine (combustion turbine) for industrial use (Škorpík, 2020b). Notation: a, air inlet; b, compressor section; c, combustion chambers; d, expander (turbine) section; e, exhaust opening; m_s [kg/s], air bypass for supplying to the turbine seals (usually labyrinth seals); m_a [kg/s], inlet air; m_{fuel} [kg/s], fuel supply; m_{co} [kg/s], cooling airflow.

The supplied fuel – usually gaseous or liquid – is burned inside the combustion chamber, within the airflow, supplied by the compressor. The combustion releases reaction products, which – together with the combustion air – form the flue gas. The flue gas from the combustion chamber enters the expander (turbine) section. Part of the power output of the turbine section is used by the compressor section and the remainder is supplied to an electric generator (generation configuration) or another device (direct-drive configuration).

Gas turbines are generally installed in various types of applications (Soares, 2020):
–   standalone configurations for power generation;
–   for direct drive, for example, for natural gas liquefaction;
–   Combined Heat and Power (CHP) generation plants;
–   for driving ships, and as jet engines for propulsion of airplanes.

https://doi.org/10.1515/9783110630091-007

The term "direct-drive" has different interpretations. In some cases, it is applied to the prime mover; in this case, a gas turbine, directly powering the equipment, for example, a compressor or a pump. However, in a US technology overview (Soares, 2020), the term is used to refer to a gas turbine directly driving a power generator. For the other context – directly driving process equipment – they use the term "mechanical drive".

7.1 Fundamentals

Gas turbines are based on the Brayton cycle (Razak, 2007). They can be built as compact units with rated power generation capacity from 30 kW – called microturbines (Capstone, 2019) – and the largest units reach 400 MW (Siemens, 2020), and up to 500 MW and more (Mitsubishi Hitachi Power Systems Ltd, 2020). Gas turbines are widely used for propulsion (Mattingly et al., 2016), especially for airplanes because of the favourable ratio of power to the equipment size (weight).

7.1.1 Brayton cycle (Joule cycle)

The Brayton cycle was designed by George Brayton in the United States on a piston engine (Kroos and Potter, 2015, p. 341). Currently, it typically implemented in turbines and is performed through several pieces of equipment (Fig. 7.2) – a compressor, a heater or combustion chamber and an expander. The working fluid in the Brayton cycle is a gas. There are closed-cycle (Fig. 7.2a) and open-cycle (Fig. 7.2b) gas turbines.

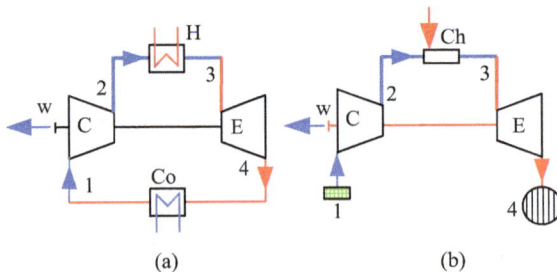

Fig. 7.2: A principle diagram of a gas turbine (combustion turbine) for industrial use (Škorpík, 2020b). (a) Closed cycle and (b) open cycle. Notation: C, compressor; H, heater; E, expander; Co, cooler; Ch, combustion chamber with fuel inlet; w [kJ/kg], specific work of the cycle.

In the closed-cycle type, the working fluid circulates entirely within the equipment items constituting the machine. It has a heater, supplying high-temperature heat from an external source and a cooler, taking away low-temperature heat to an external medium. The heat is supplied to the working gas through a heat exchanger at high pressure. Such heat exchangers for gas heating at high pressures and temperatures are associated with problems – large size and the use of special materials – giving rise to high investment cost.

An open-cycle gas turbine takes ambient air and exhausts the expanded gas back to the environment. Instead of a heater, it has a combustion chamber and uses the surroundings as a large cooler. This is the more common type of gas turbines, mainly due to the avoidance of the large heat exchangers associated with the closed-cycle implementations.

Assuming the ideal case (reversible changes), the following thermodynamic processes take place in the individual parts of the gas turbine flowsheet (see Fig. 7.2):

- **Compression.** Isentropic compression of the working gas from state 1 to state 2.
- **Heating of the working gas.** The working gas is heated (in the heater H or the combustion chamber Ch) from the temperature of state 2 to that of state 3 – an isobaric process.
- **Expansion.** The isentropic expansion of the working gas from state 3 to state 4 takes place in the expander (turbine section). During the expansion, the working gas performs work, which powers the rotation of the machine shaft. Typically, the turbine and compressor are connected to the same shaft. The compressor consumes part of the shaft work of the turbine. The remaining part of the shaft work can be used to rotate an electric generator or another piece of equipment.
- **Cooling of the working gas inside the cooler.** The isobaric cooling of the working gas on temperature T_1 is performed inside the cooler. The cycle can be repeated from this state.

In the case of an open cycle, atmospheric air is sucked and compressed to the pressure at state 2 by the compressor. Next, the pressurised air is mixed with fuel (flammable fluid) at pressure p_2 inside the combustion chamber, where it continuously burns and is then expanded in the turbine part. The exhaust gas is sent to a chimney (stack) or to a heat recovery steam generator (HRSG).

The temperature–entropy diagram of a reversible Brayton cycle for an ideal gas is shown in Fig. 7.3. The data assumed in constructing the diagram include specific heat capacity $c_p = 1.004$ kJ/kg- K (assuming dry air without CO_2, at atmospheric conditions); $k = 1.402$ [1] – adiabatic index for air; inlet pressure $p_1 = 1$ atm = 1.01325 bar; high pressure of the cycle $p_2 = 1$ MPa = 10 bar, inlet temperature $t_1 = 20$ °C, temperature after the combustion chamber $t_3 = 1,300$ °C. The assumptions for ideal conditions include a closed cycle, for which the composition of the working gas does not change. For the case of an open cycle, the fuel combustion in the chamber changes the composition of the working gas.

Referring to the flowsheet (Fig. 7.2) and to the $T–s$ diagram (Fig. 7.3), the energy balance of the Brayton cycle can be written as follows. The heat supplied to the cycle (in the heater or the combustion chamber):

$$q_{2-3} = h_{3c} - h_{2c} \qquad\qquad (7.1)$$

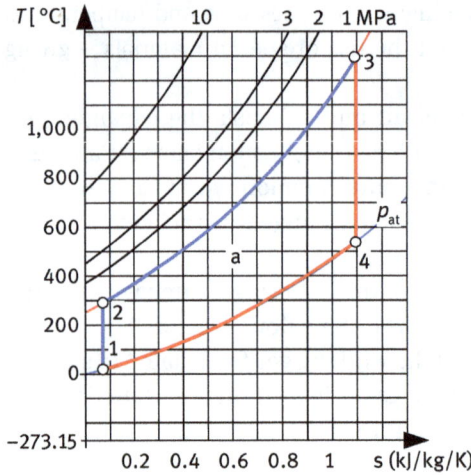

Fig. 7.3: The Brayton cycle in the $T-s$ diagram of an ideal gas.

Heat rejected from the cycle (after the expander):

$$q_{4-1} = h_{1c} - h_{4c} \tag{7.2}$$

Specific work of the working gas in the expander

$$w_e = h_{3c} - h_{4c} \tag{7.3}$$

Specific work performed by the compressor

$$w_c = h_{1c} - h_{2c} \tag{7.4}$$

Net specific work of the cycle

$$w = w_e + w_c = q_{2-3} + q_{4-1} = (h_{3c} - h_{2c}) + (h_{1c} - h_{4c}) \tag{7.5}$$

Cycle efficiency:

$$\eta_{cycle} = 1 + \frac{h_{1c} - h_{4c}}{h_{3c} - h_{2c}} \tag{7.6}$$

The notation in Eqs. (7.1)–(7.6) includes
- q_{2-3} [kJ/kg] – heat supplied to the cycle
- q_{4-1} [kJ/kg] – heat rejection from the cycle
- $h_{<x>c}$: $x = [1..4]$ [kJ/kg] – specific stagnation enthalpies (the sum of static enthalpy and kinetic energy – see Balachandran, 2013) at states "x"
- w, w_c, w_e [kJ/kg] – specific work rates of the cycle – net, compression, expansion
- η_{cycle} [1] – efficiency of the cycle

7.1.2 The gas turbine as a process unit

The gas turbine is a separate compact process unit, implementing the entire Brayton cycle Fig. 7.4. This is contrary to steam turbines, which are only parts of the Rankine cycle. The gas turbine only requires a connection to the fuel supply, a connection to a driven machine or electricity users, air intake (with filters) and flue gas extraction. The compressor part of a combustion turbine can be equipped with the coolers for intercooling of the air being compressed (e.g. Schobeiri, 2018, p. 20). In the industry or in thermal power plants, gas turbines almost always contain heat exchangers (referred to as HRSGs; Ganapathy, 2003) that utilise the heat contained in the exhaust. This heat can be used for heating water, steam generation, or supplying heat directly to processes. This type of comprehensive utilisation of the primary heat source is called cogeneration (Schobeiri, 2018), emphasising that power and heat are generated simultaneously. More complex arrangements are also possible – including trigeneration (Aviso and Tan, 2018), which is the simultaneous generation of power, heating and cooling utilities. In case the gas turbine shaft drives an electric generator, the useful output is in the form of electrical power supplied to an industrial site or the central grid.

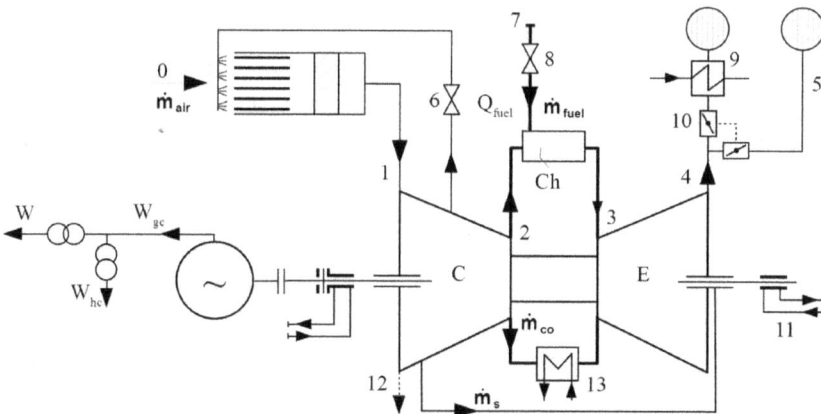

Fig. 7.4: Connection of a gas turbine within a power plant. Notation: 0, air state before the suction filters and the silencer; 1, air state at the compressor compartment inlet; 2, compressed air for combustion; 3, the exhaust gas state at the combustion chamber outlet and the turbine inlet; 4, exhaust state at the turbine outlet; 5, flue gas outlet to the stack; 6, warm air intake to heat the inlet air for prevention of icing; 7, fuel supply; 8, fuel valve (before it is a shut-off valve and a safety valve); 9, waste heat boiler; 10, bypass boiler cap; 11, lubrication system of bearings; 12, drainage of the compressor part; 13, cooling air cooler. W_{gc} [MW], the power output of the system at the generator contacts; W_{hc} [MW], internal power use; W [MW], power supplied to the users or to the grid.

Anti-icing systems are sometimes used (General Electric Company, 2015) depending on the climatic conditions of the turbine installation. At cold ambient conditions and high humidity levels, the suction filter may become clogged with frost (position 6 in Fig. 7.4). The air for cooling the thermally stressed parts of the turbine is cooled in external heat exchangers (m_{co}, position 13 in Fig. 7.4). This reduces air consumption. It is then directed to flow through the fine filter to prevent clogging of the nozzles. Outside the blades, this air flow is used to cool and flood the seals at the point between the rotor and the combustion chamber.

7.1.3 Real gas turbine cycle

The actual gas turbine cycle features various types of losses in the individual parts of the machine. This includes pressure losses in the paths of the fluid flows and mechanical losses in the mechanisms – transmissions, shafts, bearings and generator parts. Examples of pressure losses include those incurred in the turbocharger suction, noise dampers, the air inlet, combustion chambers, exhaust outlet and the stack. These losses have a direct impact on the specific internal work of the gas, as shown in the T–s diagram (Fig. 7.5). In the figure, Δp is the pressure loss in the combustion chamber. For this case, in the real cycle, the composition of the working gas changes – in the combustion chamber. Therefore, the T–s diagram of the real Brayton cycle consists of segments calculated using different data about the gas properties. The numbering of the working gas flows and states in Fig. 7.5 corresponds to the notation in Fig. 7.3.

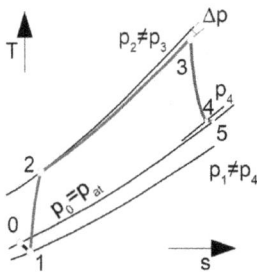

Fig. 7.5: Real gas turbine cycle T–s diagram.

The efficiency of a gas turbine is defined as the ratio of the delivered power related to the amount of usable energy supplied with the fuel to the combustion chamber. This is also referred to as fuel efficiency or power generation efficiency. Referring to the contours in Fig. 7.4, the power delivered to the customers is formed by subtracting the internal power use from the power generation:

$$W = W_{gc} - W_{hc} \; (MW) \tag{7.7}$$

The efficiency to the point after the power generator is then the ratio of the power generation to the heat flow supplied with the fuel:

$$\eta_{\text{gc}} = \frac{W_{\text{gc}}}{Q_{\text{fuel}}} \, [1] \tag{7.8}$$

The net efficiency of the gas turbine plant accounts for the internal power use

$$\eta_{\text{gt}} = \frac{W_{\text{gt}}}{Q_{\text{fuel}}} \, [1] \tag{7.9}$$

7.1.4 Operation characteristics of gas turbines

For gas turbines, the control of power generation is primarily driven by varying the fuel injection rate, thereby changing temperature T_3 (state 3 in Fig. 7.5) after the combustion chamber. For example, by lowering the amount of fuel, temperature T_3 decreases, thereby decreasing the volumetric flow of the turbine section (the flow cross sections of the turbine are constant), which means that the pressure p_2 downstream of the compressor section also decreases. It can also be controlled by manipulating the compressor part.

A significant influence on the power output of a gas turbine is also caused by the pressure and the temperature of the ambient air. In the case of jet engines, the power changes caused by the change of the pressure and temperature due to the adjustment of the outlet nozzle geometry can be partially compensated.

Power, efficiency and fuel consumption are interdependent, and if the turbine speed can be varied, the resulting characteristic is relatively complex. A typical set of characteristic curves, created in Škorpík (2011b) from data of Pratt & Whitney ST6 gas turbine, is shown in Fig. 7.6. This is a typical set of control characteristic curves for a combustion turbine at speed and load changes. It shows how the thermal efficiency and maximum load of the turbine decreases as the speed decreases from nominal speed. Figure 7.6 shows an example of using the diagram, traced with red arrows. If the fuel consumption is kept the same (here $\varepsilon_2 = 1$), the power decreases ($\varepsilon_1 \approx 0.64$) as the engine speed decreases (here $n/n_n = 0.4$). This means that efficiency decreases about $\varepsilon_3 = 1.54$ (efficiency decreased by approximately 35 %).

The construction of gas turbine characteristics is very demanding because it is a compressor and a turbine set. This means that it is necessary to construct the characteristics of both the compressor part and the turbine and to connect them.

If the gas turbine is installed in a set with an electrical generator connected to the mains, the speed at any operating point is kept constant. Under such circumstances, the power drop is approximately proportional to the drop in fuel input (Kim and Hwang, 2006). Figure 7.7 shows a qualitative plot of such a part-load performance characteristic.

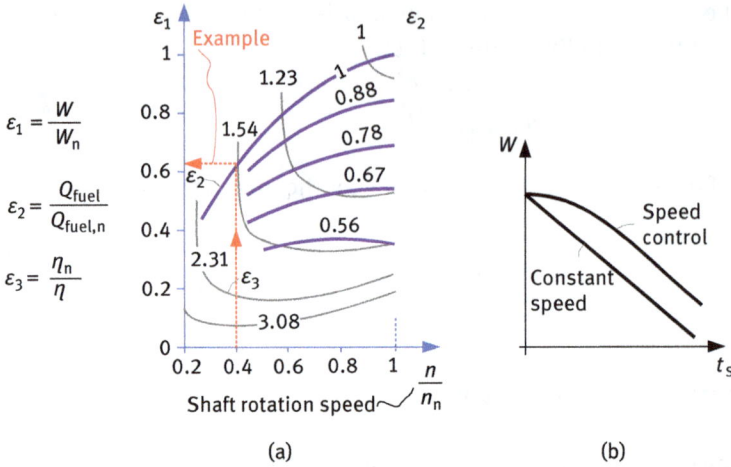

$$\varepsilon_1 = \frac{W}{W_n}$$

$$\varepsilon_2 = \frac{Q_{fuel}}{Q_{fuel,n}}$$

$$\varepsilon_3 = \frac{\eta_n}{\eta}$$

Fig. 7.6: Characteristic curves (Škorpík, 2011b) for a combustion gas turbine: (a) full characteristics; (b) power dependence on air temperature in compressor intake. Notation: t_s [°C], air temperature at the compressor suction point; W [MW], power generation by the gas turbine (without electrical part); index "n" denotes nominal parameters.

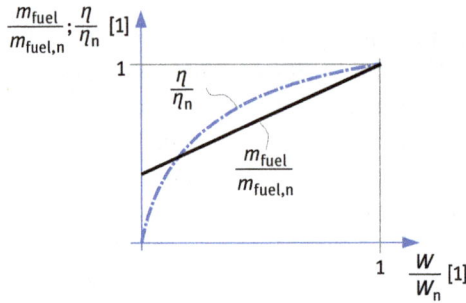

Fig. 7.7: Part-load performance characteristic of a single-shaft gas turbine with an electrical generator connected to the grid.

7.1.5 Main producers and typical performance

Gas turbines are offered by a variety of manufacturers, but for the large part, the market is dominated by the same technology companies as for steam turbines. According to a recent market analysis (Envision Intelligence, 2020), the main market players are:

- General Electric, share 49 %
- Siemens, share 29 %
- Mitsubishi-Hitachi, share 11 %
- Ansaldo, share 8 %
- The remaining share of 3 % is held by other suppliers

A good source of overview for gas turbines on offer and their performance is the periodical "Gas Turbine World" (Pequot Publishing Inc, 2019), which lists various manufacturers and the gas turbines currently available to customers. The catalogue and the magazine are published several times per year, keeping the information up-to-date. From the provided listing, the electrical efficiency of the gas turbine options varies in the interval from about 24 % for the smaller models about 1–2 MW offered by Dresser-Rand (a Siemens subsidiary) up to 41–44 % for the larger models above 100 MW, offered by General Electric, Siemens and Ansaldo. It has to be pointed out that the referenced efficiency rates are based on the power generation rate relative to the fuel use (heat rate) only. As such, some of the cited models can be enhanced gas turbines with heat recuperation before the combustion chambers.

The potential for heat recovery for use in combined cycles or cogeneration is considered in other sources, mainly presented by the individual equipment suppliers. For instance, General Electric (2020a) provides an overview of its line of combined cycle gas turbine (CCGT) and simple cycle gas turbines with a portfolio ranging from 30 to 50 MW up to 1.6 GW power plant capacity. An overview of the main technologies and suppliers involving gas turbines in combined cycles can be found in Soares (2020). The source also discusses all application areas, including power generation, CHP and propulsion.

A separate segment on the gas turbine market is that of microturbines. These are small gas turbines rated in the range from 10 to 15 kW up to a few hundreds of kW (do Nascimento et al., 2013). The review points out as the leading manufacturers on the microturbine market AlliedSignal, Elliott Energy System, Capstone, ABB, citing units rated from 30 to 200 kW.

The sub-market of microturbines has been valued to more than 360 M USD for the year 2018 (Fortune Business Insights, 2019). The analysis is based on data analysis about manufacturers such as Capstone Turbine Corporation, Bladon Micro Turbine, UAV Turbine, Ansaldo Energia, Aurelia Turbine, MTT Microturbine, FlexEnergy Inc., ICR Turbine Engine Corporation, Dresser-Rand, Turbo Tech Precision Engineering Pvt. Ltd. and Brayton Energy LLC. The report indicates that approximately 2/3 of the microturbine applications are for CHP and the remaining are standby (backup) machines.

7.2 Overview of gas turbine applications

Although gas turbines have many advantages, thermal power plants with combustion turbines alone are not widespread. Gas turbines are often used as backup power generators, for example there are products by Kawasaki (2020) and General Electric (2020b), and for peak-shaving, due to their quick start capabilities. As mentioned in the previous section, microturbines also fill that market niche.

Special materials and alloys are used in constructing gas turbines, which allows achieving high firing temperature and longer equipment life, but the machine parts are susceptible to damage (Igoe and Welch, 2014). This makes it necessary to use very clean fuels without particulates or other impurities, as well as to properly filter the incoming air. The usual fuels used include natural gas and petroleum derivatives (Basha et al., 2012), various biofuels (Gupta et al., 2010), for example biogas biodiesel, and synthesis gas after cleaning. Gas turbines can also burn lower-grade fuels, but it is associated with problems.

Those higher costs give additional reasons for extracting the maximum possible energy from the used fuel. Related to gas turbines, recovered energy can be used either to reduce the fuel consumption for generating the same power, or for integrating the gas turbine with other process units. The typical arrangements for making those happen are illustrated in Fig. 7.8. One way to improve energy utilisation is through heat regeneration in a modification of the Brayton cycle, where the exhaust gas is used for preheating the compressed air before entering the combustion chamber (Fig. 7.8a). Another option is to recover the exhaust heat in a heat exchanger (Fig. 7.8b) for direct process heating, or an **HRSG**, for raising steam. In turn, the steam can be used in steam turbines or for process heating. Combinations of these arrangements are also possible.

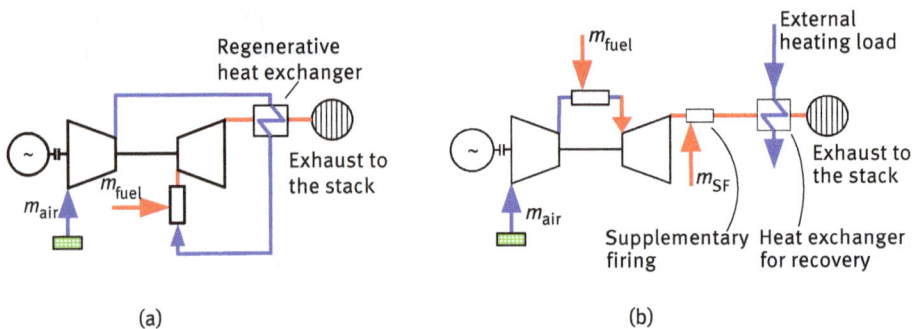

Fig. 7.8: Integration options for heat recovery and reuse for gas turbines: (a) heat regeneration and (b) process heating. Notation: m_{air} [kg/s], mass flow rate of incoming air; m_{fuel} [kg/s], mass flow rate of the fuel burned in the turbine; m_{SF} [kg/s], mass flow rate of the fuel for supplementary firing.

The temperature of hot exhaust gas from a gas turbine can be up to 600 °C without heat regeneration and up to 1,177 °C for regenerative gas turbines (Razak, 2007). Depending on the needs of the served processes, the exhaust gas can be treated in different ways. If no advanced materials are available, or they are too expensive, while the served process can tolerate lower temperature heating, the exhaust can be mixed with cold air, also increasing its volume flow rate of the mixture, before passing it to the process heaters. Another common arrangement is supplementary firing in the HRSG for increasing the steam generation rate, pressure and temperature. Supplementary firing, however, reduces the overall thermal efficiency of the process.

The steam, generated in an HRSG, can be used for power generation, bottoming a steam cycle that complements the gas turbine. The combination of a Brayton cycle and a steam (Rankine) cycle is called a CCGT. This combination features a significantly higher fuel utilisation efficiency than the gas turbine alone – reaching 55–60 %. There are examples – the alliance of General Electric and Électricité de France claiming a 62.22 % efficient CCGT plant in 2016 (General Electric, 2020c) and another comes from Japan (Proctor, 2018) – again powered by a General Electric gas turbine.

According to the thermodynamic properties of heat cycles, to obtain maximum thermal efficiency, the heat input should be provided at the highest possible temperature, and the heat rejection should take place at the lowest possible temperature. Within the CCGT arrangement, the Brayton cycle fits in the upper temperature range, while the Rankine cycle fits in the lower temperature range, whereby the combination of both provides a large temperature span. A typical CCGT flowsheet is illustrated in Fig. 7.9a.

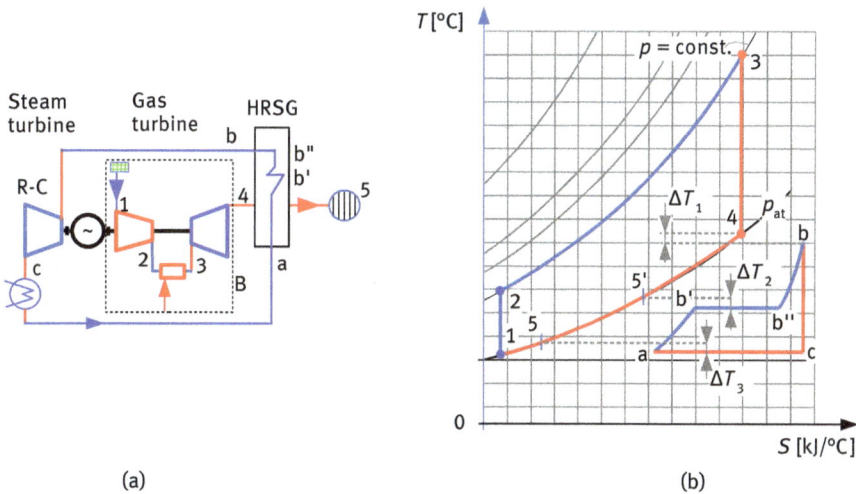

(a)   (b)

Fig. 7.9: Gas turbine powered combined cycle: (a) flowchart; (b) T–s diagram.

The flowsheet in Fig. 7.9a features a gas turbine implementing the Brayton cycle (denoted as B), a HRSG and a steam turbine, implementing the Rankine–Clausius cycle (denoted as R-C). The steam in the HRSG is generated from the heat carried with the hot exhaust gas from the gas turbine, in this way, cooling the exhaust.

To extract the maximum power from the fuel in the single-pressure steam cycle from Fig. 7.9, it is necessary to synchronise both thermodynamic cycles. The steam cycle part of the CCGT is adjusted to match the heat available from the exhaust gas at the outlet of the gas turbine expander. In Fig. 7.9a and b, this corresponds to state 4 and temperature T_4. From this description, it is evident that the maximum temperature of the steam cycle T_b must be lower than the exhaust temperature T_4, at a temperature difference ΔT_1, to ensure efficient heat transfer and acceptable size of the HRSG. The saturation temperature of the steam T_b' is another constraint, which can form a Pinch Point with the gas turbine exhaust inside the HRSG evaporator section, at temperature difference ΔT_2. The temperature difference between exhaust gas exit from the HRSG and the boiler feed water forms another potential constraint point ΔT_3. For maximum power output, usually, a multi-pressure steam cycle is used in CCGT plants. They are popular worldwide due to their high efficiency and their short construction time. CCGT units are also used to drive ships.

An example flowsheet with multiple steam pressure levels is shown in Fig. 7.10. It presents the layout of the steam cycle in the "Red Mill" thermal power plant in

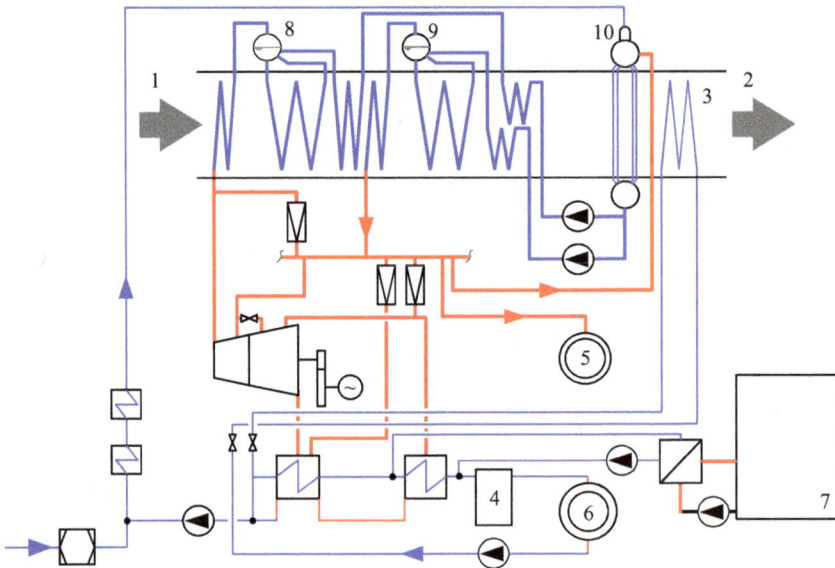

Fig. 7.10: Flowchart of steam part of CCGT power plant Red Mill (Brno, Czech Republic), amended from Škorpík (2020a). Notation: 1, input of hot exhaust gas from combustion turbine to steam recovery boiler; 2, exhaust output to the chimney (stack); 3, heat exchanger of the exhaust gas/hot water circuit (economiser); 4, natural gas water boiler; 5, steam user connection (0.9 MPa, 200 °C); 6, hot water user connection (130/70 °C); 7, hot water accumulator, non-pressurised storage tank (5,600 m³); 8, boiler drum of the high-pressure circuit; 9, boiler drum of the low-pressure circuit; 10, mixing of steam with condensate and pre-heat of the boiler feed water.

Brno – Czech Republic (Allforpower.cz., 2017) – as an example of a CCGT with a two-pressure-level cycle. The exhaust of the combustion turbine is connected with the steam recovery boiler through an exhaust pipe, which contains its own stack. Thanks to that, it is possible to operate the gas turbine without the steam part. This would become necessary in the case if there is a requirement for a faster start-up of power generation production.

The total electric power capacity of this plant is 94 MW, of which the gas turbine has a capacity of 70 MW. The plant also has a heating output capacity of 140 MW. The steam generation rating is 100 t/h, at 6.84 MPa (68.4 bar), 500 °C. The steam is supplied to the users at 0.92 MPa (9.2 bar) and 200 °C. The heat from the system is available as hot water (125 MW) and as steam (15 MW). The electrical efficiency of the system is 47.5 %, the overall (cogeneration) efficiency is 89.0 % and the power-to-heat ratio is 0.68.

The steam turbine in the "Red Mill" power plant contains one controlled steam extraction and one uncontrolled steam extraction. The steam from the controlled extraction is used for feeding steam to the users, but if steam pressure inside the steam pipe is higher, then the steam flows through a valve of controlled extraction to turbine and mass flow rate of steam is maximal. The uncontrolled steam extraction is used for feeding a heater of hot water during periods of high heating demand.

The small combustion turbines with power output to around 500 kW, as mentioned in the previous section, are referred to as *microturbines*. They are typically used in smaller scale CHP units and typically have a short start time, minimum maintenance requirements and small size.

Figure 7.11 shows the flowsheet of the Capstone C30 gas turbine model (Capstoneturbine.com., 2020). The notation includes: 1, high-frequency generator; 2, inverter (50/60 Hz); 3, power supply of accessories of the unit; 4, starter unit. The microturbine module is composed of a single-stage compressor and a single-stage turbine with a radial design. The power output of the unit is 30 kWe, and its maximum

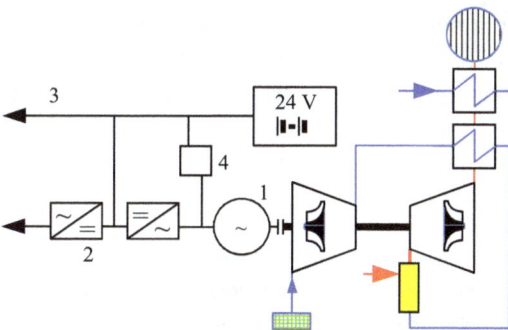

Fig. 7.11: Single-line diagram of the small combustion turbine Capstone C30 (amended from Škorpík, 2011a).

electric efficiency is 26 % (the best value in microturbines category). The NO_x emission level is 9 ppm; the exhaust temperature before the heat exchanger is 275 °C. The weight of the unit is 405 kg, and the height is 1.5 m.

A disadvantage of microturbines is their relatively high rotation speed (20,000–150,000 min^{-1}). This poses a high load on the turbine and generator bearings and requires expensive equipment for the inverter. Those properties make microturbines relatively expensive (capital cost per unit capacity), and they burn only high-grade fuels. Therefore, they are used most frequently as backup power generators. An advantage is that they are lighter than the backup units with piston combustion engines of similar capacities. The small units can also be assembled in a group that has a common control system, and it optimises their operation.

Gas turbine expanders without a compression section are usually called **turboexpanders**. They are frequently single-stage turbines. In the case of higher working gas inlet temperature, turboexpanders with axial stages are more suitable, due to the potential better cooling. Turboexpanders have niche applications:
- Reduction of gas pressure in pipelines (Kuczyński et al., 2019)
- Liquefaction of gases and gas mixtures with a follow-up extraction of valuable components – a typical application is the recovery of Natural Gas Liquids (Li et al., 2017)
- Re-gasification processes in LNG reception terminals (Mokhatab et al., 2014)

Combustion turbines are also used as direct drives. A typical application is to drive compressors of compressor stations on gas transportation pipelines for compensating pressure drop along with the transportation distance. Figure 7.12 shows an example flowsheet for the gas turbine driving the compressor station Werne in Germany (DEHN + SÖHNE, 2018). For transit pipelines, compressor stations are typically installed every 100–150, where the mean pressure inside the pipelines is about 7.5 MPa and requirement of power input of the compressors is up to 40 MW/segment. Fuel for the compressor stations is usually taken from the gas being transported through the pipeline. If the power output of the combustion turbine is higher than power demand by the compressor, a power generator can also be linked to the shaft, resulting in a hybrid type of installation – direct drive with a generator. The installation contains a two-shaft combustion turbine Pratt & Whitney FT8-55 with an approximate power output of 26 MW.

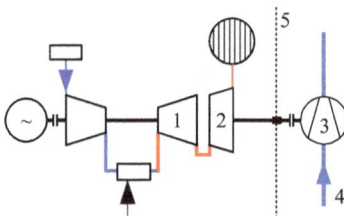

Fig. 7.12: Flowsheet of a gas turbine with a turbocompressor. Notation: 1, high-pressure turbine; 2, low-pressure turbine; 3, compressor to be driven; 4, pipeline; 5, a wall that separates the natural gas compressor from the machine room of the combustion turbine.

Gas turbines are also used as components of compressed air energy storage (CAES) systems and their extensions (RWE Power AG, 2010). This type of unit contains a turbocompressor powered by an electric motor, a storage tank for compressed air (it can be an excavated underground mine or other types of underground space) and a gas turbine with a combustion chamber. It functions as follows. In periods of a surplus of electricity in the electrical grid, the compressor is switched on, and it compresses air to the storage, powered by electricity from the grid. The air storage is used in periods with high electricity demands and grid deficit. This works by feeding the stored compressed air to the combustion chamber of the turbine and firing fuel. The exhaust from combustion chambers is expanded in the turbine which drives the electric generator. The flow of air from the tank can tear off dust and dirt, which can damage the blades of the turbine, and these blades have to be specially modified and often controlled. This system has a quick start of accumulation mode, even production mode. The CAES units for the accumulation of electricity have approximately about 50 % round-trip efficiency. The efficiency can be increased by building additional thermal energy storage and integrating it with the main CEAS system. During the charging of the air storage, the heat from the compression is directed to the heat storage and during the air release from the air storage, heat is released from the heat storage and provided to the outgoing air stream. This scheme is termed adiabatic CAES. It can achieve round-trip efficiency up to 70–75 %.

Figure 7.13 shows a principle flowsheet of the ADELE cycle. During storage, the compressed air is cooled by a heat carrier. In the process, the heat carrier is pumped from tank b to tank c, which is thermally isolated. During the retrieval from the storage, the compressed air is heated by the heat carrier retrieved from the thermal storage (tank c) before passing the air to the turbine (with an eventual combustion chamber). In the process of the second heat exchange, the heat carrier is pumped from tank c to tank b.

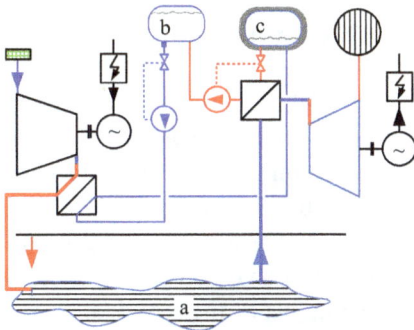

Fig. 7.13: Compressed air energy storage flowsheet. Notation: a, storage tank of compressed air; b, storage tank of cold accumulation liquid (oil); c, storage tank of hot accumulation liquid (regenerated heat).

Gas turbines for aircraft use are under strong and regular control. They also have a limited service life due to strict regulations for reliability reasons. This service life for propulsion is very short compared to the technical service life of the machine. Therefore, these turbines are in good condition after their removal from the airplane, and they are usually adapted for further service in the power industry. The jet engines have to be modified, to be suitable for stationary operation. The issues corrected include problems of high shaft rotation speed, removal of bypasses, the addition of stages to the turbine and the turbocompressor. Due to their origin and the necessary modification process, such turbines are called aeroderivative turbines. Figure 7.14 shows one possible arrangement for an aeroderivative gas turbine, where the additional expander is connected to a separate shaft. More details on these designs can be found in the dedicated literature (e.g. Schobeiri, 2018).

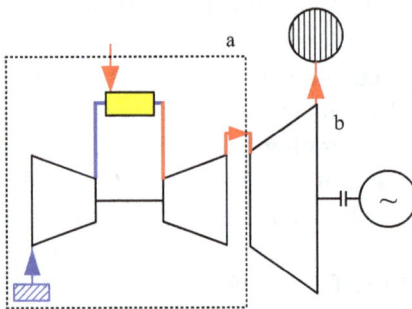

Fig. 7.14: Flowchart of a simple aeroderivative gas turbine. Notation: a, core of the aeroderivative (combustion turbine from a jet engine); b, expansion turbine for the complete utilisation of the thermal and kinetic energy components contained in the exhaust after the first expander.

All major players on the power machinery market offer aeroderivative gas turbines (Global Market Insights, Inc, 2020). The available models cover a wide range of rated capacities – from less than 50 kW up to 70–100 MW. The main advantages of aeroderivative gas turbines are they start quickly, thanks to the lightweight rotors and the lower cost for the cases of small power output (if used aircraft engine parts are used).

7.3 Gas turbine models for utility network optimisation

For designing and operating gas turbines, detailed models, like the ones discussed in Schobeiri (2018), can be used. However, for screening among many gas turbines for utility system design, operational simulation and optimisation of the overall utility networks, simpler models are needed, balancing the complexity and fidelity. This means that the models should be sufficiently accurate, as to trust the computation results, but their complexity should not cause excessive computation loads and delays. It has to be stressed that a gas turbine is not just a single device. It consists of a number of units, forming a flowsheet of its own, as shown in Section 7.1.

This section starts with an overview of the modelling developments for gas turbines, related to the formulation of a balanced model relating to performance at full and part loads, as well as cost. This is followed by the presentation of the main model from Varbanov et al. (2004), which provides the basis for understanding the main trends and for further customisation of the model to specific problems.

7.3.1 Modelling developments

There have been previous gas turbine models used for network analysis and optimisation. An attempt was made for a simple model in Maréchal and Kalitventzeff (1998), where a chart of typical gas turbine full-load performance characteristics was presented. The chart is based on an in-house data set, from a project performed by the authors. The efficiency plot is used by them in an optimisation procedure for optimal utility system selection. While this gives some idea of gas turbine performance, it is not suitable for computational models.

Shang (2000) developed a gas turbine model, allowing part-load modelling. It allows the construction of simplified utility system models for the purposes of gas turbine screening and system design. It features a variation of the Willans Line equation, where the power generation is related to the difference of a term proportional to the fuel feed to the gas turbine and a loss term.

That model has been thoroughly revised by Varbanov et al. (2004). Based on rigorous mass and energy balances, it relates the main properties – flow rates of the fuel, inlet air, injected steam and exhaust. It demonstrates that equating the slope and intercept coefficients of the model in Shang (2000) with incremental gas turbine power generation and losses leaves both terms varying – as functions of the current part load, which invalidates the assumption made in that source on the constant proportionality of the loss term to the turbine size. Varbanov et al. (2004) further provide a model, consisting of a set of linear equations, clearly showing that the coefficients in the provided Gas Turbine Willans Line (GTWL) are linearisation parameters, only loosely related to the rigorous power equation.

Based on the consistent gas turbine model in Varbanov et al. (2004), a customisation has been developed in Aguilar et al. (2007), focusing on utility network synthesis for flexible operation, including gas turbine flexibility. This involves a selection of site gas turbines from a large number of possible candidates, also accounting for part-load performance variations.

For those interested in detailed modelling of gas turbines, including dynamics in the performance, a detailed model of a microgas turbine is developed and discussed in (Rahman and Malmquist, 2016). That work also refers to previous models, also accounting for details of the gas turbine flowsheet and dynamic performance.

7.3.2 The balanced gas turbine model

A gas turbine model has been developed in a similar way to that for steam turbines
(Chapter 6). Configurations allow steam injection (Fig. 7.15).

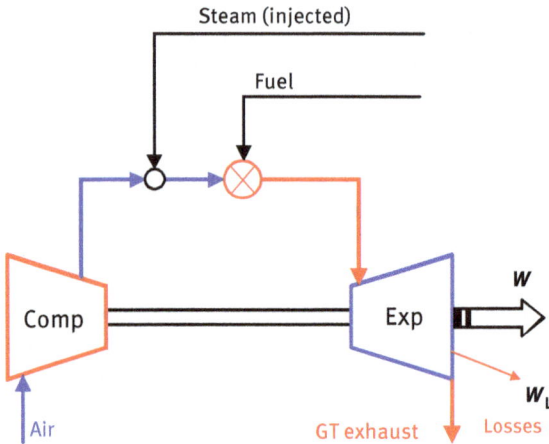

Fig. 7.15: Configuration of a gas turbine with optional steam injection.

7.3.2.1 Mass balance
The fuel consumption and the steam injection can be characterised by their ratios
(fuel-to-air ratio f and steam-to-air ratio s) with respect to the airflow:

$$f = \frac{m_f}{m_a} \tag{7.10}$$

$$s = \frac{m_{st}}{m_a} \tag{7.11}$$

The fuel-to-air ratio f usually changes with part load, depending on the chosen control system for the gas turbine, as shown in a conference paper on single-shaft
heavy-duty machines (Facchini, 1993) and in a more comprehensive evaluation including emissions (Andreini and Facchini, 2004). From the point of view of the
mass balance, the gas turbine can be viewed as a sequence of mixers:

$$m_a + m_{st} + m_f - m_{ex} = 0 \tag{7.12}$$

All mass flows can be expressed as functions of the fuel consumption:

$$m_a = \frac{1}{f} \times m_f \tag{7.13}$$

$$m_{st} = \frac{s}{f} \times m_f \tag{7.14}$$

$$m_{ex} = \frac{f+s+1}{f} \times m_f \tag{7.15}$$

7.3.2.2 Energy balance

A relationship of the shaft power and the fuel flow for the gas turbine is needed. It can be derived on the basis of an energy balance around the turbine, and the expressions for the mass flows from Eqs. (7.13) to (7.15). The standard form of the energy balance is

$$m_a \times h_a + m_{st} \times h_{st} + m_f \times NHV - m_{ex} \times h_{ex} - W_{gt} - W_{L,gt} = 0 \tag{7.16}$$

After the mass flows of air, steam and the exhaust are substituted with their equivalent in terms of the fuel flow, the resulting form of the balance is solved for the shaft power:

$$W_{gt} = w_{sp} \times m_f - W_{L,gt} \tag{7.17}$$

where w_{sp} is the specific shaft power generation – a net amount, accounting for the compressor load:

$$w_{sp} = \frac{1}{f} \times h_a + \frac{s}{f} \times h_{st} + h_f + NHV - \frac{f+s+1}{f} \times h_{ex} \tag{7.18}$$

Both the slope and the intercept in Eq. (7.17) vary with the fuel flow, and hence, the resulting performance line is curved. The actual form of the performance line will be considered later. The expression, obtained in Eq. (7.17) fits the form of the Willans Line. However, its coefficients have a direct physical meaning, unlike the linearisation for steam turbines. The slope represents the specific energy available in the gas turbine. The intercept represents energy losses. They are mainly mechanical losses for the compressor transmission, other friction losses, casing heat losses and kinetic energy losses from the exhaust. The efficiency of mechanical transmission is very high; around 98 % and higher. The casing losses are also relatively small.

7.3.2.3 Factors determining gas turbine performance

Similar to steam turbines, a combination of factors determines the performance of gas turbines. The factors considered here are the turbine size, the current load and the ambient temperature.

7.3.2.3.1 Effect of the gas turbine size on the performance and full-load model

One of the major factors affecting gas turbine performance is its size. It is usually measured in terms of rated power production at ISO rating conditions, which refer to $P = 1$ atm, $T = 15$ °C and 60 % relative humidity ('Gas Turbines. Procurement. Part 2: Standard Reference Conditions and Ratings' 2008). In general, gas turbine efficiency changes non-linearly with size. Turbines from different manufacturers are likely to feature different efficiencies for the same size. Maréchal and Kalitventzeff (1998) presented a curve of gas turbine efficiencies at full load and ISO conditions. A similar, more elaborate set of curves can also be found in Razak (2007). A qualitative illustration of this relationship is given in Fig. 7.16. Such a plot represents data from many gas turbines at their rated design points and features a monotonic non-linear shape.

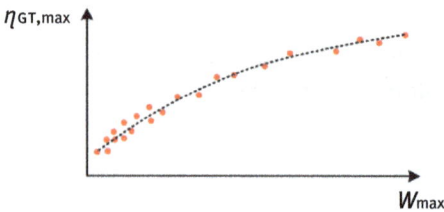

Fig. 7.16: Efficiency versus rated power for gas turbines.

Equivalent information can be obtained if the relationship illustrated in Fig. 7.16 is interpreted as one between the heat flow from fuel combustion (*fuel heat* in short) and the shaft power, at full load. The relationship between the fuel heat and the turbine shaft power can also be expressed via its efficiency or heat rate:

$$Q_f = \frac{W_{gt}}{\eta_{gt}} \text{ or } Q_{f,\max} = \frac{W_{gt,\max}}{\eta_{gt,\max}} \text{ (for full load)} \tag{7.19}$$

$$\mathrm{HR} = \frac{Q_f}{W_{gt}} = \frac{1}{\eta_{gt}} \tag{7.20}$$

Data on gas turbine performance is published by manufacturers. For example, in the General Electric Reference Library, there is a document designated "GER-3567 H" (Brooks, 2000). A plot of the gas turbine fuel heat consumption versus their rated power at full load, obtained from this source, is given in Fig. 7.17.

The trend of the plot in Fig. 7.17 is, essentially, linear. Similar plots, but from a different data source, are given in Manninen (1999). Therefore, the following equation can be formulated to describe the relation between fuel heat flow and power generation of a set of gas turbines, at full load:

$$Q_{f,\max} = A_{gt} + B_{gt} \times W_{gt,\max} \tag{7.21}$$

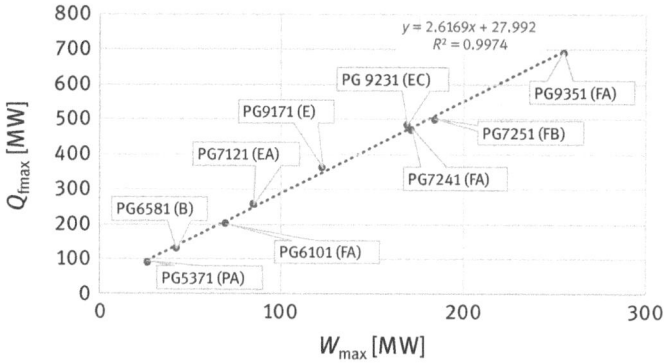

Fig. 7.17: Fuel heat for industrial gas turbines at full load obtained from GER-3567 H (Brooks, 2000).

Values for the coefficients of this relationship can be obtained by regression. It is important to emphasise that these coefficients differ between different gas turbine manufacturers and even different gas turbine types from the same manufacturer. The following values for the coefficients A_{gt} and B_{gt} have been obtained from the data set in Brooks (2000) plotted in Fig. 7.17:

- $A_{gt} = 21.9917$ MW
- $B_{gt} = 2.6683$ [1]
- 26.1 MW $\leq W_{gt,max} \leq 255.6$ MW

7.3.2.3.2 Performance trends at part load

Regarding part-load operation, the gas turbine efficiency follows a trend, similar to that of steam turbines. However, the curve for gas turbines is somewhat steeper in comparison with steam turbines, due to the different working fluid (flue gas). The works by Andreini and Facchini (2004) investigate the part-load performance of gas turbines in detail. A qualitative sketch, representing the gas turbine power efficiency at part load, is given in Fig. 7.18a. The behaviour illustrated in Fig. 7.18a needs to be modelled. It is desirable to produce a linear model, to allow for simplicity and convexity of the overall optimisation model. The rigorous relationship of the gas turbine shaft power versus the mass flow of fuel is a curve with a slope that increases with increasing the fuel flow (Fig. 7.18b).

The most convenient approach to modelling the gas turbine part load is to introduce a piecewise-linear approximation of the actual performance. The overall load range can be partitioned into modelling intervals and inside each interval, the power–fuel flow relationship can be approximated with a single straight line. This provides the optimal trade-off between accuracy and simplicity of the model for the purposes of simulation and optimisation.

(a)

(b)

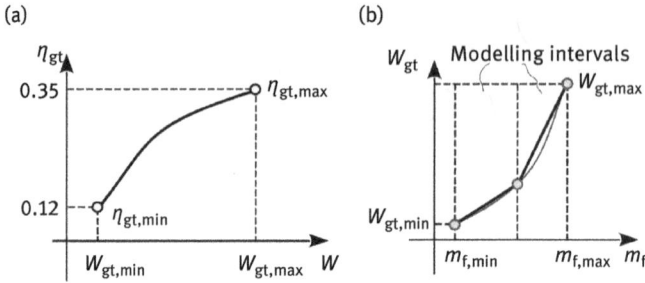

Fig. 7.18: Gas turbine efficiency trends and gas turbine Willans line piece-wise Linearisation: (a) efficiency variation at part load and (b) power generation as a function of steam flow rate.

7.3.2.3.3 Part-load model equations

The gas turbine part-load operation is expressed as GTWL. Such a linear equation can be written for a single modelling interval, as well as for the point of maximum load:

$$W_{gt} = n_{gt} \times m_f - W_{gt,\,int} \qquad (7.22)$$

$$W_{gt,\,max} = n_{gt} \times m_{f,\,max} - W_{gt,\,int}$$

Equation (7.22) can be directly applied to the existing gas turbines by obtaining operating point data from measurements or vendor catalogues and subjecting them to regression. On the other hand, since Eq. (7.22) is used for utility system design and network optimisation, then it is usually incorporated within a larger model, combining full-load and part-load performance. The slope and intercept of GTWL can be related to the full-load performance coefficients as follows (Varbanov et al., 2004):

$$n_{gt} = \frac{L_{gt} + 1}{B_{gt}} \times \left(NHV - \frac{A_{gt}}{m_{f,\,max}} \right) \qquad (7.23)$$

$$W_{gt,\,int} = \frac{L_{gt}}{B_{gt}} \times \left(NHV \times m_{f,\,max} - A_{gt} \right) \qquad (7.24)$$

7.3.2.3.4 Accounting for the ambient temperature

The model developed so far accounts only for the gas turbine size and part load. In order to make it valid for a range of ambient temperatures, temperature correction plots, provided by gas turbine manufacturers, are commonly used. According to the manufacturer information (Brooks, 2000), the correction plots are individual for each gas turbine model. Consequently, the ambient temperature can be taken into account for an individual gas turbine. Example temperature correction plots for GE gas turbine, model PG5371(PA), are given in Fig. 7.19.

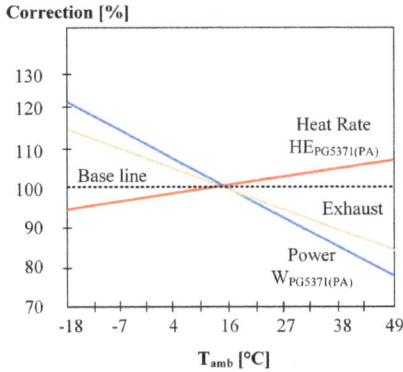

Fig. 7.19: Temperature correction plots for PG5371(PA) (after Brooks, 2000).

If the correction curves in the figure are assumed as straight lines, the correction equations for rated power and for heat rate can be written as follows (measurement units in square brackets):

$$W_{\text{PG5371(PA)}}[\text{MW}] = 29.036\ [\text{MW}] - 0.1894[\text{MW}/^{\circ}\text{C}] \times T_{\text{amb}}[^{\circ}\text{C}] \tag{7.25}$$

$$\text{HR}_{\text{PG5371(PA)}}[1] = 3.4094\ [1] + 0.0087[1/^{\circ}\text{C}] \times T_{\text{amb}}[^{\circ}\text{C}] \tag{7.26}$$

In Eq. (7.26), the heat rate here is dimensionless, with the meaning of MW/MW. Based on the correction plots for the rated power variation with the ambient temperature in Fig. 7.19, the fuel heat at full load can be expressed as a quadratic function of the ambient temperature in °C:

$$Q_{f,ma} = A_{\text{gt},t} + B_{\text{gt},t} \times T_{\text{amb}} + C_{\text{gt},t} \times T_{\text{amb}}^{2} \tag{7.27}$$

However, the coefficient before the second-order term is insignificant compared to the linear coefficients, as given in Tab. 7.1. The presented values indicate that the relationship, given in Eq. (7.27), is essentially linear.

Tab. 7.1: Coefficients for the temperature correction, for Eq. (7.27).

Coefficient	Value
$A_{\text{gt},t}$ (MW)	80.8917
$B_{\text{gt},t}$ (MW/°C)	0.695
$C_{\text{gt},t}$ (MW/°C^2)	4.033×10^{-4}

7.3.2.3.5 Estimating the condition of the exhaust gas

The gas turbine model described in the previous sections is capable of estimating the power generation by gas turbines having the fuel flow specified. However, the gas turbine exhaust is often connected to an HRSG. This requires that the model also estimates the mass flow rate, the temperature and the specific heat capacity of the exhaust stream.

The actual value of the specific heat capacity of a gas turbine exhaust depends on the used fuel and the fuel-to-air ratio, which determine the composition and the temperature of the stream. In general, the specific heat capacity of gas turbine exhausts is around 1–1.1 kJ/(kg ×°C) (Konstantin and Konstantin, 2018) for a wide variety of gas turbines and fuels. For most preliminary calculations, this value is precise enough. However, if the energy losses from a gas turbine need to be properly estimated, a more precise value of the exhaust gas heat capacity may be required. This may be necessary for simulation or operational optimisation of existing systems or at the phase of detailed system design and specification, when the main topology and equipment types have already been selected. The C_p value of the gas turbine exhaust can be estimated using process simulators.

Relatively small variations in the values of this parameter may result in significant differences in the calculated gas turbine energy losses. The reason for this is that in the energy balance equation, the term for the exhaust heat includes a product of the specific heat capacity and the mass flow rate of the exhaust. The former is a much smaller value, very close to unity. The latter is a value two orders of magnitude larger.

There are two possible ways to obtain an estimate of the exhaust gas mass flow rate. The first is to use the mass balance of the gas turbine and the other is to develop an explicit set of regression equations. Both approaches have their advantages and disadvantages. These options are discussed next.

Case 1: Exhaust mass flow from the mass balance

The mass flow rate of the exhaust can be calculated on the basis of the gas turbine mass balance. This modelling case requires the fuel-to-air ratio f to be provided as a specification. Its value may be constant or variable with respect to the current gas turbine load (Facchini, 1993). If the turbine efficiency has to be maximised, this requires a control strategy with variable fuel-to-air ratio (Kim and Hwang, 2006). The following equation gives the mass balance, solved for the exhaust flow:

$$m_{ex} = \frac{f+1}{f} \times m_f + m_{st} \tag{7.28}$$

This modelling approach requires knowledge of the values of f for any load. There is a danger if the precision of the data for the fuel-to-air ratio is poor. Because the fuel flow m_f is much smaller than the air flow m_a and the total flow of the exhaust m_{ex}, any error in the value of f may be multiplied by larger numbers leading to appreciable imprecision.

Case 2: Independent regression equations

The second option is to perform a separate regression analysis. In this way, the mass flow of the exhaust is directly calculated as a function of the rated power and the ambient temperature. First, the mass flow of the exhaust is related to the current shaft power. Facchini (1993) presents a plot of the mass flow rate of air through the compressor of a gas turbine, which is linear. Taking into account that the fuel flow rate is small compared to that of the air, it is reasonable to assume that the exhaust flow will also depend on the power generation at part-load linearly, where both coefficients in the equation depend on the ambient temperature:

$$m_{ex} = A_{mex} + B_{mex} \times W_{gt} \tag{7.29}$$

This modelling approach has the potential for greater precision. However, it has the disadvantage that it is more complex to implement. The first difficulty is that the model becomes bilinear in terms of the ambient temperature T_{amb} and the gas turbine shaft power W_{gt}, and the latter is itself dependent on T_{amb}. Another issue is the increased data requirement, to perform such a regression analysis.

The exhaust temperature can be calculated from the overall energy balance. While obtaining a precise estimate of the specific heat capacity of the exhaust is impossible using general manufacturer data ('Powering the World', 2016), an approximate estimate of the balance is possible, which in most cases yields satisfactory precision.

If the terms in the gas turbine energy balance are weighted, the most significant terms are the heat released by the fuel combustion, the heat contained in the exhaust and the produced shaft power (Fig. 7.20). All other terms are relatively less

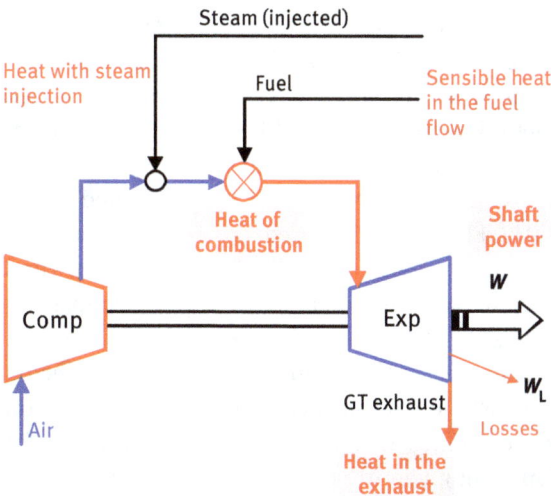

Fig. 7.20: Significance screening of the terms in the energy balance.

significant. The energy balance, consisting only of the significant terms, can be solved for the heat in the exhaust. This can provide an estimated imprecision of about 3 %. Better precision is possible, if data for the less significant terms is available. Using the simplified energy balance, the heat flow of the exhaust can be calculated from the following:

$$Q_{ex} = m_f \times NHV - W_{gt} \tag{7.30}$$

The temperature of the exhaust can be further calculated from the heat flow of the exhaust and the ambient temperature:

$$T_{ex} = T_{amb} + \frac{m_f \times NHV - W_{gt}}{m_{ex} \times C_{p,ex}} \tag{7.31}$$

7.3.2.4 Estimation of the greenhouse gas emissions

The fuel combustion in the gas turbine and the HRSG produces emissions of CO_2, SO_2 and NO_x (mainly NO and NO_2) (Habib et al., 2010). In the view of the ultimate necessity to reduce the global emissions of these gases, the mechanisms of their generation need to be understood.

The formation of carbon dioxide (CO_2) and sulphur dioxide (SO_2) during fuel combustion can be modelled as a complete conversion of the fuel carbon and sulphur contents. For these gases, constant emission factors are defined. Thus, the emissions are estimated as directly proportional to the amounts of fuel burnt.

The case with the nitrogen oxides is more complicated. These can be formed from fuel-bound nitrogen as well as by oxidation of the nitrogen contained in the combustion air. In this case, the amount of the formed NO_x depends largely on the firing temperature and the applied combustion and NO_x reduction technologies. The latter is an integral part of the gas turbine machine design. It is common for gas turbine manufacturers to specify guaranteed NO_x levels for gas turbines and combined cycle systems – for example for General Electric (Chase and Kehoe, 2000). However, due to the unknown composition of the exhaust gas, the estimate of the emissions of nitrogen oxides is not very precise. A methodology adopted by the United States Environmental Protection Agency (US EPA 1993) is based on experimental fuel-related coefficients. This is used to calculate the NO_x emissions from gas turbines and HRSGs. Any estimates of other fuels than hydrocarbon-based ones – for example biogas, containing some carbon monoxide (CO), must be treated with caution. However, as these fuels have much lower heating values (LHVs), it is expected that they should not produce higher NO_x emission levels than the hydrocarbon fuels. This NO_x estimation procedure defines NO_x emission indices, specific for each fuel, which relate the emission level given in *ppmvd* to mass of NO_x per unit burnt fuel (e.g. [kg NO_x]/[1,000 kg fuel]).

7.3.2.5 Summary of the model

The elements of the gas turbine model should be applied systematically, to produce consistent estimates. One possible algorithm, employing the gas turbine model, is shown in Fig. 7.21. The model requires the various regression coefficients and specifications of the ambient temperature and the fuel properties – including heating value. Finally, as it provides one degree of freedom, it requires either the fuel flow or the shaft power to be specified. The calculation estimates the intermediate parameters and produces a final estimate of the mass flow and temperature of the exhaust and an estimate of either the fuel flow or the shaft power, depending on the chosen specification.

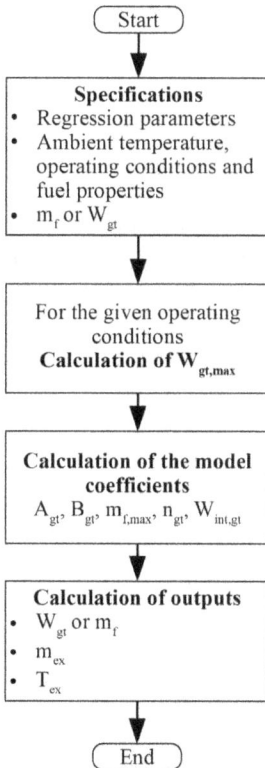

```
                    ┌─────────┐
                    │  Start  │
                    └────┬────┘
                         ▼
       ┌─────────────────────────────────┐
       │         Specifications          │
       │  • Regression parameters        │
       │  • Ambient temperature,         │
       │    operating conditions and     │
       │    fuel properties              │
       │  • m_f or W_gt                  │
       └────────────────┬────────────────┘
                        ▼
       ┌─────────────────────────────────┐
       │    For the given operating      │
       │           conditions            │
       │    Calculation of W_{gt,max}    │
       └────────────────┬────────────────┘
                        ▼
       ┌─────────────────────────────────┐
       │    Calculation of the model     │
       │           coefficients          │
       │  A_{gt}, B_{gt}, m_{f,max},     │
       │  n_{gt}, W_{int,gt}             │
       └────────────────┬────────────────┘
                        ▼
       ┌─────────────────────────────────┐
       │      Calculation of outputs     │
       │  • W_{gt} or m_f                │
       │  • m_{ex}                       │
       │  • T_{ex}                       │
       └────────────────┬────────────────┘
                        ▼
                    ┌─────────┐
                    │   End   │
                    └─────────┘
```

Fig. 7.21: Algorithm for gas turbine calculation.

7.4 Gas turbine performance evaluation at full load: a worked example

Gas turbines can use different gaseous and liquid fuels. They have different properties. From the point of view of gas turbine performance, the most significant property is the fuel heating value. It has been found that different fuel heating values influence gas turbine performance both at full load and at part load. Carcasci et al. (2000) provide such type of evaluation of single-shaft gas turbines. A similar evaluation has been presented in Colitto Cormacchione and Facchini (2001) but for gas turbines with multiple shafts.

This section presents an evaluation example for single-shaft gas turbines, based on the data from Carcasci et al. (2000). The paper investigates the variations of gas turbine power efficiency, power output, pressure ratio and exhaust temperature of four gas turbines. It relates these properties to the LHV of the fuel used. The study covers both full-load and part-load performance. The load ranges considered for each turbine are different, but do not fall below 69 % of the full load. The most likely reason for this may be that below this value the efficiency of gas turbines drops sharply and there is no interest in lower partial loads.

7.4.1 Data and calculation

The data specification starts with the heating value of the nominal fuel. It is selected as natural gas with $LHV_{Case1} = 48,608.7$ kJ/kg. For this fuel, the gas turbines have been simulated and evaluated. Table 7.2 lists the efficiency and fuel heat use at full load for the evaluated models. This is labelled as "Case 1".

Tab. 7.2: Gas turbine calculations for natural gas as fuel (Case 1).

Gas turbine model	W_{size}	Efficiency	Q_f
	[MW]	[1]	[MW]
GE MS9001E	127.22	0.338	376.4
KWU V64.3A	70	0.368	190.2
ABB GTX100	43	0.349	123.2
NP PGT10B	11.15	0.322	34.6

At the next step, the fuel heating value was reduced to 34 % of that of natural gas, corresponding approximately to the heating value of biogas, resulting in $LHV_{Case2} = 16,526.9$ kJ/kg. The simulations resulted in the values listed in Tab. 7.3. Cases 3 and 4 have been evaluated for 20 % and 15.22 %. The evaluation results are listed in Tabs. 7.4 and 7.5.

Tab. 7.3: Gas turbine calculations for Case 2 – fuel with 34 % of the heating value of Case 1.

Gas turbine model	W_{size}	Efficiency	Q_f
	[MW]	[1]	[MW]
GE MS9001E	133.6	0.345	386.7
KWU V64.3A	73.5	0.376	195.4
ABB GTX100	45.2	0.357	126.5
NP PGT10B	11.7	0.329	35.6

Tab. 7.4: Gas turbine calculations for Case 3 – fuel with 20 % of the heating value of Case 1.

Gas turbine model	W_{size}	Efficiency	Q_f
	[MW]	[1]	[MW]
GE MS9001E	142.5	0.354	402.2
KWU V64.3A	78.4	0.386	203.3
ABB GTX100	48.2	0.366	131.6
NP PGT10B	12.5	0.338	37.0

Tab. 7.5: Gas turbine calculations for Case 4 – fuel with 15.22 % of the heating value of Case 1.

Gas turbine model	W_{size}	Efficiency	Q_f
	[MW]	[1]	[MW]
GE MS9001E	142.5	0.3542	402.2
KWU V64.3A	78.4	0.3857	203.3
ABB GTX100	48.2	0.3659	131.6
NP PGT10B	12.5	0.3376	37.0

7.4.2 Full-load performance trends with different fuels

The data from Carcasci et al. (2000) have been analysed, plotting the gas turbine performance data at full load (Fig. 7.22), for the investigated values of the fuel heating value. They have been regressed for obtaining the coefficients for the full-load performance of gas turbines according to Eq. (7.21). The regression equations are shown in the figure. The values of the regression coefficients show some minor variation depending on the fuel LHV.

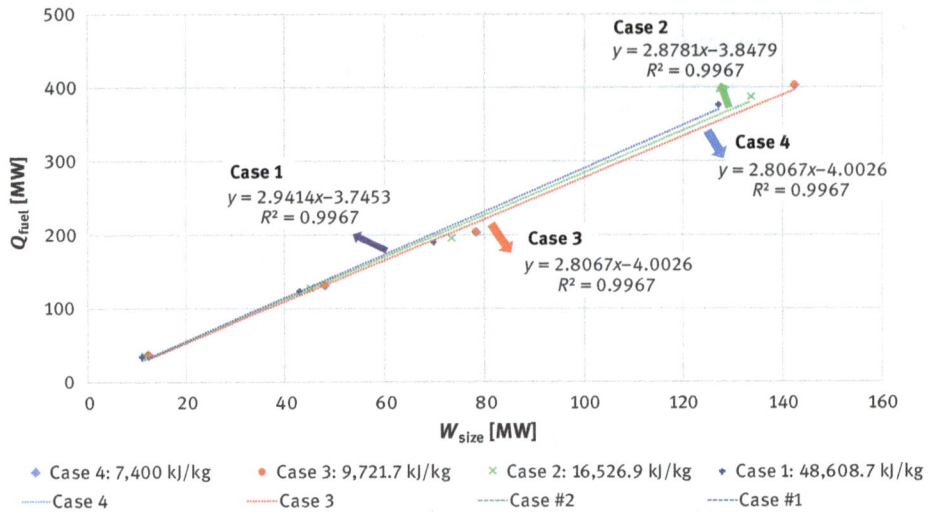

Fig. 7.22: Regression lines correlating the intercept ratio to full-load power generation, for single-shaft gas turbines simulated in Carcasci et al. (2000).

7.4.3 Part-load performance with different fuels

Another important issue for the gas turbine model is to characterise the part-load performance. To a large degree, this is determined by the gas turbine intercept ratio L_{gt}. The data for the gas turbines from Carcasci et al. (2000) have been analysed. The intercept ratios can be related to the gas turbine size or to the turbine pressure ratio. For this example, the first option has been chosen, as manufacturers tend to use quite different pressure ratios for turbines of similar sizes. The analysis results are shown in Tab. 7.6 and Fig. 7.23.

Tab. 7.6: Analysis of the intercept ratio based on the data in Carcasci et al. (2000).

GT model	W (ISO) [MW]	Pressure ratio [1]	L_{gt} (natural gas) [1]	L_{gt} (biomass synthesis gas) [1]
GE MS9001E	127.22	12.33	0.06805	0.06494
KWU V64.3A	70	16.6	0.08128	0.07336
ABB GTX100	43	20	0.09428	0.07988
NP PGT10B	11.15	15.6	0.13591	0.11138

Fuel heating values [kJ/kg]	
Natural gas	Biomass synthesis gas
48,608.7	7,340.0

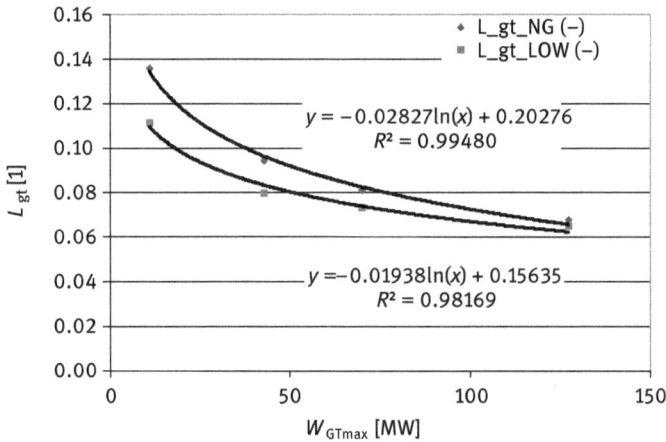

Fig. 7.23: Regression lines correlating the intercept ratio to full-load power generation, for single-shaft gas turbines simulated in Carcasci et al. (2000).

The results indicate that the gas turbine intercept ratio L_{gt} changes non-linearly with the turbine size. Moreover, a different curve is obtained for every fuel, based on its heating value. The data in Tab. 7.6 and the curves in Fig. 7.23 show that the influence of different fuels on the gas turbine intercept ratio is significant.

7.5 Summary

This chapter presents gas turbines – a class of machines that have both firing and fluid expansion in a turbine section. The main principle of operation and types of gas turbine arrangements – closed and open – which implement the Brayton cycle have been presented. The thermodynamic basics of the Brayton cycle have been explained. This is followed by a presentation of the main performance trends of gas turbines, and an overview of the market – suppliers and typical performance of the offered units. The various gas turbine applications have been discussed – dedicated power applications with heat regeneration for maximum power extraction, as well as the cogeneration connection mode and combined cycles. The main use of gas microturbines is for CHP and backup power units. An advanced cycle for power storage, based on gas turbine technology, has also been presented. This is followed by the presentation of the gas turbine model for utility system simulation and optimisation.

Further information on gas turbines can be found in dedicated books such as the one on industrial gas turbines (Razak, 2007), detailed component consideration and machine design (Schobeiri, 2018), propulsion applications (Mattingly et al., 2016) and HRSGs (Ganapathy, 2003). One can also refer to the chapter on gas turbines in the *Energy Conversion* book by Goswami and Kreith (2017). Supplier catalogue data can be also consulted, for example General Electric (2020a) or Capstone (Capstoneturbine.com., 2020) as well as the dedicated magazine-type periodicals of market data like *Gas Turbine World* (Pequot Publishing Inc, 2019).

Nomenclature

Symbol	Measurement unit	Description
A_{gt}	MW	Regression parameter (fixed term) correlating the full-load performance across a selection of gas turbines
$A_{gt,t}$	MW	Fixed term coefficient in the regression equation of gas turbine performance with ambient temperature variation
A_{mex}	kg/s	Fixed term coefficient in the correlation equation for the gas turbine exhaust flow
B_{gt}	[1]	Regression parameter (slope) correlating the full-load performance across a selection of gas turbines
$B_{gt,t}$	MW/°C	Coefficient of the linear term in the regression equation of gas turbine performance with ambient temperature variation
B_{mex}	kg/MJ	Coefficient in of the linear term in the correlation equation for the gas turbine exhaust flow
$C_{gt,t}$	MW/°C^2	Coefficient of the quadratic term in the regression equation of gas turbine performance with ambient temperature variation
c_p	kJ/kg/K; kJ/kg/°C	Specific heat capacity
$c_{p,ex}$	kJ/kg/K; kJ/kg/°C	Specific heat capacity of the gas turbine exhaust
F	[1]	Fuel-to-air ratio
h_{1c}	kJ/kg	Specific stagnation enthalpy before the compressor
h_{2c}	kJ/kg	Specific stagnation enthalpy after the compressor
h_{3c}	kJ/kg	Specific stagnation enthalpy after the heater or the combustion chamber
h_{4c}	kJ/kg	Specific stagnation enthalpy after the expander
h_a	kJ/kg	Specific enthalpy of the air entering the gas turbine
h_{ex}	kJ/kg	Specific enthalpy of the gas turbine exhaust flow
h_{st}	kJ/kg	Specific enthalpy of the steam injected to the gas turbine
$h_{\{#\}c}$	kJ/kg	Specific stagnation enthalpies
HR	[1]	Heat rate of a gas turbine
$HR_{PG5371(PA)}$	[1]	Heat rate of the gas turbine model PG5371(PA) by General Electric
k	[1]	Adiabatic index for air
L_{gt}	[1]	Gas turbine intercept ratio
LHV. NHV	kJ/kg	Lower/Net Heating Value
m_a, m_{air}	kg/s	Mass flow rate of air intake to a gas turbine
m_{co}	kg/s	Cooling airflow of a gas turbine
m_{ex}	kg/s	Exhaust flow rate of the gas turbine

(continued)

Symbol	Measurement unit	Description
m_f, m_{fuel}	kg/s	Fuel consumption of the gas turbine
$m_{f,max}$	kg/s	Maximum fuel consumption of the gas turbine
m_s	kg/s	Mass flow rate of the air bypass to the turbine seals
m_{st}	kg/s	Mass flow rate of the steam injection
m_{SF}	kg/s	Mass flow rate of the fuel for supplementary firing
n_{gt}	MJ/kg, MWh/t	Slope coefficient of the Gas Turbine Willans Line
p_1	bar	Inlet pressure of the Brayton cycle (Fig. 7.3, State 1)
p_2	bar	High pressure of the Brayton cycle (Fig. 7.3, State 2)
q_{2-3}	kJ/kg	Specific heat supplied to the Brayton cycle
q_{4-1}	kJ/kg	Specific heat rejected from the Brayton cycle
Q_{ex}	kW; MW	Heat flow carried with the gas turbine exhaust
Q_f	kW; MW	Heat supplied by burning the gas turbine fuel at any load
$Q_{f,max}$	kW; MW	Heat supplied by burning the gas turbine fuel at full load
s	kJ/kg/K	Specific entropy
s	[1]	Steam-to air ratio
t_1	°C	Inlet air temperature of the Brayton cycle (Fig. 7.3 State 1)
t_3	°C	Air temperature after the combustion chamber of Brayton cycle (Fig. 7.3 State 1)
t_s	°C	Air temperature at the compressor suction point (Fig. 7.6)
t_4	°C	Air temperature at state 4 of the gas turbine powered combined cycle (Fig. 7.9)
T_{amb}	°C	Ambient temperature
T_b	°C	Maximum air temperature of the steam cycle (Fig. 7.9)
$T_{b'}$	°C	Constraint temperature can form a Pinch Point C with the gas turbine exhaust inside the HRSG evaporator section (Fig. 7.9)
T_{ex}	°C	Temperature of the exhaust
w, w_{sp}	kJ/kg	Net specific work of the Brayton Cycle
W	kW; MW	Power delivered to the users (Fig. 7.4)
w_c	kJ/kg	Specific work performed by the compressor
w_e	kJ/kg	Specific work of the working gas in the expander
W_{gc}	kW; MW	Power output of the gas turbine system at the generator contacts
W_{gt}	kW; MW	Gas turbine shaft power
$W_{gt,int}$	kW; MW	Intercept coefficient of the Gas Turbine Willans Line
$W_{gt,max}$	kW; MW	Maximum shaft power of gas turbine
W_{hc}	kW; MW	Internal power use of the gas turbine system for driving auxiliary equipment
$W_{PG5371(PA)}$	MW	Rated power of the model PG5371(PA) by General Electric
W_{size}	MW	Rated power of a gas turbine – in the case study
ΔT_1	°C	Temperature difference between Brayton Cycle exhaust temperature (T_4) and the superheated steam temperature of the Rankine cycle (Fig. 7.9)
ΔT_2	°C	Temperature difference under the condition of $T_{b'}$ of the Rankine cycle and the gas turbine exhaust profile (Fig. 7.9)

(continued)

Symbol	Measurement unit	Description
ΔT_3	°C	Temperature difference between exhaust gas exit from the HRSG and the boiler feed water inlet (Fig. 7.9)
η_{cycle}	[1]	Efficiency of the Brayton cycle
η_{gc}	[1]	Efficiency to the point after the power generator
η, η_{gt}	[1]	Efficiency of gas turbines
$\eta_{gt,max}$	[1]	Gas turbine efficiency at full load

References

Aguilar, O., Perry, S.J., Kim, J.-K., and Smith, R. 2007. Design and Optimization of Flexible Utility Systems Subject to Variable Conditions: Part 1: Modelling Framework. Chemical Engineering Research and Design 85(8): 1136–1148. https://doi.org/10.1205/cherd06062.

Allforpower.cz. 2017. Siemens: Red Mill technologies overhaul (Siemens: Generálka technologií Červeného mlýna). 1 June 2017. http://www.allforpower.cz/clanek/siemens-generalka-techno logii-cerveneho-mlyna/, accessed 24/03/2020.

Andreini, A. and Facchini, B. 2004. Gas Turbines Design and Off-Design Performance Analysis With Emissions Evaluation. Journal of Engineering for Gas Turbines and Power 126(1): 83–91. https://doi.org/10.1115/1.1619427.

Aviso, K.B. and Tan, R.R. 2018. Fuzzy P-Graph for Optimal Synthesis of Cogeneration and Trigeneration Systems. Energy 154 (July): 258–268. https://doi.org/10.1016/j. energy.2018.04.127.

Balachandran, P. 2013. Engineering Fluid Mechanics. New Delhi, India: PHI Learning Private Ltd.

Basha, M.S., Shaahid, M., and Al-Hadhrami, L. 2012. Impact of Fuels on Performance and Efficiency of Gas Turbine Power Plants. Energy Procedia, 2011 2nd International Conference on Advances in Energy Engineering (ICAEE), 14 (January): 558–565. https://doi.org/10.1016/j. egypro.2011.12.975.

Brooks, F.J. 2000. GE Gas Turbine Performance Characteristics. GER-3567H, https://www.ge.com/ content/dam/gepower-pgdp/global/en_US/documents/technical/ger/ger-3567h-ge-gas-turbine-performance-characteristics.pdf, accessed 11/04/2019.

Capstone. 2019. Capstone Microturbines to Convert Biogas Into Heat and Power for Irvine Water Recycling Plant Capstone Turbine Corporation (CPST). https://www. capstoneturbine.com/news/press-releases/detail/3406/, accessed 22/08/2019.

Capstoneturbine.com. 2020. C30: Capstone Turbine Corporation (CPST). https://www.capstonetur bine.com/products/c30, accessed 24/03/2020

Carcasci, C., Colitto Cormacchione, N.A., and Facchini, B. 2000. Single Shaft Gas Turbine Comparison Using Low BTU Fuel (Bio-Fuel) and Part Load Control Systems. In: Proceedings of Powergen Europe 2000 Congress, June 2000, Helsinki, Finland.

Chase, D.L. and Kehoe, P.T. 2000. GE Combined Cycle Product Line and Performance. General Electric Reference Library – GER-3574G. October 2000. https://www.ge.com/content/dam/ gepower-pgdp/global/en_US/documents/technical/ger/ger-3574g-ge-cc-product-line-performance.pdf, accessed 12/04/2019

Colitto Cormacchione, N.A. and Facchini, B. 2001. Multiple Shafts Gas Turbines Comparison Using Low Btu Fuel (Bio-Fuel) and Part Load Control Systems. In Proceedings of Powergen Europe 2001 Congress, 29-31 May 2001, Brussels, Belgiu Brussels, Belgium.

DEHN + SÖHNE. 2018. DEHN Protects. Open Grid Europe's Compressor Station in Werne. 2018. https://www.dehn-international.com/sites/default/files/media/files/2018-08/ref051-e-werne-realisierung.pdf, accessed 22/10/2020.

do Nascimento, M.A.R., de Oliveira Rodrigues L., dos Santos E.C., Gomes E.E.B., Dias F.L.G., Velásques E.I.G., and Carrillo R.A.M. 2013. Micro Gas Turbine Engine: A Review. In: Progress in Gas Turbine Performance, Benini E., London. UK: IntechOpen. https://doi.org/10.5772/54444.

Envision Intelligence. 2020. Gas Turbine Manufacturers Market Share. Envision Intelligence. https://www.envisioninteligence.com/blog/gas-turbine-manufacturers-market-share/. accessed 06/03/2020

Facchini, B. 1993. A Simplified Approach to Off-Design Performance Evaluation of Single-Shaft Heavy Duty Gas Turbines. In: Proceedings of the 7th Congress on gas turbines in cogeneration and utility; industrial and independent power generation, COGEN TURBO POWER '93; 1993; Bournemouth, UK, 8:189–98. ASME -PUBLICATIONS- IGTI.

Fortune Business Insights. 2019. Microturbine Market Size, Share, Industry Report – 2026. July 2019. https://www.fortunebusinessinsights.com/industry-reports/microturbine-market-100514, 23/03/2020.

Ganapathy, V. 2003. Industrial Boilers and Heat Recovery Steam Generators: Design, Applications, and Calculations. New York, USA: Marcel Dekker.

Gas Turbines. Procurement. Part 2: Standard Reference Conditions and Ratings. 2008. London, United Kingdom: BSI British Standards. https://doi.org/10.3403/BSISO3977.

General Electric. 2020a. Combined & Simple Cycle Power Plant Solutions | GE Power. https://www.ge.com/power/gas/power-plants, accessed 23/03/2020.

General Electric. 2020b. Mobile Power Plants & Temporary Power Generation | GE Power. https://www.ge.com/power/applications/fast-power, accessed 23/03/2020.

General Electric. 2020c. World's Most Efficient Combined-Cycle Power Plant | GE Power. 23 March 2020. https://www.ge.com/power/about/insights/articles/2016/04/power-plant-efficiency-record, accessed 22/03/2020.

General Electric Company. 2015. Anti-Icing System, GEA32069A. August 2015. https://www.ge.com/content/dam/gepower-pgdp/global/en_US/documents/technical/upgrade-documents/GEA32069A-AntiIcing-US-R1-LR.pdf, accessed 22/03/2020.

Global Market Insights, Inc. 2020. Aeroderivative Gas Turbine Market Size and Share | Statistics – 2026. Global Market Insights, Inc. February 2020. https://www.gminsights.com/industry-analysis/aeroderivative-gas-turbine-market, accessed 24/03/2020.

Goswami, D.Y. and Kreith, F. 2017. Energy Conversion. 2nd ed. *Mechanical and Aerospace Engineering Series*. Boca Raton, FL, USA: CRC Press, Taylor & Francis Group.

Gupta, K.K., Rehman, A., and Sarviya, R.M. 2010. Bio-Fuels for the Gas Turbine: A Review. Renewable and Sustainable Energy Reviews 14(9): 2946–2955. https://doi.org/10.1016/j.rser.2010.07.025.

Habib, Z., Parthasarathy, R., and Gollahalli, S. 2010. Performance and Emission Characteristics of Biofuel in a Small-Scale Gas Turbine Engine. Applied Energy 87(5): 1701–1709. https://doi.org/10.1016/j.apenergy.2009.10.024.

Igoe, B. and Welch, M. 2014. Fuels, Combustion & Environmental Considerations in Industrial Gas Turbines. Power Engineering 16 (May): 2014. https://www.power-eng.com/2014/05/16/fuels-combustion-environmental-considerations-in-industrial-gas-turbines/, accessed 23/03/2020.

Kawasaki Gas Turbine Standby Generator Sets https://global.kawasaki.com/en/energy/pdf/20141030Standby.pdf, accessed 22/03/2020.

Kim, T. and Hwang, S. 2006. Part Load Performance Analysis of Recuperated Gas Turbines Considering Engine Configuration and Operation Strategy. Energy 31(2–3): 260–277. https://doi.org/10.1016/j.energy.2005.01.014.

Konstantin, P. and Konstantin, M. 2018. The Power Supply Industry: Best Practice Manual for Power Generation and Transport, Economics and Trade. Cham, Switzerland: Springer International Publishing AG.

Kroos, K. and Potter, M.C. 2015. Thermodynamics for Engineers. Stamford. CT, USA: Cengage Learning.

Kuczyński, Szymon, Mariusz Łaciak, Andrzej Olijnyk, Adam Szurlej, and Tomasz Włodek. 2019. 'Techno-Economic Assessment of Turboexpander Application at Natural Gas Regulation Stations'. Energies 12 (4): 755. https://doi.org/10.3390/en12040755.

Li, Y., Xu, F., and Gong, C. 2017. System Optimization of Turbo-Expander Process for Natural Gas Liquid Recovery. Chemical Engineering Research and Design 124 (August): 159–169. https://doi.org/10.1016/j.cherd.2017.06.001.

Manninen, J. 1999. Flowsheet Synthesis and Optimisation of Power Plants. PhD Thesis, Manchester, UK: University of Manchester Institute of Science and Technology.

Maréchal, F. and Kalitventzeff, B. 1998. Process Integration: Selection of the Optimal Utility System. Computers & Chemical Engineering, 22 (March): S149–56. https://doi.org/10.1016/S0098-1354(98)00049-0.

Mattingly, J.D., Boyer, K.M., and Ohain, H.V. 2016. Elements of Propulsion: Gas Turbines and Rockets. Second ed. AIAA Education Series. Reston, VA, USA: American Institute of Aeronautics and Astronautics, Inc.

Mitsubishi Hitachi Power Systems Ltd. 2020. 'Product Lineup | Gas Turbines | Mitsubishi Hitachi Power Systems Ltd. Mitsubishi Hitachi Power Systems Official Website. https://www.mhps.com/products/gasturbines/lineup/, accessed 13/03/2020.

Mokhatab, S., Mak, J.Y., Valappil, J.V., and Wood, D.A., eds. 2014. Chapter 7 – LNG Plant and Regasification Terminal Operations. In: Handbook of Liquefied Natural Gas, 297–320. Boston, USA: Gulf Professional Publishing. https://doi.org/10.1016/B978-0-12-404585-9.00007-6.

Pequot Publishing Inc. 2019. 2019 Performance Specs, 35th ed. Gas Turbine World, July-August 2019. https://gasturbineworld.com/shop/performance-specs/2019-performance-specs-35th-edition/, accessed 23/03/2020.

'Powering the World'. 2016. 2016. https://www.ge.com/content/dam/gepower-pgdp/global/en_US/documents/product/2016-gas-power-systems-products-catalog.pdf, accessed 22/10/2020.

Proctor, D. 2018. Another World Record for Combined Cycle Efficiency. POWER Magazine (blog). 1 October 2018. https://www.powermag.com/another-world-record-for-combined-cycle-efficiency/, accessed 23/03/2020..

Rahman, M. and Malmquist, A. 2016. Modeling and Simulation of an Externally Fired Micro-Gas Turbine for Standalone Polygeneration Application. Journal of Engineering for Gas Turbines and Power 138(11): 112301–112301–15. https://doi.org/10.1115/1.4033510.

Razak, A.M.Y. 2007. Industrial Gas Turbines: Performance and Operability. Boca Raton, FL, USA: Woodhead Publishing Limited.

RWE Power AG. 2010. ADELE – Adiabatic Compressed-Air Energy Storage for Electricity Supply http://www.rwe.com/web/cms/mediablob/en/391748/data/235554/1/rwe-power-ag/press/company/Brochure-ADELE.pdf, accessed 24/03/2020.

Schobeiri, M.T. 2018. Gas Turbine Design, Components and System Design Integration. Cham: Springer International Publishing. https://doi.org/10.1007/978-3-319-58378-5.

Shang, Z. 2000. Analysis and Optimisation of Total Site Utility Systems. PhD Thesis, Manchester, UK: University of Manchester Institute of Science and Technology.

Siemens, AG 2020. Gas Turbines | Power Generation | Siemens. https://new. siemens.com/cn/en/products/energy/power-generation/gas-turbines.html, accessed 13/03/2020.

Škorpík, J. 2011a. Heat Turbines and Turbocompressors. Transformační Technologie, February. https://www.transformacni-technologie.cz/en_23.html, accessed 25/03/2020.

Škorpík, J. 2011b. Gas Turbine as Part of Process Units (Plynová Turbína v Technologickém Celku). Transformační Technologie. https://www.transformacni-technologie.cz/27.html, accessed 22/03/2020 (in Czech).

Škorpík, J. 2020a. Steam turbines as part of overall plants (Parní turbína v technologickém celku). Energy Conversion Technologies (Transformační technologie), January. https://www. transformacni-technologie.cz/25.html, accessed 26/02/2020 (in Czech).

Škorpík, J. 2020b. Thermal turbines and turbocompressors (Tepelné turbíny a turbokompresory). Transformační technologie, March. https://www.transformacni-technologie.cz/23.html, accessed 09/03/2020 (in Czech).

Soares, C.M. 2020. Gas Turbines in Simple Cycle & Combined Cycle Applications https://netl.doe. gov/sites/default/files/gas-turbine-handbook/1-1.pdf, accessed 10/03/2020.

US EPA, Emission Standards Division. 1993. Alternative Control Techniques Document NO_x Emissions From Stationary Gas Turbines, EPA 453/R-93-007. US EPA. https://nepis.epa.gov/Exe/ZyNET. exe/2000HING.TXT? ZyActionD=ZyDocument&Client=EPA&Index=1991+Thru+1994&Docs= &Query=&Time=&EndTime=&SearchMethod=1&TocRestrict=n&Toc=&TocEntry=&QField= &QFieldYear=&QFieldMonth=&QFieldDay=&IntQFieldOp=0&ExtQFieldOp=0&XmlQuery=&File=D %3A%5Czyfiles%5CIndex%20Data%5C91thru94%5CTxt%5C00000014%5C2000HING.txt&User= ANONYMOUS&Password=anonymous&SortMethod=h%7C-&MaximumDocuments= 1&FuzzyDegree=0&ImageQuality=r75g8/r75g8/x150y150g16/i425&Display= hpfr&DefSeekPage=x&SearchBack=ZyActionL&Back=ZyActionS&BackDesc=Results% 20page&MaximumPages=1&ZyEntry=1&SeekPage=x&ZyPURL#, accessed 12/04/2019.

Varbanov, P.S., Doyle, S., and Smith, R. 2004. Modelling and Optimization of Utility Systems. Chemical Engineering Research and Design 82(5): 561–578. https://doi.org/10.1205/ 026387604323142603.

Part 3: **Utility networks as a whole – modelling and optimising utility systems**

8 Steam network modelling

With the basic understanding of steam turbines, it is possible to create simple utility network models for evaluating and optimising their performance. It is necessary to identify the degrees of freedom and trade-offs in utility plants to understand the system-level features.

8.1 Contribution to sustainability

As indicated in Chapter 3, the efficiency of operation on industrial sites is directly linked to their resource consumption and environmental impacts. The inputs to a utility system include mainly fuels and water, while some sites may also harvest renewable non-fuel energy sources like solar. The outputs are mainly flue gases, water evaporation and aqueous waste streams (purges) in the form of boiler blowdown and cooling tower blowdown. These utility system interfaces give rise to various footprints linked to the environmental impact and sustainability contribution of the industrial sites.

Sustainability in the general sense has three main aspects – environmental, economic and societal. Industry penetrates all these spheres, but specifically utility systems interact mainly with the environment. This makes utility systems important in terms of maximising their efficiency, which simultaneously minimises the resource depletion and environmental pollution.

The emissions of oxides of carbon, sulphur and nitrogen, emitted by fuel combustion in the utility systems, cause various harmful effects on the environment and the quality of human life mainly by provoking global warming and acid rain. In addition, the emissions of particulates also cause serious health issues and discomfort to the personnel and to people situated nearby the industrial sites.

In order to cope with the problem efficiently, the amounts of these emissions need to be reduced significantly. There are principally two ways of incorporating emission reduction into a process optimisation model:
(a) Impose costs related to emissions. These may be financial penalties such as carbon tax, cost of abatement (sequestration) or simply a price that an industrial company is willing to spend on dealing with these emissions.
(b) Enforce specified maximum allowed emission rates.

The resource withdrawal and pollutant impacts on the environment are quantified by relevant footprints related to industrial processes. Čuček et al. (2012) presented a comprehensive overview of the developed footprint tools, such as Greenhouse Gas Footprint (GHGFP), Water Footprint (WFP) and Nitrogen Footprint (NFP). Intensive indicators, footprints per unit generation, can also be defined.

https://doi.org/10.1515/9783110630091-008

The most common footprints are the GHGFP and the WFP, which are scalable and can be applied even for international trade evaluation (Liu et al., 2017). While the GHGFP is a well-established metric, the WFP concept is less known. The WFP of a product is the volume of freshwater used to produce it or polluted, measured over the full supply chain (Hoekstra et al., 2012). It is a multidimensional indicator, showing water consumption volumes by source and polluted volumes by type of pollution.

Maximising the system efficiency naturally acts upon reduction of the footprints. One measure tends to reduce both Greenhouse Gases (GHGs) and WFPs – limiting on-site power generation to the cogeneration mode – that is, only to the extent enabled by the process steam demands. Such a constraint should be removed only in cases when there is no grid power supply available or if the grid power supply comes with lower efficiency and higher footprint intensities.

The ultimate societal goal is to achieve low GHG emission levels (ideally zero). However, there is no consensus on the matter among experts. The discussions carry on while the preference towards the "cap-and-trade" mechanism (Carl and Fedor, 2016) prevails in existing regulations.

8.2 Optimisation of existing site utility systems

The utility system on a site needs to establish the balances of power and steam. They are separate balances, which are related to each other via the heat-based power generators (steam and gas turbines). The analysis of a utility system begins with the specifications. The latter usually includes:
- The process heating and cooling demands
- The site power demands, stemming from the process units and the need to run the utility system itself
- Pressure and temperature of the boiler steam
- The pressure of the condensing header
- For modelling purposes, usually only one deaerator is assumed. More deaerators could be selected in practice, for instance – one per boiler.
- Steam headers, steam turbine and any other equipment
- Pressures and temperatures for all remaining steam headers (mains)
- Topology connections between the elements – headers, turbines, boilers and deaerator
- Steam header, from which the steam for the deaerator is drawn. This is usually the lowest pressure (LP) steam main. This choice is supported by the frequent excess of steam on the lower pressure headers.

8.2.1 Problem description

The utility system is a key part of any processing sites. Power cogeneration is an essential feature of such systems, particularly from steam turbines or gas turbines, or a combination of both. On large processing sites, the cost of fuel and power can be significant together with the related emissions. Better management of the utility system can lead to significant cost and emission savings. Such savings can often be accomplished without capital expenditure by more effective day-to-day management of the system. Therefore, an optimisation model of utility systems can be an important tool in determining the strategy for cost and emission reduction (e.g. by improved Heat Integration or optimal distribution of the steam and fuel flows) on the site and for modifications to the utility system.

The models, described in Chapters 5–7, can be incorporated, along with those for the other system elements, into an overall model of the utility network. For existing sites, the degrees of freedom of the utility systems allow the significant opportunity for cost and footprint reduction through operational optimisation. This results from the complexity of the topology and interactions within such systems, the continually changing demands on the system and the complexity of cost tariffs for power and fuel. In this part, a procedure for optimisation of existing utility systems is presented. It uses the models discussed in Chapters 5–7 and algorithmic decomposition.

Traditionally, the objective of optimising a utility system is to minimise the system operating cost. This objective is in line with the need for emission and footprints reduction, of which the GHGs and the WFPs are the most significant. Within the context of operational optimisation, the main cost items come from the bills for fuel, water (fresh and discharged) and power import. Linked to the fuel use, a charge for the emitted GHG can be also applied. All these bills and the released emissions are proportional to the amounts of used resources – fuels and water, and to the imported power. Therefore, cost minimisation also tends to minimise the environmental impacts.

The optimisation task is carried out accounting for a number of parameter variations. The tariffs for the purchase of power from centralised power generation and the sale of the surplus power exported usually vary significantly throughout the year and during every day. In some circumstances, the fuel is also subject to variable tariffs. This means that various operating scenarios need to be considered.

8.2.2 Variation of the specifications with time

The discussed utility system models assume steady state. The set of specifications, defining the system state, includes the site operating parameters and market conditions. These conditions usually vary with time, which is the most important factor

The most important factors that change are:
- the electricity tariffs;
- the fuel feedstocks and their prices;
- the ambient conditions.

The variations of the market prices of electrical power deserve special consideration. Power supply companies tend to reflect their operating costs according to supply and demand, with the highest prices at times of highest demand. As a result, there are two types of electricity price variations:
- **Seasonal**. These variations are related to the changing ambient conditions. For temperate climate, usually there are two distinct price periods – summer and winter.
- **Current**. Supply companies usually divide every 24 h period into a number of load intensity periods – typically peak, semi-peak and off-peak time intervals, each with different power prices.
- Also, the price of imported power might differ from the price of exported power.

It is most convenient to model different tariffs by introducing a series of operating cases, usually labelled by periods (Time Slices). Each case is characterised by a different set of values for the system specifications. The impact of these varying site specifications on the various tasks concerning utility systems is different. For operational optimisation of existing utility systems, each individual operating case may be optimised separately. For synthesis and design, it is necessary to optimise the system topology, the equipment sizes and the operation specifications across all scenarios simultaneously. This results in multiperiod formulations for utility system synthesis – such as using fixed equipment efficiencies (Iyer and Grossmann, 1997) and accounting for the equipment load variation (Varbanov et al., 2005). The multiperiod optimisation for synthesis is discussed in more detail later.

8.2.3 Major degrees of freedom and trade-offs

The overall utility network can be partitioned into the main blocks, containing facilities with similar functions. These are firing machines (steam generators and gas turbines) and the steam distribution system.

8.2.3.1 Firing machines
Starting from the higher temperature parts of the system, the first degrees of freedom relate to the devices with combustion – gas turbines, boilers (fuel-based steam generators) and heat recovery steam generators (HRSGs) – see Fig. 8.1. Firing in utility systems is done mainly in gas turbines and boilers. A gas turbine can be

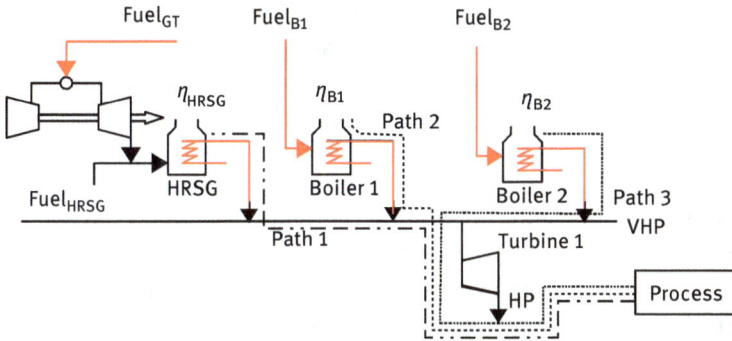

Fig. 8.1: Typical configuration of firing machines in a utility system.

connected to an HRSG, in which some supplementary firing can be applied. Every firing machine can use a different fuel or, perhaps, a different combination of fuels and has its own efficiency. That efficiency typically varies with the load. The reason for that variation lies in the inherent heat losses from a boiler regardless of its part loading rate. For a boiler to generate steam at the specified conditions, it has to be pressurised to the required pressure and heated to the required temperature regardless of whether it would generate steam at full capacity or at 50 % part load. Similar trends are at play in the efficiency relationship to the boiler size. The supplementary firing in the HRSG devices presents additional degrees of freedom.

8.2.3.2 Steam distribution system

Moving down the temperature scale, steam can be transferred between headers via let-down stations or via steam turbines (Fig. 8.2). Generally, steam turbines have different efficiencies, which depend on their sizes. As discussed in Chapter 6, most often larger turbines are more efficient than smaller ones. Also, the efficiency varies with part load.

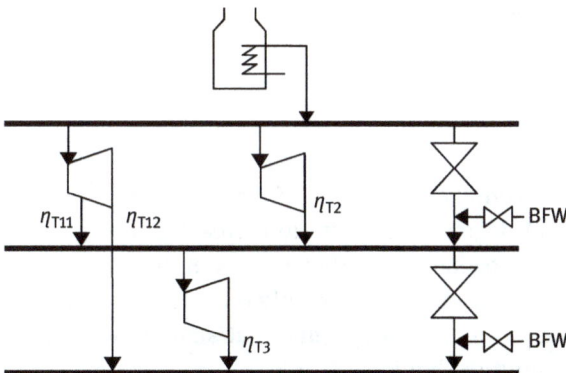

Fig. 8.2: Typical paths and trade-offs in the steam distribution system.

Letdown stations can be implemented as simple expansion valves, but due to the nearly adiabatic expansion of the steam in such valves, the usual arrangement of a letdown station is to combine the valve with boiler feedwater (BFW) injection for conditioning the steam before feeding it to the lower pressure steam main.

Given these potential steam expansion options, if there are two or more steam paths connecting two steam headers, this introduces additional degrees of freedom for internal flow distribution allowing to optimise the power generation. Usually, large letdown flows indicate missed opportunity for cogeneration. However, letdown stations play an essential role in maintaining the steam at each header at the desired (usually superheated) condition by expanding steam from higher pressure headers to lower pressure ones, nearly adiabatically. Also, letdown flows might allow a constraint in a steam turbine to be bypassed. For example, if all steam turbines between two headers are at their maximum flow rates, expanding steam through a letdown station may allow lower pressure steam turbines or process users to receive more steam.

Condensing steam turbines provide utility systems with degrees of freedom, generating extra power from higher steam flows (Fig. 8.3). However, the cogeneration efficiency of this arrangement is lower, compared with the combination of backpressure turbines and process steam use.

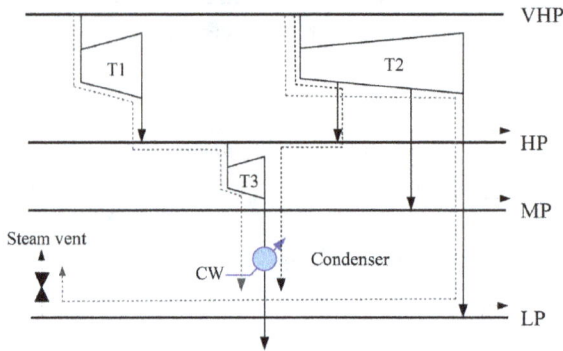

Fig. 8.3: Condensing steam turbines and vents.

As with condensing steam turbines, venting steam from LP headers (bottom-left of Fig. 8.3) also provides an additional degree of freedom to increase power generation. This option also results in lower cogeneration efficiency, even lower than that of the condensing turbine option. The reason is that not only the latent heat carried with the vented steam is lost, but also the heat of bringing that amount of water to the state of BFW up to the state of saturated liquid.

8.2.4 Indicators of environmental impact

The following equations are formulated to calculate GHG emissions from boilers. They are based on the assumption that the emissions are proportional to the fuel use, applying constant emission factors. For the CO_2 and SO_2 emissions, this is precisely the case, because the carbon and sulphur content in each fuel is usually at a fixed share:

$$m_{CO_2, BLR}(bo, btype, bf, sc) = m_{BF}(bo, btype, bf, sc) \times EF_{BF, CO_2}(bf) \qquad (8.1)$$

$$m_{SO_2, BLR}(bo, btype, bf, sc) = m_{BF}(bo, btype, bf, sc) \times EF_{BF, SO_2}(bf) \qquad (8.2)$$

where
- $m_{CO_2, BLR}$ [t/h] and $m_{SO_2, BLR}$ are the CO_2 and SO_2 emission flow rates
- m_{BF} [t/h] are the fuel flow rates
- EF_{BF, CO_2} [1] and EF_{BF, SO_2} [1] are the emission factors
- The flow rates are given over the domain (bo, btype, bf, sc) and the emission factors only over the domain (bf), where the indexes denote the following: bo is the boiler instance, btype is the boiler type, bf is the boiler fuel and sc is the scenario index, representing the various Time Slices in a multiperiod optimisation

For the NO_x emissions, the emission factors are more complicated to estimate, because their values result from both the fuel and the combustion conditions, where some of the nitrogen in the air may be oxidised. It can be, however, approximated and correlated as proportional to the fuel use:

$$m_{NO_x, BLR}(bo, btype, bf, sc) = m_{BF}(bo, btype, bf, sc) \times EF_{BF, NO_x}(bf) \qquad (8.3)$$

where $m_{NO_x, BLR}$ [t/h] is the NO_x emission flow rate and EF_{BF, NO_x} [1] is the corresponding emission factor. The domain indices are the same as for the other emissions.

8.2.5 Objective function components for the optimisation

For optimising the operation of utility systems, the main cost items are associated with
- fuel supply,
- utility cooling provided by the cooling subsystem,
- water supply and
- emissions – in the case that levies and other charges are instituted.

In the case of new system design or retrofit, the capital costs need to be accounted for. Further cost items can be added too, for example – cost of treating wastewater or for any environmental impacts according to local regulations.

Focusing on the operating costs, the most significant items are the cost of fuel, cooling water, emissions and makeup water for the boilers. The fuel cost (FC) [€/year] of all firing facilities (boilers, gas turbines, HRSG) is calculated as proportional to the consumed fuel flows m_{BF} [t/h] and summed up over those facilities, fuel types used and operating periods:

$$FC = \sum_{bo, btype} \sum_{bf} \sum_{sc} [m_{BF}(bo, btype, bf, sc) \times Price_{BF}(bf) \times YF(sc) \times HrYr] \qquad (8.4)$$

where the additional items are as follows:
- $Price_{BF}(bf)$ [€/t] is the price of the fuel denoted by index bf.
- $YF(sc)$ [1] = [y/y] is the fraction of the year within which period **sc** takes place.
- HrYr [h/y] is the number of hours per year for which the system operates. This can be equal to 8,760 h/y, which is the number of hours in a regular (non-leap) year overall, but this parameter may have a lower value if the site operates shorter during the year – for example, if there are stoppages for planned or unplanned maintenance.

The cost of cooling provided by the cooling system, assuming that all is performed using cooling water, is calculated in several steps. It starts from the estimation of the overall utility cooling demand of the site $Q_{CW}(sc)$. This can be modelled as a sum of the heat duties of the steam turbine condensers $Q_{COND}(stcond, sc)$ and all residual process cooling demands $Q_{DPC}(sc)$ that cannot be used for steam generation, calculated for each of the modelling time periods:

$$Q_{CW}(sc) = \sum_{stcond} [Q_{COND}(stcond, sc)] + Q_{DPC}(sc) \qquad (8.5)$$

Assuming that the cooling water is supplied and returned back to the cooling water system at given fixed temperatures $T_{CW, Spl}(sc)$ (°C) and $T_{CW, Ret}(sc)$ [°C], the mass flow rate of the cooling water $m_{CW}(sc)$ [t/h] is then estimated as:

$$m_{CW}(sc) = \frac{Q_{CW}(sc)}{C_{P, W} \times [T_{CW, Ret}(sc) - T_{CW, Spl}(sc)]} \qquad (8.6)$$

where $C_{P, W}$ = 4.19 kJ/kg/°C = 1.164×10^{-3} MWh/t/°C is the specific heat capacity of water.

The cost of the cooling water CWC [€/y] is estimated as the product of the cooling water flow rate, the price and the annual period fraction, summed up over all periods:

$$CWC = \sum_{sc} [m_{CW}(sc) \times HrYr \times YF(sc) \times Price_{CW}(sc)] \qquad (8.7)$$

where $Price_{CW}(sc)$ [€/t] is the price of cooling water. Note that the cooling water price is modelled as dependent on the period because the generation of cooling water has different efficiency and its own costs, which depends on the ambient conditions.

The cost of demineralised water DWC [€/y] for the make-up of losses from the steam system, mainly from boiler blowdown and condensate loss, can be calculated as a product of the unit price $Price_{DW}$ [€/t] and the annually weighted sum of the demands for demineralised water $m_{DW}(sc)$ [t/h] by the modelling periods (Time Slices):

$$DWC = Price_{DW} \times \sum_{sc} [m_{DW}(sc) \times HrYr \times YF(sc)] \qquad (8.8)$$

The items for emission costs of all firing devices (boilers, gas turbines, HRSG) are modelled as proportional to the emission flow rates defined in Eqs. (8.1)–(8.3). The emissions of each type are summed up by all firing machines and fuels, for each Time Slice, then the sum is multiplied by the corresponding charge price for the pollution and the results are summed, providing the emission cost EC(sc) [€/h] for each Time Slice and the annualised emission cost [€/y]:

$$EC(sc) = Price_{CO_2} \times \sum_{bo,btype} \sum_{bf} [m_{CO_2.BLR}(bo, btype, bf, sc)]$$

$$+ Price_{SO_2} \times \sum_{bo,btype} \sum_{bf} [m_{SO_2.BLR}(bo, btype, bf, sc)] \qquad (8.9)$$

$$+ Price_{NO_x} \times \sum_{bo,btype} \sum_{bf} [m_{NO_x.BLR}(bo, btype, bf, sc)]$$

$$EC_{Total} = HrYr \times \sum_{sc} [EC(sc) \times YF(sc)] \qquad (8.10)$$

where
- $Price_{CO_2}$ [€/t] is the price (tax) charged for CO_2 emissions
- $Price_{SO_2}$ [€/t] is the price (tax) charged for SO_2 emissions
- $Price_{NO_x}$ [€/t] is the price (tax) charged for NO_x emissions

8.2.6 Implications of the hardware models on the optimisation task and the solution procedure

The optimisation problems for existing utility systems, using the models developed in the previous sections, involve making continuous and discrete decisions. Discrete decisions relate to the operational status (on/off) of the different devices and the mode of interaction with the central electricity grid (import/export).

The energy balances of the system elements contain non-linear terms – mainly products of the specific enthalpies with mass flow rates and the non-linear correlations for estimating the water and steam properties.

Thus, the model defines the problem as one of mixed-integer non-linear programming (MINLP). If solved directly, this may lead to computation problems, inherent to MINLP. These include convergence speed and attainment of the global optimum. There are solvers on the market, such as BARON (BARON | The Optimization Firm, 2019) or SCIP (Vigerske and Gleixner, 2018), which claim to attain the global optima. They are used in research such as the industrial gas network optimisation (Puranik et al., 2016).

Using those solvers is an option, but some are commercial, and it is not always justified to obtain the licence. The free solvers are mostly experimental and not part of the widely used software. These reasons give rise to the idea of overcoming the non-linearity issues. One way to do this is the successive application of mixed-integer linear programming (MILP), as discussed next.

It is possible to decouple the overall optimisation procedure by alternating optimisation steps that use a simplified model with fixed enthalpies of the streams, and simulation steps that use the complete model to recalculate and adjust the enthalpies. The simplified model fixes the process stream enthalpies, resulting in a linear formulation. The linear optimisation is repeated, always followed by simulation and so on until convergence is achieved (Fig. 8.4). This is referred to as the successive MILP (SMILP) procedure (Varbanov et al., 2004b). The procedure in Fig. 8.4 is

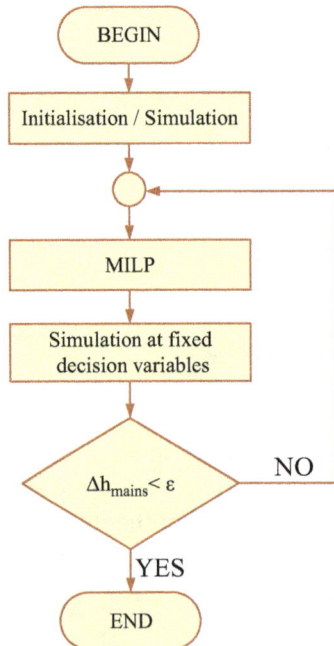

Fig. 8.4: SMILP procedure for utility system optimisation.

characterised by relatively rapid convergence. In the case of operational optimisation, it usually requires no more than five iterations to reach reasonably small error levels. The non-linearity effects, inherent to utility system parts, are discussed next.

8.2.6.1 Non-linearity resulting from firing machines

8.2.6.1.1 Gas turbines

One source of non-linearity stems from the gas turbine model. This is the calculation of the exhaust temperature from the overall energy balance of a gas turbine (eq. (7.16)). Both the mass flow and the temperature of the exhaust vary significantly at part-load. Therefore, it is not suitable to assume either of them as fixed optimisation parameters if the gas turbine load should be varied. A potential workaround stems from the practice of industrial sites to use gas turbines at maximum 2 to 3 load levels, keeping the load steady at each such regime. This allows simulating the gas turbine in advance at each potential set points and then providing these set points as possible discrete alternatives in the optimisation model.

8.2.6.1.2 Steam boilers

The steam boilers can be modelled in two alternative ways – with a constant efficiency model or with a variable efficiency model. The constant efficiency model (eq. (5.1)) is linear. However, a more realistic system model requires the variation of the boiler efficiency with the load. In order to avoid non-linearity, a different form of the aforementioned relationship is used, which is given in eq. (5.5). Instead of directly relating the efficiency to the steam generation flow rate, it calculates the fuel consumption. This results in an equation which contains terms proportional to the maximum load and the current load. The model is linear while accounting for the variation in efficiency with load, as well as for the boiler blowdown losses.

8.2.6.1.3 Non-linearity effects in the steam distribution sub-system

The main modelling challenges in this part of the system are the steam mains, the steam turbines, letdown stations and deaerators. Steam headers are modelled as a sequence of a mixer and a splitter. Both mass balances have linear forms:

$$\left[\sum_{\text{inlets}} m_{\text{inlet, hdr}}\right] - m_{\text{total, hdr}} = 0 \tag{8.11}$$

$$\left[\sum_{\text{outlets}} m_{\text{outlet, hdr}}\right] - m_{\text{total, hdr}} = 0 \tag{8.12}$$

However, the energy balance of the mixing contains bilinear terms for steam mass flows and enthalpies:

$$\left[\sum_{\text{inlets}} (m_{\text{inlet, hdr}} \times h_{\text{inlet, hdr}})\right] - m_{\text{total, hdr}} \times h_{\text{hdr}} = 0 \qquad (8.13)$$

8.2.6.1.4 Steam turbines

The steam turbine model, in its general form, features several sources of non-linearity. The first source is the calculation of the isentropic enthalpy change. This is a particular case of estimation of steam properties. The relationship between the turbine outlet pressure, the inlet steam conditions and the outlet ideal enthalpy from isentropic expansion is non-linear, usually resulting from regression equations of the steam properties. As a result, the Willans Line [Eq. (6.24)] becomes bilinear because its slope and the intercept become variables, yielding a product of two variables ($n_{\text{ST}} \times m_{\text{STEAM}}$). Also, the energy balance across a steam turbine (eq. 6.34), used to calculate the exhaust enthalpy, is bilinear.

8.2.6.1.5 Letdown valves

It is possible to implement several types of letdown valves (stations), as shown in Fig. 8.5: (a) injecting steam – for reconditioning outlets of steam turbines after exhausting superheat potential; (b) direct expansion (isenthalpic/adiabatic) – for adjusting steam mains' condition when necessary to increase the enthalpy of the lower main; (c) BFW injection valves (de-superheating). For adiabatic letdown valves, both the mass and energy balances are linear, as they are trivial – equating the inlet and outlet flows of mass and energy.

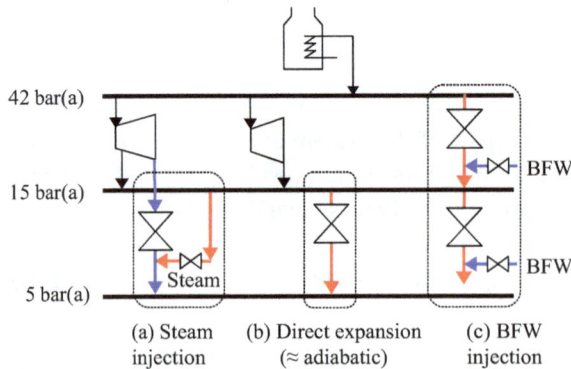

Fig. 8.5: Letdown valve arrangements.

However, for cases (a) and (c), the mixing of the main steam flow and the injected flow causes the valve energy balance to become bilinear, featuring a product of the steam mass flows and specific enthalpies:

$$m_{in, LD} + m_{inject, LD} - m_{out, LD} = 0 \qquad (8.14)$$

$$m_{in, LD} \times h_{in, LD} + m_{inject, LD} \times h_{inject, LD} - m_{out, LD} \times h_{out, LD} = 0 \qquad (8.15)$$

8.2.6.2 Deaerators

The deaerator, in general, also has a bilinear energy balance, caused by the product of the inlet steam mass flow and specific enthalpy, which have to be subjected to the SMILP decoupling scheme:

$$m_{stm, DA} + m_{DAF} - m_{BFW} - m_{V, DA} = 0 \qquad (8.16)$$

$$m_{stm, DA} \times h_{stm, DA} + m_{DAF} \times h_{DAF} - m_{BFW} \times h_{BFW} - m_{V, DA} \times - h_{V, DA} = 0 \qquad (8.17)$$

The deaerator feed m_{DAF} is formed by adding makeup water (demineralised water) to the condensate return. Additionally, the deaerator vent is assumed to be a small fraction of the steam supply – usually 3–5 %:

$$m_{V, DA} = VF_{DA} \times m_{stm, DA}, VF_{DA} \in [0.03...0.05] \qquad (8.18)$$

The vented steam is expected to be at saturation vapour condition and the BFW at the saturated liquid condition at the deaerator pressure:

$$h_{V, DA} = \text{Saturated vapour} \qquad (8.19)$$

$$h_{BFW} = \text{Saturated liquid} \qquad (8.20)$$

8.2.6.3 Simplifications applied to the optimisation model

In order to facilitate the optimisation steps and to apply the formulated successive MILP procedure, the following system properties are fixed during the optimisation steps.

1. The gas turbine exhaust temperature is fixed and the direct temperature calculations are excluded from the optimisation formulation.
2. The gas turbine performance coefficients (for the Willans Line) are fixed.
3. The temperatures of the steam headers and, consequently, their enthalpies are fixed.
4. The enthalpies and temperatures of steam turbine exhausts are fixed and the steam turbine energy balances are excluded from the optimisation formulation.
5. The enthalpies of the deaerator flows are fixed.

These alterations to the rigorous model convert it to a linear one, resulting in a MILP optimisation formulation. The non-linear effects are accounted for during the simulation steps of the overall optimisation procedure.

8.2.6.4 Feasibility and convergence of the optimisation procedure

Following the algorithm shown in Fig. 8.4, the utility network is initially simulated to determine the temperatures of the gas turbine exhaust, steam mains and steam turbine exhausts. These values are further used as parameters in the MILP formulation. After each optimisation step, the resulting flow pattern is resimulated with the rigorous model, calculating the actual enthalpies and temperatures. This process is repeated until the differences between successive values of the gas turbine exhaust and steam header temperatures are within a specified tolerance. This usually occurs within two to five iterations.

8.2.7 Additional model features

There are additional features that are useful to be included in the optimisation model. They include the internal power consumption and the handling of the resulting power balance, in relation to the power import and export decisions from and to the electricity grid.

8.2.7.1 Internal power consumption

There is some internal power consumption for running the utility system itself. The most significant items of this consumption are:
- boiler feedwater pumping;
- boiler fan operation and
- other internal power demands can be added to the account as long as they are found significant.

Boiler fans usually consume a small amount of power at approximately constant rate, regardless of the current steam load, as long as the boiler is running. This mode of operation can be modelled by introducing a parameter specification, denoted as $\mathbf{SPD_{BFan,boiler}}$ [MWh/t], which specifies the fan power consumption per unit flow for each boiler.

The pumping of boiler feed-water caters for the demands of several operations:
- lifting the water from the elevation level of the condensate line to the deaerator;
- raising the feedwater pressure from the deaerator conditions up to the boiler conditions;
- any other elevation/lifts between the deaerator and the boilers.

Additionally, there are certain pressure drops inherent to the pipes and all other devices within the utility system, which also needs to be overcome by the pumps. For a general optimisation model, an additional parameter specification can be defined: ΔP_{pipes}, [bar] allowing the design engineer to specify any additional pressure resistance to be accounted for by the utility system pumps.

For instance, the pumping pressure drop for elevation can be estimated by the rise in the height of the water. For every metre of height rise, the pressure drop is:

$$\Delta p_{specific} = \rho_{WATER} \times g = 1,000 \times 9.81 = 9,810 \left[\frac{Pa}{m}\right] \approx 10,000 \left[\frac{Pa}{m}\right] = 0.1 \left[\frac{bar}{m}\right] \quad (8.21)$$

where "g" (9.81 m/s^2) is the acceleration due to gravity.

As a result, the power consumption for feedwater pumping can be modelled as a product of the total sum of pressure drops and the feedwater volume flow:

$$PD_{FWP}[MW] = \frac{m_{BFW} \ [t/h]}{\rho_{WATER} \ [t/m^3]} \times \left\{ (P_{VHP} - P_{DA}) + \Delta P_{pipes} \right\} \ [bar] \times \frac{1}{36} \quad (8.22)$$

The total internal power demand is estimated by adding the fan requirements and the feedwater pump consumption:

$$PD_{INTERNAL}[MW] = \sum_{boilers} (SPD_{BFan, boiler} \times m_{boiler, total}) + PD_{FWP} \quad (8.23)$$

8.2.7.2 Power import and export by the site

At any given moment, the site can import or export power, but not both. First, the site power balance is established by summing up the process power demands and the internal power demand of the utility system, then subtracting the power generation by gas and steam turbines:

$$W_{Balance} = PD_{processes} + PD_{INTERNAL} - W_{GT, sum} - W_{ST, sum} \quad (8.24)$$

A positive value of the site power balance is defined as power import, and negative value is defined as power export.

Next, the balance $W_{Balance}$ is linked to the import and export flows. In an algebraic model, this would need defining an integer variable, reflecting the direction of the power flow across the site boundary. First, the site balance is connected to the power import and power export variables using the following equality constraint:

$$W_{Balance} - W_{Import} + W_{Export} = 0 \quad (8.25)$$

The remaining task is to ensure that the import and export flows are not simultaneously non-zero. The power flow direction variable y_{IMP} is defined, which takes the value of 1, if the mode is import and the value of 0, if the mode is export.

The following equation ensures that the import mode variable y_{IMP} assumes the value of 1 when the balance is positive:

$$W_{Balance} - y_{IMP} \times E_{VL} \leq 0 \tag{8.26}$$

where E_{VL} is a large value with the meaning of power upper bound. The value is usually set sufficiently higher than any value expected for the main variables of the same type. In this case, if the power balance of the site is expected within the range ±100 MW, the value of this bound would be set as $E_{VL} = 1,000$ MW.

The next item of the construct is to constrain the import mode variable y_{IMP} to assume the value of zero when the site power balance is negative:

$$W_{Balance} + (1 - y_{IMP}) \times E_{VL} \leq 0 \tag{8.27}$$

The next stage is to link the power import W_{Import} and power export W_{Export} variables to the import mode variable y_{IMP}:

$$W_{Import} - y_{IMP} \times E_{VL} \leq 0 \tag{8.28}$$

$$W_{Export} - (1 - y_{IMP}) \times E_{VL} \leq 0 \tag{8.29}$$

The parameter E_{VL} in Eqs. (8.28) and (8.29) can be replaced by the total sum of the site power demand and a power export upper limit, determined by a contract. The resulting power flows can then be linked to the relevant power import cost and power export revenue using the contractual prices.

Importing electricity also implies a certain amount of GHG emissions and other footprints caused by the equivalent generation by the electricity provider. These have to be taken into account within the overall environmental impact assessment of the site.

8.2.8 An illustrative example of utility system optimisation

The described models and reasoning can be used for optimising the operation of a utility system – be it an existing one or during process design. This section provides an example based on a real site, adapted from (Varbanov et al., 2004b). Consider the utility system in Fig. 8.6. This includes a gas turbine with a HRSG, two steam boilers, four steam turbines for electricity generation and two direct-drive steam turbines. Three fuels are available to the site, specified in Tab. 8.1. The site can import up to 50 MW of power at 0.045 $/kWh and it could export up to 10 MW of power at the price of 0.060 $/kWh. The site power demand excluding the drivers is 50 MW. Additional site configuration data are given in Tab. 8.2. Table 8.3 lists the values of the steam turbine performance parameters. Note that the steam turbines are labelled as DRV1 and DRV2. These are direct-drive turbines, which means that they have a fixed load, which cannot be changed during the optimisation.

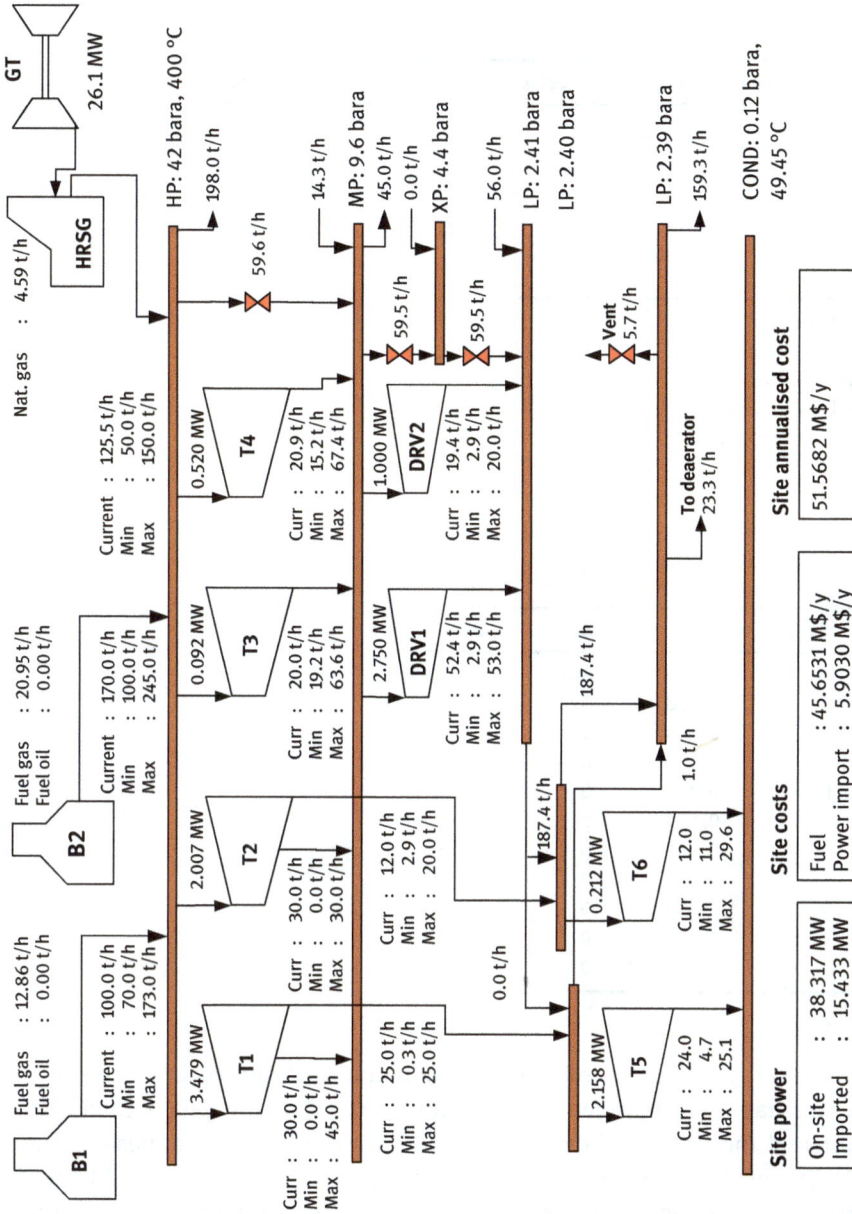

Fig. 8.6: Utility system example: initial operating point.

Tab. 8.1: Fuel data.

	Fuel 1: fuel gas	Fuel 2: fuel oil	Fuel 3: natural gas
NHV [kJ/kg]	32,502.8	40,245.0	46,151.8
Price [$/t]	103.41	70.82	159.96

Tab. 8.2: Site configuration data.

Ambient temperature	°C	25.00
Minimum stack temperature	°C	150.00
Deaerator pressure	bara	1.01325
Boiler feed-water temperature	°C	80.00
Condensate return ratio	[1]	0.5557

Tab. 8.3: Steam turbine performance parameters for the illustrative example.

Turbine: Stage	L	A	B
	[1]	MW	[1]
T1: HP-MP	0.228	0	1.96
T1: MP-LP-t1	0.010	0	3.15
T2: HP-MP	2.802	0	1.82508
T2: MP-LP-t2	0.193	0	3.15156
T3	0.429	0	1.43
T4	0.289	0	1.47
T5	0.229	0	1.46
T6	0.588	0	1.0445
DRV1	0.100	0	1.5
DRV2	0.040	0	1.53

Figure 8.6 represents the initial operating point, as identified from the existing system. This has been subjected to optimisation according to the described SMILP procedure and the result is shown in Fig. 8.7. As a result of the optimisation, turbines T2, T5 and T6 have been turned off. On the other hand, turbines T3 and T4 have been loaded to the maximum. The driver steam turbines remain in operation. Boiler B2 load is reduced to the minimum, while the HRSG flow is maximised. As a whole, the on-site power production is increased by 2.447 MW. The power import cost is reduced by 0.9359 M$/y and FCs by 6.4145 M$/y. The total operating cost of the site is reduced by 7.3508 M$/y, a 14 % reduction with respect to the initial operation.

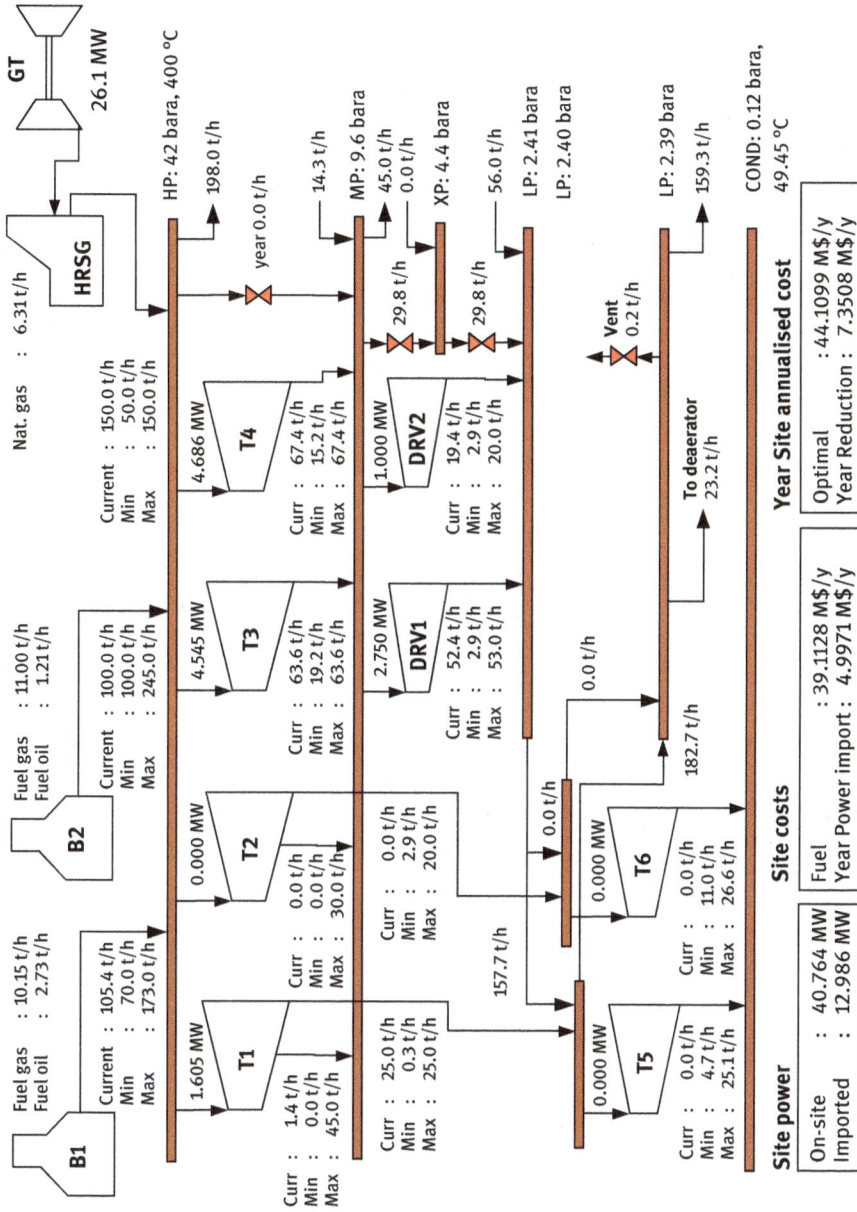

Fig. 8.7: Utility system example: optimal operation.

The reason for the cost trend shown in the optimisation is that power import is cheaper than the on-site power generation on its own. However, the much higher efficiency of the combined power and heat generation makes part of the on-site generated power even cheaper. As a result, on-site power cogeneration, alongside heat required by the site processes is maximised, but any steam turbine load above the one satisfying the steam requirement is not profitable.

8.3 Summary of utility system optimisation

This chapter presents an efficient method for operation optimisation of existing utility systems. It is based on a robust procedure using a successive MILP algorithm. The procedure features a relatively rapid convergence in most cases, achieving the solution in at most five SMILP iterations. The approach has been illustrated using an industry-scale case study. This points out the most typical feature of industrial utility systems – when the cogeneration results in effectively very low cost of the part of the on-site generated power.

The developed optimisation procedure can also be applied to other types of utility system analysis. For instance, the capabilities of an existing utility system can be explored in order to identify the best directions for energy-efficient retrofits of the existing site processes and their Heat Exchanger Networks (Varbanov et al., 2004a).

Nomenclature

Symbol	Unit	Description
B{#}	–	Boiler {#}
BFW	–	Boiler feedwater
$C_{P,W}$	kJ/kg/°C; MWh/t/°C	Specific heat capacity of water
CW	–	Cooling water
CWC	€/y	Cooling water cost
DRV{#}	–	Direct-drive steam turbine {#}
DWC	€/y	Demineralised water cost
EC(sc)	€/h	Emission cost within each Time Slice (scenario)
EC_{Total}	€/y	Annualised emission cost
EF, $EF_{BF,CO2}$, $EF_{BF,SO2}$, $EF_{BF,NOx}$	[1]	Emission factors
E_{VL}	MW	A large value with the meaning of power upper bound
FC	€/y	Fuel cost
g	m/s^2	The acceleration due to gravity

(continued)

Symbol	Unit	Description
GHG	–	Greenhouse gas
GHGFP	kg CO_2-e/{amount}; kg CO_2-e/kg product; kg CO_2-e/y	Greenhouse gas footprint. The exact measurement unit depends on the context and application.
GT	–	Gas turbine
h_{BFW}	kJ/kg; MWh/t	Specific enthalpy of the BFW
h_{DAF}	kJ/kg; MWh/t	Specific enthalpy of the deaerator feedwater
$h_{inject,LD}$	kJ/kg; MWh/t	Specific enthalpy of the BFW injected to a letdown valve
$h_{inlet,hdr}$	kJ/kg; MWh/t	Specific steam enthalpy at the inlet of the steam header
$h_{in,LD}$	kJ/kg; MWh/t	Specific steam enthalpy at the inlet of the letdown valve
h_{hdr}	kJ/kg; MWh/t	Specific enthalpy of steam in a header
$h_{out,LD}$	kJ/kg; MWh/t	Specific steam enthalpy at the outlet of the letdown valve
$h_{stm,DA}$	kJ/kg; MWh/t	Specific enthalpy of the deaerator steam feed
HP (steam)	–	High-Pressure steam
HRSG	–	Heat Recovery Steam Generator
HrYr	h/y	The number of hours per year for which the system operates
$h_{V,DA}$	kJ/kg; MWh/t	Specific enthalpy of the deaerator vent
LP (steam)	–	Low-pressure steam
m_{BF}	t/h	Fuel flow rate
m_{BFW}	t/h	Mass flow rate of the BFW
m_{CW}	t/h	The mass flow rate of the cooling water
$m_{CO_2,BLR}$	t/h	CO_2 emission flow rate
m_{DAF}	t/h	Mass flow rate of the deaerator feedwater
m_{DW}	t/h	Demineralised water flow (demand)
MILP	–	Mixed-integer linear programming
$m_{in,LD}$	t/h	Steam mass flow rate at the inlet of a letdown valve
$m_{inject,LD}$	t/h	Mass flow rate of BFW injected to a letdown valve
$m_{inlet,hdr}$	t/h	Steam mass flow rate at the inlet of the steam head mixer
MINLP	–	Mixed-integer non-linear programming
$m_{NO_x,BLR}$	t/h	NO_x emission flow rate
$m_{out,LD}$	t/h	The steam mass flow rate at the outlet of the letdown valve
$m_{outlet,hdr}$	t/h	Steam mass flow rate at the outlet of the steam head splitter
$m_{SO_2,BLR}$	t/h	SO_2 emission flow rate
$m_{boiler,total}$	t/h	Total boiler steam generation flow
MP (steam)	–	Medium-pressure steam
$m_{total,hdr}$	t/h	The total mass flow rate of the steam header
m_{STEAM}	t/h	Steam mass flow rate
$m_{stm,DA}$	t/h	Mass flow rate of the deaerator steam feed
$m_{V,DA}$	t/h	Mass flow rate of the deaerator vent
NFP	kg N/capita/y	Nitrogen footprint. The exact measurement unit can vary and depends on the context and application.
NHV	MWh/t; kJ/kg	Net heating value

(continued)

Symbol	Unit	Description
n_{ST}	MWh/t; kJ/kg	The Willans Line slope coefficient
P_{DA}	bar	The outlet pressure of the deaerator
PD_{FWP}	MW	Power consumption for feedwater pumping
$PD_{INTERNAL}$	MW	Total internal power demand
$PD_{processes}$	MW	The process power demands
$Price_{BF}$	€/t	Fuel price
$Price_{CO_2}$	€/t	Price (tax) charged for CO_2 emissions
$Price_{CW}$	€/t	Cooling water price
$Price_{DW}$	€/t	Demineralised water price
$Price_{NO_x}$	€/t	The price (tax) charged for NO_x emissions
$Price_{SO_2}$	€/t	Price (tax) charged for SO_2 emissions
P_{VHP}	bar	VHP steam pressure
Q_{COND}	MW	Heat duties of the steam turbine condensers
Q_{CW}	MW	Utility cooling demand
Q_{DPC}	MW	All residual process cooling demands that cannot be used for steam generation
SMILP	–	Successive mixed-integer linear programming
$SPD_{BFan, boiler}$	MWh/t	Specific energy demand of boiler fans per unit steam generation
$T\{\#\}$	–	Steam turbine {#}
$T_{CW,Ret}$	°C	Temperature of the cooling water returned back from the processes to the cooling water system
$T_{CW,Spl}$	°C	Temperature of the cooling water as supplied by the cooling water system to the processes
VF_{DA}	[1]	Deaerator steam vent fraction
VHP (steam)	–	Very high pressure steam
$W_{Balance}$	MW	The site power balance
W_{Export}	MW	Power export of the site
WFP	m³/{amount}; m³/kg product m³/y	Water footprint. The exact measurement unit depends on the context and application.
$W_{GT,sum}$	MW	Power generation by gas turbines
W_{Import}	MW	Power import of the site
$W_{ST,sum}$	MW	Power generation by steam turbines
XP (steam)	–	Extra pressure – an intermediate steam main (Figs. 8.6 and 8.7)
YF(sc)	[1]	The fraction of the year within which period **sc** takes place
y_{IMP}	[1]	Binary variable denoting the power flow direction
ΔP_{pipes}	bar	Pressure drop in the piping
$\Delta p_{specific}$	bar/m	Specific pressure drop per unit elevation
$\eta_{B\{\#\}}$	[1]	Efficiency of boiler {#}
η_{HRSG}	[1]	HRSG efficiency
$\eta_{T\{\#\}}$	[1]	Efficiency of steam turbine {#}
ρ_{WATER}	kg/m³	The density of water

Indices

Symbol	Description
B{#}	Index for a boiler number
bf	The boiler fuel
bo	Boiler instance (optional)
Btype	The boiler type
Sc	The scenario index

References

'BARON | The Optimization Firm'. 2019. https://www.minlp.com/baron, accessed 03/05/2019.

Carl, J. and Fedor, D. 2016. Tracking Global Carbon Revenues: A Survey of Carbon Taxes versus Cap-and-Trade in the Real World. Energy Policy 96 (September). 50–77. https://doi.org/10.1016/j.enpol.2016.05.023.

Čuček, L., Klemeš, J.J., and Kravanja, Z. 2012. A Review of Footprint Analysis Tools for Monitoring Impacts on Sustainability. Journal of Cleaner Production 34 (October). 9–20. https://doi.org/10.1016/j.jclepro.2012.02.036.

Hoekstra, A.Y., Chapagain, A.K., Aldaya, M.M., and Mekonnen, M.M. 2012. The Water Footprint Assessment Manual: Setting the Global Standard. 1st ed. London, UK: Routledge. https://doi.org/10.4324/9781849775526.

Iyer, R.R. and Grossmann, I.E. 1997. Optimal Multiperiod Operational Planning for Utility Systems. Computers & Chemical Engineering 21(8). 787–800. https://doi.org/10.1016/S0098-1354(96)00317-1.

Liu, X., Klemeš, J.J., Varbanov, P.S., Čuček, L., and Qian, Y. 2017. Virtual Carbon and Water Flows Embodied in International Trade: A Review on Consumption-Based Analysis. Journal of Cleaner Production 146: 20–28. https://doi.org/10.1016/j.jclepro.2016.03.129.

Puranik, Y., Mustafa Kilinç, N.V., Sahinidis, T.L., Gopalakrishnan, A., Besancon, B., and Roba, T. 2016. Global Optimization of an Industrial Gas Network Operation. AIChE Journal 62(9). 3215–24. https://doi.org/10.1002/aic.15344.

Varbanov, P., Perry, S., Klemeš, J., and Smith, R. 2005. Synthesis of Industrial Utility Systems: Cost-Effective de-Carbonisation. Applied Thermal Engineering 25(7). 985–1001. https://doi.org/10.1016/j.applthermaleng.2004.06.023.

Varbanov, P., Perry, S., Makwana, Y., Zhu, X.X., and Smith, R. 2004a. Top-Level Analysis of Site Utility Systems. Chemical Engineering Research and Design 82(6). 784–95. https://doi.org/10.1205/026387604774196064.

Varbanov, P.S., Doyle, S., and Smith, R. 2004b. Modelling and Optimization of Utility Systems. Chemical Engineering Research and Design 82(5). 561–78. https://doi.org/10.1205/026387604323142603.

Vigerske, S. and Gleixner, A. 2018. SCIP: Global Optimization of Mixed-Integer Nonlinear Programs in a Branch-and-Cut Framework. Optimization Methods and Software 33(3). 563–93. https://doi.org/10.1080/10556788.2017.1335312.

9 Utility system simulation: a solved case study

Establishing the steam and power balances for a utility system is a key activity for system identification. Its results can be used for tuning and optimising the operation of the involved equipment and the system as a whole, as well as for improving the system via equipment replacements or a wider retrofit. This chapter presents a worked case study on modelling a utility system. It uses a published utility system example, from (Varbanov et al., 2004), to provide an illustration of how some of the models from the previous chapters can be applied. The case study is considered in two modes. First, it is illustrated on how to model the system using simple calculations in MS-Excel (Microsoft, 2019). The Excel file can be downloaded from the book companion page. That is followed by a guide on how to model the same case in the Petro-SIM process simulator (KBC, 2019). The simulator input file is also available for download from the book companion web page.

9.1 Step-by-step steam balancing and calculation steps of a utility system with Excel

The utility network to be modelled is shown in Fig. 9.1, having placeholders for some of the stream flow rates. Each boiler (Boiler 1, Boiler 2 and HRSG) has a blowdown flow. These are collapsed (hidden) for the simplicity of the picture. However, the blowdown flows are explicitly modelled.

The task is to balance the system. The known flow rates and other properties of the utility system are shown in Tab. 9.1. There are modelling assumptions adopted for the case study:

– All steam mains do not vent steam. As a result, no vent connections are shown in Fig. 9.1 and the corresponding flowrates are specified as zeroes in Tab. 9.1. The pressure values are given as "absolute" – that is, not subtracting the atmospheric pressure.
– The condensate in the network has the temperature of saturated liquid at the given pressure.
– The deaerator uses exactly as much LP steam as needed to bring its inlet flow to the boiling point at the specified pressure.
– Each steam main is modelled as a sequence of a stream mixer, forming the content of the steam main and a stream splitter, distributing to the outlets the formed steam.
– The calculation procedure uses a customised set of measurement units. The reason for this is that on one hand, the widely accepted unit by site engineers for steam flow is "t/h" and for power or energy flow is "MW." On the other hand, the used correlations of the IAPWS'97 (IAPWS, 2020) water–steam properties use

https://doi.org/10.1515/9783110630091-009

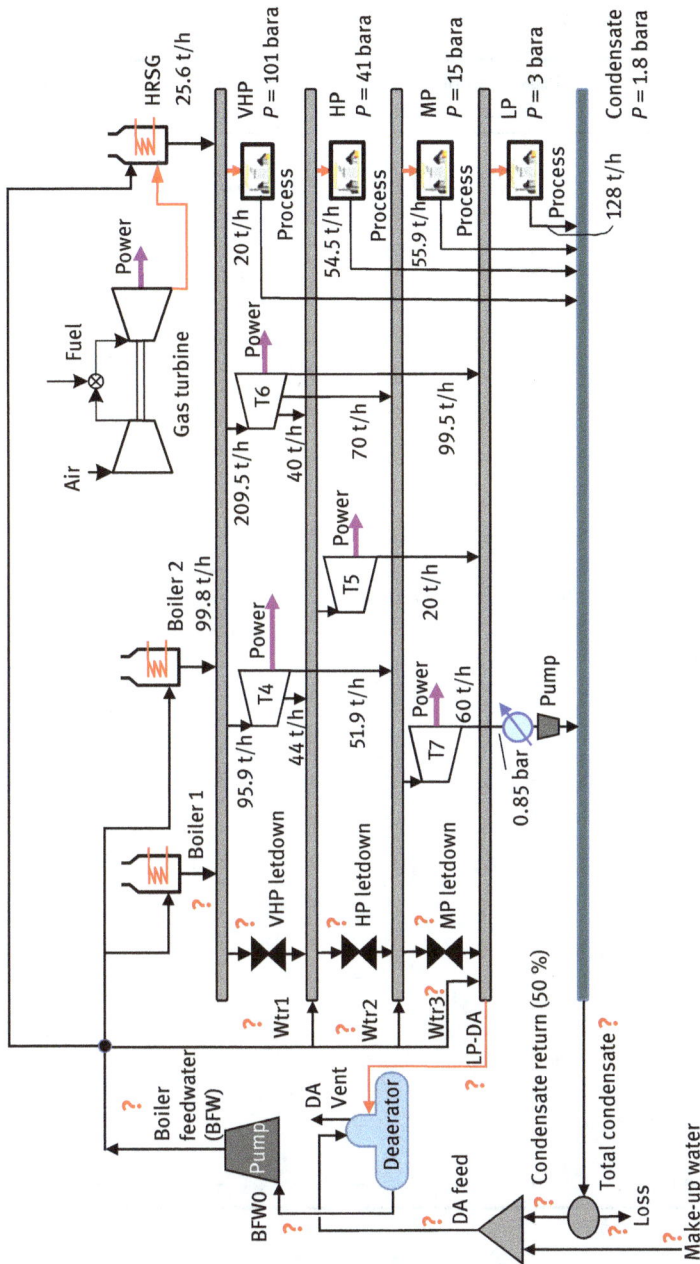

Fig. 9.1: Steam system PFD with placeholders for some flow rate values.
(Adapted from Varbanov et al., 2004)

Tab. 9.1: Known parameters of the utility system.

Parameters	Pressure [MPa] [(bar)] (absolute, not gauge)	Temperature [°C]	Mass flow [t/h]	Others [unit]
Pressure levels	–	–	–	–
VHP	10.1 (101)	510	–	–
HP	4.1 (41)	330	–	–
MP	1.5 (15)	230	–	–
LP	0.3 (3)	145	–	–
Boiler 2 steam (S_2)	–	–	99.8	–
HRSG steam (S_{HRSG})	–	–	25.6	–
VHP process steam	–	–	20	–
HP process steam	–	–	54.5	–
MP process steam	–	–	55.9	–
LP process steam	–	–	128	–
Make-up water	0.18 (1.8)	25	–	–
Condensate	0.18 (1.8)	45	–	–
Deaerator	0.15 (1.5)	–	–	–
BFW0	0.15 (1.5)	–	–	–
BFW	12 (120)	–	–	–
Turbine	–	–	–	–
T4-in	–	–	95.9	–
T4A-out	–	–	44	–
T5-in	–	–	20	–
T6-in	–	–	209.5	–
T6A-out	–	–	40	–
T6B-out	–	–	70	–
T7-in	–	–	60	–
VHP vent	–	–	0	–
HP vent	–	–	0	–
MP vent	–	–	0	–
LP vent	–	–	0	–
Boiler blowdown fraction (F_{BD}) – see Chapter 5	–	–	–	0.08 [1]
CRR	–	–	–	0.5 [1]
Pump efficiency	–	–	–	0.75 [1]
A fraction of the deaerator vent within the used steam	–	–	–	0.02 [1]
Boiler efficiency (fuel to steam)	–	–	–	0.9 [1]

SI units, except for pressure – which is in "bar." This makes it necessary to convert the specific enthalpy values between the SI unit of kJ/kg and the nonstandard unit of MWh/t. The other pair of conversions is for temperature between °C and K, for the same reason.

The given data includes:
- Steam mains specifications – pressures and temperatures
- Steam generation flow rates by Boiler 2 and the HRSG
- The process steam use flow rates
- The steam turbine steam loads and power generation rates (see Tab. 9.1)
- Correlations and regression coefficients for estimation of the steam turbine performance (see Tab. 9.2).
- Turbine steam capacities and power generation efficiency values (Tab. 9.3), using intercept ratio $L = 0.2$ [1].
- Other site parameters such as the condensate return (CRT) ratio, blowdown fraction and pump efficiency
- Pressures and temperatures of the CRT
- The tolerance for converging numeric differences is TOL $= 10^{-4}$

Tab. 9.2: Regression coefficients for steam turbine performance estimation.

Parameter	Unit	Explanation	Value
L	[1]	Intercept ratio. Specification	–
$b_{0,BP}$	MW	Backpressure steam turbine regression parameter, Chapter 6	0
$b_{1,BP}$	MW/K	Backpressure steam turbine regression parameter, Chapter 6	0.00423
$b_{2,BP}$	[1]	Backpressure steam turbine regression parameter, Chapter 6	1.155
$b_{3,BP}$	1/K	Backpressure steam turbine regression parameter, Chapter 6	0.000538
$b_{0,COND}$	MW	Condensing steam turbine regression parameter, Chapter 6	−0.463
$b_{1,COND}$	MW/K	Condensing steam turbine regression parameter, Chapter 6	0.00353
$b_{2,COND}$	[1]	Condensing steam turbine regression parameter, Chapter 6	1.57
$b_{3,COND}$	1/K	Condensing steam turbine regression parameter, Chapter 6	0.0007

Tab. 9.3: Maximum steam flows and power generation efficiencies.

		Backpressure		Backpressure	Backpressure			Condensing
		Turbine T4: two stages		Single stage	Turbine T6: three stages			Single stage
		T4A	T4B	T5	T6A	T6B	T6 C	T7
m_{max}	t/h	165	90	35	335	220	150	90
η_{mech}	[1]	0.95	0.95	0.95	0.95	0.95	0.95	0.95

It is then required to obtain estimates of:
- The overall Boiler feedwater (BFW) flow rate and its constituents – CRT flow, make-up water and deaerator feed
- The throughput of Boiler 1

- The BFW supply to the steam mains below the VHP main
- The letdown flow rates
- The missing temperatures and stream conditions

Next follows a description of the key modelling techniques applied, to estimate the performance of the steam turbines, as well as the letdown expansion behaviour. The solution steps are described after that, providing guidance.

9.1.1 Performance modelling of steam turbines

Steam turbines are modelled using the Willans Line equation (see Chapter 6) as follows:

$$W = n_{st} \cdot m_{steam} - W_{int} \tag{9.1}$$

where W represents power generation by the turbine, MW; m_{steam} represents steam flow across each turbine, t/h.

Equation (9.1) is applied to T5 and T7 directly. Turbines T4 and T6 are of the extraction type. They are decomposed and the Willans Line is applied to each of its the component turbines. This decomposition scheme is illustrated in Fig. 9.2

9.1.2 Estimation of properties of inlet and outlet streams of letdown valves

All the letdown valves are modelled as isenthalpic, as shown in Eqs. (9.2)–(9.4):

$$\text{VHP – HP letdown: } h_{SD1in} = h_{SD1out} = h_{VHP} \tag{9.2}$$

$$\text{HP – MP letdown: } h_{SD2in} = h_{SD2out} = h_{HP} \tag{9.3}$$

$$\text{MP – LP letdown: } h_{SD3in} = h_{SD3out} = h_{MP} \tag{9.4}$$

The enthalpies of the inlet and outlet streams of all valves can be calculated using Eqs. (9.2)–(9.4). Since the pressures of the inlet and outlet streams are known, the outlet steam temperature of all valves can be obtained, according to Eq. (9.5), which uses a VBA macroimplementation of the IAPWS 1997 industrial formulation for the thermodynamic properties of water and steam (Wagner and Kretzschmar, 2008):

$$T[K] = t\text{Water(Pressure [bar], enthalpy [kJ/kg])} \tag{9.5}$$

The spreadsheet calculations adhere to the following colour coding convention:
- Generally, the cells are with white background.
- Green fill of a cell indicates a specification.
- Orange fill of a cell indicates a link to another cell.

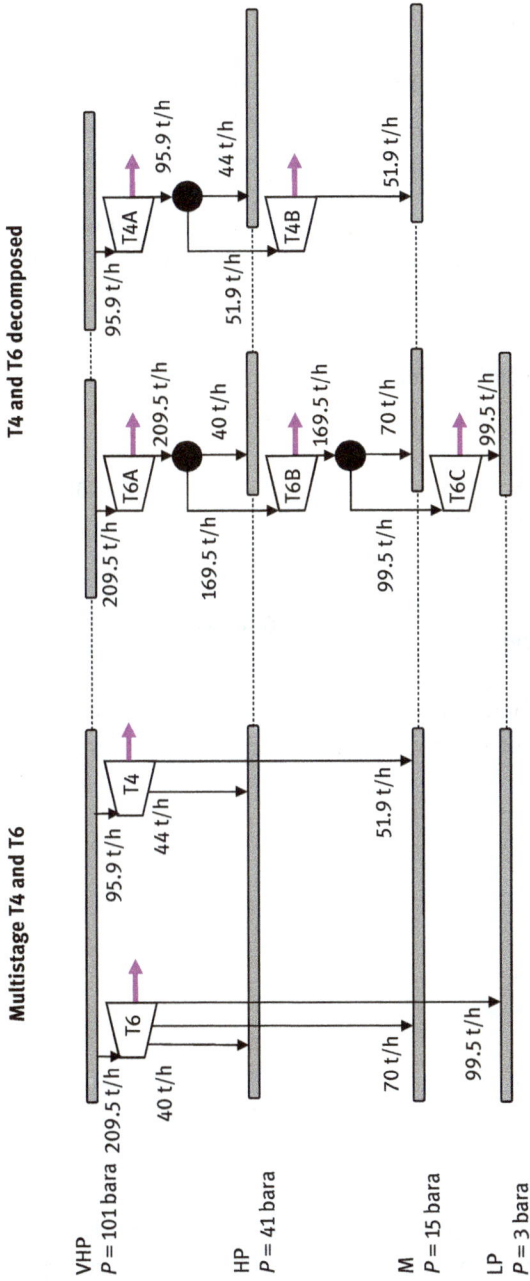

Fig. 9.2: Steam turbine decomposition scheme.

9.1.3 Parameter assignments and one-off calculations

The problem specifications are given in Tab. 9.1. The pressures of the steam mains (P_{VHP}, P_{HP}, P_{MP} and P_{LP}) are specified in bar and converted to MPa for displaying in SI units. The temperatures of the steam mains (T_{VHP}, T_{HP}, T_{MP} and T_{LP}) are specified in °C and converted to K.

The condensate pressure is specified in a similar way – in bar and converted to MPa. For the condensate, no direct temperature specification is given. Following the assumption in the problem setup, the temperature is calculated as that of saturated liquid:

$$T_{cond}\ [K] = tSatW(P_{cond}\ [bar]) = 368.28\,K \tag{9.6}$$

$$T_{cond}\ [°C] = T_{cond}\ [K] - 273.15 = 95.13\,°C \tag{9.7}$$

Using the IAPWS'97 correlations, the enthalpies of the steam mains and the condensate are calculated in kJ/kg and converted to MWh/t:

$$h_{\#LEVEL}\ [kJ/kg] = enthalpyW(T_{\#LEVEL}\ [K], P_{\#LEVEL}\ [bar]) \tag{9.8}$$

$$h_{\#LEVEL}\ [MWh/t] = h_{\#LEVEL}\ [kJ/kg] \times \frac{1}{3,600}\left[\frac{h}{s}\right] \times \frac{1,000\ [kg/t]}{1,000\ [kWh/MWh]} \tag{9.9}$$

where $\#LEVEL \in [VHP, HP, MP, LP, cond]$. The labels in this enumeration represent the steam mains and the condition of the condensing turbine (T7). The resulting values are summarised in Tab. 9.4.

Tab. 9.4: Steam main specific enthalpies.

Parameter	Value	
	[kJ/kg]	[MWh/t]
h_{VHP}	3,399.6	0.9443
h_{HP}	3,039.8	0.8444
h_{MP}	2,875.5	0.7987
h_{LP}	2,750.3	0.7640
h_{cond}	398.5	0.1107

Additionally, for calculating the heat supply by fuel or gas turbine exhaust to the steam boilers, it is necessary to know the value of the specific enthalpy of liquid water at the VHP level. This is calculated using the IAPWS'97 correlations as follows:

$$h_{VHP,SatLiq}\ [kJ/kg] = enthalpySatLiqPW(P_{VHP}\ [bar]) \tag{9.10}$$

$$h_{\text{VHP, SatLiq}} \text{ [MWh/t]} = h_{\text{VHP, SatLiq}} \text{ [kJ/kg]} \times \frac{1}{3,600} \left[\frac{h}{s}\right] \times \frac{1,000 \text{ [kg/t]}}{1,000 \text{ [kWh/MWh]}} \qquad (9.11)$$

The CRT flow is specified as having a higher pressure than the condensate collection, which implies that pumps are installed between those parts of the network. Using the CRT temperature and pressure specifications, the specific enthalpy is calculated:

$$h_{\text{CRT}} \text{ [kJ/kg]} = \text{enthalpy}W(T_{\text{CRT}} \text{ [K]}, P_{\text{CRT}} \text{ [bar]}) = 188.6 \text{ kJ/kg} \qquad (9.12)$$

$$h_{\text{CRT}} \text{ [MWh/t]} = h_{\text{CRT}} \text{ [kJ/kg]} \times \frac{1}{3,600} \left[\frac{h}{s}\right] \times \frac{1,000 \text{ [kg/t]}}{1,000 \text{ [kWh/MWh]}} = 0.0524 \text{ MWh/t} \qquad (9.13)$$

Next is the calculation of the enthalpy of the deaerator vent, using the IAPWS'97 correlations and assuming saturated vapour condition:

$$h_{\text{DV}} \text{ [kJ/kg]} = \text{enthalpySatVapPW}(P_{\text{DA}} \text{ [bar]}) = 2,693.1 \text{ kJ/kg} \qquad (9.14)$$

$$h_{\text{DV}} \text{ [MWh/t]} = h_{\text{DV}} \text{ [kJ/kg]} \times \frac{1}{3,600} \left[\frac{h}{s}\right] \times \frac{1,000 \text{ [kg/t]}}{1,000 \text{ [kWh/MWh]}} = 0.7481 \text{ MWh/t} \qquad (9.15)$$

The BFW pressure after the deaerator (P_{BFWO}) is also 1.5 bar. That pressure value is converted to kPa for further applying the estimation expression for the BFW enthalpy change as a result of the pumping to the boiler pressure. It is assumed that the enthalpy of the BFWO stream is the same as the saturated liquid in the deaerator, resulting in a small vapour fraction (about 1 %). The enthalpy calculation is as follows:

$$h_{\text{BFWO}} \text{ [kJ/kg]} = \text{enthalpySatLiqPW}(P_{\text{BFWO}} \text{ [bar]}) = 467.1 \text{ kJ/kg} \qquad (9.16)$$

$$h_{\text{BFWO}} \text{ [MWh/t]} = h_{\text{BFWO}} \text{ [kJ/kg]} \times \frac{1}{3,600} \left[\frac{h}{s}\right] \times \frac{1,000 \text{ [kg/t]}}{1,000 \text{ [kWh/MWh]}} = 0.1297 \text{ MWh/t} \qquad (9.17)$$

Since the deaerator steam (LPDA) is drawn from the LP steam main, it is assigned the specific enthalpy using the temperature and pressure specifications for that main:

$$h_{\text{LPDA}} \text{ [kJ/kg]} = \text{enthalpy}W(T_{\text{LP}} \text{ [K]}, P_{\text{LP}} \text{ [bar]}) = 2,750.3 \text{ kJ/kg} \qquad (9.18)$$

$$h_{\text{LPDA}} \text{ [MWh/t]} = h_{\text{LPDA}} \text{ [kJ/kg]} \times \frac{1}{3,600} \left[\frac{h}{s}\right] \times \frac{1,000 \text{ [kg/t]}}{1,000 \text{ [kWh/MWh]}} = 0.7640 \text{ MWh/t} \qquad (9.19)$$

The make-up water specific enthalpy is calculated as follows:

$$h_{\text{MuW}} \text{ [kJ/kg]} = \text{enthalpy}W(T_{\text{MuW}} \text{ [K]}, P_{\text{MuW}} \text{ [bar]}) = 105.0 \text{ kJ/kg} \qquad (9.20)$$

$$h_{\text{MuW}} \ [\text{MWh/t}] = h_{\text{MuW}} \ [\text{kJ/kg}] \times \frac{1}{3,600} \left[\frac{\text{h}}{\text{s}}\right] \times \frac{1,000 \ [\text{kg/t}]}{1,000 \ [\text{kWh/MWh}]} = 0.0292 \ \text{MWh/t}$$

(9.21)

Using the specified pressures and the given pump efficiency, the BFW specific enthalpy is calculated as follows:

$$h_{\text{BFW}} \left[\frac{\text{kJ}}{\text{kg}}\right] = h_{\text{BFW0}} \left[\frac{\text{kJ}}{\text{kg}}\right] + \frac{P_{\text{BFW}} - P_{\text{BFW0}}}{\rho \cdot \eta} \left[\frac{\text{kN} \times \text{m}^3}{\text{kg} \times \text{m}^2}\right] = 483.7 \ \text{kJ/kg} \qquad (9.22)$$

$$h_{\text{BFW}} \ [\text{MWh/t}] = h_{\text{BFW}} \ [\text{kJ/kg}] \times \frac{1}{3,600} \left[\frac{\text{h}}{\text{s}}\right] \times \frac{1,000 \ [\text{kg/t}]}{1,000 \ [\text{kWh/MWh}]} = 0.1344 \ \text{MWh/t}$$

(9.23)

The blowdown flows of Boiler 2 and the HRSG are calculated using the specified blowdown fractions:

$$\text{BD}_{\text{\#BOILER}} \ [\text{t/h}] = S_{\text{\#BOILER}} \ [\text{t/h}] \times \frac{F_{\text{BD}}}{1 - F_{\text{BD}}} \ [1] \qquad (9.24)$$

where $\text{\#BOILER} \in [2, \text{HRSG}]$. The calculated values are $\text{BD}_2 = 8.678$ t/h and $\text{BD}_{\text{HRSG}} = 2.226$ t/h.

The fuel heat used in Boiler 2 is estimated by first calculating the BFW taken into the boiler and then applying the enthalpy balance of BFW preheat and steam generation:

$$\text{BFW}_2 \ [\text{t/h}] = S_2 + \text{BD}_2 = 108.478 \ [\text{t/h}] \qquad (9.25)$$

$$Q_{\text{BF2}} \ [\text{MW}] = \text{BFW}_2 \ [\text{t/h}] \times \left(h_{\text{VHP, SatLiq}} - h_{\text{BFW}}\right) \ [\text{MWh/t}] + S_2 \ [\text{t/h}]$$
$$\times \left(h_{\text{VHP}} - h_{\text{VHP, SatLiq}}\right) [\text{MWh/t}] = 83.073 \ [\text{MW}] \qquad (9.26)$$

Similarly, the heat absorbed from the gas turbine exhaust for steam generation in the HRSG is estimated as follows:

$$\text{BFW}_{\text{HRSG}} \ [\text{t/h}] = S_3 + \text{BD}_{\text{HRSG}} = 27.826 \ [\text{t/h}] \qquad (9.27)$$

$$Q_{\text{HRSG}} [\text{MW}] = \text{BFW}_{\text{HRSG}} \ [\text{t/h}] \times \left(h_{\text{VHP, SatLiq}} - h_{\text{BFW}}\right) \ [\text{MWh/t}] + S_3 \ [\text{t/h}]$$
$$\times \left(h_{\text{VHP}} - h_{\text{VHP, SatLiq}}\right) \ [\text{MWh/t}] = 21.309 \ [\text{MW}] \qquad (9.28)$$

The flow rate of steam expanding through the second stage of T4 is calculated as the difference of the turbine inlet flow rate and that of the extraction to the HP main:

$$S_{\text{4Bout}} = S_{\text{4in}} - S_{\text{4Aout}} = 51.9 \ \text{t/h} \qquad (9.29)$$

Similarly, using the turbine decomposition scheme from Fig. 9.2, the flow rates of steam expanding through the second and the third stages of turbine T6 are calculated as $S_{\text{6B}} = 169.5$ t/h and $S_{\text{6C}} = 99.5$ t/h.

The next step is the calculation of the overall condensate collection. Condensate is formed from the process steam use (steam drawn from the VHP, HP, MP and LP mains), plus the steam condensed at the outlet of turbine T7 (Fig. 9.3).

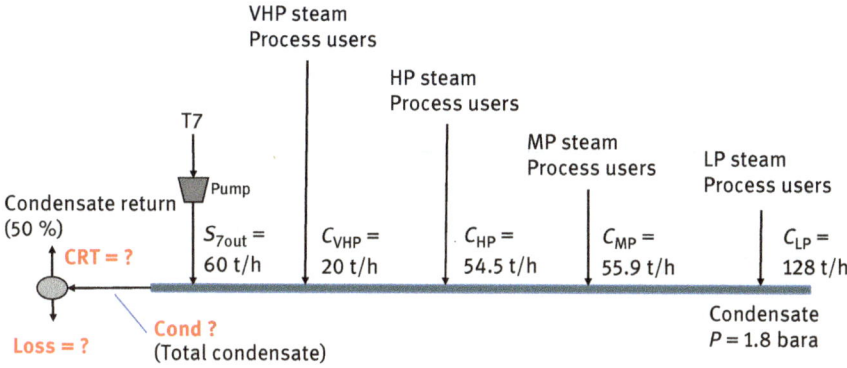

Fig. 9.3: Condensate collection in the example utility system.

This configuration is reflected by the mass balances of the condensate collection and the condensate loss as follows:

$$\text{Cond} = C_{VHP} + C_{HP} + C_{MP} + C_{LP} + S_{7out} = 318.4 \text{ t/h} \tag{9.30}$$

$$\text{CRT} = \text{CRR} \times \text{Cond} = 159.2 \text{ t/h} \tag{9.31}$$

$$\text{Loss} = \text{Cond} - \text{CRT} = 159.2 \text{ t/h} \tag{9.32}$$

The variables associated with the steam turbines include regression coefficients, Willans Line coefficients, power generation and properties of the exhaust steam flows. These all are calculated only once. The reason for this is that all turbines draw steam from the mains at pressures and temperatures fixed by specifications and the flow rates of the expanding steam are also specified. The calculations are performed for each of the component/simple turbines in the system. For each component turbine, the following calculation sequence is applied.

The pressure values for inlets to/and outlets from the turbines, as well as the inlet steam temperature values, are taken from the specifications and assigned as shown in Tab. 9.5. Note that the inlet temperatures for T4B, T6B and T6C are not assigned in Tab. 9.5. The corresponding enthalpy values are calculated later as a result of turbine stages calculation and the entropy, and related calculations are performed based on those enthalpies.

Tab. 9.5: Pressure values assigned to steam turbine inlets and outlets.

		Backpressure		Backpressure	Backpressure			Condensing
		Turbine T4: two stages		Single stage	Turbine T6: three stages			Single stage
		T4A	T4B	T5	T6A	T6B	T6C	T7
P_{In}	bar (a)	101	41	41	101	41	15	15
P_{Out}	bar (a)	41	15	3	41	15	3	0.85
T_{In}	K	783.15	–	603.15	783.15	–	–	503.15

The saturation temperatures at the turbine inlets and outlets are calculated using the IAPWS'97 correlations:

$$T_{Sat, In, \#TURBINE} \ [K] = tSatW(P_{In, \#TURBINE} \ [bar]) \tag{9.33}$$

$$T_{Sat, Out, \#TURBINE} \ [K] = tSatW(P_{Out, \#TURBINE} \ [bar]) \tag{9.34}$$

$$\Delta T_{Sat, \#TURBINE} \ [K] = T_{Sat, In, \#TURBINE} \ [K] - T_{Sat, Out, \#TURBINE} \ [K] \tag{9.35}$$

where #TURBINE $\in$ {T4A, T4B, T5, T6A, T6B, T6C, T7}.

From those values, the parameters A_{ST} and B_{ST} are calculated for each turbine:

$$A_{ST, \#TURBINE} \ [MW] = b_{0, BP} + b_{1, BP} \times \Delta T_{Sat, \#TURBINE} \ [K] \tag{9.36}$$

for #TURBINE $\in$ {T4A, T4B, T5, T6A, T6B, T6C}

$$B_{ST, \#TURBINE} \ [1] = b_{2, BP} + b_{3, BP} \times \Delta T_{Sat, \#TURBINE} \ [K] \tag{9.37}$$

for #TURBINE $\in$ {T4A, T4B, T5, T6A, T6B, T6C}

$$A_{ST, \#TURBINE} \ [MW] = b_{0, COND} + b_{1, COND} \times \Delta T_{Sat, \#TURBINE} \ [K] \tag{11.38}$$

for #TURBINE $\in$ {T7}

$$B_{ST, \#TURBINE} \ [1] = b_{2, COND} + b_{3, COND} \times \Delta T_{Sat, \#TURBINE} \ [K] \tag{9.39}$$

for #TURBINE $\in$ {T7}

Then, the specific enthalpy and entropy of the inlet steam are calculated:

$$h_{, \#TURBINE} \ [kJ/kg] = enthalpyW(T_{in, \#TURBINE} \ [K], P_{In, \#TURBINE} \ [bar]) \tag{9.40}$$

$$s_{In, \#TURBINE} \left[\frac{kJ}{kg \times K}\right] = entropyW(T_{in, \#TURBINE} \ [K], P_{In, \#TURBINE} \ [bar]) \tag{9.41}$$

where #TURBINE $\in$ {T4A, T5, T6A, T7}.

For #TURBINE $\in$ {T4B, T6B, T6C}, the assignment of the specific enthalpies is done in following eqs. (9.42) to 9.44 after calculating the power generation and exhaust enthalpy values of the upper stages from eqs. (9.53) and (9.54):

$$h_{\text{In, T4B}} \ [\text{kJ/kg}] := h_{\text{Out, T4A}} \ [\text{kJ/kg}] \tag{9.42}$$

$$h_{\text{In, T6B}} \ [\text{kJ/kg}] := h_{\text{Out, T6A}} \ [\text{kJ/kg}] \tag{9.43}$$

$$h_{\text{In, T6C}} \ [\text{kJ/kg}] := h_{\text{Out, T6B}} \ [\text{kJ/kg}] \tag{9.44}$$

This follows the decomposition scheme from Fig. 9.2.

The specific entropy at the inlets of the component turbines #TURBINE $\in$ {T4B, T6B, T6C} is calculated as follows:

$$s_{\text{In, #TURBINE}} \left[\frac{\text{kJ}}{\text{kg} \times \text{K}}\right] = s\text{SteamPH}(P_{\text{in, #TURBINE}} \ [\text{bar}], h_{\text{In, #TURBINE}} \ [\text{kJ/kg}]) \tag{9.45}$$

For calculation of the Willans Line coefficients, the outlet steam enthalpy values for isentropic expansion for each component turbine are calculated, followed by calculation of the isentropic enthalpy change values, also converted from kJ/kg to MWh/t:

$$h_{\text{Out, IS, #TURBINE}} \ [\text{kJ/kg}] = \text{enthalpyWPS}\left(P_{\text{Out, #TURBINE}} \ [\text{bar}], s_{\text{In, #TURBINE}} \left[\frac{\text{kJ}}{\text{kg} \times \text{K}}\right]\right) \tag{9.46}$$

$$\Delta h_{\text{IS, #TURBINE}} \ [\text{kJ/kg}] = h_{\text{In, #TURBINE}} \ [\text{kJ/kg}] - h_{\text{Out, IS, #TURBINE}} \ [\text{kJ/kg}] \tag{9.47}$$

$$\Delta h_{\text{IS, #TURBINE}} \ [\text{MWh/t}] = \Delta h_{\text{IS, #TURBINE}} \ [\text{kJ/kg}] \times \frac{1}{3,600} \left[\frac{\text{h}}{\text{s}}\right] \times \frac{1,000 \ [\text{kg/t}]}{1,000 \ [\text{kWh/MWh}]} \tag{9.48}$$

where #TURBINE $\in$ {T4A, T4B, T5, T6A, T6B, T6C, T7}.

The Willans Line coefficients are then calculated using the equations derived in Chapter 6:

$$n_{\text{ST, #TURBINE}} \ [\text{MWh/t}] = \frac{L+1 \ [1]}{B_{\text{ST, #TURBINE}} \ [1]} \times \left(\Delta h_{\text{IS, #TURBINE}} \ [\text{MWh/t}] - \frac{A_{\text{ST, #TURBINE}} \ [\text{MW}]}{m_{\text{max, #TURBINE}} \ [\text{t/h}]}\right) \tag{9.49}$$

$$W_{\text{INT, #TURBINE}} \ [\text{MW}] = \frac{L \ [1]}{B_{\text{ST, #TURBINE}} \ [1]} \times (\Delta h_{\text{IS, #TURBINE}} \ [\text{MWh/t}] \\ \times m_{\text{max, #TURBINE}} \ [\text{t/h}] - A_{\text{ST, #TURBINE}}) \ [\text{MW}] \tag{9.50}$$

where #TURBINE $\in$ {T4A, T4B, T5, T6A, T6B, T6C, T7}.

The power generation and the total energy extraction from the passing steam flow for each turbine are calculated using the Willans Line and using the power generation efficiency, followed by the specific enthalpy and dryness fraction of the exhaust steam:

$$W_{\text{#TURBINE}} \ [\text{MW}] = n_{\text{ST, #TURBINE}} \ [\text{MWh/t}] \times m_{\text{#TURBINE}} \ [\text{t/h}] - W_{\text{INT, #TURBINE}} \ [\text{MW}] \tag{9.51}$$

$$W_{\text{total, #TURBINE}} \ [\text{MW}] = \frac{W_{\text{#TURBINE}} \ [\text{MW}]}{\eta_{\text{#TURBINE}} \ [1]} \tag{9.52}$$

$$h_{\text{Out}, \#\text{TURBINE}} \, [\text{MWh/t}] = h_{\text{In}, \#\text{TURBINE}} \, [\text{MWh/t}] - \frac{W_{\text{total}\#\text{TURBINE}} \, [\text{MW}]}{m_{\#\text{TURBINE}} \, [\text{t/h}]} \quad (9.53)$$

$$h_{\text{Out}, \#\text{TURBINE}} \, [\text{kJ/kg}] = h_{\text{Out}, \#\text{TURBINE}} \, [\text{MWh/t}] \times 3,600 \, \left[\frac{\text{s}}{\text{h}}\right] \times \frac{0.001 \, [\text{t/kg}]}{0.001 \, [\text{MWh/kWh}]} \quad (9.54)$$

$$X_{\text{Out}, \#\text{TURBINE}} \, [1] = x\text{Steam}(P_{\text{Out}, \#\text{TURBINE}}, h_{\text{out}, \#\text{TURBINE}} \, [\text{kJ/kg}]) \quad (9.55)$$

where $\#\text{TURBINE} \in \{\text{T4A, T4B, T5, T6A, T6B, T6C, T7}\}$.

The described calculations produce the numerical values given in Tab. 9.6. In addition, turbine T7 has a condenser. The condenser duty is calculated as follows:

$$Q_{\text{Cond}, T7} \, [\text{MW}] = S_{7\text{in}} \, [\text{t/h}] \times (h_{\text{Out}, T7} - h_{\text{Cond, SatVap}}) \, [\text{MWh/t}] = 36.426 \, \text{MW} \quad (9.56)$$

Tab. 9.6: Steam turbine calculation results.

		Backpressure		Backpressure	Backpressure			Condensing
		Turbine T4: two stages		Single stage	Turbine T6: three stages			Single stage
		T4A	T4B	T5	T6A	T6B	T6C	T7
$T_{\text{sat,in}}$	K	584.88	524.98	524.98	584.88	524.98	471.45	471.45
$T_{\text{sat,out}}$	K	524.98	471.45	406.68	524.98	471.45	406.68	368.28
ΔT_{sat}	K	59.91	53.53	118.30	59.91	53.53	64.77	103.17
A_{ST}	MW	0.2534	0.2264	0.5004	0.2534	0.2264	0.2740	−0.0988
B_{ST}	[1]	1.1872	1.1838	1.2186	1.1872	1.1838	1.1898	1.6422
h_{in}	kJ/kg	3,399.6	3,196.7	3,039.8	3,399.6	3,188.7	2,975.2	2,875.5
se_{in}	kJ/kg/K	6.6266	6.7340	6.4865	6.6266	6.7221	6.8071	6.6166
$h_{\text{out,IS}}$	kJ/kg	3,126.7	2,936.0	2,519.5	3,126.7	2,929.7	2,649.9	2,374.3
Δh_{IS}	kJ/kg	272.9	260.7	520.3	272.9	259.0	325.3	501.1
Δh_{IS}	MWh/t	0.0758	0.0724	0.1445	0.0758	0.0720	0.0904	0.1392
n_{st}	MWh/t	0.0751	0.0709	0.1282	0.0759	0.0719	0.0893	0.1025
W_{int}	MW	2.065	1.063	0.748	4.236	2.636	2.233	1.538
m	t/h	95.900	51.900	20.000	209.500	169.500	99.500	60.000
W	MW	5.135	2.614	1.817	11.657	9.550	6.653	4.613
W_{total}	MW	5.405	2.752	1.912	12.271	10.053	7.003	4.856
h_{out}	MWh/t	0.8880	0.8349	0.7488	0.8858	0.8265	0.7561	0.7178
h_{out}	kJ/kg	3,196.7	3,005.8	2,695.6	3,188.7	2,975.2	2,721.8	2,584.1
X_{out}	[1]	1.0000	1.0000	0.9865	1.0000	1.0000	0.9986	0.9631

9.1.4 Initialise the computational loops

Utility systems form networks, which usually result in computational loops. This is also the case with the current system, as can be seen in Fig. 9.4. A computational loop is formed within a system of equations when no direct calculation sequence can be established, and the values of the unknown variables can be calculated only by solving the equations together. Within the domain of process systems engineering, any flowsheet with a topological loop also inevitably forms computational loops in the underlying mathematical model. More details on the mathematical modelling of the process networks and the sequential method of chemical system simulation can be found in the famous book on chemical engineering calculations (Himmelblau and Riggs, 2012).

Figure 9.4 highlights the main topology loops formed by the unknown stream properties in the considered example. That involves the variables related to the BFW – mass flow rate and specific enthalpy. Since the BFW specific enthalpy is fixed by specifying its pressure and temperature, the remaining computation loop related to the BFW is tied to the mass flow rate. This means that to simulate the utility system using the sequential-modular method, one has to assume a value for the BFW flow rate, calculate the process units along the path and then end up calculating the same BFW flow rate. At that point, the assumed and the calculated values are compared. If the difference is smaller than a preset tolerance, the iterations stop and it is deemed that calculation convergence is achieved. In the opposite case, the calculated value of the BFW flow rate is used as the new iteration estimate and the loop is repeated again. This procedure is known as the "method of successive substitutions" (Himmelblau and Riggs, 2012). It is followed in the current section for implementing the principle of sequential-modular simulation on the utility system example.

The initialisation of the computational loops starts with obtaining initial estimates of the flow rates of the BFW and the deaerator steam. It is important to obtain estimates that would be as close to the solution values as possible. The estimation starts by assuming initial values of the BFW injection flows to the steam mains as zero:

$$\mathrm{Wtr}1 = \mathrm{Wtr}\,2 = \mathrm{Wtr}\,3 = 0\,\mathrm{t/h} \tag{9.57}$$

At the next step, the approximation of the deaerator feed stream with the sum of the VHP steam main outlet flows, increased by using the boiler blowdown fraction is performed:

$$\mathrm{DAF} = (S_{4\mathrm{in}} + S_{6\mathrm{in}} + \mathrm{VHP_{out}}) \times \frac{1}{1 - F_{\mathrm{BD}}} + \mathrm{Wtr}\,1 + \mathrm{Wtr}\,2 + \mathrm{Wtr}\,3 =$$

$$= (95.9 + 209.5 + 20) \times \frac{1}{1 - 0.08} + 0 = 353.7\,\mathrm{t/h} \tag{9.58}$$

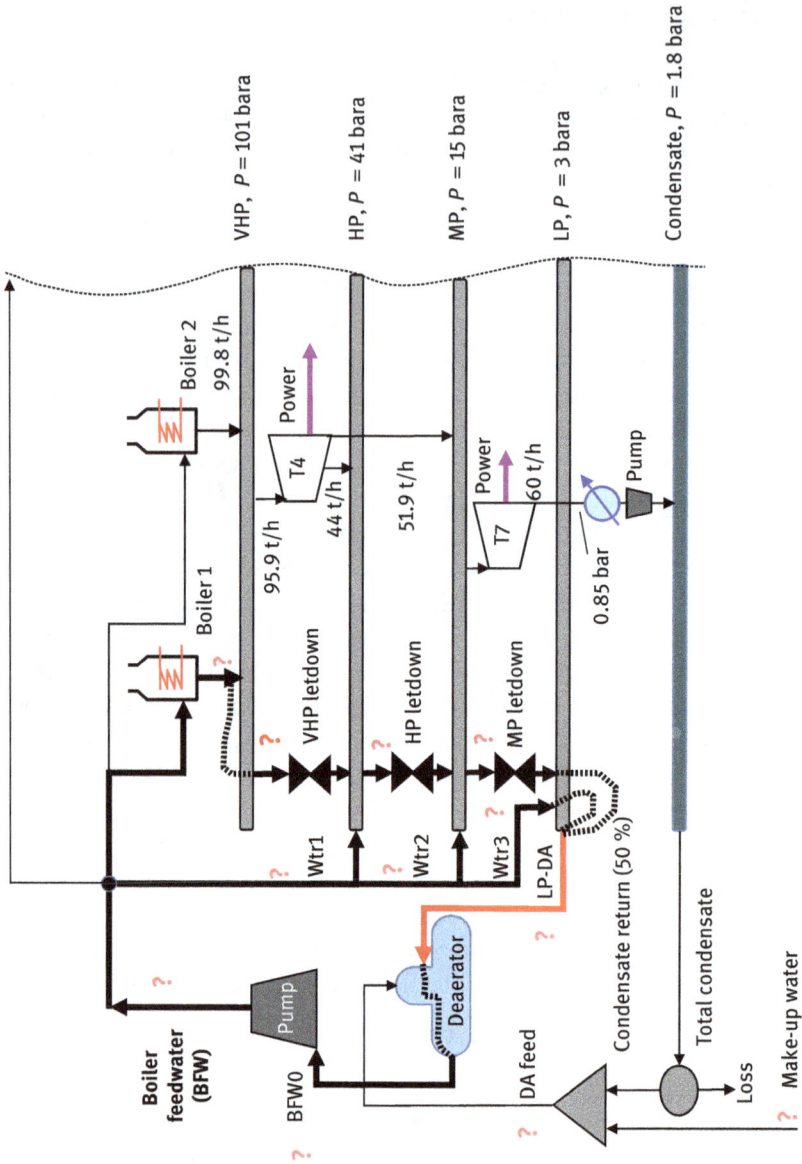

Fig. 9.4: The computational loop highlighted.

Further, the make-up water flow rate is approximated with the sum of the water losses from the system – total blowdown flows and condensate losses:

$$\text{TBDF} = \text{DAF} - (S_{4\text{in}} + S_{6\text{in}} + \text{VHP}_{\text{out}}) = 353.7 - 325.4 = 28.3 \text{ t/h} \quad (9.59)$$

$$\text{MuW} = \text{Loss} + \text{TBDF} = 159.2 + 28.3 = 187.496 \text{ t/h} \quad (9.60)$$

This is necessary for combining the BFW and deaerator steam initialisations. Having an estimate for the flow rate of the make-up water allows estimating the specific enthalpy of the deaerator feed:

$$h_{\text{DAF}} = \frac{\text{MuW} \cdot h_{\text{MuW}} + \text{CRT} \cdot h_{\text{CRT}}}{\text{DAF}} = 0.0390 \text{ MWh/t} \quad (9.61)$$

$$h_{\text{DAF}} \text{ [kJ/kg]} = h_{\text{DAF}} \text{ [MWh/t]} \times 3{,}600 \left[\frac{\text{s}}{\text{h}}\right] \times \frac{0.001 \text{ [t/kg]}}{0.001 \text{ [MWh/kWh]}} = 140.5 \text{ kJ/kg} \quad (9.62)$$

The deaerator steam flow rate and the BFW flow rate are then estimated by iteratively solving the mass and enthalpy balances of the deaerator together, following the configuration shown in Fig. 9.5. This involves manipulating the value of the deaerator steam flow rate (LPDA, Eq. (9.63)), calculating from that the flow rate of the deaerator vent (DAV, Eq. (9.64)), calculating then the estimate of the feedwater flow rate leaving the deaerator (BFW0, Eq. (9.65)) from the deaerator mass balance. This is followed by the calculation of the specific enthalpy of the BFW0 flow (h_{BFW0}, Eq. (9.66)) and ensuring that the difference between the calculated value and the specific enthalpy of saturated liquid at the deaerator pressure is smaller than the calculation tolerance – according to Eq. (9.68).

$$\text{LPDA : manupulated (initialised at 60 t/h)} \quad (9.63)$$

$$\text{DAV} = \text{VF}_{\text{DA}} \times \text{DAS} \quad (9.64)$$

$$\text{BFW} = \text{BFW0} = \text{DAF} + \text{LPDA} - \text{DAV} \quad (9.65)$$

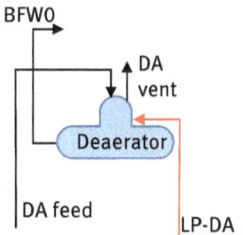

Fig. 9.5: Deaerator configuration.

$h_{\text{BFWO, CALC}}$ [MWh/t]

$$= \frac{\text{DAF } [t/h] \times h_{\text{DAF}} \text{ [MWh/t]} + \text{LPDA } [t/h] \times h_{\text{DAS}} \text{ [MWh/t]} - \text{DAV } [t/h] \times h_{\text{DAV}} \text{ [MWh/t]}}{\text{BFWO } [t/h]}$$

$$(9.66)$$

$$h_{\text{BFWO, CALC}} \text{ [kJ/kg]} = h_{\text{BFWO, CALC}} \text{ [MWh/t]} \times 3,600 \left[\frac{s}{h}\right] \times \frac{0.001 \text{ [t/kg]}}{0.001 \text{ [MWh/kWh]}} \quad (9.67)$$

$$|h_{\text{BFWO, CALC}} - \text{enthalpySatLiq}(P_{\text{DA}})| \leq \text{TOL} \quad (9.68)$$

This iterative process of adjusting the LPDA value produces the values listed in Tab. 9.7.

Tab. 9.7: Deaerator initialisation.

Variable	Unit	Value
LPDA	t/h	55.897
DAV	t/h	1.118
BFW(Loop 1), BFWO	t/h	408.475
MuW	t/h	187.496
$h_{\text{BFWO, CALC}}$	kJ/kg (MWh/t)	0.1363 (490.67)

To balance the enthalpy of the steam mains, a modelling decomposition is used as illustrated in Fig. 9.6. This approach allows calculating the average specific enthalpy of the steam contained in the steam main.

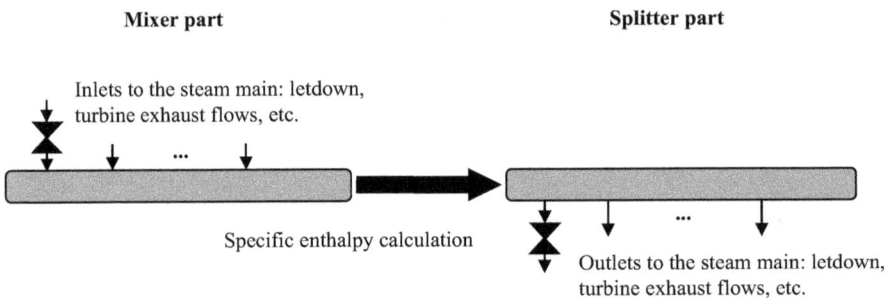

Fig. 9.6: Steam main decomposition into a mixer and a splitter.

The next steps in the initial (first) loop use the mass balances of the splitter of each steam main in the set {LP, MP, HP} to calculate the flow rate of the letdown valve, exhausting steam from the upper steam main. The enthalpy balance of the mixer part is then used for calculating the amount of BFW injected to the steam main.

The process starts with the LP main. The flow rate of the BFW (Wtr3), injected to the LP main, is manipulated to adjust the calculated specific enthalpy of the LP main to the specification. Using the splitter mass balance (Eq. (9.70)) provides an estimate of the sum of the outlet flow rates for the main ($LP_{outlets,Total}$), while the mixer mass balance (Eq. (9.71)) allows estimating the flow rate of the MP-LP letdown (SD3).

$$Wtr3 : \text{manupulated (initialised at 0 t/h)} \tag{9.69}$$

$$LP_{outlets,\,Total} = LPDA + LP_{out} \tag{9.70}$$

$$SD3 = LP_{outlets,\,Total} - Wtr3 - S_{5out} - S_{6Cout} \tag{9.71}$$

The Wtr3 flow, originating from the BFW (see Fig. 9.4), has the same enthalpy, that is $h_{Wtr3} = h_{BFW} = 0.1409\,\text{MWh/t}$ (see Eq. (9.22)). Similarly, the enthalpy of the incoming letdown flow is $h_{SD3} = h_{MP} = 0.7987\,\text{MWh/t}$. Using these values and the calculated specific enthalpies of the exhausts from turbines T5 and T6C allows estimating the calculated value of the LP main specific enthalpy:

$$h_{LP,\,main,\,calc} = \frac{Wtr3 \times h_{Wtr3} + SD3 \times h_{SD3} + S_{5out} \times h_{T5out} + S_{6Cout} \times h_{T6Cout}}{LP_{outlets,\,Total}} \tag{9.72}$$

where all specific enthalpies are in [MWh/t] and all flow rates in [t/h]:

$$h_{LP,\,main,\,calc}\ [\text{kJ/kg}] = h_{LP,\,main,\,calc}\ [\text{MWh/t}] \times 3,600\ \left[\frac{s}{h}\right] \times \frac{0.001\ [\text{t/kg}]}{0.001\ [\text{MWh/kWh}]} \tag{9.73}$$

The calculated estimate is then compared with the specification:

$$\left| h_{LP,\,main,\,calc}\ [\text{kJ/kg}] - h_{LP,\,main,\,spec}\ [\text{kJ/kg}] \right| \le \text{TOL} \tag{9.74}$$

This iterative process of adjusting the Wtr3 value results in satisfying the inequality in eq. (9.74) and produces the values listed in Tab. 9.8.

Tab. 9.8: LP steam main calculation results (Loop 1).

Variable	Unit	Value
Wtr3	t/h	1.628
SD3	t/h	87.769
$LP_{outlets,Total}$	t/h	183.897

Next is balancing the MP steam main using equations of the same structure and meaning as Eqs. (9.69) to (9.74) for the LP steam main. This iterative process of adjusting the Wtr2 value results in minimising the difference between the calculated and the specified specific enthalpy values for the MP steam main below the required tolerance and produces the values listed in Tab. 9.9.

Tab. 9.9: MP steam main calculation results (Loop 1).

Variable	Unit	Value
Wtr2	t/h	4.460
SD2	t/h	52.308
MP$_{\text{outlets,Total}}$	t/h	203.669

The HP steam main is balanced in the same way, providing the estimates listed in Tab. 9.10.

Tab. 9.10: HP steam main calculation results (Loop 1).

Variable	Unit	Value
Wtr1	t/h	9.772
SD1	t/h	33.036
HP$_{\text{outlets,Total}}$	t/h	126.808

The balance of the VHP steam main is calculated in Eq. 9.75.:

$$\text{VH}_{\text{outlets, Total}} = \text{SD1} + S_{4\text{in}} + S_{6\text{in}} + \text{VHPout} = 358.436 \, \text{t/h} \tag{9.75}$$

This allows calculating the estimates of steam generation and blowdown flow for Boiler 1:

$$S_1 = \text{VHP}_{\text{outlets, Total}} - S_2 - S_3 = 233.036 \, \text{t/h} \tag{9.76}$$

$$\text{BD}_1 \, [\text{t/h}] = S_1 \, [\text{t/h}] \times \frac{F_{\text{BD}}}{1 - F_{\text{BD}}} \, [1] = 20.264 \, \text{t/h} \tag{9.77}$$

The BFW and fuel heat for Boiler 1 are then estimated as follows:

$$\text{BFW}_1 = S_1 + \text{BD}_1 = 245.307 \, [\text{t/h}] \tag{9.78}$$

$$Q_{\text{BF1}} \, [\text{MW}] = \text{BFW}_1 \, [\text{t/h}] \times \left(h_{\text{VHP, SatLiq}} - h_{\text{BFW}} \right) [\text{MWh/t}] + S_1 \, [\text{t/h}]$$
$$\times \left(h_{\text{VHP}} - h_{\text{VHP, SatLiq}} \right) [\text{MWh/t}] = 187.857 \, [\text{MW}] \tag{9.79}$$

This calculation completes Loop 1 and further starts Loop 2 by calculating an estimate for the BFW flow rate and comparing it with the value of the variable from Loop 1:

$$\text{BFW(Loop2)} = S_1(\text{Loop1}) + S_2 + S_3 + \text{BD}_{1(\text{Loop1})} + \text{BD}_2 + \text{BD}_{\text{HRSG}}$$
$$+ \text{Wtr} \, 1(\text{Loop1}) + \text{Wtr} \, 2(\text{Loop1}) + \text{Wtr} \, 3(\text{Loop1}) = 405.465 \tag{9.80}$$

$$\Delta BFW(Loop\,1 \rightarrow Loop\,2) = BFW(Loop\,2) - BFW(Loop\,1) = 405.103 - 408.475 = -3.009\,t/h$$

$$(9.81)$$

The discrepancy resulting in Eq. (9.81) indicates the need to carry on further loops. The calculations resume with calculating again the deaerator, then the LP steam main, MP steam main, HP steam main and the VHP steam main – all for Loop 2. The loops are repeated until convergence is achieved at Loop 8, achieving difference between the BFW estimates smaller than the specified tolerance. The evolution of the values of the variables for all loops is traced in Tab. 9.11.

Tab. 9.11: Evolution of the network estimates until the convergence at Loop 8.

Variable	Unit	Values							
		Loop 1	Loop 2	Loop 3	Loop 4	Loop 5	Loop 6	Loop 7	Loop 8
LPDA	t/h	51.589	50.784	50.664	50.646	50.644	50.643	50.643	50.643
DAV	t/h	1.032	1.016	1.013	1.013	1.013	1.013	1.013	1.013
MuW	t/h	187.496	191.549	190.808	190.697	190.681	190.678	190.678	190.678
BFW	t/h	404.253	400.518	399.659	399.531	399.512	399.509	399.508	399.508
Wtr3	t/h	1.501	1.459	1.453	1.452	1.452	1.451	1.451	1.451
SD3	t/h	58.588	57.825	57.712	57.695	57.692	57.692	57.692	57.692
$LP_{outlets,\,Total}$	t/h	179.589	178.784	178.664	178.646	178.644	178.643	178.643	178.643
Wtr2	t/h	8.760	8.711	8.703	8.702	8.702	8.702	8.702	8.702
SD2	t/h	43.829	43.115	43.008	42.993	42.990	42.990	42.990	42.990
$MP_{outlets,\,Total}$	t/h	174.488	173.725	173.612	173.595	173.592	173.592	173.592	173.592
Wtr1	t/h	8.647	8.558	8.545	8.543	8.543	8.543	8.543	8.543
SD1	t/h	25.682	25.056	24.963	24.949	24.947	24.947	24.947	24.947
$HP_{outlets,\,Total}$	t/h	118.329	117.615	117.508	117.493	117.490	117.490	117.490	117.490
$VHP_{outlets,\,Total}$	t/h	351.082	350.456	350.363	350.349	350.347	350.347	350.347	350.347
$S1$	t/h	225.682	225.056	224.963	224.949	224.947	224.947	224.947	224.947
BD1	t/h	19.625	19.570	19.562	19.561	19.561	19.561	19.561	19.561
Q_{BF1}	MW	208.730	208.151	208.065	208.052	208.050	208.050	208.050	208.050

9.2 Using a process simulator for steam balancing

This section describes the steam balancing and utility system simulation using the Petro-SIM process simulator (KBC, 2019). The main goals of this working session are to provide the readers with an extended set of skills for modelling utility systems in an efficient way that enables the interaction with partners – other researchers, customers or experts. The same example as in Section 9.1 is used, with the same specifications.

9.2.1 Process preview

The PFD preview of the utility system is shown in Fig. 9.7. The flowsheet contains four steam mains (VHP, HP, MP and LP), two steam boilers (B1 and B2) as well as a gas turbine (omitted from the flowsheet for simplicity) and a heat recovery steam generator (HRSG), two single-stage turbines [of which one back-pressure turbine (T5) and one condensing turbine (T7)], two multistage turbines (T4, T6), three letdown valves (V1, V2, V3), four steam users (Use-VHP, Use-HP, Use-MP and Use-LP), one cooler emulating condensate heat loss (E-HL), one deaerator (Deaerator) and pumps (P). There are also other auxiliary process units and streams. The whole flowsheet is shown in Fig. 9.7.

The instructions below show screenshots of the simulator views and configuration windows. The screenshots are from the converged simulation. The Petro-SIM convention is that the specified values are rendered in blue font.

9.2.2 Simulation basis

1. Open the Petro-SIM. Create a new case, "File" → "New" → "Case from scratch;" and save the file as "Case Study Utility System;"
2. Select the NBS-Steam of property package;
3. Add components: H_2O;
4. Enter the simulation environment.

9.2.3 Set up the simulation process

9.2.3.1 Add boiler B1, B2 and HRSG1
Add a water stream naming it BFW1. This is done by clicking once on the blue block arrow in the component palette (Fig. 9.8a) on the top-left of the simulator window and then clicking on the sheet. Then double-click on the stream icon on the sheet, which opens the property window (Fig. 9.8b), where on the "Composition" page

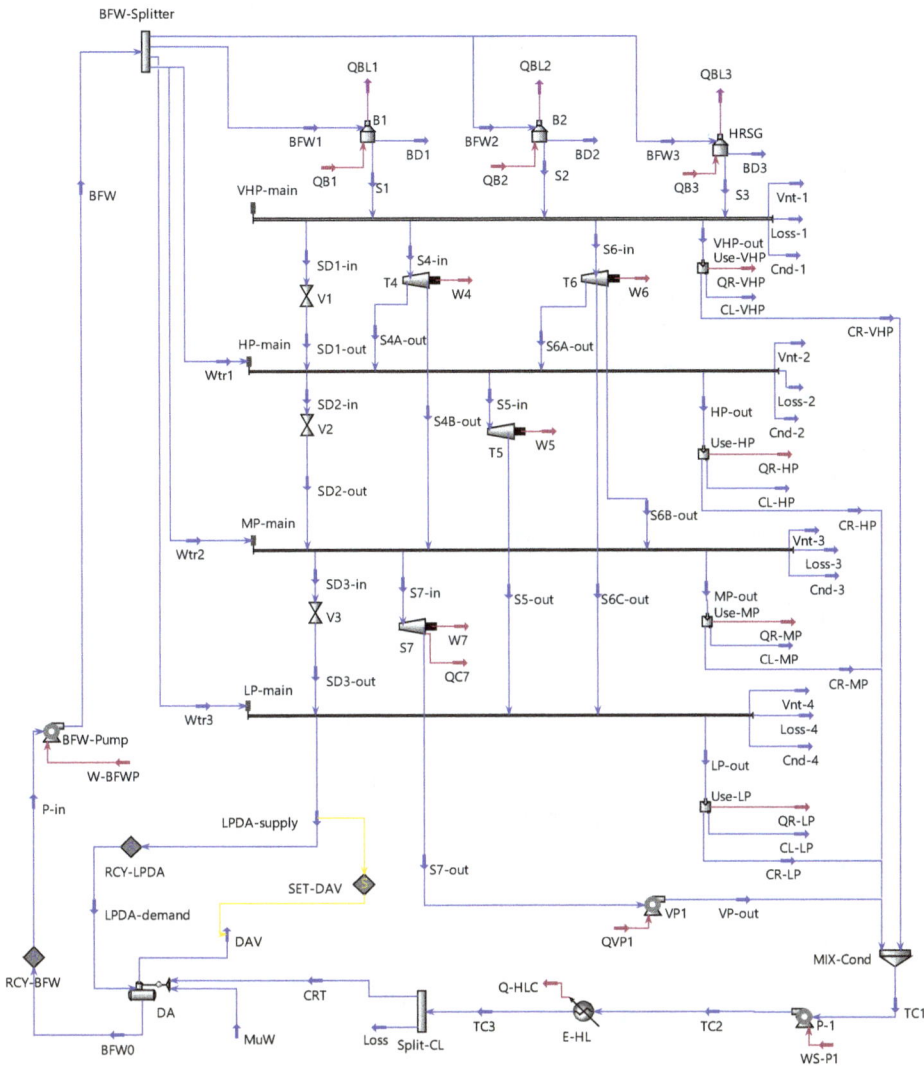

Fig. 9.7: The utility system flowsheet in Petro-SIM.

specify 1 as the mole fraction of water. Similarly, add two more water streams named S1 (steam generation by Boiler 1) and BD1 (blowdown stream of boiler 1), then add two energy streams: QB1 (fuel heat for boiler 1) and QBL1 (heat losses from boiler 1). Set the temperature of S1 to 510 °C on the "Conditions" page. Pressure is not specified at this point. It will be propagated from the pressure specification of the VHP main when it is added to the flowsheet. After that, add a boiler named B1.

Next, it is necessary to provide the settings for boiler B1. Enter the information as shown in Fig. 9.9 on the connections page. The "Boiler feedwater" connection of

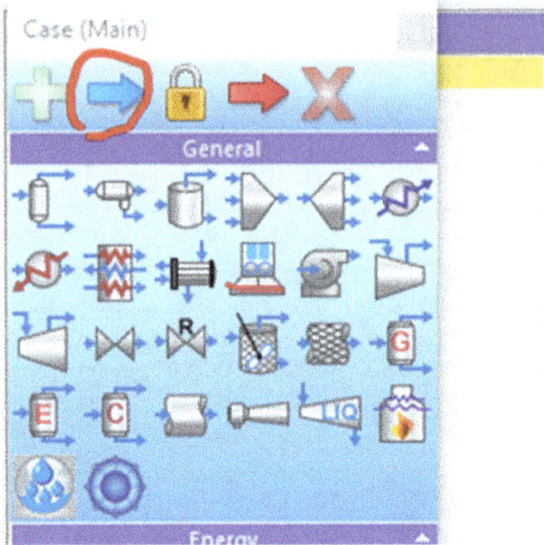

(a) Component palette

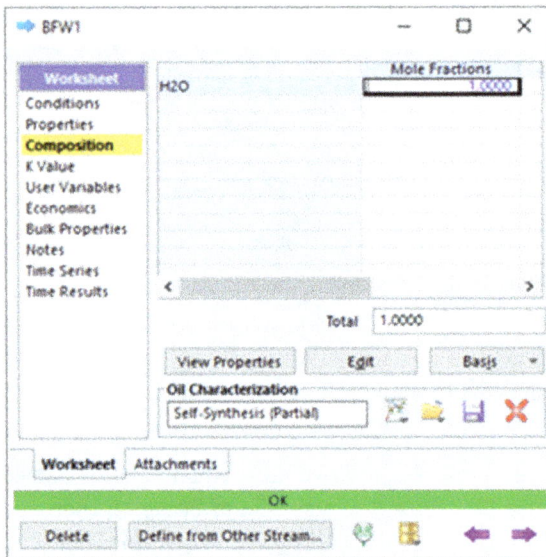

(b) Specification window for BFW1

Fig. 9.8: Adding stream BFW1 (a) component palette and (b) specification window for BFW1.

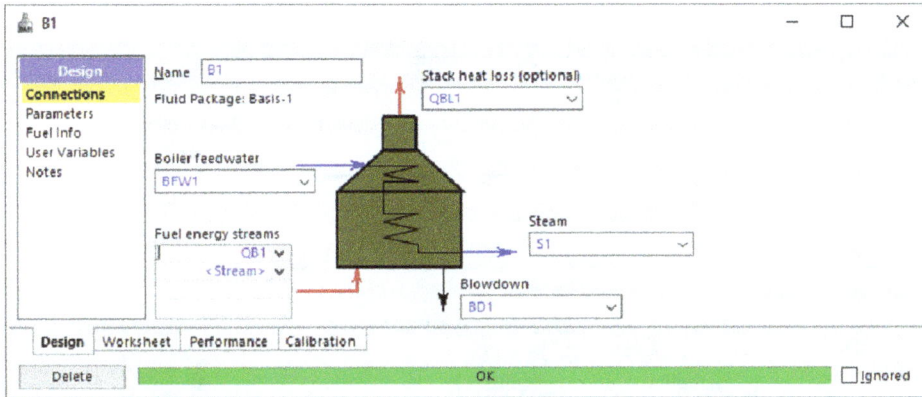

Fig. 9.9: Connections of Boiler B1.

B1 should be set as BFW1, the stream "Fuel energy streams" is QB1, the stream "Stack heat loss" is QBL1, the stream "Steam" is S1, the stream "Blowdown" is BD1. On the parameters page of boiler B1, set the blowdown ratio to 8 % and accept the default setting for the radiant losses, as shown in Fig. 9.10.

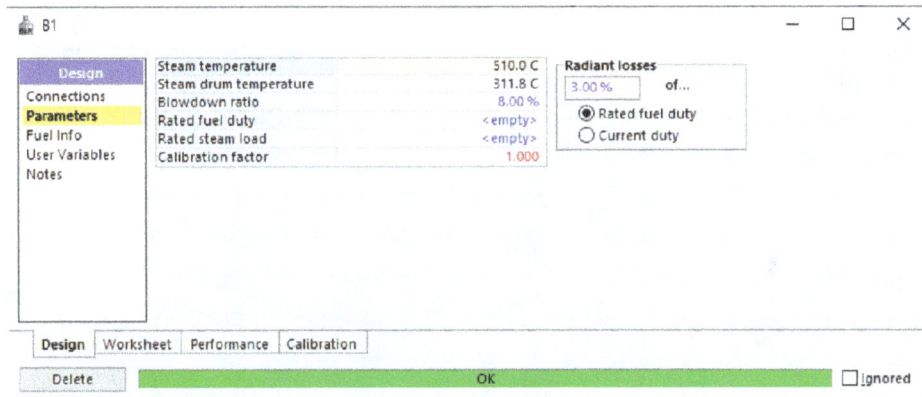

Fig. 9.10: Connections of Boiler B1.

The connections and parameters of B2 and HRSG are set the same as B1, as shown in Tab. 9.12. BFW2, QB2, QBL2, S2 and BD2 are the streams connected with boiler B2. BFW3, QB3, QBL3, S3 and BD3 are the streams connected with the HRSG. On the "Parameters" pages of B2 and HRSG, the blowdown ratio is set to 8 % and the default setting for the radiant losses is accepted. Then, for the steam stream S2, the temperature is set to 510 °C and mass flow is set to 99.8 t/h on the "Worksheet"

Tab. 9.12: Settings of the B2 and HRSG1.

Boiler	Connections					Parameters
	Boiler feedwater	Fuel energy streams	Stack heat loss	Steam	Blowdown	Blowdown ratio [%]
B2	BFW2	QB2	QBL2	S2	BD2	8.00
HRSG1	BFW3	QB3	QBL3	S3	BD3	8.00

page of the "Connections" section. For the steam stream S3, the temperature is set to 510 °C and "Mass flow" is set to 25.6 t/h.

9.2.3.2 Add steam headers: VHP, HP, MP and LP

Start with the VHP main by adding a steam main component from the palette, placing it below the boilers on the flowsheet. This action inserts three steam streams in addition to the main: Vnt-1, Loss-1 and Cnd-1. Specify the molar fraction of water for each of them as 1. Name the object "VHP-main." After that, insert several steam streams named SD1-in, S4-in, S6-in and VHP-out (water molar fraction is also 1). These are the inlets to the VHP-HP letdown station, turbines T4 and T6 and the stream representing the process steam use. Connect the VHP steam main, using the "Connections" sheet as shown in Fig. 9.11. The inlet streams are S1, S2 and S3. The outlet streams are SD1-in, S4-in, S6-in and VHP-out. The auxiliary streams, inserted automatically by the simulator, are also automatically connected to the steam main. The vent stream is Vnt-1, the mass loss stream is Loss-1 and the condensate stream is Cnd-1.

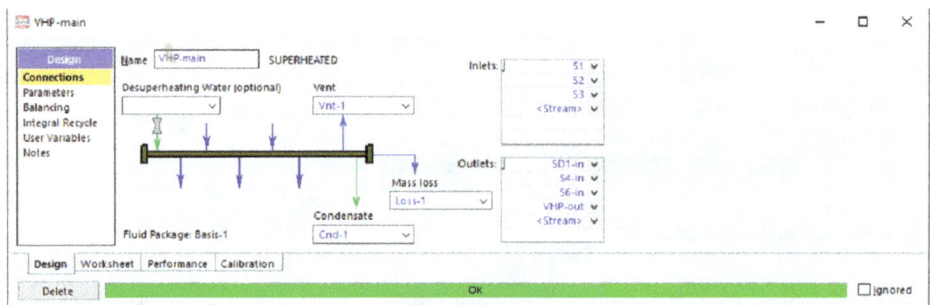

Fig. 9.11: Connections of the VHP steam main.

The parameters of the VHP steam main are shown in Fig. 9.12. Use the data in the figure and set the header pressure as 10.1 MPa (equals to 101 bar(a)) and accept the other default specifications.

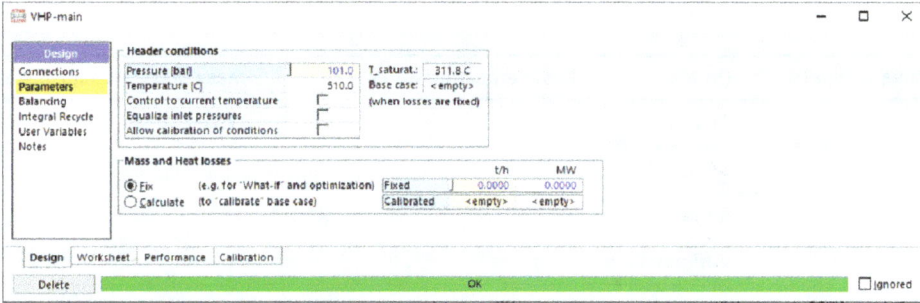

Fig. 9.12: Parameters of the VHP steam main.

The balancing of steam header VHPS is shown in Fig. 9.13. In that screen, for stream S1, in the column "Balance method," choose "Control flow." This setting tells the simulator that the VHP steam header object will adjust the flow rate of stream S1 in order to balance the steam between all inlets and outlets for the header. The specifications of the VHP main at this stage are completed by setting the "Mass flow" values of streams S4-in and S6-in as 95.9 t/h and 209.5 t/h. Remember that the flow rate values of S2 and S3 have already been specified during the setup of the boilers. This leaves the flow rates of the VHP steam use and the stream SD1-in undetermined. The former will be specified later and the latter will be calculated once the flowsheet is completed and solved.

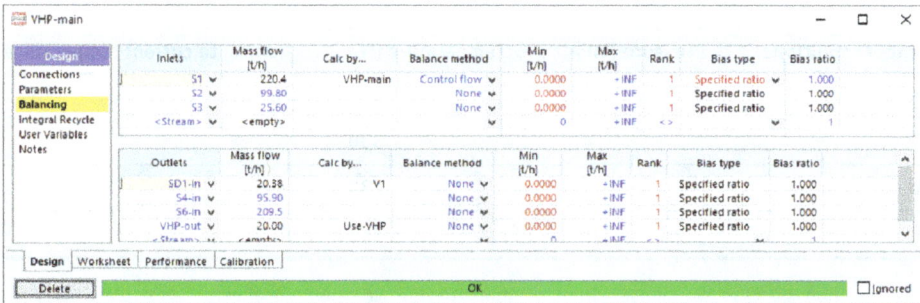

Fig. 9.13: Balancing of the VHP main.

Then, the connections, parameters and balancing of HPS, MPS and LPS are set as same VHPS, as shown in the following tables. Table 9.13 shows the settings of corresponding streams connected with HPS, MPS and LPS. The pressure and temperature settings for all steam mains are shown in Tab. 9.14 and the balance control settings are given in Tab. 9.15. Please use the data in these tables for specifications of the remaining steam mains.

Tab. 9.13: Settings of the Connections page of VHPS, HPS, MPS and LPS.

Name	Inlets	Outlets	Desuperheating water	Vent	Mass loss	Condensate
VHP	S1	SD1-in	–	Vnt-1	Loss-1	Cnd-1
	S2	S4-in				
	S3	S6-in				
		VHP-out				
HP	SD1-out	S5-in	Wtr1	Vnt-2	Loss-2	Cnd-2
	S6A-out	SD2-in				
	S4A-out	HP-out				
MP	S6B-out	S7-in	Wtr2	Vnt-3	Loss-3	Cnd-3
	S4B-out	SD3-in				
	SD2-out	MP-out				
LP	S6C-out	LPDA-supply	Wtr3	Vnt-4	Loss-4	Cnd-4
	S5-out	LP-out				
	SD3-out					

Tab. 9.14: Settings of the Parameters page of VHPS, HPS, MPS and LPS.

Header conditions	Pressure [MPa]	Temperature [°C]	Control to current temperature
VHP	10.1	–	–
HP	4.1	330	√
MP	1.5	230	√
LP	0.3	145	√

9.2.3.3 Add steam turbines: T4, T5, T6 and T7; and a cooler: E1

The settings of connections, parameters, stages and balancing pages of turbine T4 are specified next. The connections of turbine T4 are shown in Fig. 9.14. The "Inlets" list of stream turbine T4 has one entry: S4-in. The "Outlets" list contains S4A-out, S4B-out. The power stream of T4 is W4.

The "Parameters" sheet of turbine T4 is shown in Fig. 9.15. That is used to specify that the Willans Line coefficients will be used for each stage of turbine T4. For all other specifications, use the default values.

Tab. 9.15: Settings of the Balancing page of VHPS, HPS, MPS and LPS.

Name	Inlets			Outlets		
	Streams	Mass flow [t/h]	Balance method	Streams	Mass flow [t/h]	Balance method
VHP	S1	<empty>	Control flow	SD1-in	<empty>	None
	S2	99.8	None	S4-in	95.9	None
	S3	25.9	None	S6-in	209.5	None
				VHPS-out	<empty>	None
HP	SD1-out	<empty>	Control flow	S5-in	20	None
	S6A-out	40	None	SD2-in	<empty>	None
	S4A-out	<empty>	None	HP-out	<empty>	None
MP	S6B-out	<empty>	None	S7-in	60	None
	S4B-out	51.9	None	SD3-in	<empty>	None
	SD2-out	<empty>	Control flow	MP-out	<empty>	None
LP	S6C-out	99.5	None	LP-DA0	<empty>	None
	S5-out	<empty>	None	LP-out	<empty>	None
	SD3-out	<empty>	Control flow			

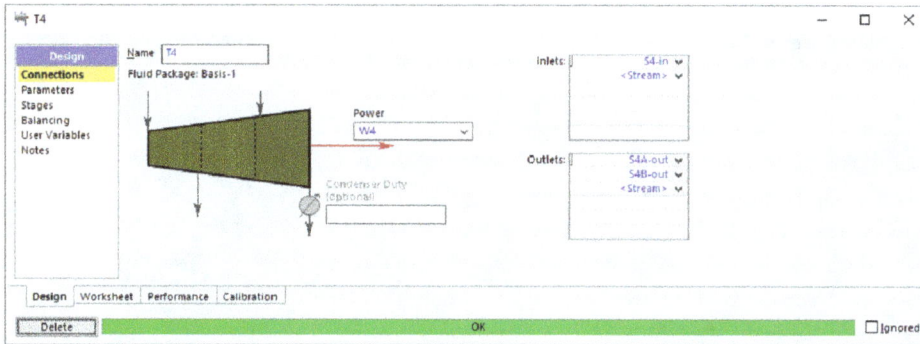

Fig. 9.14: Connections of T4.

The "Stages" sheet of turbine T4 is shown in Fig. 9.16. Insert two stages. In the first stage, stream IN is S4-in, stream OUT is S4A-out and the coefficients are Willans-A = 0.9903, Willans-C = −2.065 MW. For the second stage, stream OUT is S4B-out, and the coefficients are Willans-A = 0.9785 [1], Willans-C = −1.063 MW.

The "Balancing" sheet of turbine T4 is shown in Fig. 9.17. In the section "Inlets," it lists stream S4-in with mass flow as 95.9 t/h, as specified during setting

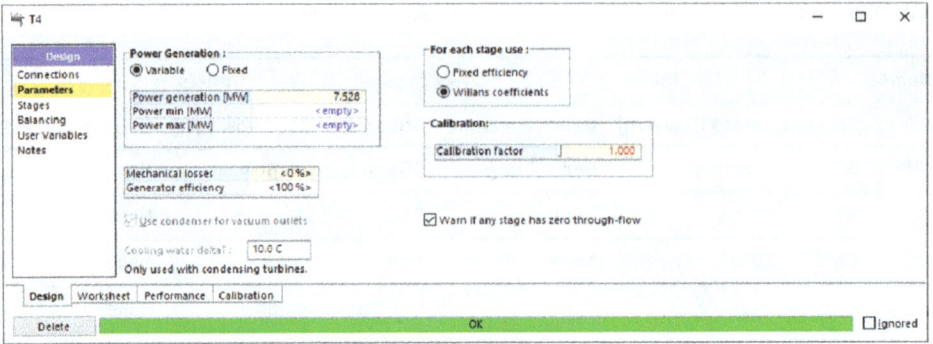

Fig. 9.15: Parameters of T4.

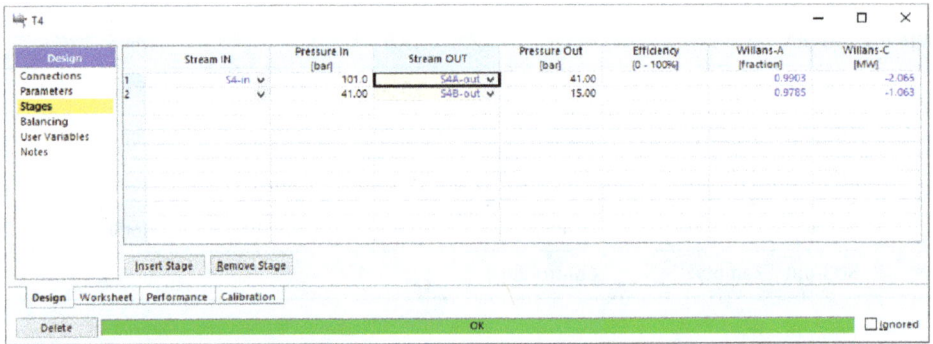

Fig. 9.16: Stages of T4.

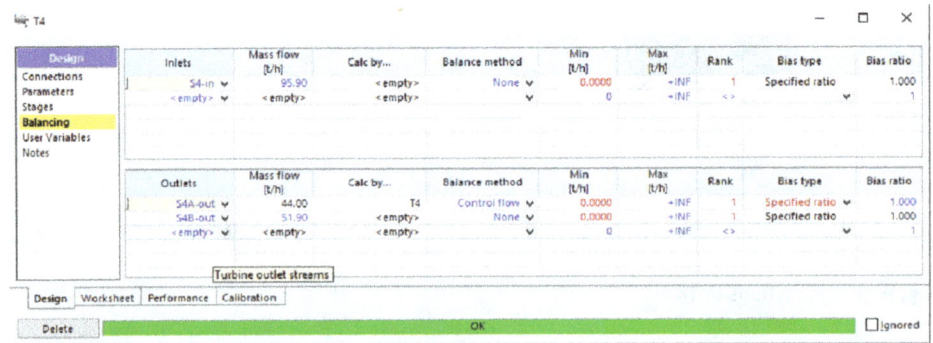

Fig. 9.17: Balancing of T4.

up the VHP main. S4A-out and S4B-out are "Outlets." Specify S4A-out as calculated by selecting the "Control flow" option. The mass flow of S4B-out has already been set to 51.9 t/h through the specifications of the MP steam main.

In a similar way, turbines (T5, T6 and T7) are added and connected to the relevant streams, specifying that the Willans Line coefficients are used in the model. The specifications for the "Connections," "Stages" and "Balancing" pages of T4, T5, T6 and T7 are shown in Tabs. 9.16–9.18.

Tab. 9.16: Settings of the Connections page of T4, T5, T6 and T7.

Turbine	Stream IN	Stream OUT	Power
T4	S4-in	S4A-out	W4
		S4B-out	
T5	S5-in	S5-out	W5
T6	S6-in	S6A-out	W6
		S6B-out	
		S6C-out	
T7	S7-in	S7-out	W7

Tab. 9.17: Settings of the Stages page of T4, T5, T6 and T7.

Turbine	Stream IN	Stream OUT	Willans-A [1] $\in$ [0. .1]	Willans-C [MW]
T4	S4-in	S4A-out	0.9903	−2.065
		S4B-out	0.9785	−1.063
T5	S5-in	S5-out	0.8873	−0.7480
T6	S6-in	S6A-out	1.000 *	−4.236
		S6B-out	0.9992	−2.636
		S6C-out	0.9882	−2.233
T7	S7-in	S7-out	0.7365	−1.538

* The conversion equation produces 1.0007. However, the simulator expects only values in the interval [0. .1]

Tab. 9.18: Settings of the Balancing page of T4, T5, T6 and T7.

Turbine	Inlets			Outlets		
	Streams	Mass flow [t/h]	Balance method	Streams	Mass flow [t/h]	Balance method
T4	S4-in	95.9	None	S4A-out	<empty>	Control flow
			None	S4B-out	51.9	None
T5	S5-in	20	None	S5-out	<empty>	Control flow
T6	S6-in	209.5	None	S6A-out	40	None
				S6B-out	<empty>	Control flow
				S6C-out	99.5	None
T7	S7-in	60	None	S7-out	<empty>	Control flow

The Willans Line coefficients for the simulator specifications (Tab. 9.17) are obtained from the Willans Line coefficients of the Excel model by conversion. The conversion equations are derived by comparing the model from Chapter 7 with the Petro-SIM documentation, and the resulting expressions are as follows:

$$\{"\text{Willan} - A"\}_{,\#\text{TURBINE}}[1] = \frac{n_{,\#\text{TURBINE}}[\text{MWh/t}]}{\Delta h_{\text{IS},\#\text{TURBINE}}[\text{MWh/t}]} \tag{9.82}$$

$$\{"\text{Willans} - C"\}_{,\#\text{TURBINE}}[\text{MW}] = -W_{\text{INT},\#\text{TURBINE}}[\text{MW}] \tag{9.83}$$

The connections of the condensing turbine T7 are shown in Fig. 9.18. The inlet stream is S7-in and the outlet stream is S7-out, whose pressure is set to 0.085 MPa (0.85 bar). The power is specified as energy stream W7. The condenser duty is directed to energy stream QC7.

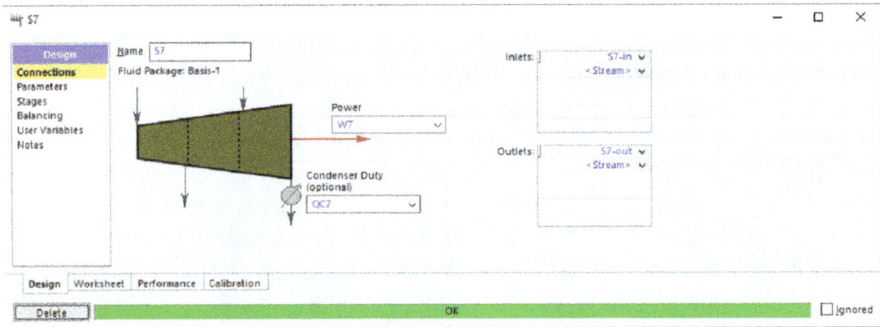

Fig. 9.18: Connections of T7.

The outlet stream S7-out is further connected to a vacuum condensate pump VP1. The connections of the pump are shown in Fig. 9.19. The inlet stream is S7-out and the outlet stream is VP-out (set as a water stream with pressure equal to that of the collected condensate: 0.101325 MPa = 1.01325 bar). The energy stream is QVP1. Then, on the parameters page (same window, next item in the list), the adiabatic efficiency value is set 75 %.

9.2.3.4 Add steam use units

"Steam Use" is the name of the operating unit in the simulator that represents the process steam use. It is found in the "Energy" section in the PFD component palette. Add four instances of that unit operation named Use-VHP, Use-HP, Use-MP and Use-LP. The "Connections" sheet for Use-VHP is shown in Fig. 9.20. The "Steam IN" connection of Use1 is selected as VHP-out. The "Condensate recovered" connection is CR-VHP, "Condensate lost" is CL-VHP and "Heat removed" is QR-VHP. Enter the settings shown in the figure.

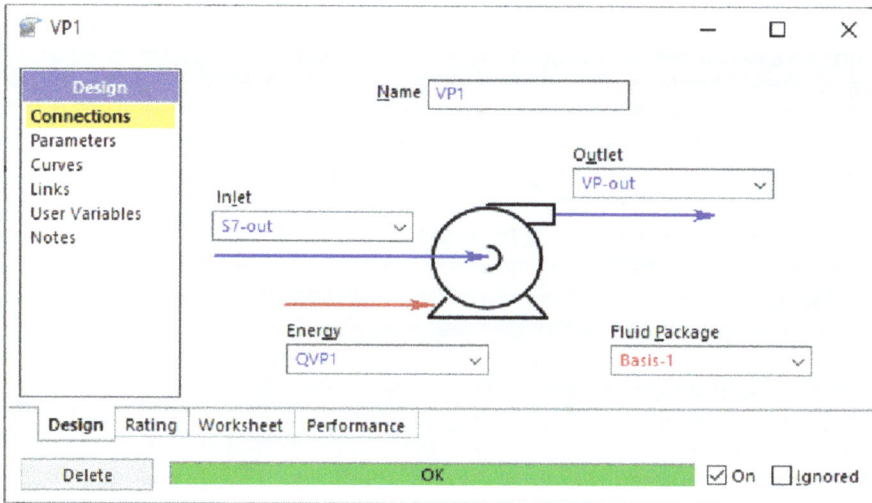

Fig. 9.19: Connections of VP1.

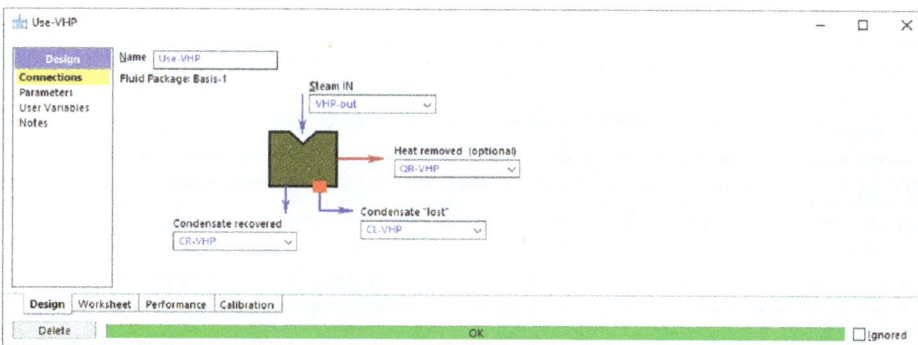

Fig. 9.20: Connections of Use-VHP.

On the "parameters" page of Use-VHP, the mass flow is set 20 t/h. Since the CRT and loss are modelled separately in the steam use units, the "Cond. Recov." is set 100% (Fig. 9.21). The condensate from the process steam users is assumed to be collected at atmospheric pressure. For this reason, the condensate pressure is set as 0.101325 MPa (1.01325 bar), as shown in Fig. 9.21. Use-HP, Use-MP and Use-LP units are set up in the same way as Use-VHP, using the specifications in Tabs. 9.19 and 9.20.

9.2.3.5 Add the letdown valves: V1, V2 and V3
Next, the letdown valves are added: V1 between the VHP and HP mains, V2 between HP and MP mains and V4 between MP and LP mains. First, add valve V1. Use the

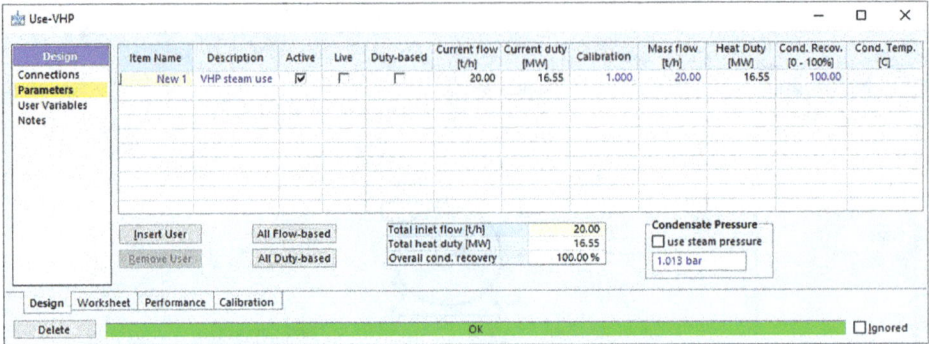

Fig. 9.21: Parameters of Use-VHP.

Tab. 9.19: Settings for the connections page of Use2, Use3 and Use4.

Steam use	Steam IN	Heat removed	Condensate recovered	Condensate "lost"
Use-HP	HP-out	QR-HP	CR-HP	CL-HP
Use-MP	MP-out	QR-MP	CR-MP	CL-MP
Use-LP	LP-out	QR-LP	CR-LP	CL-LP

Tab. 9.20: Settings for the parameters page of use-HP, use-MP and use-LP.

Steam use	Calibration	Mass flow [t/h]	Cond. recov. [%]	Condensate pressure [MPa] (bar)
Use-HP	1	54.5	100	0.101325 (1.01325)
Use-HP	1	55.9	100	0.101325 (1.01325)
Use-HP	1	128	100	0.101325 (1.01325)

information from the "Connections" page as shown in Fig. 9.22. The "Inlet" connection of valve V1 is selected as SD1-in and the "Outlet" as SD1-out. Then, the pressure of SD1-out is set 4.10 MPa (41 bar).

The connections of V2 and V3 are set in a similar way, using the data in Tab. 9.21. Then, the pressure of SD2-out is set to 1.50 MPa (15 bar) and that of SD3-out is set to 0.30 MPa (3 bar).

9.2.3.6 Add the other units

The next steps involve connecting the condensate from the process steam use to the modelling elements for the condensate loss. Add a mixer and name it MIX-Cond. This connects the condensate coming from the process steam users and from turbine T7 with the intermediate stream for total condensate (TC1), before modelling the heat

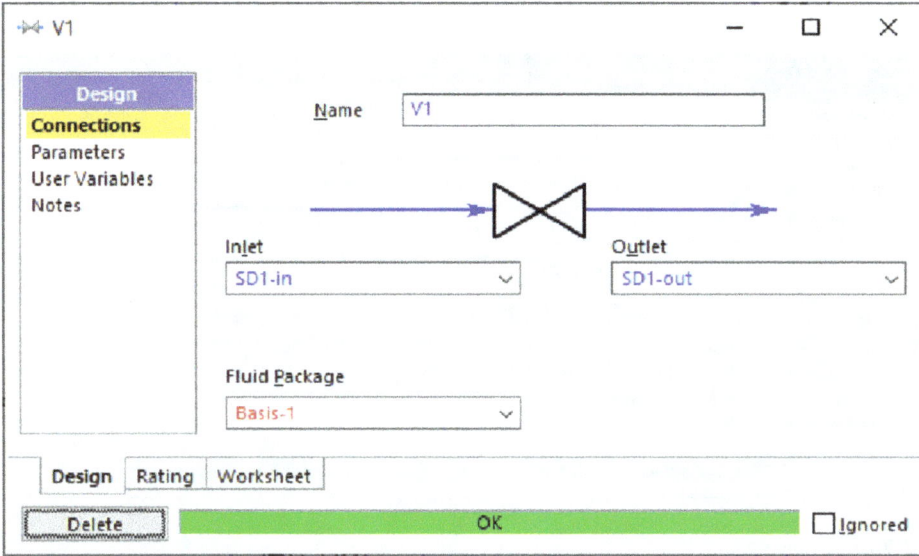

Fig. 9.22: Connections of V1.

Tab. 9.21: Settings of the Connections page of V2 and V3.

Isenthalpic valve	Inlet	Outlet
V2	SD2-in	SD2-out
V3	SD3-in	SD3-out

and mass loss. As inputs to MIX-Cond, set streams CR-VHP, CR-HP, CR-MP, CR-LP and VP-out. As the outlet stream, define the new water stream named TC1. This configuration is shown in Fig. 9.23.

The collected condensate comes at atmospheric pressure, while the CRTto the steam system comes at 0.18 MPa (1.8 bar). The pressure increase is performed by a pump, which is added to the flowsheet as a pump unit named P1. The inlet to the pump is the initial TC1, and the outlet is the TC1 stream after the pressure increase, it is named TC2. These settings are shown in Fig. 9.24 For the outlet TC2 of P1, the set the pressure to 0.18 MPa (1.8 bar).

At this point, the process units simulating the heat loss and the condensate mass loss are added. Add a cooler named E-HL, which takes as inlet stream TC2 and has as an outlet a new stream named TC3. As the energy connection specification of E-HL, define a new energy stream named Q-HLC. Set TC to be a water stream with pressure 0.18 MPa (1.8 bar) and temperature 45 °C. The connections for E-HL are shown in Fig. 9.25.

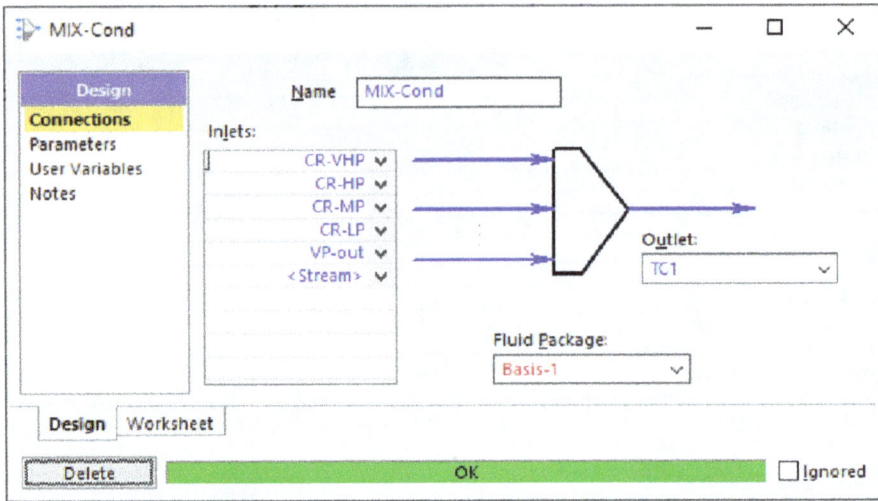

Fig. 9.23: Connections of MIX-Cond.

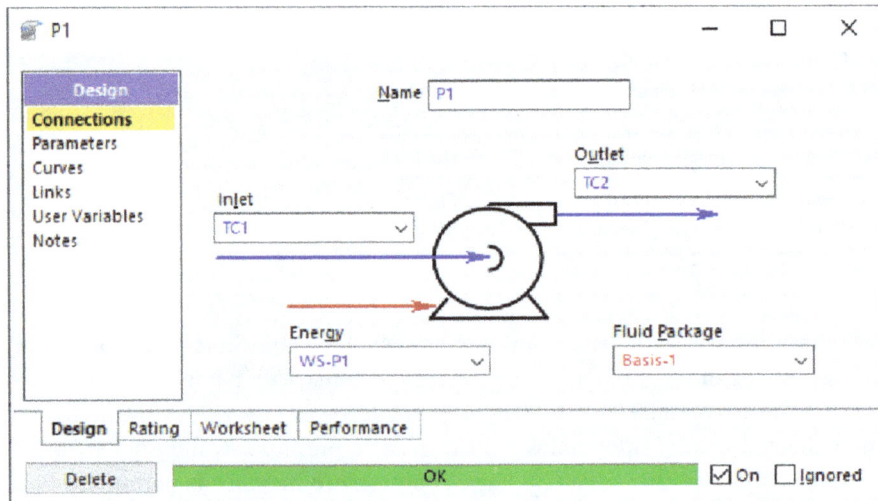

Fig. 9.24: Connections of P1.

The mass loss of condensate is modelled by a splitter unit. In Petro-SIM, this type of unit is labelled as "TEE." Place one splitter (TEE) on the flowsheet next to the cooler E-HL, name it "Split-CL" and connect its inlet to stream TC3. The outlet connections are specified as water streams named CRT and loss (condensate loss). The connections page should look like the screenshot in Fig. 9.26. For Split-CL, on the page design → parameters → splits, the flow ratio of CRT is set to 0.5, as shown in Fig. 9.27 and the fraction for loss is calculated then also to 0.5.

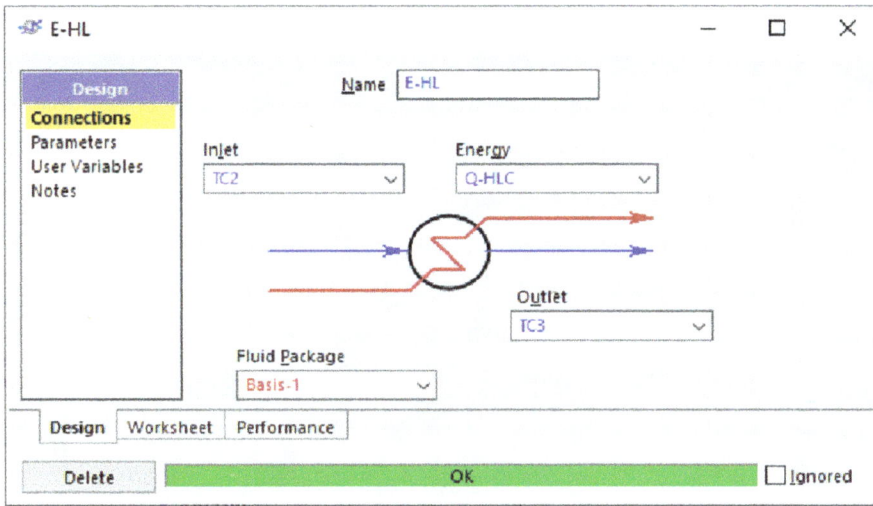

Fig. 9.25: Connections of E-HL.

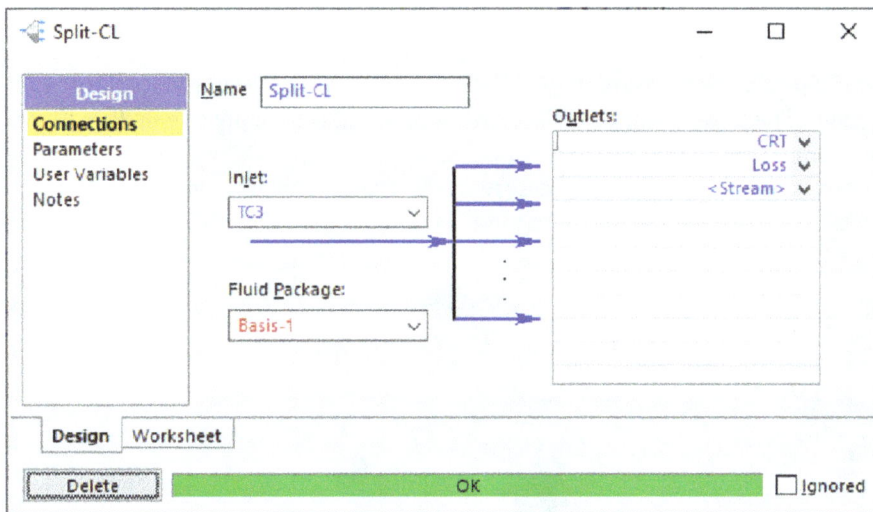

Fig. 9.26: Connections of Split-CL.

Add a Deaerator unit naming it DA. Name its vent stream, that appears automatically, as DAV and specify its composition as 100% water (molar fraction of 1). Route stream CRT to the "Condensate" connection of DA. As a Make-up water connection, define a new water stream named MuW. The pressure and the temperature of MuW are set as 0.18 MPa (1.8 bar) and 25 °C. The "Boiler feedwater connection" of DA is set

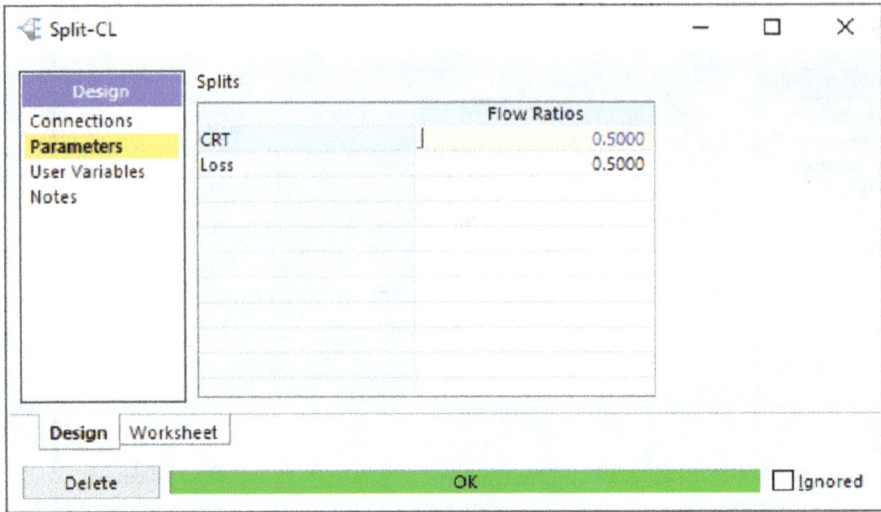

Fig. 9.27: Parameters of Split-CL.

as a new water stream named BFW0, specified as a water stream of 0.15 MPa (1.5 bar) pressure. As the "Steam IN" connection, specify a new water stream named "LPDA demand." Specify the pressure of "LPDA-demand" as 3 bar – equal to that of the LP steam main. The stream will be further linked to the LPDA-supply stream that leaves the LP steam main and is defined earlier in the current workflow. For DA, on the Parameters page, the Operating pressure in the group named "Vent Data" is set to 0.15 MPa (1.5 bar). The connections and parameters of the de-aerator are shown in Figs. 9.28 and 9.29

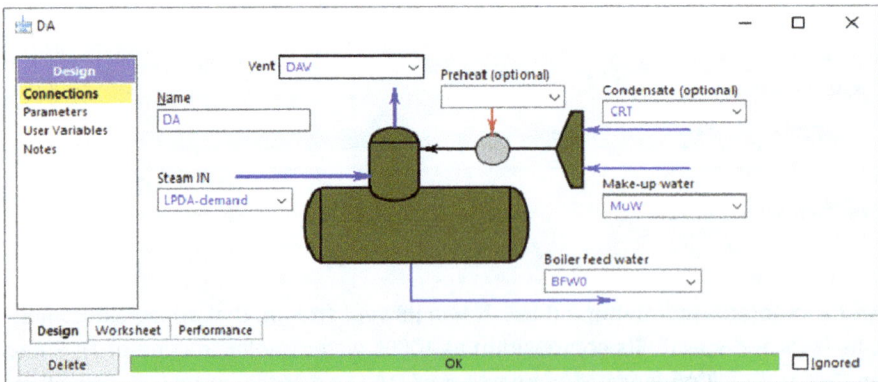

Fig. 9.28: Connections of the deaerator.

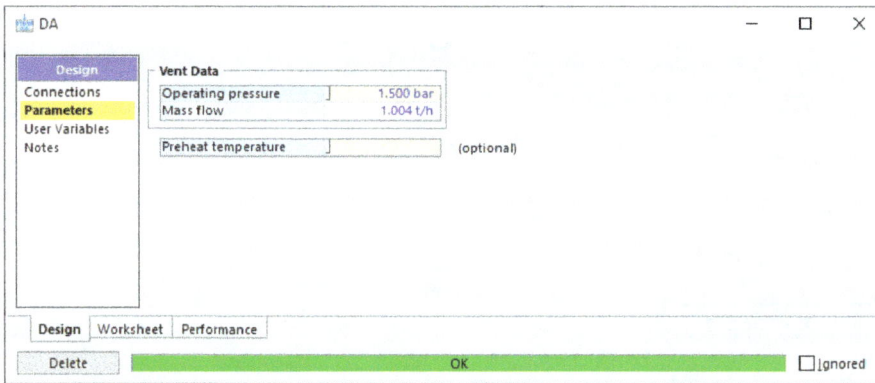

Fig. 9.29: Parameters of the deaerator.

At this point, it is important to realise, referring to Fig. 9.1, that there are two stream loops mapping to two computational loops. One is formed by the steam use unit operations, which then returns some of the condensate to the de-aerator. This loop is implicit in the manual calculations, as shown in Excel. In the Petro-SIM simulation, this loop is closed by using a recycle unit operation tearing into supply and demanding the BFW stream, leaving the deaerator. That recycle unit operation links the pair of streams BFW0 and P-in.

The second loop involves the LP steam used for the deaerator and in the manual calculations is explicitly iterated, see Fig. 9.4. For this reason, there are two simulator streams corresponding to the single flowsheet stream of the LP steam used for the deaerator. The LP-DA stream from Fig. 9.1 is represented by two simulator streams: LPDA-supply and LPDA-demand (Fig. 9.7).

The flow rate (property "mass flow") of the deaerator vent DAV is set to 2 % of the flow rate of stream LPDA. Such a proportional setting cannot be done directly from the properties of the stream itself. To do this, add a SET unit from the section "Logical" of the component palette (named SET-DAV). The object in the "Target Variable" group is selected as DAV, and the setting "Variable" is selected as "Mass Flow." The "Source" property of SET-DAV is selected as LPDA-supply in the connections page. It is important to use "LPDA-supply" because of the direction, in which the computational loops are performed. These selections are illustrated in Fig. 9.30, which reflects the connections of the SET-DAV unit, and in Fig. 9.31, which shows the window for detailed selection of the data source when the "Select Variable" button of the main properties window is clicked. In the parameters page of SET-DAV, set "Multiplier" to 0.02 and "Offset" to 0, as shown in Fig. 9.32.

Next is the addition of the BFW splitter, that connects the BFW total flow to the individual connections. Add a "TEE" unit in the upper left corner of the

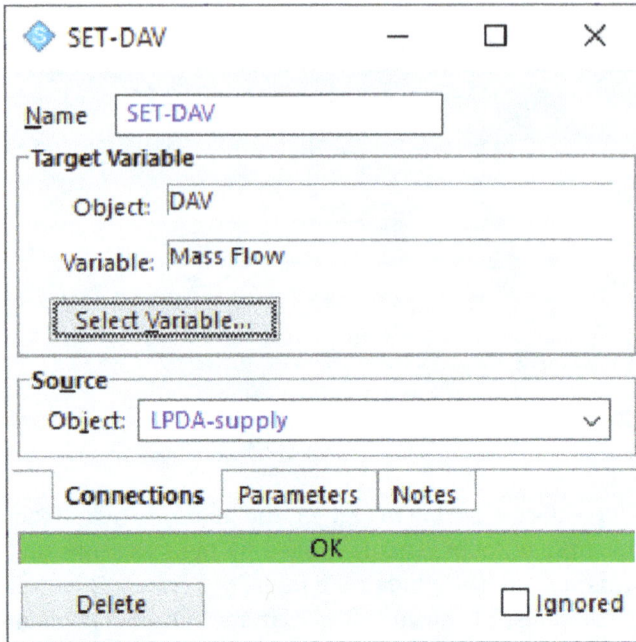

Fig. 9.30: Connections page of the SET-DAV unit.

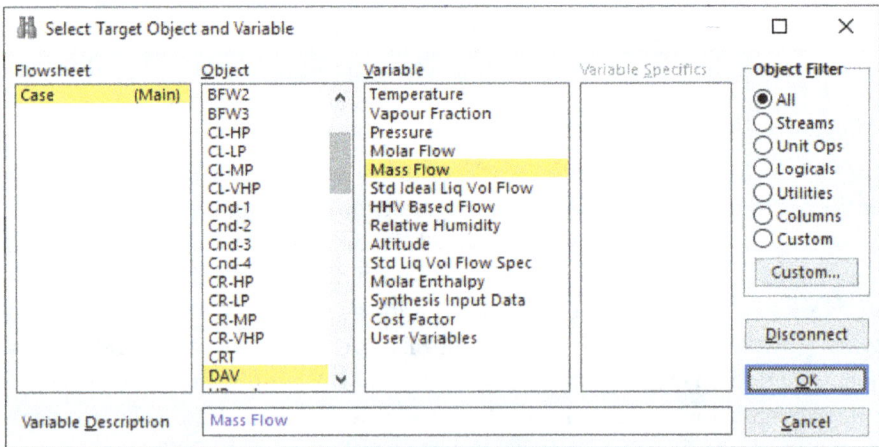

Fig. 9.31: Detailed selections window for specifying the object and variable of the data source for the SET-DAV unit.

flowsheet, close to the boiler icons and name it "BFW-Splitter" (Fig. 9.33). As an inlet stream, create a new water stream named BFW. As outlets select BFW1, BFW2, BFW3, Wtr1, Wtr2 and Wtr3. The splitter outlet flows are either specified

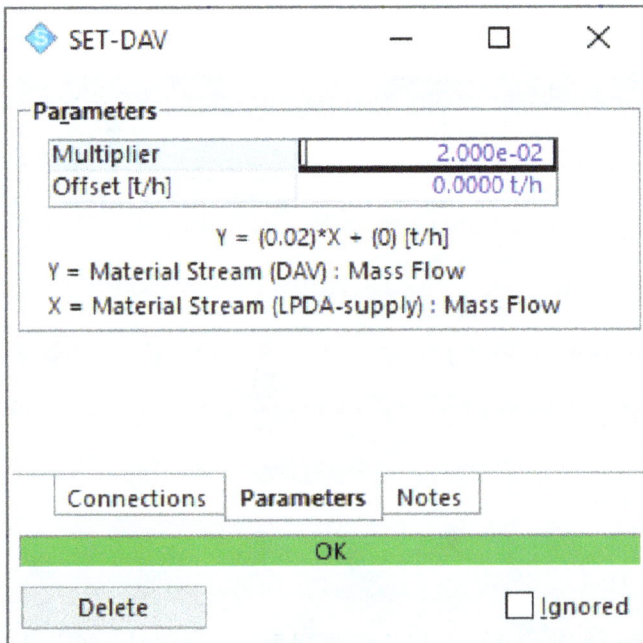

Fig. 9.32: Parameters page of the SET-DAV unit.

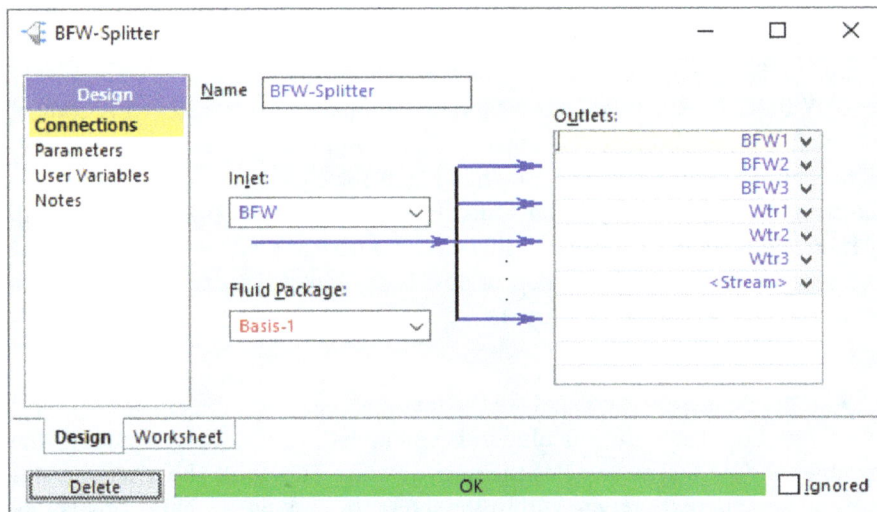

Fig. 9.33: Connections of BFW-splitter.

or could be be calculated by the simulation, so the split ratio specifications should be left empty. Next, add a pump, name it as "BFW-Pump" (Fig. 9.34) and specify for outlet BFW, then for inlet a water stream, P-in. Then, for the inlet stream

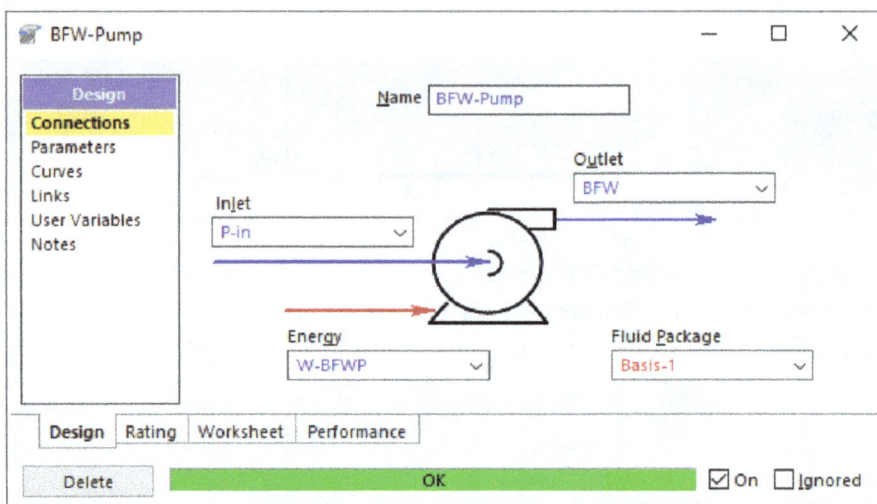

Fig. 9.34: Connections of BFW-pump.

P-in, specify the composition as 100% water, the pressure as 0.15 MPa (1.5 bar) and the vapour/phase fraction as 0. The energy stream for BFW-Pump is specified in the "Connections" page as "W-BFWP."

To complete the flowsheet topology, it is necessary to add two "Recycle" logical units from the component palette. Add one such unit and name it "RCY-BFW" (see Fig. 9.35). Set the connections as: Inlet: BFW0, outlet: P-in.

For RCY-BFW, on the tolerances of parameters page, the transfer directions of pressure and flow are both set backwards, and the others are set not transferred, and all tolerances should be set as shown in Fig. 9.36.

The second "Recycle" unit to be added links LPDA-supply and LPDA-demand. Add it from the palette and name as RCY-LPDA (Fig. 9.37). For RCY-LPDA, on the tolerances page, the transfer direction of flow is set backwards, and the others are set not transferred, as shown in Fig. 9.38.

9.2.3.7 Run the simulation and analyse the results

Run the simulation. Once the calculation is completed, the resulting values of the key variables should be the same as the ones given in Tab. 9.22. The discrepancies between the two simulations are within the 4.832 t/h maximum value. This is due to the differences between the thermodynamic property packages used for the Excel file and in the simulator. The specific enthalpies of the steam turbine exhausts and the steam mains are calculated at values differing by up to nearly 70 kJ/kg for the LP main. These differences are smaller for the higher pressure exhausts – from close to zero and up to 9 kJ/kg for the HP main. A summary of the differences are

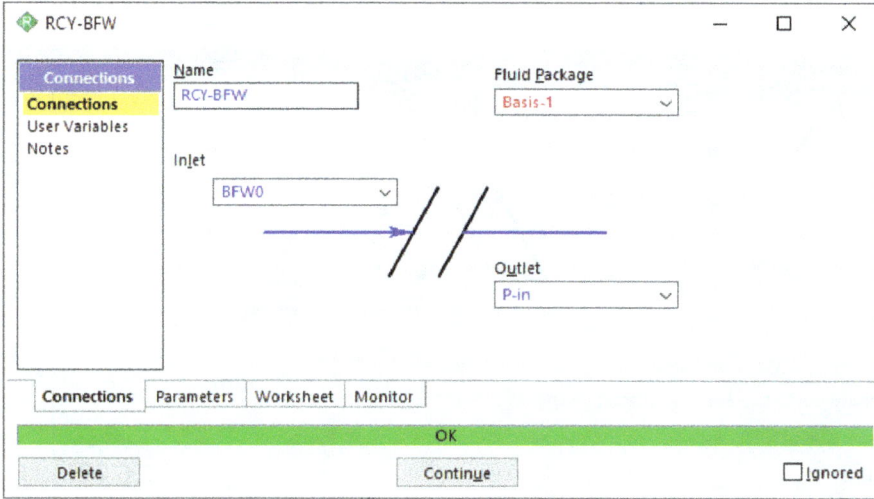

Fig. 9.35: Connections of RCY-BFW.

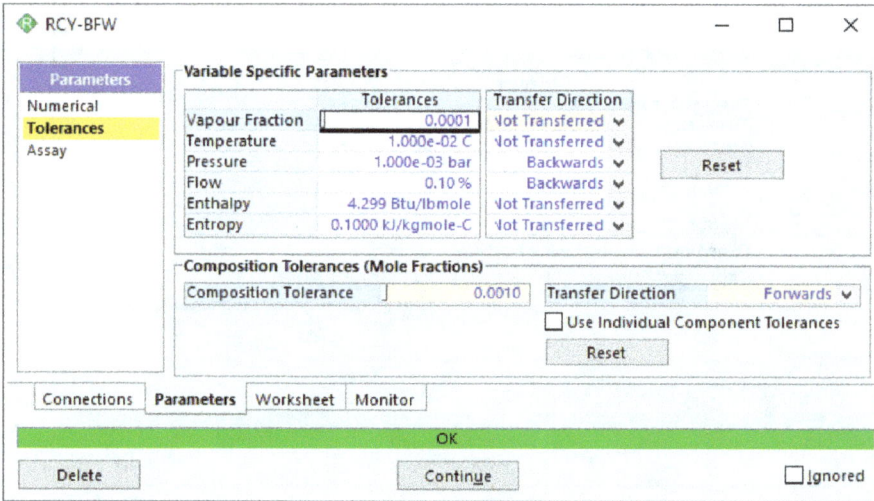

Fig. 9.36: The tolerances/parameters page of RCY-BFW.

provided in Tab. 9.23. These differences in the enthalpy calculations can be connected to the differences in the letdown flows, and by implication over the loop – to the deaerator flows. For reference to the readers, Tab. 9.24 lists the essential properties of the material and energy streams.

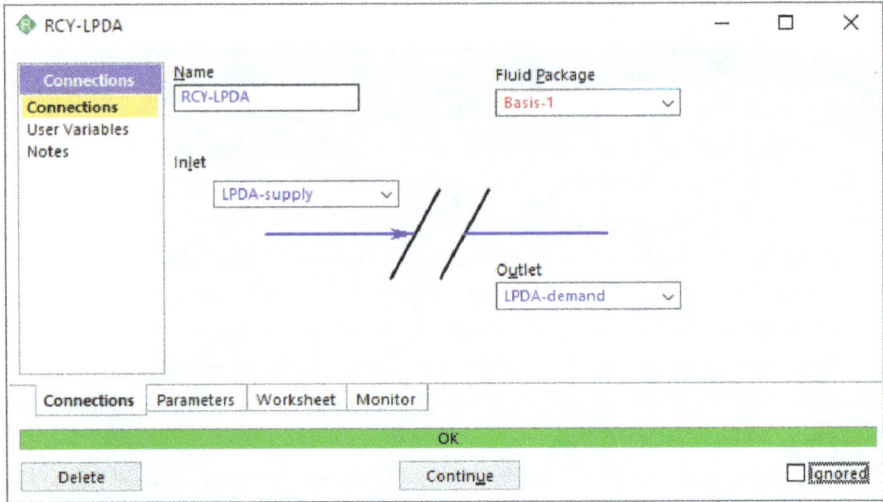

Fig. 9.37: Connections of RCY-LPDA.

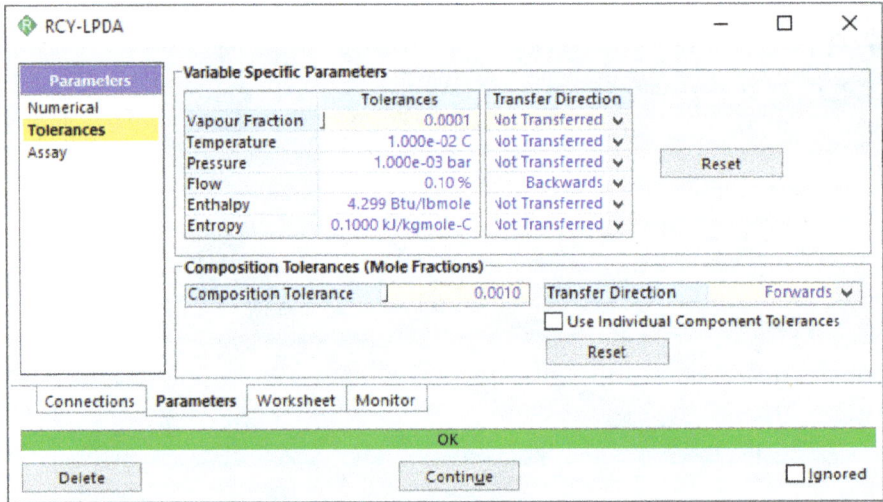

Fig. 9.38: The tolerances/parameters page of RCY-LPDA.

9.3 Summary

This chapter shows two possible ways of performing steam balances and simulating a utility system – explicit modelling using spreadsheet software and automating the simulation using a commercial process simulator. The first part is important for

Tab. 9.22: Simulation results from the spreadsheet model and the Petro-SIM model: a comparison.

Variable	Unit	Values		Difference = Excel–{Petro-SIM}	
		Excel model	Petro-SIM model	[t/h] or [MW]	%
LPDA	t/h	50.643	50.183	0.46	0.91
DAV	t/h	1.013	1.004	0.009	0.89
MuW	t/h	190.678	187.668	3.01	1.58
BFW	t/h	399.508	396.246	3.262	0.82
Wtr3	t/h	1.451	4.340	−2.889	−199.10
SD3	t/h	57.692	54.343	3.349	5.80
Wtr2	t/h	8.702	10.185	−1.483	−17.04
SD2	t/h	42.990	38.158	4.832	11.24
Wtr1	t/h	8.543	8.276	0.267	3.13
SD1	t/h	24.947	20.382	4.565	18.30
S1	t/h	224.947	220.382	4.565	2.03
BD1	t/h	19.561	17.631	1.930	9.87
Q_{BF1}	MW	208.050	203.307	4.743	2.28

Tab. 9.23: Differences between the specific enthalpies of the two models, kJ/kg.

Variable		Excel model	Petro-SIM model	Difference = Excel–{Petro-SIM}
$h_{T5,out}$	Exhaust of T5	2,695.6	2,711.5	−15.9
$h_{T6C,out}$	LP exhaust of T6	2,721.8	2,789.1	−67.2
$h_{LP,calc}$	Enthalpy at the LP main	2,750.33	2,750.15	0.18
$h_{T4B,out}$	MP exhaust of T4	3,005.8	3,031.5	−25.7
$h_{T6B,out}$	MP extraction of T6	2,975.2	3,003.1	−27.9
$h_{MP,calc}$	Enthalpy at the MP main	2,875.46	2,874.1	1.4
$h_{T4A,out}$	HP extraction of T4	3,196.7	3,205.8	−9.1
$h_{T6A,out}$	HP extraction of T6	3,188.7	3,198.3	−9.5
$h_{HP,calc}$	Enthalpy at the HP main	3,039.8	3,039.8	−9.5

learning the principles of steam balancing and process simulation, providing deeper understanding of modelling and simulation techniques including the principle of sequential-modular simulation. The second part, using the Petro-SIM simulator, is important for acquiring skills in using process simulators, as tools that can save the

Tab. 9.24: Complete list of stream properties from the Petro-SIM simulation.

Object	DAV	Wtr1	Wtr2	Wtr3	BD1	BD2	BD3	CR-VHP	S5-in	S5-out
Mass flow [t/h]	1.00366	8.27618	10.1853	4.33981	17.6305	7.984	2.048	20	20	20
Temperature [°C]	111.4	113.3	113.3	113.3	311.8	311.8	311.8	100	330	133.6
Pressure [bar]	1.5	120	120	120	101	101	101	1.013	41	3
Vapour fraction	1	0	0	0	0	0	0	0	1	0.9936
Mass enthalpy [kJ/kg]	2,693	483.8	483.8	483.8	1,412	1,412	1,412	419.1	3,038	2,711

Object	VHP-out	SD1-in	SD1-out	S3	BFW3	SD2-in	SD2-out	S6A-out	S4A-out	LPDA-supply
Mass flow [t/h]	20	20.3818	20.3818	25.6	27.648	38.158	38.158	40	44	50.183
Temperature [°C]	510	510	480.1	510	113.3	330	300.5	394.5	397.6	145
Pressure [bar]	101	101	41	101	120	41	15	41	41	3
Vapour fraction	1	1	1	1	0	1	1	1	1	1
Mass enthalpy [kJ/kg]	3,399	3,399	3,399	3,399	483.8	3,038	3,038	3,198	3,206	2,750

Object	LPDA-demand	S4B-out	SD3-in	SD3-out	CR-HP	HP-out	CR-MP	MP-out	S7-in	S7-out
Mass flow [t/h]	50.1866	51.9	54.3432	54.3432	54.5	54.5	55.9	55.9	60	60
Temperature [°C]	145	301.3	230	204.4	100	330	100	230	230	95.15
Pressure [bar]	3	15	15	3	1.013	41	1.013	15	15	0.85
Vapour fraction	1	1	1	1	0	1	0	1	1	0

Mass enthalpy [kJ/kg]	2,750	3,040	2,874	2,874	419.1	3,038	419.1	2,874	2,874	398.6
Object	VP-out	S6B-out	S4-in	S6 C-out	S2	BFW0	BFW1	BFW2	CR-LP	LP-out
Mass flow [t/h]	60	70	95.9	99.5	99.8	396.051	238.012	107.784	128	128
Temperature [°C]	95.15	287.8	510	163.3	510	111.4	113.3	113.3	100	145
Pressure [bar]	1.013	15	101	3	101	1.5	120	120	1.013	3
Vapour fraction	0	1	1	1	1	0	0	0	0	1

Mass enthalpy [kJ/kg]	398.6	3,010	3,399	2,789	3,399	467.2	483.8	483.8	419.1	2,750
Object	CRT	Loss	MuW	S6-in	S1	TC1	TC2	TC3	BFW	P-in
Mass flow [t/h]	159.2	159.2	187.668	209.5	220.382	318.4	318.4	318.4	396.246	396.246
Temperature [°C]	45	45	25	510	510	99.09	99.1	45	113.3	111.4
Pressure [bar]	1.8	1.8	1.8	101	101	1.013	1.8	1.8	120	1.5
Vapour fraction	0	0	0	1	1	0	0	0	0	0

Mass enthalpy [kJ/kg]	188.6	188.6	104.9	3,399	3,399	415.2	415.3	188.6	483.8	467.2
Object	Q-HLC	QB1	QB2	QB3	QC7	QR-HP	QR-LP	QR-MP	QR-VHP	QVP1
Heat/power Flow [MW]	20.056	203.307	92.0677	23.6166	36.6487	39.6494	82.8829	38.1208	16.5527	3.77E-04

Object	W-BFWP	W4	W5	W6	W7	WS-P1
Heat/power flow [MW]	1.83073	7.5278	1.81478	26.6313	4.60892	9.67E-03

modelling time and enable larger-scale studies. It is important to stress that having performed both studies, allows comparing the results and verifying the modelling assumptions. In this particular case, the importance of selecting the right steam property package has been illustrated.

Put together, the two parts equip the readers with the skills and knowledge necessary to complete tasks relevant to practical industrial problems. At this point, the readers should be able to describe industrial utility systems, compose flowsheet models and solve them for evaluating the performance of the utility systems.

Nomenclature

Symbol	Measurement unit	Description
A_{ST}; $A_{ST\#TURBINE}$	MW	Steam turbine performance regression parameter
$b_{0,BP}$	MW	Backpressure steam turbine regression parameter
B1, B2	–	Steam boilers in the Petro-SIM flowsheet
$b_{1,BP}$	MW/K	Backpressure steam turbine regression parameter
$b_{2,BP}$	[1]	Backpressure steam turbine regression parameter
$b_{3,BP}$	1/K	Backpressure steam turbine regression parameter
$b_{0,COND}$	MW	Condensing steam turbine regression parameter
$b_{1,COND}$	MW/K	Condensing steam turbine regression parameter
$b_{2,COND}$	[1]	Condensing steam turbine regression parameter
$b_{3,COND}$	1/K	Condensing steam turbine regression parameter
$BD_{\#BOILER}$	t/h	Mass flow rates of the boiler blowdown of Boiler 2 and the HRSG
BD_1	t/h	The blowdown flows of Boiler B1
BD_2	t/h	Mass flowrate of the blowdown from Boiler B2
BD_{HRSG}; BD3	t/h	Mass flow rate of the blowdown from the HRSG
BFW	–	Boiler Feedwater
BFW{#}	–	BFW streams in the flowsheet, {#} = {1. . .3}
BFW	t/h	The mass flow of BFW from the pump
BFW0	t/h	The mass flow of BFW from the deaerator
BFW_1	t/h	The mass flow of Boiler B1 feed water
BFW_2	t/h	Mass flow rate of BFW to Boiler B2
BFW_{HRSG}	t/h	Mass flow rate of the BFW to HRSG
B_{ST}; $B_{ST,\#TURBINE}$	[1]	Steam turbine regression performance parameter
Cond	t/h	The mass flow of condensate collection
C_{HP}	t/h	The mass flow of condensate recovered of HP steam through the process
C_{LP}	t/h	The mass flow of condensate recovered of LP steam through the process
C_{MP}	t/h	The mass flow of condensate recovered of MP steam through the process

(continued)

Symbol	Measurement unit	Description
C_{VHP}	t/h	The mass flow of condensate recovered from the VHP steam process users
CRR	[1]	Condensate return ratio
CRT	–	Condensate return
CRT	t/h	The mass flow of Condensate ReTurn
DA	–	Deaerator
DAF	t/h	Mass flow rate of the deaerator feed stream
DAS	t/h	Mass flow rate of the LP steam drawn to feed the deaerator
DAV	t/h	Mass flow rate of the deaerator vent
E-HL	–	Cooler flowsheet unit emulating condensate heat loss
F_{BD}	[1]	The blowdown fractions of boilers and the HRSG
$h_{\#LEVEL}$	kJ/kg; MWh/t	Enthalpy of steam main {#LEVEL} – including condensate
h_{BFW}	kJ/kg	Enthalpy of the BFW from the pump
h_{BFW0}	kJ/kg	The BFW enthalpy after the deaerator
$h_{BFW0,CALC}$	MWh/t	The calculated specific enthalpy of the BFW0 flow
h_{cond}	kJ/kg; MWh/t	Specific enthalpy if the condensate main
$h_{Cond,SatVap}$	MWh/t	The condenser specific enthalpy of outlet steam of turbine T7
h_{CRT}	kJ/kg	Specific enthalpy of condensate Return
h_{DAF}	MWh/t	Specific enthalpy of the deaerator feed
h_{DAS}	MWh/t	Specific enthalpy of the drawn from the LP main
h_{DAV}	MWh/t	Specific enthalpy of the deaerator vent
h_{DV}	kJ/kg; MWh/t	Specific enthalpy of the deaerator vent
h_{HP}	kJ/kg; MWh/t	Specific enthalpy of HP steam pipe network
h_{HPout}	kJ/kg	Specific enthalpy of HP steam supplying the process
h_{in}	kJ/kg	Specific enthalpy of steam turbine inlet
$h_{In,T4B}$	kJ/kg	Specific enthalpy of inlet steam of turbine T4B
$h_{In,T6B}$	kJ/kg	Specific enthalpy of inlet steam of turbine T6B
$h_{In,T6C}$	kJ/kg	Specific enthalpy of inlet steam of turbine T6C
$h_{In,\#TURBINE}$	kJ/kg	Specific enthalpy of the inlet steam for each turbine
h_{LP}	kJ/kg	Specific enthalpy of LP steam pipe network
h_{LPDA}	kJ/kg	Specific enthalpy of the deaerator steam (LPDA)
$h_{LP,main,calc}$	MWh/t	The calculated value of the LP main enthalpy
$h_{LP,main,spec}$	MWh/t	The specificied value of the LP main enthalpy
h_{LPout}	kJ/kg	Specific enthalpy of LP steam supplying the process
h_{MP}	kJ/kg; MWh/t	Specific enthalpy of MP steam pipe network
h_{MPout}	kJ/kg	The enthalpy of MP steam supplying the process
h_{out}	kJ/kg; MWh/t	The enthalpy of a steam turbine outlet
h_{MuW}	kJ/kg	Specific enthalpy of the Make-up Water
$h_{out,is}$; $h_{Out,IS,\#TURBINE}$	kJ/kg	The outlet steam enthalpy for isentropic expansion for each component turbine
$h_{Out,T4A}$	kJ/kg	Specific enthalpy of outlet steam of turbine T4A
$h_{Out,T6A}$	kJ/kg	Specific enthalpy of outlet steam of turbine T6A
$h_{Out,T6B}$	kJ/kg	Specific enthalpy of outlet steam of turbine T6B

(continued)

Symbol	Measurement unit	Description
$h_{Out,T7}$	kJ/kg	Specific enthalpy of outlet steam of turbine T7
$h_{Out,\#TURBINE}$	kJ/kg	Specific enthalpy of a steam turbine outlet
HP (steam)	–	High-pressure steam
HP_{out}	t/h	The mass flow of HP steam supplying the process
$HP_{outlets,Total}$	t/h	Total mass flow rate of steam leaving the HP main
HP vent	t/h	Mass flow rate of the vent on the HP main
HRSG	–	Heat recovery steam generator
h_{T4Aout}	kJ/kg	Specific enthalpy of outlet A steam of turbine T4
h_{T4Bout}	kJ/kg	Specific enthalpy of outlet B steam of turbine T4
h_{T5in}	kJ/kg	Specific enthalpy of inlet steam of turbine T5
h_{T5out}	MWh/t	Specific enthalpy of outlet steam of turbine T5
h_{T6Aout}	kJ/kg	Specific enthalpy of outlet A steam of turbine T6
h_{T6Bout}	kJ/kg	Specific enthalpy of outlet B steam of turbine T6
h_{T6Cout}	MWh/t	Specific enthalpy of outlet C steam of turbine T6
h_{T7in}	kJ/kg	Specific enthalpy of inlet steam of turbine T7
h_{SD1in}	kJ/kg; MWh/t	Enthalpy of inlet steam of letdown valve VHP letdown
h_{SD1out}	kJ/kg; MWh/t	Enthalpy of outlet steam of letdown valve VHP letdown
h_{SD2in}	kJ/kg; MWh/t	Enthalpy of inlet steam of letdown valve HP letdown
h_{SD2out}	kJ/kg; MWh/t	Enthalpy of outlet steam of letdown valve HP letdown
h_{SD3}	MWh/t	The enthalpy of MP letdown
h_{SD3in}	kJ/kg; MWh/t	Enthalpy of inlet steam of letdown valve MP letdown
h_{SD3out}	kJ/kg; MWh/t	Enthalpy of outlet steam of letdown valve MP letdown
h_{VHP}	kJ/kg; MWh/t	Enthalpy of VHP steam main
$h_{VHP,SatLiq}$	kJ/kg	Enthalpy of liquid water at the VHP level
h_{Wtr1}	kJ/kg	The enthalpy of Wtr1
h_{Wtr2}	kJ/kg	The enthalpy of Wtr2
h_{Wtr3}	kJ/kg	The enthalpy of Wtr3
IAPWS	–	The International Association for the Properties of Water and Steam
L	[1]	Intercept ratio
Loss	t/h	The mass flow of condensate loss
LP (steam)	–	Low-pressure steam
LPDA	t/h	The mass flow of LP steam into the deaerator
LP_{out}	t/h	Mass flow of the LP steam supplied to the processes
$LP_{outlets,Total}$	t/h	The sum of the outlet flow rates for the LP main
LP vent	t/h	Mass flow rate of the vent on the LP main
m	t/h	Steam flow through a turbine
m_{max}	t/h	Maximum steam flows for each turbine
$m_{(max,\#TURBINE)}$	t/h	Maximum steam flow for a turbine
MP (steam)	–	Medium Pressure steam
MP_{out}	t/h	The mass flow of MP steam supplying the process
$MP_{outlets,Total}$	t/h	The sum of the outlet flow rates for the MP main
MP vent	t/h	Mass flowr ate of the vent on the MP main

(continued)

Symbol	Measurement unit	Description
m_{steam}	t/h	Steam flow across each turbine (stage)
MuW	t/h	Mass flow rate of make-up water
$m_{\#TURBINE}$	t/h	Mass flow rate of steam passing through a turbine
n_{st}	MWh/t	The slope of the Willans Line
$n_{ST,\#TURBINE}$	MWh/t	The slope of the Willans Line of {#TURBINE}
$P_{\#LEVEL}$	bar	The pressure of the steam mains and the condensate
P_{BFW}	bar	The pressure of BFW from the pump
P_{BFW}	bar	The BFW pressure after the deaerator
P_{cond}	bar	The condensate pressure
P_{CRT}	bar	Pressure of the condensate return flow
P_{DA}	bar	The pressure of the deaerator
PFD	–	Process Flow Diagram
P_{HP}	bar	Steam pressure at the HP level
P_{in}	bar	The inlet pressure of steam turbine
$P_{In,\#TURBINE}$	bar	The pressure at the turbine inlets
P_{LP}	bar	Steam pressure at the LP level
P_{MP}	bar	Steam pressure at the MP level
P_{MuW}	bar	Pressure of the make-up water
P_{out}	bar	The outlet pressure of steam turbine
$P_{Out,\#TURBINE}$	bar	The pressure at the turbine outlets
P_{VHP}	bar	Steam pressure at the VHP level
Q_{BF1}	MW	Fuel heat required by Boiler B1
Q_{BF2}	MW	Fuel heat required by Boiler B2
$Q_{Cond,T7}$	MW	The condenser duty of turbine T7
QB{#}	MW	Fuel heat stream for Boiler {#} = {1..3}
QBL{#}	MW	Heat loss stream for Boiler {#} = {1..3}
Q_P	MW	The power required by the pump
$S_{\#BOILER}$	t/h	Mass flow rates of steam generation by Boiler 2 and the HRSG
S_1	t/h	The mass flow of steam generated by Boiler B1
S_2	t/h	Mass flow rate of steam generated by Boiler B2
S_3	t/h	Mass flow rate of steam generated by HRSG
S_{4in}	t/h	The mass flow of inlet steam of turbine T4
S_{4Aout}	t/h	The mass flow of outlet A steam of turbine T4
S_{4Bout}	t/h	The mass flow of outlet B steam of turbine T4
S_{5in}	t/h	The mass flow of inlet steam of turbine T5
S_{5out}	t/h	The mass flow of outlet steam of turbine T5
S_{6B}	t/h	Mass flow rate of steam expanding through the second stage of turbine T6
S_{6C}	t/h	Mass flow rate of steam expanding through the third stage of turbine T6
S_{6in}	t/h	The mass flow of inlet steam of turbine T6
S_{6Aout}	t/h	The mass flow of outlet A steam of turbine T6
S_{6Bout}	t/h	The mass flow of outlet B steam of turbine T6

(continued)

Symbol	Measurement unit	Description
S_{6Cout}	t/h	The mass flow of outlet C steam of turbine T6
S_{7in}	t/h	The mass flow of inlet steam of turbine T7
S_{7out}	t/h	The mass flow of outlet steam of turbine T7
se_{in}	kJ/kg/K	Specific entropy if steam turbine inlet
$SD_{\{\#letdown\}}$	t/h	The mass flow of steam through a letdown valve
$s_{In,\#TURBINE}$	kJ/kg/K	The entropy of the inlet steam for each turbine
T	K	Temperature
$T\{\#\}$	–	Steam turbine number $\{\#\}$
$T_{\#LEVEL}$	K;°C	The temperature of the steam mains and the condensate
T4-in	t/h	Mass flow rate of the Inlet to steam turbine T4
T4A	–	HP extraction connection of turbine T4/VHP-HP stage of turbine T6
T4A-out	t/h	Mass flow rate of HP extraction from turbine T4
T4B	–	Backpressure (MP) exhaust connection of turbine T4/HP-MP stage of turbine T6
T5-in	t/h	Mass flow rate of the Inlet to steam turbine T5
T6A	–	HP extraction connection of turbine T6/VHP-HP stage of turbine T6
T6-in	t/h	Mass flow rate of the Inlet to steam turbine T6
T6A-out	t/h	Mass flow rate of HP extraction from turbine T6
T6B-out	t/h	Mass flow rate of MP extraction from turbine T6
T6B	–	MP extraction connection of turbine T6/HP-MP stage of turbine T6
T6C	–	Backpressure (LP) exhaust connection of turbine T6/MP-LP stage of turbine T6
T7-in	t/h	Mass flow rate of the Inlet to steam turbine T6
TBDF	t/h	The sum of the water losses from the system with blowdown and condensate loss
T_{cond}	K; °C	The condensate temperature
T_{CRT}	K; °C	Temperature of the condensate return flow
T_{in}	K	The inlet temperature of steam turbine
$T_{in,\#TURBINE}$	K	The temperatures at the turbine inlets
T_{HP}	K; °C	The temperature of HP steam main
T_{LP}	K; °C	The temperature of LP steam main
T_{MP}	K; °C	The temperature of LP steam main
TOL	Multiple units	Tolerance for converging numeric differences
$T_{Sat,in}$	K	The saturation temperatures at the turbine inlets
$T_{Sat,out}$	K	The saturation temperatures at the turbine outlets
T_{MuW}	K	The temperature of make-up water
$T_{Sat,In,\#TURBINE}$	K	The saturation temperatures at the turbine inlets
$T_{Sat,Out,\#TURBINE}$	K	The saturation temperatures at the turbine outlets
T_{VHP}	K; °C	Temperature of steam at the VHP main
Use-$\{\#main\}$	–	Steam users for $\{\#main\}$ = {VHP, HP, MP, LP}
$V\{\#\}$	–	Letdown valves, $\{\#\}$ = {1..3}
VHP (steam)	–	Very high-pressure steam
VHP_{out}	t/h	Mass flow rate of the VHP steam supplied to the processes

(continued)

Symbol	Measurement unit	Description
VHP$_{outlets,Total}$	t/h	The sum of the outlet flow rates for the VHP main
VHP vent	t/h	Mass flow rate of the vent on the VHP main
VF$_{DA}$	[1]	Deaerator vent fraction
W	MW	Turbine power generation
W_{int}; $W_{INT, \#TURBINE}$	MW	The intercept of the Willans Line
{"Willans–A"}$_{\#TURBINE}$	[1]	Willans-A coefficient specification for steam turbines in Petro-SIM
{"Willans–C"}$_{\#TURBINE}$	MW	Willans-C coefficient specification for steam turbines in Petro-SIM
W_{total}; $W_{total, \#TURBINE}$	MW	The total energy extraction from the passing steam flow for each turbine
Wtr{#}	t/h	The mass flow of BFW injected into the steram mains, to control the temperature of steam pipe network. {#} = {1, 2, 3} = {HP, MP, LP}
$W_{\#TURBINE}$	MW	The power generation of a steam turbine
X	[1]	The dryness fraction of the exhaust steam
X_{out}; $X_{Out, \#TURBINE}$	[1]	The dryness fraction a steam turbine exhaust
ΔBFW	t/h	The BFW flow difference between the calculated estimate with the specification
Δh_{IS}; $\Delta h_{IS, \#TURBINE}$	kJ/kg; MWh/t	Steam enthalpy difference for isentropic expansion for each component turbine
ΔP	Pa	Pump pressure difference
ΔT_{sat}, $\Delta T_{Sat, \#TURBINE}$	K	Saturation temperature difference across a steam turbine
η	[1]	The efficiency of the BFW pump
η_{mech}	[1]	Power generation efficiencies of the turbines
$\eta_{\#TURBINE}$	[1]	The power generation efficiency for each turbine
ρ	kg/m^3	The density of BFW through the pump

References

Himmelblau, D.M. and Riggs, J.B. 2012. Basic Principles and Calculations in Chemical Engineering. 8th ed. Prentice Hall International Series in the Physical and Chemical Engineering Sciences, Upper Saddle River, NJ, USA: Prentice Hall.

IAPWS. 2020. International Association for the Properties of Water and Steam. http://www.iapws.org/, accessed 19/04/2020.

KBC. 2019. Petro-SIM | Process Simulation Software | KBC. https://www.kbc.global/software/process-simulation-software, accessed 19/08/2020.

Microsoft. 2019. Spreadsheet Software – Excel Free Trial – Microsoft Excel. https://products.office. com/en-us/excel, accessed 28/08/2019.

Varbanov, P., Perry, S., Makwana, Y., Zhu, X.X., and Smith, R. 2004. Top-Level Analysis of Site Utility Systems. Chemical Engineering Research and Design 82(6): 784–795. https://doi.org/ 10.1205/026387604774196064.

Wagner, W. and Kretzschmar, H.-J., eds 2008. IAPWS Industrial Formulation 1997 for the Thermodynamic Properties of Water and Steam. In: International Steam Tables: Properties of Water and Steam Based on the Industrial Formulation IAPWS-IF97, 7–150. Berlin, Heidelberg, Germany: Springer Berlin Heidelberg. https://doi.org/10.1007/978-3-540-74234-0_3.

10 Macro-analyses

Employing the models of utility system components, steam turbines (Chapter 6), gas turbines (Chapter 7) for the simulation (Chapter 9) and optimisation (Chapter 8) of the utility networks, are important activities for studying existing sites or designing new ones. In some cases, other tasks also become necessary. Often it is required to improve the efficiency of site processes with regard to utility heating and cooling. The selection of processes to examine, seeking good energy retrofit options, can be helped by Top-Level Analysis (TLA). Alternatively, it may be desired to evaluate and improve the performance of the utility system itself, taking into consideration several alternative policies and equipment options. This is the kind of task where *R*-curve analysis can be the right tool. Both such analyses require the simulation and optimisation of entire utility systems, varying the evaluated conditions and scenarios. This is why they are referred to as "macro-analyses".

10.1 Top-Level Analysis

Often, industrial utility systems need to be analysed, to determine the true value of steam savings and to identify beneficial improvement directions for the considered utility system. When site energy managers are seeking for good projects for the energy retrofit, usually the attention is directed to the site processes – their heating and cooling demands, which result from the combination of the main process flowsheet and its heat recovery network. The reason for this is simple, the major share of the energy consumed on industrial sites, up to 70–80%, are due to the process energy use, as can be seen in the Singapore energy efficiency technology roadmap (NCCS, 2019) and an analysis of the energy demand reduction opportunities in the UK (Chowdhury et al., 2018).

The potential benefits to be identified can be used to determine the economics of retrofitting the processes on a site by means of decreasing their energy consumption. This can be done by applying the optimisation model for industrial utility systems, described in Chapter 8. In Varbanov et al. (2004a), based on a similar model, a stepwise optimisation procedure is applied to determine the marginal price of steam at the various steam mains with their pressure levels. Such an analysis inherently accounts for the constraints on steam savings imposed by the utility system. That allows the application of automated tools for utility system optimisation. Steam at a given level (steam main) can have different prices depending on how much is saved, due to the variation of the energy conversion efficiency with changing loads of the components – most notably, the steam turbines.

This is a much more practical approach than attributing a single value to steam at a given level, which can be valid only in a close vicinity around the current operating

https://doi.org/10.1515/9783110630091-010

point of the utility system. The procedure, described here, is built on the principle of identifying the steam main with the highest marginal price of steam in the utility system, and then in a step-wise manner exploring the what-if scenarios of reduction of steam demand, also checking for potential changes of the marginal price. This is referred to as "TLA" because it concerns the site utility system, which is situated at the top of the hierarchy of energy supply and used for an industrial site. TLA provides a strategy for targeting and prioritising steam saving on Total Sites and modifications to utility systems.

10.1.1 Energy-saving strategies for an industrial site

A typical industrial site comprises process production units linked to a common (centralised) utility system. The centralised utility system meets the process demands for heat and power, creating indirect links between the processes. Generally, energy savings in industrial sites can be achieved through:
– retrofit of site processes to increase energy efficiency
– utility system improvements
– efficiency audits and operational optimisation of existing processes or utility systems

Each of these actions requires knowledge of the economic benefits that can be expected as a result of their implementation that could justify any identified changes. The method, described in this chapter, is on the process and Heat Exchanger Network (HEN) retrofits targeting for the purpose of saving hot utility – mainly steam.

Considerable amounts of steam can often be saved by optimising and retrofitting site processes and their HENs. A good example of HEN retrofit is the Network Pinch method, first published by Asante and Zhu (1996). A well-documented implementation has been discussed in Al-Riyami et al. (2001). The study has clearly demonstrated that retrofitting a process HEN features different benefits at different steps and that the energy savings obtained come at usually increasing investment cost with every further step. The particular study identified approximately 27 % utility reduction scope for the HEN of a fluid catalytic cracking (FCC) plant to be economical, for the specified steam prices.

Although such projects might have the potential to save steam at various pressure levels, establishing the economic value of such savings is far from straightforward. This results from the complex interactions within site utility systems, which supply the steam to the processes. Thermodynamic analysis can be used to identify which energy-saving projects on existing industrial sites can be beneficial. This can be done by using, for example, fixed estimates of the steam price, based on the burned fuel and the boiler efficiency. However, such analyses do not take into account the structure and the power cogeneration inside the utility system, let alone the varying the performance of their components and the constraints imposed by existing equipment.

The initial idea of the TLA was developed by Makwana (1998) in his PhD thesis. This was inspired by the need for an approach to screen energy projects without collecting data from all the site processes. The reason for such a prioritisation stems from the high cost and the long time for data measurement and processing – including data reconciliation as can be seen on the examples for the HENs (Ijaz et al., 2013) and for the site utility systems (Yong et al., 2016).

The initial procedure for TLA (Makwana, 1998) demonstrated the benefit of prioritising data acquisition. It required only collection and analysis of basic data relating to the central utility system. The analysis was based on the concept of heat flow paths through the utility system, distinguishing between current and optional paths. Current paths were analysed to find the scope for steam saving through the reduction in steam consumption by processes. Optional paths were identified to utilise any steam surplus provided by the current paths more efficiently for the generation of additional power.

The analysis recognised that each path contributed differently to the overall power generation efficiency and attempted to pass any excess steam, resulting from a decrease in process consumption through more efficient paths, where this was cost-effective. As a final result, the analysis produced steam marginal prices for different steam headers at different rates of potential steam saving. That procedure (Makwana, 1998) required explicit enumeration and calculation of the utility paths and was complex when applied to real sites, which becomes a limiting issue when it should be implemented in a software tool for efficient automation.

The current chapter presents an advanced procedure for TLA, which obtains the marginal steam prices by applying a utility system optimisation software tool. The knowledge for performing this analysis builds on the hardware models for steam and gas turbines (Chapters 6 and 7) and the successive mixed-integer linear programming procedure for optimisation of existing utility systems, described in Chapter 8.

10.1.2 Interactions within the utility system when optimising under variable steam loads

It is apparent that if steam savings are to be made within processes on a site, this would change the demands on the utility system. Changing the steam demands sets a different operating point than the initial one. This results in a different optimal distribution of the steam flowing through the utility system. Hence, the utility system optimisation needs to be considered in the context of variable steam demands.

Figure 10.1 shows a site utility configuration with three steam generation devices. Boiler 1 fires coal to generate very high pressure (VHP) steam, while Boiler 2 fires fuel oil to generate VHP steam. A gas turbine firing natural gas produces VHP steam from a heat recovery steam generator (HRSG), which can use supplementary

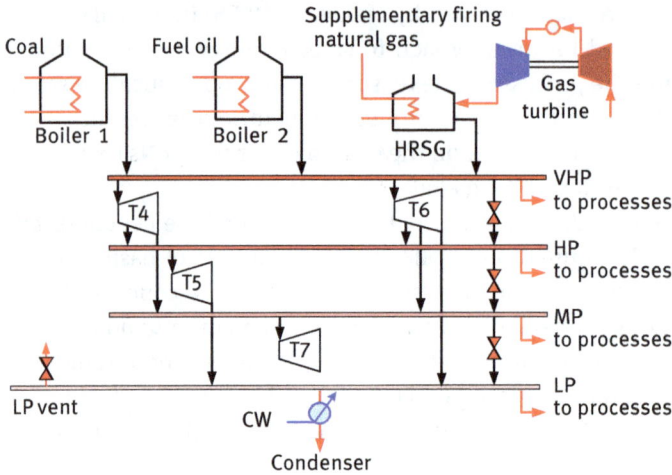

Fig. 10.1: Utility system example for Top-Level Analysis.

firing using natural gas. There has been a demand for VHP steam from various processes on the site to drive steam turbines on fixed drives with the condensed exhaust. In addition, steam from the VHP main is expanded through turbines T4 and T6. Turbine T4 exhausts to the medium pressure (MP) main also supplying steam to the high pressure (HP) header through the extraction outlet. Turbine T6 has extractions to the HP and MP mains and exhausts to LP steam. Turbine T5 expands steam from the HP to the low pressure (LP) main. Finally, turbine T7 expands steam from the MP main through to condensation. There are further steam demands from the processes for HP, MP and LP steam for process heating.

Consider the hypothetical case if it becomes possible to reduce the consumption of HP steam by processes. This can be immediately valorised by reducing the expansion of VHP steam via the letdown and, in turn, reducing the load on either of the boilers. However, turning down the generation of VHP steam and saving fuel is only one option. Another option could be to use the excess VHP steam, created by the reduction in HP steam demand, to generate extra power. The excess VHP steam could be expanded through turbine T4 and then T7 to generate extra power.

The excess VHP steam could be alternatively expanded through turbine T6 and then T7 to generate extra power. A third option would be to pass the surplus VHP steam through turbine T4 and T5 and then release it through the vent valve of the LP main. Finally, the excess VHP steam could be expanded through turbine T6 to the LP main and vented. The benefit to be gained by such options depends not only on the cost of fuel and power but also the performance and the constraints on the individual components in the utility system, as well as the external parameters – as the prices of power import from and export to the central electricity grid. Each steam turbine in the utility system generally has different efficiency and flow rate constraints. Consequently, each option is likely to result in a different economic benefit.

To summarise, if a surplus of steam is created because of energy reduction projects in site processes, in principle, there are two alternatives to exploit that:
- turn down the utility steam generation and save fuel, or
- use the surplus to generate power more efficiently.

10.1.3 Stepwise optimisation procedure

Even for a relatively simple utility system such as the one shown in Fig. 10.1, there are many complex interactions to be explored in order to determine the true economic value of steam saved at any one of the pressure levels. Therefore, the use of utility system optimisation in a macro-procedure can enable the use of software tools for performing the analysis more efficiently. The complex interactions and trade-offs inside the utility system can be accounted for by exploiting the utility system optimisation for a set of different operating conditions.

10.1.3.1 Steam marginal price
The concept of steam marginal price is used as an indicator in TLA. It was defined in Varbanov et al. (2004a) as the change in utility system operating cost per unit change in steam demand of a given header (change in the steam header balance):

$$PM_{header} = \frac{\Delta Cost}{\Delta m_{stm, header}} \tag{10.1}$$

Note that existing and running site utility systems are not always operated at the optimal conditions. Often, the current operation is shifted from the optimum, due to periodic discrepancies of the operation and the variations of the external conditions. To provide a fair basis for comparison of the operation conditions, the utility system operation at the current process steam demands is first optimised.

Only after that, the what-if analysis is started. The change in the operating cost is then taken between the optimal operation before the steam demand change and the optimal operation after the steam demand change for the current step. Obviously, the result is context-specific and, for different operating conditions, each steam header will have a different marginal steam price. Even, as will be demonstrated in the examples, the marginal steam price can change for the same header in the course of the potential steam demand variation.

In general, the steam balance of a header may be changed by the following:
- increase/decrease in process steam demand
- increase/decrease in process steam generation

- switching a process demand from one steam header to another
- change to the utility system, for example, shutdown of a boiler and steam turbine

The approach used in TLA has the advantage that it uses a single criterion – the steam marginal price. This criterion allows to assess the potential economic benefits from steam savings on the site using the same procedure regardless of the site scale, the nature of the evaluated changes to the site or the modelling tools employed.

10.1.3.2 Procedure description

TLA follows a stepwise optimisation procedure. As reasoned earlier, it is assumed that the objective of the analysis is to reduce the steam demand through energy conservation in the site processes. However, the algorithm is sufficiently general to be applied to cases of process steam generation, or production expansion, resulting in higher utility demands or different operating cases.

The overall algorithm is shown in Fig. 10.2. The initial step is to optimise the operation for the initial steam demands, using any available tool. This provides a basis for calculating the real steam prices in the subsequent steps, by eliminating any inefficient effects built in the current operation, which have to be corrected in any case.

Next, a loop is performed, comprising two major blocks. The first block determines the header with the highest steam marginal price for the current steam demands and selects it for analysis. The second block exploits the capacity for a potential decrease in demand at the selected header. The demand is lowered in a stepwise manner with operational optimisation is performed at each step. If a constraint on usage (or generation) is reached or the marginal price for the header changes, the steps of demand decrease (generation increase) are terminated. At this point, a test is performed to determine whether further decrease of process steam demand (increase in generation) at any mains is possible. If so, the loop moves to the next iteration. Otherwise, the procedure is terminated.

It is important to note that in this description, it is meant to detect the significant changes in the marginal steam price. Small-scale fluctuations may occur due to number rounding and the solution tolerances of the used optimisation tools. The numerical tools used for optimisation usually produce certain smaller oscillations in the steam marginal price values. These small fluctuations are not a result of any real change in the system behaviour and for the purposes of the procedure should be neglected.

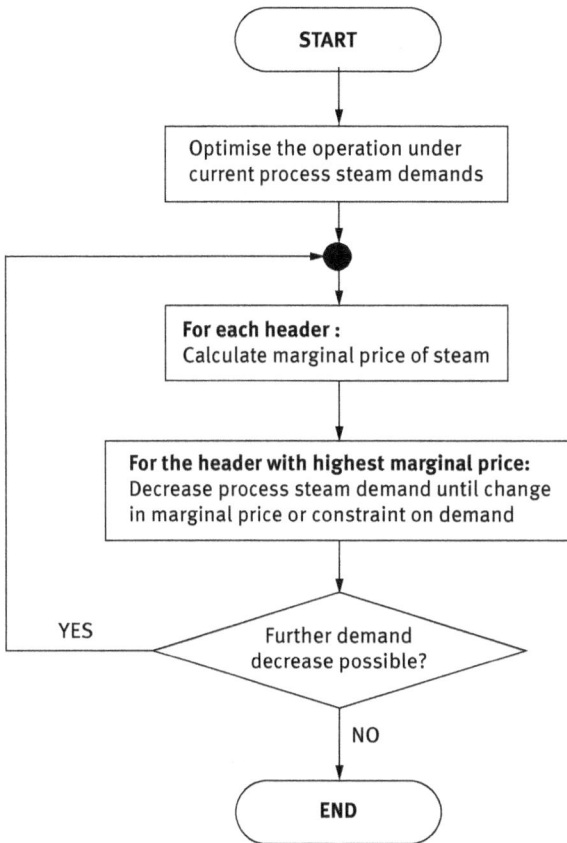

Fig. 10.2: The algorithm of the stepwise optimisation procedure.

10.1.4 Top-Level Analysis of generator-only utility systems

A system containing only steam turbine generators is considered. The steam flow rates through turbines of this type can be varied smoothly between their minimum and maximum limits because the shaft power is not fixed but is allowed to vary. This results in a continuous decrease in the site operating cost. In the case of analysis for steam saving potential, this means a monotonic profile of the marginal steam prices.

Figure 10.3 gives details of the conditions for the initial operating point and constraints on the flows in an existing utility system. All steam turbines are assumed to have machine efficiency $\eta_m = 0.95$, intercept ratio $L = 0.176$ and performance coefficients according to Tab. 10.1. The fuels in use are summarised in Tab. 10.2 and the site configuration data are given in Tab. 10.3. The current process steam requirements are to be examined for potential benefits of saving steam.

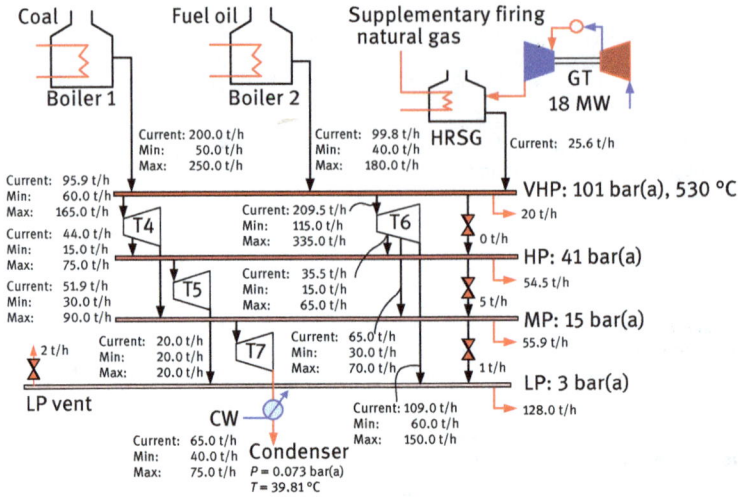

Fig. 10.3: Top-Level Analysis of generator-only systems. The initial operation for initial steam demands.

Tab. 10.1: Steam turbine regression coefficients (after Varbanov et al., 2004b) – Chapter 6.

Validity ranges	Backpressure turbines		Condensing turbines	
	$W_{max} \leq 2$ MW	$W_{max} > 2$ MW	$W_{max} \leq 2$ MW	$W_{max} > 2$ MW
b_0 [MW]	0	0	0	−0.463
b_1 [MW/°C]	0.00108	0.00423	0.000662	0.00353
b_2 [1]	1.097	1.155	1.191	1.220
b_3 [1/°C]	0.00172	0.000538	0.000759	0.000148

Tab. 10.2: Fuel data for the generator-only and driver-only case studies (after Varbanov et al., 2004a).

Notes		Fuel 1	Fuel 2	Fuel 3
		Solid (coal)	Liquid (fuel oil)	Gaseous (natural gas)
NHV	MJ/kg	28	40	52
Cost	$/kg	0.065	0.12	0.22
	$/t	65.0	120.0	220.0
	$/kWh	0.0084	0.0108	0.0152

Tab. 10.3: Site configuration data for the generator-only case study (after Varbanov et al., 2004a).

Ambient temperature	°C	25.0
BFW temperature	°C	45.0
Cooling water temperature	°C	25.0
Site power demand	MW	68
Minimum power import	MW	0
Maximum power import	MW	50
Minimum power export	MW	0
Maximum power export	MW	0
Unit power cost (import)	$/kWh	0.06
Unit power value (export)	$/kWh	0.05
Unit water cost	$/kWh	0.005

The curve development starts by first optimising the operation point for the initial process of steam demands. Figure 10.4 shows the optimised flowsheet.

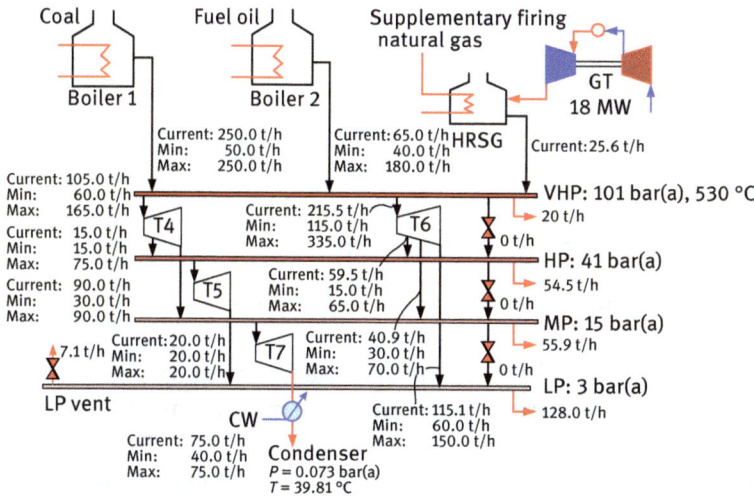

Fig. 10.4: Generator-only case study. The optimal operation for the initial steam demand.

Figure 10.5 gives a plot of steam marginal price versus the potential steam savings for the considered system and Fig. 10.6 shows how the steam turbine flows vary with the development of the analysis. The plot in Fig. 10.5 features six segments:

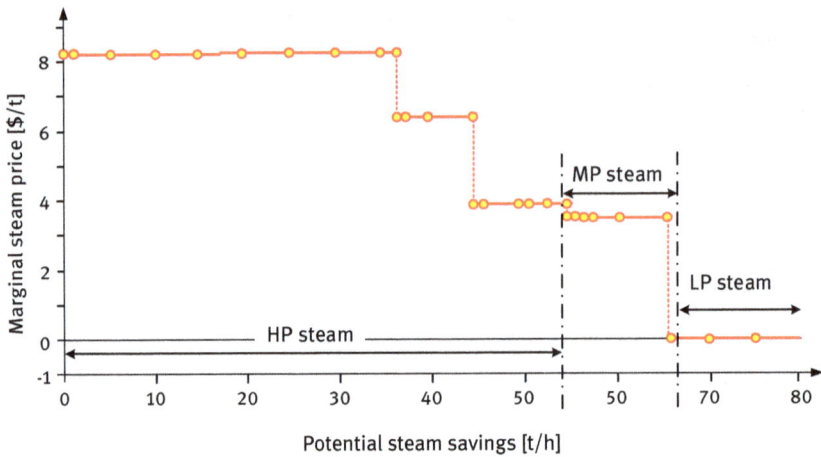

Fig. 10.5: Steam marginal prices for the generator-only case study.

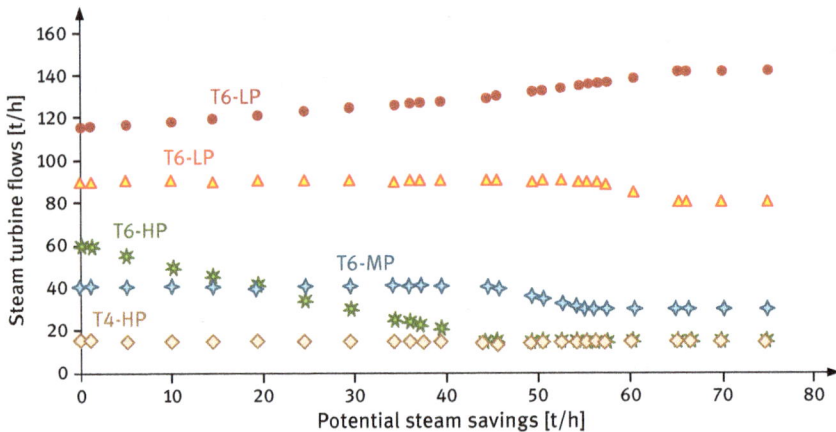

Fig. 10.6: Generator-only case study: steam turbine flows versus steam savings.

three for HP steam savings, two MP segments and one LP steam saving segment. Note that within each segment, there are slight marginal price variations, resulting from the numerical fluctuations.

At the initial point, the HP extraction of turbine T4 is set to its minimum of 15 t/h. The first HP segment of the curve (Fig. 10.5) reflects the steam saving by reduction of the HP extraction of turbine T6. Part of this saving is used by the reduction in steam generation in Boiler 2 and another by increasing the LP exhaust of turbine T6 and the LP vent. This segment is terminated by Boiler 2, reaching its minimum capacity. After this point, any boiler steam savings are realised in Boiler 1. The second HP segment features further gradual decrease in the HP extraction from T6 and is terminated at its minimum of 15 t/h. The third HP segment in Fig. 10.5 exhibits interesting behaviour. As can be seen in Figs. 10.4 and 10.6, at this point both direct paths for saving HP steam are already exhausted after the first two segments. The HP outlet flows of turbines T4 and T6 are at their minimum limits of 15 t/h and cannot be reduced further. However, because eventual letdown of HP steam to the MP main increases the enthalpy of the MP steam, the power generated by the condensing steam turbine T7 is always larger without this letdown. This, in turn, makes it more profitable to save further HP steam, passing the saved HP steam to steam turbine T7 instead of moving to MP steam savings.

After the HP steam demand is reduced to zero, there are two MP steam saving segments. The first one saves steam through the MP extraction of turbine T6 and is quite short. The second MP segment reduces the MP exhaust of turbine T4 until the minimum on the inlet steam for T4 is reached. Any further saving of MP or LP steam cannot be exploited for load reduction in the boilers and is vented.

An interesting feature is that the marginal price of the LP steam is zero throughout the analysis, not only in the last segment of the curve. This can be explained with the relatively high efficiency of power generation through the path *LP main →*
T6 → VHP main and the price structure of the site energy resources. The effect causes the site to be in power balance for any amount of steam savings.

Figure 10.5 illustrates that the value to steam at a given pressure level can significantly vary with the rate of energy savings in site processes. The value at steam main depends on how much is being consumed, as well as on the fuel cost, power cost and so on. This indicates that, for internal accounting inside a site and for steam exchange between sites, the use of a single value of steam for energy management and distribution of costs can be only valid within certain narrow operational ranges. Any more sizeable operation variations or retrofits on a site on the site should trigger a recalculation of the steam prices.

10.1.5 Systems with direct-drive turbines

Direct driver steam turbines introduce additional constraints in utility system operation. Each steam turbine used as a direct driver has a specification of fixed power to be generated. As a result, the steam flow rates through driver turbines are virtually constant while they are in operation. However, if the utility system under consideration features low steam demands, then powering certain drivers with steam turbines may become less economical than to switch them to electric motors running on imported or site-generated power. As a result, the profiles of marginal steam prices and the steam flows through turbines take rectangular shapes and the marginal price curves lose their monotonicity. The case study in Fig. 10.7 is introduced to illustrate the utility system behaviour under such constraints.

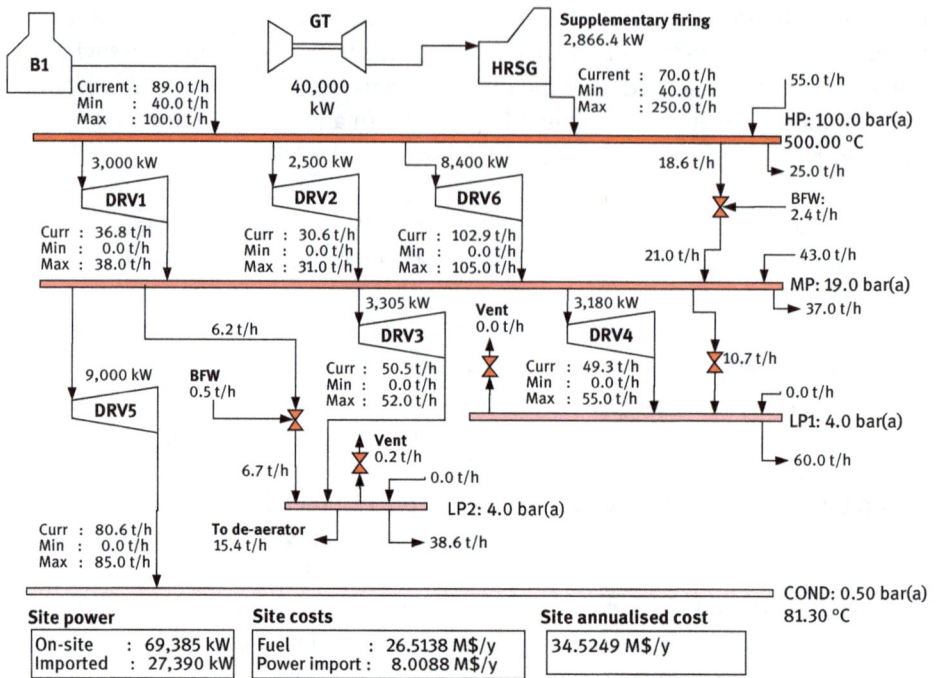

Fig. 10.7: A case study with driver steam turbines: flowsheet and existing operation.

In this example, gaseous fuel is used with a cost of 0.017 \$/kWh. The cost of electricity for import or export is 0.034 \$/kWh. All steam turbines feature machine efficiency $\eta_m = 0.95$, intercept ratio $L = 0.176$ and performance coefficients according to Tab. 10.2. Site configuration data are given in Tab. 10.4.

Tab. 10.4: Site configuration data for the case of driver-only steam turbines (after Varbanov et al., 2004a).

Ambient temperature	°C	25.00
BFW temperature	°C	99.99
Cooling water temperature	°C	25.00
Site power demand	MW	67.390
Unit power cost (import)	\$/kWh	0.034
Unit power value (export)	\$/kWh	0.034
Unit water cost	\$/kWh	0.005

The flowsheet in Fig. 10.7 shows the existing operation point under the initial steam demands and system constraints. The driver steam turbines can be turned on or off. In the case of switching a turbine off, the relevant drive is switched to an electric motor. Such action adds to the site power demand, which in this case is satisfied by import from the grid.

The TLA procedure has been applied to this system (Fig. 10.7), which results in the marginal steam price profile shown in Fig. 10.8. The initial optimisation switches off boiler B1 and shifts all boiler steam production to the HRSG. The operating cost of the utility system is reduced from 34.5249×10^6 \$/y to 34.1262×10^6 \$/y (1.2 % with respect to the initial operation). The small magnitude of the cost savings can be explained by the fact that all steam turbines are direct drives. As a result, any savings are realised through the steam load switch from the stand-alone boiler to the HRSG and by minimising the flows through letdown stations and vents. The subsequent steps of reduction of process steam demand cause a gradual decrease in the site operating cost. The cost reduction for saving HP steam (Fig. 10.8) is again caused by lower fuel consumption in the HRSG (for supplementary firing) as a result of a direct decrease of the HP steam demand.

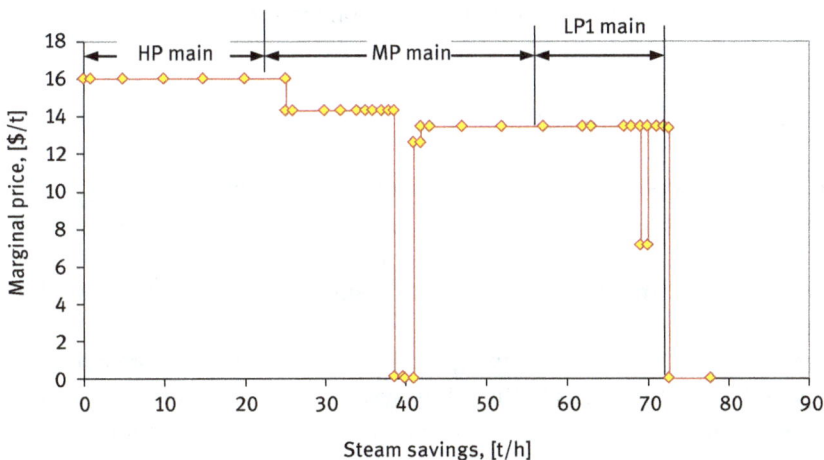

Fig. 10.8: A case study with driver steam turbines: marginal steam prices.

After the HP demand reaches zero, initially the MP steam savings generate further cost reduction by reducing the flow through the HP–MP letdown, until the letdown flow becomes zero. This corresponds to 38.76 t/h cumulative steam savings (Fig. 10.8). After this point, follows a segment with a zero steam marginal price. The reason for this is that there is no further scope for reduction in the HP–MP letdown flow, which is zero, and it is not yet economical to turn off any of the driver turbines DRV1, DRV2 or DRV6, passing their duties to electric motors.

However, if the MP steam demand is further decreased, a breakpoint is reached at 41.0 t/h total steam savings (Fig. 10.8), at which turbine DRV2 is turned off, and its duty is replaced by a backup electric motor. This results in a sharp increase of both the steam flow through the HP–MP letdown and power import. This switch of the operational mode of DRV2 gives rise to the short segment with the marginal steam price of 12.63 $/t between steam savings of 41.0 and 42.0 t/h (Fig. 10.8) However because the HP–MP letdown at that point has got some scope for steam flow reduction, this causes the marginal steam price to jump up to the level of 13.41 $/t.

Further reduction of steam demand continues until the complete exhaustion of the MP demand and after that switches to steam savings through the LP1 main Fig. 10.8. These savings cause further change in the operating status of the high-pressure steam turbine drivers. At total steam savings of 69.0 t/h (Fig. 10.8), turbine DRV2 is turned on again, and DRV1 turns off by switching the drive duty to an electric motor. This causes the marginal steam price to jump for a short segment to 7.12 $/t and immediately after that is restored to the previous value of 13.41 $/t. The scope for an economic decrease in steam demands extends further to 72.675 t/h. After this point, until eventual complete exhaustion of all the remaining steam demands on the headers LP1 and LP2, the marginal steam price is zero. The reason is that all further steam savings result in equal increase in the steam vent flows with no impact on the site cost.

10.1.6 Summary of Top-Level Analysis

The marginal steam prices, calculated by TLA, can be used as indicators for process modifications. They reflect the true value of steam saved from energy conservation projects in the site processes. These marginal prices, together with the limits on steam saving, provide a strategy for energy conservation on the site without modification to the utility system.

However, the step changes in the marginal steam prices also can provide indications for potential modifications to the utility system itself. For example, the case study with generator-only steam turbines (Section 10.1.5) features a significant LP vent flow at some points of the analysis. No use of any LP steam savings under the current configuration of the utility system is also detected. If the saved LP steam can be utilised in some way on the site or if it can be sold to a neighbouring site, LP steam savings may become economical, aligning the economics with thermodynamic logic. For instance, it may be beneficial to indirectly utilise LP steam heating potential by preheating the boiler feed water.

The TLA procedure provides marginal steam prices and allows us to make retrofit and operational decisions regarding site processes. The procedure allows automated simulation/optimisation tools to be exploited. It decouples the low-level system modelling tasks from the high-level system analysis and allows the procedure to remain the same as the models evolve. The stepwise optimisation procedure enables the analysis of systems with complex configurations, involving gas turbines, HRSGs, multiple boilers and multiple steam turbines. Even the most complex system can be subjected to the analysis, once its model is implemented in a steam network optimisation tool.

10.2 Site power-to-heat ratio

For energy systems, on both the sides, that of demands and that of generation, the ratio between the energy flow components, heat and power, is important. This is caused mainly by the thermodynamic limit of converting energy from one form to another (Potter and Somerton, 2014). Generating power always results in both heat and power outputs, in various ratios, where the degree of usefulness of the outlet heat flow varies from one technology to another. The outlet heat can be in the form of high-temperature exhaust gas as from gas turbines (see Chapter 7) or high-temperature fuel cells – such as molten carbonate fuel cells (Ahn et al., 2018) and solid oxide fuel cells (Eveloy et al., 2017). There can also be medium-temperature exhaust from diesel engines (Quintana et al., 2019) and phosphoric-acid fuel cells (Ito, 2017), as well as it can be of lower temperature heat (down to 90–100 °C) released by low-temperature fuel cells – for example proton exchange membrane fuel cell as well as Stirling engines (Alberti and Crema, 2014). It can be summarised that each energy conversion technology is characterised by the input energy (E_{IN}) and the flows of the power output (W), that of the useful heat (Q_{USE}) and that of the waste heat (Q_{WASTE}). The outputs then form the power-to-heat ratio of the conversion technology.

Table 10.5 shows a comparison of several energy conversion technologies in terms of typical efficiency ranges and power-to-heat ratios. It has to be noted that the power-to-heat ratio is usually evaluated for a complete technology cycle. The issue is that often the heat output of an equipment set cannot be quantified precisely. The heat output comes with a hot fluid at a certain temperature. However, to what temperature would that fluid be cooled down for useful heating is a specification made by the site energy managers and varies. Therefore, the amount of heat that can be made available to users depends on the temperature level specifications. This is the reason to evaluate the power-to-heat ratio for the technologies for complete cycles, as is in the case of the steam turbine-based combined heat and power (CHP) generation and the gas turbine-based cycles in Tab. 10.5.

It has to be noted that even within the same type of heat-and-power related equipment, there can be a substantial variation in the power-to-heat ratio. Gas turbines are a good example of equipment that can be used to cogenerate power and heat. The power-to-heat ratio of gas turbines varies with their size and current load. The variation is within a wide interval, typically 0.58 to 1.03 (US DOE, 2016a). The general trend is that the larger turbines feature higher power-to-heat ratios, mainly due to their higher

Tab. 10.5: Typical efficiency ranges and approximate power-to-heat ratios of energy technologies.

Technology	Typical efficiency [1]	Power-to-heat ratio, R [1] $= Q_{USE}/E_{IN}$	Remarks	References
PEMFC	0.40–0.50	~1	Temperature 50–100 °C	Kirubakaran et al. (2009)
PAFC	0.40	~1	Temperature ~ 200 °C	Kirubakaran et al. (2009)
Molten carbonate fuel cell (MCFC)	>0.50	~0.7	Temperature ~ 650 °C	Kirubakaran et al. (2009)
Solid oxide fuel cell (SOFC)	>0.50	~0.7	Temperature 800–1,000 °C	Kirubakaran et al. (2009)
Steam Turbine-based CHP	0.063–0.073 0.796–0.797	0 0.086–0.101	Electrical efficiency CHP efficiency	US DOE (2016b)
Gas turbines	0.23–0.35	0.3–0.7	US DOE Technology Summary	US DOE (2016a)
Combined cycle (GT + ST) – electricity generation	0.56–0.58	0	Italian natural gas-powered power plants	Jarre et al. (2016)
Combined cycle (GT + ST) – CHP	0.87–0.90	1.238–1.545	Italian natural gas-powered power plants	Jarre et al. (2016)

power efficiency. The same effect can be observed when comparing industrial with aero-derivative gas turbines. Higher power-to-heat ratios can make very large gas turbines unfavourable for industrial sites with substantial heating demand. However, on some sites (e.g. for natural gas liquefaction plants), there is a large and prevailing power demand and almost no use for the generated waste heat.

10.2.1 Background

The fuel utilisation curve (Kenney, 1984) and the R-Curve (Kimura and Zhu, 2000) are analysis tools that estimate the upper bounds of the cogeneration efficiency by a utility system. Assuming an idealised utility system configuration, the cogeneration efficiency is estimated for a given range of site power-to-heat ratios. In the initial version of the analysis (Kenney, 1984), various configurations within utility systems were evaluated, producing plots of cogeneration efficiency versus site power-to-heat ratio. This plot was referred to as a "fuel utilisation curve" (Fig. 10.9). In the figure, the term boiler means a standalone boiler, while the other references

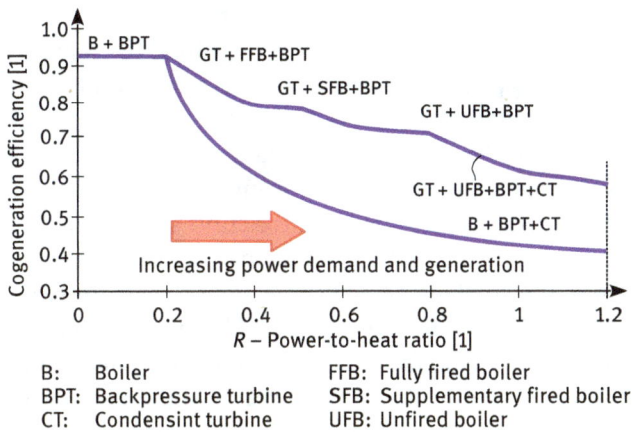

Fig. 10.9: Qualitative sketch of the fuel utilisation curve by Kenney (1984).

to boilers, fully fired, supplementary fired and unfired, mean a HRSG in several operating modes, using the heat from a gas turbine exhaust. That analysis was intended to provide quantitative design guidelines for selection of equipment, aiming for maximum fuel efficiency for a utility system. Those findings were based on the assumptions of constant power-to-heat ratio (R-ratio) for gas turbines and fixed utility system configuration.

Later, aiming to provide better insights to site operational management and de-bottlenecking, this concept was extended by Kimura and Zhu (2000). They employed the site cogeneration targeting methodology by Mavromatis and Kokossis (1998) to estimate the power generation of an ideal utility system. This allowed Kimura and Zhu (2000) to develop the "ideal R-Curve". By applying a path efficiency analysis, they attempted to redistribute the steam turbine flows in an existing utility system so that it operated at maximum fuel efficiency. This allowed the construction of the "actual R-Curve", which gives a tighter bound on the utility system efficiency.

Both of the discussed approaches assume that the utility system would generate exactly as much power as required by the site processes. For the specified site heat demands, they gradually varied the power requirement and generation (together), and for each of the resulting power-to-heat ratio points, they estimated the maximum system cogeneration efficiency. The plot (Fig. 10.9) starts from $R = 0$ (see the segment labelled "B + BPT"), which corresponds to zero power demand. At this point, they assumed cogeneration efficiency equal to that with backpressure steam turbines exploiting the steam fully flows, targeted for the satisfaction of the site heating and cooling demands by the cogeneration targeting model (Varbanov et al., 2004b). Thus, the second key point in the plot is obtained for the latter system state. The resulting cogeneration efficiency is very high since there are only backpressure steam turbines in the configuration and the heat losses are minimal.

After this point, for any larger value of the power-to-heat ratio R, the authors assume that condensing steam turbines are added and run (Fig. 10.9, the segment "B + BPT + CT"). The only useful energy they can generate is the power, while their steam outlet is condensed in utility condensers, that is, that heat is lost. The energy contained in their exhausts is rejected to the ambient. This leads to significantly lower cogeneration efficiency. At the beginning of that segment, the condensing power generation has a very small share. This segment of the curve starts from zero condensing turbine load, which is the point of full exploitation of the system cogeneration potential. As the share of the condensing power generation increases with increasing the site power demands, the system cogeneration efficiency, represented by the curve, drops from the maximum one to the values typical for standalone condensing steam turbines. The upper curve in Fig. 10.9 shows how the curve with steam turbines only ("B + BPT" and "B + BPT + CT") would shift if a gas turbine and a HRSG are added to the system.

However, this analysis has several limitations. It uses simplistic models for steam and gas turbines, fixing their efficiencies with load variation. This limits the applicability of the model to very narrow load intervals, where the actual efficiency remains within acceptably close values. Also, power import and export and steam generation by processes are not taken into account.

Balanced models for steam and gas turbines, capable of sufficient accuracy at minimal computational load, have been presented in Chapters 6 and 7. The current chapter presents one method for applying those models to the generation of the ideal R-Curve and analysing the potential for improving the efficiency of utility systems by adjusting the power-to-heat ratios. This method also explicitly distinguishes between the power-to-heat ratios of the generation and the demands, allowing to explore the scenarios including power export from the site.

10.2.2 Improved R-Curve analysis

The analysis is based on necessary definitions of the energy flow ratios and efficiencies, used in the system model. These concern
- the power-to-heat ratios of the process energy demands and of the utility system generation
- the cogeneration efficiency of the utility system

10.2.2.1 The R-ratio and the cogeneration efficiency
Thermodynamically, the cogeneration efficiency of a site can be defined as a ratio of energy flows. It is obtained from the sum of energy flows, utilised by the system, divided by the sum of energy flows supplied by primary sources, such as fuels. For an industrial utility system, the energy inputs are the fuels burned in boilers and gas turbines, while

the outputs are the heat and power, supplied to the processes. Figure 10.10 illustrates this on the example of a simple, idealised utility system, which is implied by constructing the Utility Grand Composite Curve (UGCC) (Raissi, 1994) for heat and power cogeneration targeting. More information on targeting heat and power cogeneration is provided in Chapter 11 dealing with Total Site Integration.

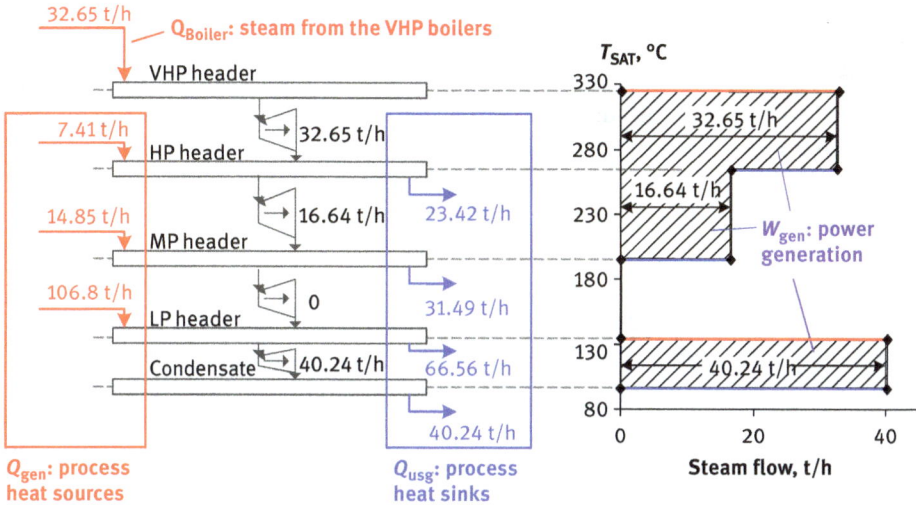

Fig. 10.10: The UGCC and the equivalent ideal utility system.

Referring to the diagram in Fig. 10.10, the overall heat supplied to the site utility system can be expressed as the sum of total process heat supplied for steam generation and the total heat produced by fuel combustion in the steam boilers:

$$Q_{supply} = Q_{gen} + Q_{Boiler} \qquad (10.2)$$

Using that, the cogeneration efficiency of the system can be defined as

$$\eta_{cogen} = \frac{W_{gen} + Q_{usg}}{Q_{supply}} \qquad (10.3)$$

In turn, the power-to-heat ratio R can be defined for the inputs to a user (for instance, a chemical process) or for the outputs of an energy conversion facility (for instance, a steam/gas turbine). The latter variant can be extended to modelling combined generation systems – for example, a steam turbine network, a site utility system burning fuels and delivering heat and power to the processes. For the utility system in Fig. 10.10, the power-to-heat ratio for the site processes as users is defined as

$$R_{usg} = \frac{W_{proc}}{Q_{usg}} \tag{10.4}$$

The ratio can also be defined for the utility system:

$$R_{gen} = \frac{W_{gen}}{Q_{usg}} \tag{10.5}$$

In the previous work (Kimura and Zhu, 2000), only the fuel heat was considered, focusing on simpler systems without process heat reuse. The definitions, provided here, are based on Varbanov et al. (2004a) and allow steam generation from process waste heat to be taken into account, alongside the heat obtained by fuel combustion in the boilers. The R-ratios, defined in Eqs. (10.4) and (10.5), account only for the steam usage. The heat from fired heaters that are not part of the utility system (i.e. not used for steam generation or boiler feed water preheat) should also be included in the calculations.

The operation of an existing site utility system needs to be optimised, in order to best utilise the primary resources. One of the criteria for the degree of utilisation of resource inputs is the cogeneration efficiency η_{cogen} (Eq. (10.3)). In order to improve site efficiency, there is a need for a targeting tool. The ideal R-Curve, introduced next, provides this functionality.

R-Curves are built assuming specific utility system configurations. In particular, the ideal R-Curve assumes the steam system to comprise one steam turbine per expansion zone (the pressure span between two consecutive steam headers), as illustrated on the left of Fig. 10.10. However, different configurations of firing equipment result in different R-Curves. The traditional utility system configuration is to combine steam boilers and backpressure steam turbines. It is possible to add condensing steam turbines. Another option is to consider a combination of a gas turbine with a HRSG, instead of, or in addition to, the steam boilers.

Having the ideal R-Curve constructed, the operation of the existing utility system can be superimposed on it by denoting the current operation as a point on the same plot, in coordinates ($R_{existing}$, $\eta_{cogen,\ existing}$). The difference in the efficiency between this current point and the ideal R-Curve for the same R-ratio provides the target scope for efficiency improvement.

The R-Curve analysis presents a targeting tool for the cogeneration efficiency for a given site power-to-heat ratio. However, it should be emphasised that this assessment is based on efficiency and not the economics of the system. Thus, it should be considered as a tool, revealing the capabilities of the utility systems and not as a means of rigorous optimisation of their economic performance.

10.2.2.2 *R*-Curve construction – case 1: power generation equal to the demands
The first case of R-Curve construction is the same as the case developed by Kimura and Zhu (2000). In the light of the definitions given in Eqs. (10.3) to (10.5), this is a

case, where the utility system generates exactly as much power as required by the site processes (Fig. 10.11).

It is assumed at this stage, that the utility system contains only steam boilers, steam turbines, letdown stations and steam mains. The curve is constructed, starting at the point where all steam, required by the site processes, is generated in steam boilers only and expanded via letdown stations. This point corresponds to zero power generation, resulting in $R = 0$. At that point, the cogeneration efficiency is equal to the efficiency of steam generation and delivery to the processes. Departing from that point, steam turbines, configured for an ideal steam cascade, are assumed. This configuration contains one simple backpressure steam turbine per expansion zone. Those turbines can be loaded with steam up to the amounts of steam, outlined in the UGCC, which corresponds to the Total Site Pinch (Klemeš et al., 1997). The space between zero power generation and filling up the steam flow capacities of the expansion zones represents the potential for power cogeneration, also referred to as CHP generation. In fact, the first utility system cogeneration model, applied to Total Sites (Raissi, 1994) has shown that the area enclosed by the UGCC is approximately proportional to the magnitude of the power cogeneration potential.

The general shape of the R-Curve, corresponding to this procedure, is shown in Fig. 10.12. It should be noted that at the R-ratio corresponding to the Site Pinch (designated as R_{PINCH}), all the steam is expanded through turbines and all the letdown valve flows are zero. The steam flows after the R-ratio for the Site Pinch can be thought of as a sum of the flows at pinched condition, plus the flows through a condensing turbine (Fig. 10.12).

10.2.2.3 *R*-Curve construction – case 2: account for power import and export

The basic case of R-Curve construction is useful for understanding and analysing the behaviour of utility systems comprising boilers and steam turbines, related to the process of heating and cooling demands. That case, however, by tying the power generation to the power demands of the site processes, leaves out the systematic consideration of power import and export options for a site. If the amount of the power generated by the utility system is differentiated from that required by the site, it is possible to adopt a more flexible strategy for site energy management, which has the potential to further improve the overall efficiency. This case, first discussed in Varbanov et al. (2004a), is illustrated in the current section.

For low power-to-heat ratios, which is the case of low on-site power demand, the system can generate extra power above that required by the site (i.e. $W_{gen} > W_{usg}$). This excess power can be exported. For very high power-to-heat ratios, the site cogeneration efficiency may drop below the average efficiency of the electricity grid as discussed in the initial work on TLA (Makwana, 1998) and in the pioneering work on the R-Curve (Kimura and Zhu, 2000). In such cases, the cogeneration efficiency analysis

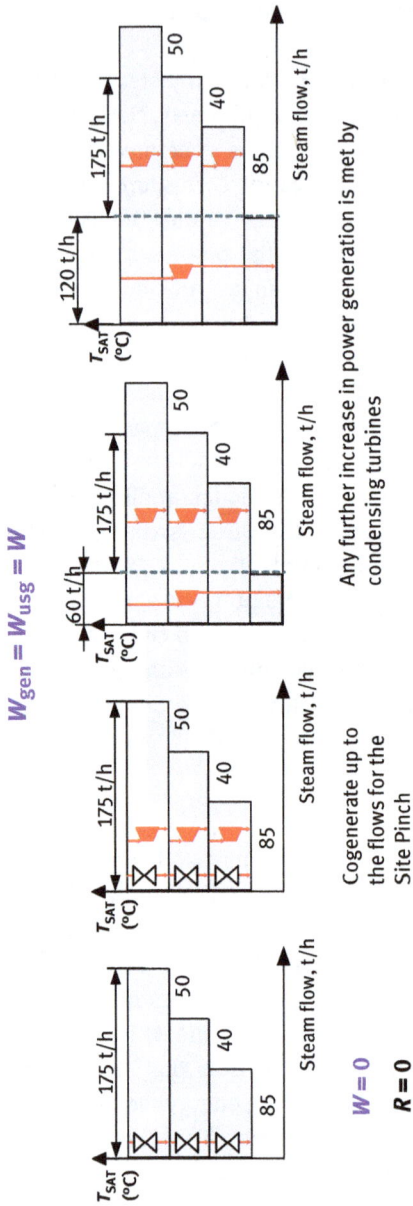

Fig. 10.11: UGCC representation of the case of power generation equal to the process power demand.

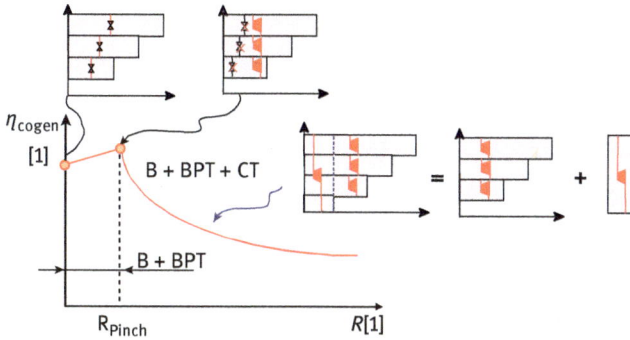

Fig. 10.12: *R*-Curve for Case 1 ($W_{gen} = W_{usg}$).

suggests power import, because the efficiency of generation of imported power becomes greater than that on the site.

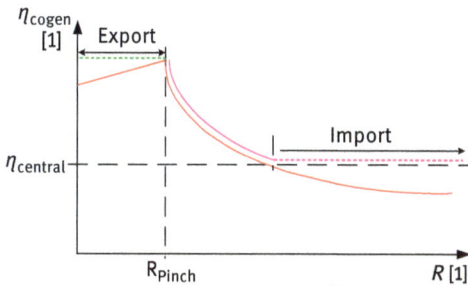

Fig. 10.13: *R*-Curve for Case 2 (flexible generation).

Applying a flexible on-site power generation strategy, that allows modelling power import and export, results in the *R*-Curve shape, shown in Fig. 10.13. The efficiency improvement resulting from this flexible energy management is clearly seen in the plot in Fig. 10.13. The power export at low *R*-ratios and import at high ratios introduce the flat segments of the new *R*-Curve. The incentives suggested for energy management policy are as follows:

- Below R_{PINCH}, the generated power is at high efficiency and it may be a good option to export power.
- After the on-site generation efficiency drops below that of the centralised power generation (the line for $\eta_{central}$ in Fig. 10.13), it may be a good opportunity to import power.

10.2.3 *R*-Curve example

A case study is used to demonstrate the *R*-Curve analysis (Varbanov et al., 2004a). The basic data, used for the calculations, are given in Tabs. 10.6 and 10.7. The steam turbine model, presented in Chapter 7, is used. There are four steam headers and a condensing (vacuum) main. An ideal utility system is assumed, which features one steam turbine in each expansion zone. The HRSG is assumed to deliver HP steam at 50 bar(a).

Tab. 10.6: Steam header data for the *R*-Curve example.

Header	P	$Q_{\text{generation}}$	Q_{usage}
	bar (a)	MW	MW
VHP	120	0.00	0.00
HP	50	5.00	15.00
MP	14	10.00	20.00
LP	3	70.00	40.00
COND	0.85	0.00	0.00

Tab. 10.7: Site configuration data for the *R*-Curve example.

$P_{\text{BFW}} = 120$ bar	ΔT_{min} for HRSG: 30 °C
$T_{\text{BFW}} = 85.0$ °C	Minimum stack temperature: 140 °C
Boiler efficiency: 0.92	Fuel cost: 0.015 $/kWh
$A_{\text{st}} = 0.5 + 0.008\ \Delta T_{\text{sat}}$ [MW]	Price power import: 0.050 $/kWh
$B_{\text{st}} = 1.18 + 0.03\ \Delta T_{\text{sat}}$ [−]	Price power export: 0.040 $/kWh
Gas turbine power: 25 MW	Rate of condensate return: 0.5
Ambient temperature: 15 °C	

The resulting set of plots is shown in Fig. 10.14. First, a curve for the simple case of balanced power generation and a boiler is considered (B + BPT + CT). Here, condensing steam turbines are only used for values of *R* greater than R_{PINCH}. This can be compared with the flexible generation strategy for the same system configuration (B + BPT + CT). Next, a gas turbine with a HRSG is considered, applying the flexible power generation strategy for unfired (B + GT + UFHRSG + CT) and supplementary fired (B + GT + SFHRSG + CT) modes. It can be seen that the kink point for R_{PINCH} for curves with a gas turbine is shifted to higher *R*-ratios. The reason for this is that for the selected gas turbine size of 18 MW if the efficiency is to be maximised,

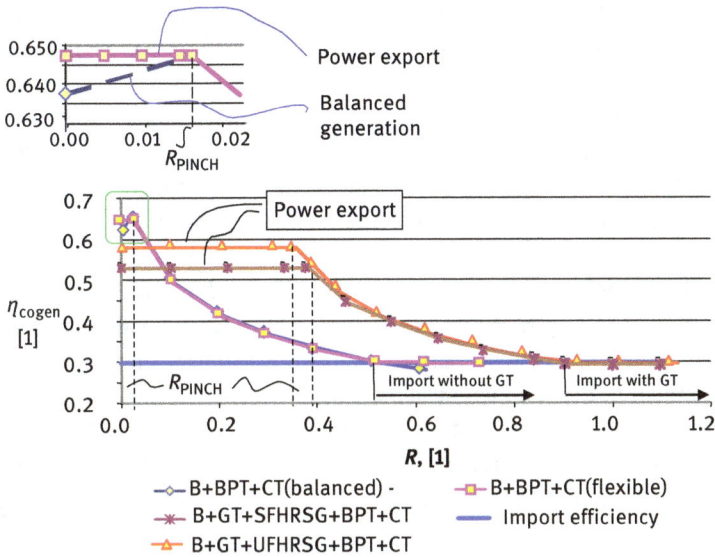

Fig. 10.14: *R*-Curve example results.

all the steam from the HRSG has to be utilised, which increases the cogeneration potential. For all configurations, the flexible power generation strategy can offer benefits in efficiency by importing or exporting power where appropriate.

10.2.4 Further developments in *R*-Curve analysis

The fundamentals of *R*-Curve analysis, presented in the previous section, have been further investigated. The developments and applications, found in the literature, illustrate the usefulness of the concept and the potential directions on how to incorporate it into practical analyses and site optimisation methods.

There have been industrial applications. Matsuda et al. (2009) published a study on a very large-scale industrial area in Japan, consisting of 31 individual sites. They have analysed the possibilities for energy savings by applying *R*-Curve analysis (described in this chapter) and Total Site Profiles (TSP) (Chapter 11). While the efficiency of each of the sites was already high before the study, the authors identified, by application of *R*-Curve analysis, potential energy-saving measures, amounting to a total of 31 % of the energy demand of the cluster. The parallel analysis for heat reuse, using the TSP, identified a further 5 % improvement in scope.

There have also been variations of the analysis – for example, exploring the site total annualised cost and emissions with varying the *R*-ratio (Karimkashi and Amidpour, 2012). A follow-up of that work (Ghaebi et al., 2012) demonstrated how to integrate an absorption chiller into an industrial site, which allows seeing the thermodynamic and cost advantages of such an action – potentially decreasing the site cost by up to 20 %.

The *R*-Curve tool can also be used for evaluating the potential integration of renewable energy sources into industrial energy systems. Depending on the site energy demands, it may be beneficial to integrate solar thermal heat, wind- or solar-generated electricity. Such a type of analysis is offered in Baniassadi et al. (2016); in the case study, the implications of integrating a solar power plant within an existing site utility system are analysed. They considered two scenarios with different *R*-ratios. It has been shown that the integration of solar power into the scenario with higher *R*-value is more beneficial than integrating it within the scenario with lower *R*-value. A similar type of analysis can also be applied to desalination systems (Salimi and Amidpour, 2017). The specific feature of this type of systems is that, depending on the choice of desalination technology, the final energy demands can feature radically different ratios between the heating and the power components of the demand.

10.2.5 Summary of *R*-Curve analysis

The *R*-Curve approach allows performing thermodynamic analysis of utility systems. The most appropriate utility system for a site power-to-heat ratio can be suggested based on the cogeneration efficiency.

R-Curve analysis can provide some guidelines of where to seek eventual cost savings by providing insights into efficiency trends and the potential improvements in utility systems, based on the inherent cogeneration potential and the efficiency capabilities of the available equipment. However, the use of this tool has its inherent limitations. Being a purely thermodynamic technique, it does not account for costs. For instance, it does not account for different fuels with different prices or perhaps low-efficiency boilers burning cheaper fuels and more efficient boilers burning expensive fuels. In this context, the most efficient operation of the utility system does not always mean the lowest cost operation.

Nomenclature

Symbol	Measurement unit	Description
A_{st}	MW	The parameters A_{st} for each turbine
B1	–	Steam boiler
BFW	t/h	The mass flow of boiler feed water from the pump
B_{st}	[1]	The parameters B_{st} for each turbine
b_0	MW	Steam turbine performance regression parameter, Chapter 6
b_1	MW/°C	Steam turbine performance regression parameter, Chapter 6
b_2	[1]	Steam turbine performance regression parameter, Chapter 6
b_3	1/°C	Steam turbine performance regression parameter, Chapter 6
CHP	–	Combined Heat and Power generation
DRV{#}	–	Direct-drive steam turbines
GT	–	Gas turbine
HP (steam)	–	High-Pressure steam
HRSG	–	Heat recovery steam generator
L	[1]	The intercept ratio
LP, LP1, LP2 (steam)	–	Low-Pressure steam
MCFC	–	Molten Carbonate Fuel Cell
MP (steam)	–	Medium-Pressure steam
NHV	kJ/kg	Net Heating Value
PM_{header}	€/t	The steam marginal price
P_{BFW}	bar	Pressure of the Boiler Feed Water
PAFC	–	Phosphoric Acid Fuel Cell
PEMFC	–	Proton-Exchange Membrane Fuel Cell
Q_{supply}	MW	The overall heat supplied to the site utility system
Q_{gen}	MW	The total process heat supplied for steam generation (site heat sources)
Q_{boiler}	MW	The total heat produced by fuel combustion in the steam boilers
Q_{usg}	MW	Process heating demands (site heat sinks) satisfied by the utility system
R	[1]	Power-to-heat ratio
R_{usg}	[1]	Power-to-heat ratio for the site process users
R_{gen}	[1]	Power-to-heat ratio of the utility system generation
$R_{existing}$	[1]	Power-to-heat ratio of an existing system
R_{PINCH}	[1]	Power-to-heat ratio at the pinched condition of the UGCC
SOFC	–	Solid Oxide Fuel Cell
ST	–	Steam turbine
T{#}	–	Steam turbine {#}
T{#}-{@main}	–	Outlet to a steam main {@main} of turbine {#}
T_{BFW}	°C	The temperature of the boiler feed water
TLA	–	Top-Level Analysis
ΔT_{min}	°C	Minimum allowed temperature difference

(continued)

Symbol	Measurement unit	Description
TSP	–	Total Site Profiles
VHP (steam)	–	Very high pressure steam
W	MW	Turbine power generation
W_{gen}	MW	Power generation by steam turbines
W_{max}	MW	Steam turbine power rating
W_{use}	MW	Power demand by the site processes
UGCC	–	Utility Grand Composite Curve
$\Delta Cost$	€/h	Changes in utility system operating cost
$\Delta m_{stm,header}$	t/h	Steam flow changes in a given header
η_m	[1]	The machine efficiency of the steam turbine
η_{cogen}	[1]	Cogeneration efficiency
$\eta_{cogen,\ existing}$	[1]	Cogeneration efficiency of an existing system
$\eta_{central}$	[1]	The on-site generation efficiency of the centralised power generation

References

Ahn, J.H., Hun Jeong, J., and Kim, T.S. 2018. Performance Enhancement of a Molten Carbonate Fuel Cell/Micro Gas Turbine Hybrid System With Carbon Capture by Off-Gas Recirculation. Journal of Engineering for Gas Turbines and Power 141(4): 041036. https://doi.org/10.1115/1.4040866.

Alberti, F. and Crema, L. 2014. Design of a New Medium-Temperature Stirling Engine for Distributed Cogeneration Applications. Energy Procedia 57: 321–330. https://doi.org/10.1016/j.egypro.2014.10.037.

Al-Riyami, B.A., Klemeš, J., and Perry, S. 2001. Heat Integration Retrofit Analysis of a Heat Exchanger Network of a Fluid Catalytic Cracking Plant. Applied Thermal Engineering 21 (13): 1449–1487. https://doi.org/10.1016/S1359-4311(01)00028-X.

Asante, N.D.K. and Zhu, X.X.. 1996. An Automated Approach for Heat Exchanger Network Retrofit Featuring Minimal Topology Modifications. Computers & Chemical Engineering, European Symposium on Computer Aided Process Engineering-6, 20 (January): S7–12. https://doi.org/10.1016/0098-1354(96)00013-0.

Baniassadi, A., Momen, M., Shirinbakhsh, M., and Amidpour, M. 2016. Application of R-Curve Analysis in Evaluating the Effect of Integrating Renewable Energies in Cogeneration Systems. Applied Thermal Engineering 93 (January): 297–307. https://doi.org/10.1016/j.applthermaleng.2015.09.101.

Chowdhury, J.I., Yukun, H., Haltas, I., Balta-Ozkan, N., Matthew, G. Jr., and Varga, L. 2018. Reducing Industrial Energy Demand in the UK: A Review of Energy Efficiency Technologies and Energy Saving Potential in Selected Sectors. Renewable and Sustainable Energy Reviews 94 (October): 1153–1178. https://doi.org/10.1016/j.rser.2018.06.040.

Eveloy, V., Rodgers, P., and Al Alili, A. 2017. Multi-Objective Optimization of a Pressurized Solid Oxide Fuel Cell – Gas Turbine Hybrid System Integrated with Seawater Reverse Osmosis. Energy 123 (March): 594–614. https://doi.org/10.1016/j.energy.2017.01.127.

Ghaebi, H., Karimkashi, S., and Saidi, M.H. 2012. Integration of an Absorption Chiller in a Total CHP Site for Utilizing Its Cooling Production Potential Based on R-Curve Concept. International Journal of Refrigeration 35(5): 1384–1392. https://doi.org/10.1016/j.ijrefrig.2012.03.021.

Ijaz, H., Ati, U.M.K, and Mahalec, V. 2013. Heat Exchanger Network Simulation, Data Reconciliation & Optimization. Applied Thermal Engineering 52(2): 328–335. https://doi.org/10.1016/j.applthermaleng.2012.11.033.

Ito, H. 2017. Economic and Environmental Assessment of Phosphoric Acid Fuel Cell-Based Combined Heat and Power System for an Apartment Complex. International Journal of Hydrogen Energy 42(23): 15449–15463. https://doi.org/10.1016/j.ijhydene.2017.05.038.

Jarre, M., Noussan, M., and Poggio, A. 2016. Operational Analysis of Natural Gas Combined Cycle CHP Plants: Energy Performance and Pollutant Emissions. Applied Thermal Engineering 100 (May): 304–314. https://doi.org/10.1016/j.applthermaleng.2016.02.040.

Karimkashi, S. and Amidpour, M. 2012. Total Site Energy Improvement Using R-Curve Concept. Energy 40(1): 329–340. https://doi.org/10.1016/j.energy.2012.01.067.

Kenney, W.F. 1984. Energy Conservation in the Process Industries. Energy Science and Engineering. Orlando, FL, USA: Academic Press.

Kimura, H. and Zhu, X.X. 2000. R – Curve Concept and Its Application for Industrial Energy Management. Industrial & Engineering Chemistry Research 39(7): 2315–2335. https://doi.org/10.1021/ie9905916.

Kirubakaran, A., Jain, S., and Nema, R.K. 2009. A Review on Fuel Cell Technologies and Power Electronic Interface. Renewable and Sustainable Energy Reviews 13(9): 2430–2440. https://doi.org/10.1016/j.rser.2009.04.004.

Klemeš, J.J., Dhole, V.R., Raissi, K., Perry, S.J., and Puigjaner, L. 1997. Targeting and Design Methodology for Reduction of Fuel, Power and CO_2 on Total Sites. Applied Thermal Engineering 17(8–10): 993–1003. https://doi.org/10.1016/S1359-4311(96)00087-7.

Makwana, Y. 1998. Energy Retrofit and Debottlenecking of Total Sites. Manchester, UK: University of Manchester Institute of Science and Technology.

Matsuda, K., Hirochi, Y., Tatsumi, H., and Shire, T. 2009. Applying Heat Integration Total Site Based Pinch Technology to a Large Industrial Area in Japan to Further Improve Performance of Highly Efficient Process Plants. Energy 34(10): 1687–1692.

Mavromatis, S.P. and Kokossis, A.C. 1998. Conceptual Optimisation of Utility Networks for Operational Variations – I. Targets and Level Optimisation. Chemical Engineering Science 53 (8): 1585–1608. https://doi.org/10.1016/S0009-2509(97)00431-4.

NCCS. 2019. Industry Energy Efficiency Technology Roadmap. Singapore National Climate Change Secretariat. https://www.nccs.gov.sg/docs/default-source/default-document-library/industry-energy-efficiency-technology-roadmap.pdf, accessed 22/10/2020.

Potter, M.C. and Somerton, C.W. 2014. Schaum's Outline of Thermodynamics for Engineers. Third ed. Schaum's Outline. New York, USA: McGraw-Hill Education.

Quintana, S.H., Castaño Mesa, E.S., and Bedoya, I.D. 2019. DECOG – A Dual Fuel Engine Micro-Cogeneration Model: Development and Calibration. Applied Thermal Engineering 151 (March): 272–282. https://doi.org/10.1016/j.applthermaleng.2019.02.008.

Raissi, K. 1994. Total Site Integration. PhD Thesis, Manchester, UK: University of Manchester Institute of Science and Technology.

Salimi, M. and Amidpour, M. 2017. Investigating the Integration of Desalination Units into Cogeneration Systems Utilizing R-Curve Tool. Desalination 419 (October): 49–59. https://doi.org/10.1016/j.desal.2017.06.008.

US DOE. 2016a. DOE/EE-1330. Combined Heat and Power Technology Fact Sheet Series: Gas
Turbines. US Department of Energy. https://www.energy.gov/sites/prod/files/2016/09/f33/
CHP-Gas%20Turbine.pdf, accessed 31/05/2020.
US DOE. 2016b. DOE/EE-1334. Combined Heat and Power Technology Fact Sheet Series: Steam
Turbines. US Department of Energy. https://www.energy.gov/sites/prod/files/2016/09/f33/
CHP-Steam%20Turbine.pdf, accessed 03/06/2019.
Varbanov, P., Perry, S., Makwana, Y., Zhu, X.X., and Smith, R. 2004a. Top-Level Analysis of Site
Utility Systems. Chemical Engineering Research and Design 82(6): 784–795. https://doi.org/
10.1205/026387604774196064.
Varbanov, P.S., Doyle, S., and Smith, R. 2004b. Modelling and Optimization of Utility Systems.
Chemical Engineering Research and Design 82(5): 561–578. https://doi.org/10.1205/
026387604323142603.
Yong, J.Y., Nemet, A., Varbanov, P., Kravanja, Z., and Klemeš, J.J. 2016. Data Reconciliation for Total
Site Integration. Chemical Engineering Transactions 52(October): 1045–1050. https://doi.org/
10.3303/CET1652175.

11 Total Site Integration

Total Site Integration has been developed as an extension of process-level Heat Integration. Its fundamentals have been covered in a previous book (Klemeš et al., 2018). In the current chapter, the fundamentals of Total Site Integration are summarised, referring the readers to the previous book. Key extensions of the Total Site concept are introduced, that relax some of the original method assumptions, making it more realistic and more widely applicable. These extensions concern handling variable energy supply and demand, embedding heat transfer properties of individual processes into the Total Site Targeting, as well as including the sensible heat of steam generation in the targeting model.

The concept of integrating the heating and cooling demands of several processes was introduced by Dhole and Linnhoff (1993), who defined the cluster of processes related to each other as a "Total Site." Klemeš et al. (1997) made further advances in the field by adding targets for power cogeneration. The Total Site concept has been used in various industrial implementations including a large-scale Japanese site (Matsuda et al., 2009), intercompany collaboration for reducing energy cost (Hackl et al., 2011), a study on improving the flexibility of combined heat and power systems (Botros and Brisson, 2011), an example from sugar and ethanol production (Morandin et al., 2011) and a study from Qatar on integrating industrial zones (Stijepovic and Linke, 2011).

The methodology has also been further extended by adding residential services, building management processes (hospitals, hotels and offices), low-grade industrial heat, waste to heat and renewables. For instance, a work comprising such extensions to the general framework has been presented by Perry et al. (2008). A case study on the integration of solar thermal energy into food processes has been one of the implementation examples (Atkins et al., 2010b). Integrating renewables with varying availability has been researched by Varbanov and Klemeš (2011) and is discussed at length in Section 11.2. Gearing Total Site Integration closer to practical implementation has also been the goal of Hackl and Harvey (2015) in developing a procedure for the composition of investment roadmaps.

The recent developments in the area include short- and mid-term planning of the Total Site heat recovery (Liew et al., 2017) based on the multiperiod targeting (Liew et al., 2018), the HEN synthesis in the part of delivering the utilities to the processes (Tarighaleslami et al., 2018), the simultaneous synthesis of the heat recovery networks for whole sites (Nemet and Kravanja, 2017) and consideration of pressure drops (Faramarzi et al., 2019).

https://doi.org/10.1515/9783110630091-011

11.1 Fundamentals of Total Site Integration

A key concept to understand is that, the Total Site refers to the collection of processes united by a common utility system. This is introduced first, followed by a summary of the site-level targeting procedure and references to the basic book on Process Integration and Intensification (Klemeš et al., 2018).

11.1.1 The site steam system as a utility exchange marketplace – steam generation as a cooling utility

It is rare to have single, isolated industrial processes. The usual situation is to have clusters of processes, interlinked with one another by various lines for exchanging raw materials, side products, intermediates and, most importantly, energy carriers. The latter usually represents the most significant links.

A chemical or another industrial site (Fig. 11.1) usually consists of a number of production and auxiliary processes. These processes require a supply of utilities to carry out their functions. These include process heating – most often using steam because of its high specific heat content in the form of latent heat and superheat. High-temperature processes, however, may require heating with hot oil or directly with flue gas in furnaces. Process cooling is performed by using cooling water, ambient air or refrigeration. Power demands arise from the need for driving process equipment, lighting and electric heating. Other utilities also include water – mainly the supply of freshwater, as well as the wastewater treatment, recycling and disposal.

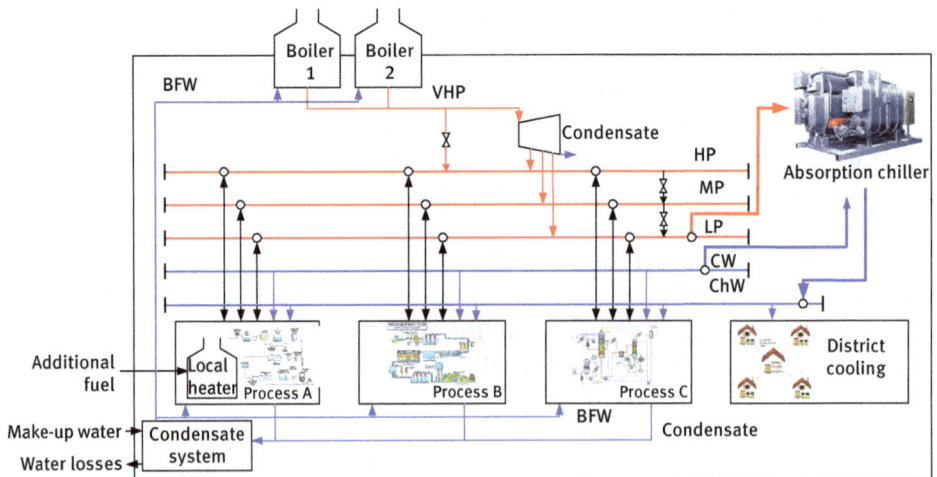

Fig. 11.1: A Total Site with a utility system seen as a marketplace for exchanging heat (amended from Klemeš et al., 2018).

Within the energy context, the utility system is considered as supplying the site processes with heating and cooling utilities as well as satisfying their power demands.

An important advantage is that a common utility system allows the processes to exchange utilities. High-temperature cooling demand of a process may be used for utility (e.g. steam) generation instead of serving the demand with utility cooling. Lower-temperature cooling demands in other processes can sometimes be served by that generated utility instead of applying boiler steam, utility hot oil or furnace flue gas. In this way, a central utility system provides a kind of a market place for exchanging utilities and enabling Heat Integration between individual processes, via the intermediate energy carriers. Having the common marketplace as an important degree of freedom in interprocess energy integration allows considering various options. When all processes on a site are considered together, the industrial site is termed a Total Site – implying that this is not just a collection of elements but an integrated system.

11.1.2 Process utility interfaces: heat sources and heat sinks

To perform an analysis of interprocess energy integration, the Total Site methodology defines very clear interfaces. These are the heat sinks and heat sources of the processes. A heat sink is a representation of heating demand. In this, it is similar to a cold process stream (Klemeš et al., 2018). It also has a starting temperature (T_S), a final/target temperature (T_T) and an enthalpy change (ΔH). A heat source is defined in a similar way with a starting temperature (T_S), a final/target temperature (T_T) and an enthalpy change (ΔH), which is analogous to a hot process stream. heat sources and sinks may often represent individual process streams – including residual heating or cooling demands, which process managers would like to expose to the exchange via the utility system. However, this is a more flexible concept, as the site-level heat sources and heat sinks may also represent composite cooling and heating demands, as is discussed later in this chapter.

11.1.3 Total Site Data Extraction

For Total Site Heat Integration targeting, the heat sources and sinks are identified using the Grand Composite Curve (GCC) (Klemeš et al., 2018) of each process. The GCC represents the net heating and cooling demands of a process after internal heat recovery has taken place. The temperature scale of the GCC is on the scale of shifted temperature T^*. The shifted temperatures are produced by shifting downwards (to lower values) the supply and target temperatures of hot streams by $0.5 \times \Delta T_{min}$, and shifting upwards (to higher values) the supply and target temperatures of the cold streams by $0.5 \times \Delta T_{min}$, embedding into the curve sufficient temperature difference for feasible heat exchange. The GCC shows the Pinch location, the minimum heating demands ($Q_{H,MIN}$) and the minimum cooling demands ($Q_{C,MIN}$) to be supplied from sources external to each process.

Obtaining the data for the heat source and sink specifications is referred to as "Total Site Data Extraction." The procedure for this is shown in Fig. 11.2. In short, the procedure starts from the GCC or problem tables of the site processes, accounts for the intraprocess heat recovery by removing the GCC pockets, identifies the segments with net heating demands (heat sinks) and net cooling demands (heat sources), and in the end constructs lists of both types of utility requirements. For a more detailed explanation of the procedure, the reader is referred to Chapter 2 of Klemeš et al. (2018).

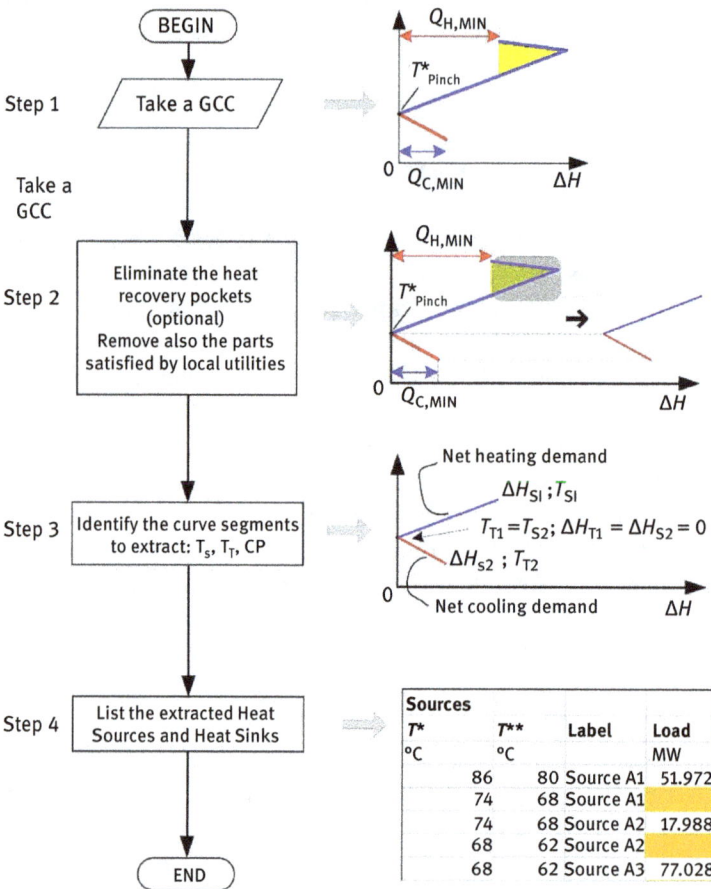

Fig. 11.2: Total Site Data Extraction procedure.

As an illustration of the Total Site Data Extraction, consider Total Site Example 2 from Chapter 2 of Klemeš et al. (2018) – describing an industrial site with two processes. The specifications are:

- $\Delta T_{min} = 10\ °C$
- Boiler (HP) steam at saturation temperature 270 °C (pressure 55.0 bar)
- LP steam at saturation temperature 140 °C (pressure 3.6 bar)
- Cooling water supplied at 15 °C and returned to the utility system at 20 °C
- The stream data for process A are given in Tab. 11.1 and for Process B – in Tab. 11.2.

Tab. 11.1: Total Site Example 2 – process A streams.

Stream	Type	T_S [°C]	T_T [°C]	CP [MW/°C]
A1	Cold	50	140	5
A2	Hot	100	30	6
A3	Cold	100	140	2

Tab. 11.2: Total Site Example 2 – process B streams.

Stream	Type	T_S [°C]	T_T [°C]	CP [MW/°C]
B1	Hot	190	120	6
B2	Cold	105	245	4
B3	Hot	80	60	2

Performing Pinch Analysis (Klemeš et al., 2018) on the two processes produces the GCCs in Figs. 11.3 and 11.4. The heat sources from both processes are listed in Tabs. 11.3 and 11.4. The tables show the T^{**} temperatures of the heat source and sink segments.

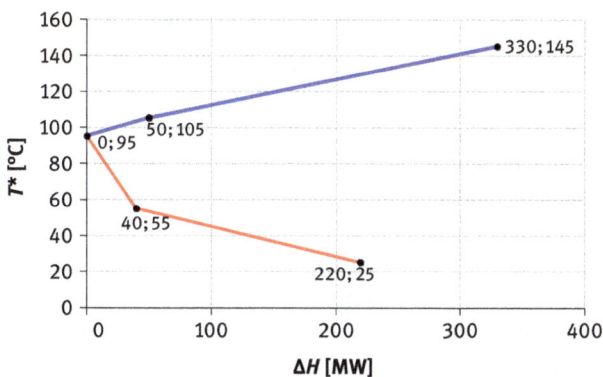

Fig. 11.3: Total Site Example – GCC for process A.

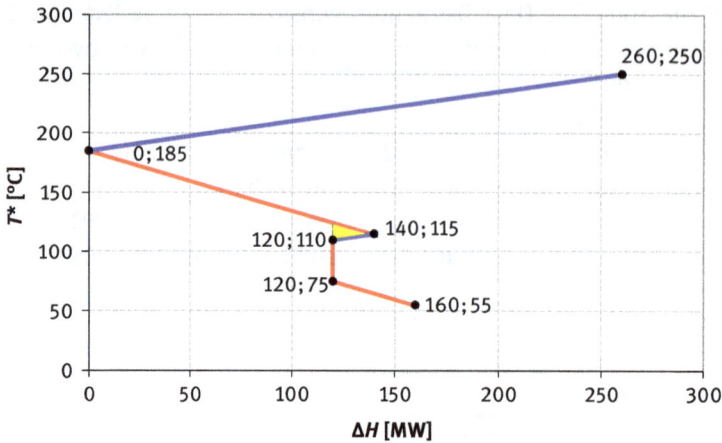

Fig. 11.4: Total Site Example 2 – GCC for process A.

Tab. 11.3: Total Site Example 2: site heat sources.

Segment	T^*_{start} [°C]	T^*_{end} [°C]	ΔH [MW]	T^{**}_{start} [°C]	T^{**}_{end} [°C]
Source A1	95	55	40	90	50
Source A2	55	25	180	50	20
Source B1	185	125	120	180	120
Source B2	75	55	40	70	50

Tab. 11.4: Total Site Example 2: site heat sinks.

Segment	T^*_{start} [°C]	T^*_{end} [°C]	ΔH [MW]	T^{**}_{start} [°C]	T^{**}_{end} [°C]
Sink A1	95	105	50.00	100.00	110.00
Sink A2	105	145	280.00	110.00	150.00
Sink B1	185	250	260.00	190.00	255.00

These are derived from the T^* values by further lowering the heat sources by $0.5 \times \Delta T_{min}$ and lifting up the heat sinks by $0.5 \times \Delta T_{min}$. This second shift results in ensuring sufficient temperature differences between the heat source and sink segments on the one hand and the site utilities on the other hand. In terms of the plots of the heat sources and sinks with T^{**} temperature representation and site utilities at their actual temperatures, when a heat source or a heat sink plot touches the plot of the utility, they are visually at the same temperature level, but at the same time, the corresponding heat carriers represented by the plots fulfil the ΔT_{min} constraints.

11.1.4 Total Site heat recovery targets

The next stage in the Total Site targeting procedure involves the combination of the heat sources from all site processes (A, B) into the composite Site Source Profile and the combination of all heat sinks to produce the Site Sink Profile. The procedure is completely analogous to the one for constructing the Composite Curves in process-level Pinch Analysis (Klemeš et al., 2018). In the considered example, the data from Tabs. 11.3 and 11.4 are taken. Applying the curve composition principles from Pinch Analysis, the temperature intervals of each data set are identified and heat interval tables are constructed calculating the sums of heat capacity flow rates and the total enthalpy changes in each temperature interval. The heat source interval table and the heat sink interval table for the illustrative example are shown in Figs. 11.5 and 11.6.

TB** °C	Source A1 CP MW / °C	Source A2 CP MW / °C	Source B1 CP MW / °C	Source B2 CP MW / °C	ΔT interval °C	ΣCP interval MW / °C	ΔH interval MW
180							
			2		60	2	120
120							
					30	0	0
90							
	1				20	1	20
70							
	1			2	20	3	60
50							
		6			30	6	180
20							

Fig. 11.5: Combining the heat sources.

TB** °C	Sink A1 CP MW / °C	Sink A2 CP MW / °C	Sink B1 CP MW / °C	ΔT interval °C	ΣCP interval MW / °C	ΔH interval MW
255						
			4.00	65	4	260
190						
				40	0	0
150						
		7.00		40	7	280
110						
	5.00			10	5	50
100						

Fig. 11.6: Combining the heat sinks.

The data from the interval tables in Figs. 11.5 and 11.6 are used to plot the Total Site Profiles (TSP), as shown in Fig. 11.7.

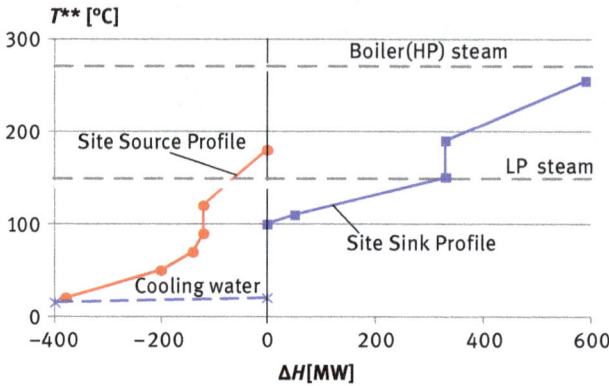

Fig. 11.7: Total Site Profiles with the HP and LP steam levels (utilities).

Using the steam cascading principle, the Site Source Composite Curve and the Site Sink Composite Curve are constructed from the profiles, as shown in Fig. 11.8. Putting the two Composite Curves together and removing the TSP, produces the plot in Fig. 11.9, which also shows the Total Site Pinch.

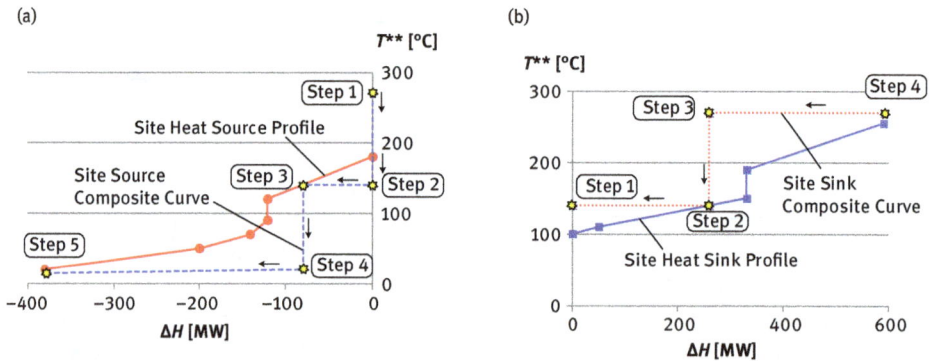

Fig. 11.8: Construction of Site Source CC and Site Sink CC (Total Site Example 2): (a) Site Source Composite Curve and (b) Site Sink Composite Curve.

A more advanced technique is the construction of the Site Utility Grand Composite Curve – Fig. 11.10 (Klemeš et al., 2018), which allows evaluating both the heat recovery targets and those for the possible power cogeneration by the utility system integrating the given processes. This is illustrated in Fig. 11.11. For evaluating such a

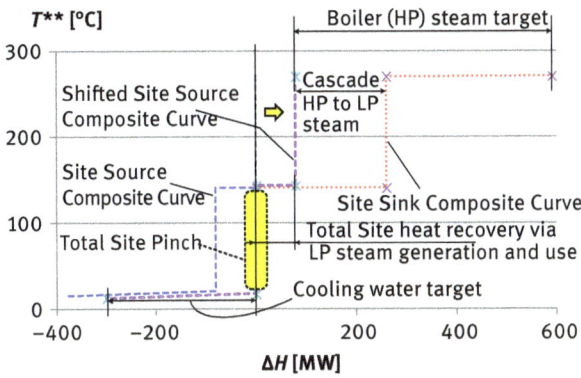

Fig. 11.9: Shift of the Site Source Composite Curve toward the Site Sink Composite Curve and identification of the Total Site Pinch.

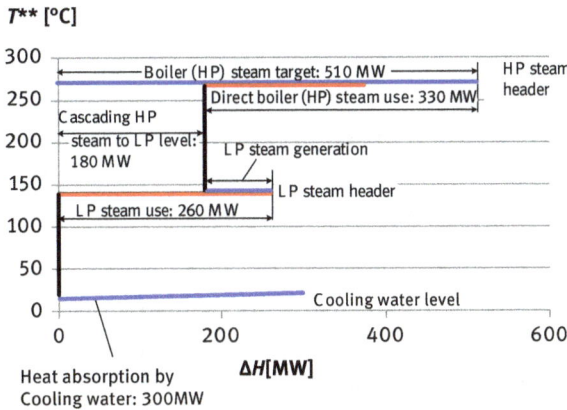

Fig. 11.10: The Site Utility Grand Composite Curve (SUGCC).

target, it is possible to apply several steam turbine models. The model, presented in Chapter 6 in detail, is frequently applied for this purpose.

It is possible to evaluate further targets based on these fundamental calculations. As an example, the power cogeneration target can be used to evaluate the target for steam turbine capital, and the targets for the hot and cold utilities can be used to evaluate the relevant greenhouse gas and water footprints (Ren et al., 2017). This basis also allows evaluating the possibilities for the reduction of the footprints exploiting the energy-water nexus as a synergy mechanism (Varbanov et al., 2018).

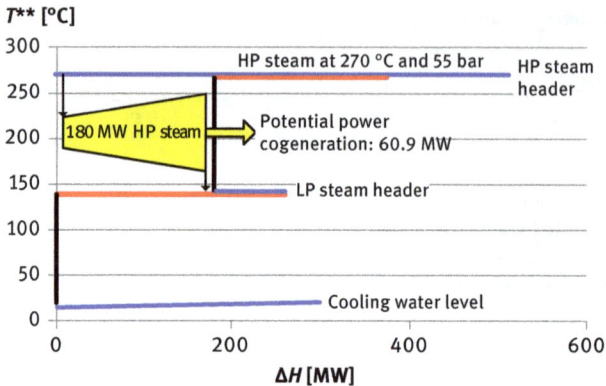

Fig. 11.11: Power cogeneration targeting using the Site Utility Grand Composite Curve (SUGCC).

11.2 Total Site with varying supply and demand including renewables

Reducing CO_2 emissions could be achieved by maximising heat recovery – which is the subject of process-level Heat Integration and Total Site (Heat) Integration (Klemeš et al., 2018) – see Section 11.1. Increasing the share of renewables in the primary energy mix supplied to the energy systems is another option. Process Integration can be used for facilitating this and maximising the efficiency of the resulting systems. This section describes the extension of the Total Site methodology to be capable of modelling the incorporation of renewable energy sources (solar, wind, biomass and some types of waste), accounting for the often substantial variability on the supply and demand sides and for the use of non-isothermal utilities.

11.2.1 Background

Although commonly described using steady-state models, in reality, energy systems, including Total Sites, are subject to variations on both the supply and demand sides. The types of variations involve both regular changes in the rates and property of the energy carriers referred to as patterns, as well as random, transient variations referred to as fluctuations. Therefore, what energy system designers and planners face, is a problem of satisfying energy demands with varying rate and of desired constant quality using energy sources with varying availability and quality. Another important feature of this problem is the multitude of energy sources (fossil fuels, nuclear, biomass, wind, solar, etc.) combined with the multitude of energy users (industrial, residential, large commercial complexes, hotels and hospitals).

There is an apparent trend of harvesting ever-increasing amounts of energy from renewable sources – such as wind and solar. Adopting renewable energy resources and maximising their utilisation acts simultaneously on reducing CO_2 emissions and energy dependencies. This brings up the need for models and algorithms to be used for designing and operating energy conversion and supply systems that work in conditions of the variable supplies and demands.

Renewable resources are usually available on smaller-scale distributed over a given area. Their availability (with the exception of biomass) is usually well below 100 %. The resource availability varies significantly with time and location. This is caused by the changing weather and geographic conditions. The energy demands (heating, cooling and power) of industrial sites and other users vary significantly with time of the day and period of the year. The variations of the renewable supplies and the demands are partly predictable, and some are changing in very regular time patterns, for instance, day and night for solar energy.

However, the availability of other renewables, such as wind-generated energy, can be less predictable. As a result, optimising the design of energy systems serving variable customer demands and using renewable resources is a complex task compared to using just fossil fuels to satisfy a constant demand.

In seeking the optimal solution for such energy systems, one would like to provide adequate tools and ensure the necessary degrees of freedom to satisfy the demands with the available supplies at minimum cost or maximum profit (Varbanov et al., 2005). The optimisation objectives can be purely economic or include environmental aspects resulting in the criteria of eco-cost or eco-profit (Čuček et al., 2012). Two tools play a significant role in this regard: energy storage on the demand side and establishment of a hierarchy for utilisation of the available energy sources concerning the supply. Installation of buffering vessels as part of the utility system can help in smoothing the fluctuations in the supplies and demands and facilitate the system control.

11.2.2 Properties of energy demand and supply

To understand the problem better, a characterisation of supply and demand types is needed. Both the supply and the demand for energy can vary with time and location. Accounting for temporal variations introduces dynamic modelling elements and concepts very similar to those used for batch processes – including terms such as horizon and Time Slice (period).

11.2.2.1 Energy supply sources – classification
The sources of energy for the considered users are in most part common:
- **Fossil fuels** currently dominate the energy markets. They can be used in all three site categories – residential, industrial sites and service building complexes.

- **Solar radiation** can be captured into thermal energy carriers (water, steam, antifreeze, etc.) or to generate electricity. A combination of both is a possibility as well, however not very much developed so far.
- **Wind.** It is used mostly for electricity generation with future potential for generating H_2 for the hydrogen economy.
- **Waste biomass and Energy crop biomass.** They can be directly utilised on-site for larger consumers as industrial sites, building complexes and farms or in district heating plants.
- **Hydropower.** It is harnessed for electricity generation. This, however, is performed mostly in centralised larger-scale power stations on dams. Microhydropower technologies are available, but they are mainly suitable for remote locations, which generally imply less energy integration.
- **Geothermal energy.** It is harnessed at the locations where it is available or close nearby.
- **Ground heat or cold.** Heat pumps are considered as renewable sources of energy by most classifications.

11.2.2.2 Energy users – classification

The energy demands vary with the various types of end-users as well as with the time schedules. Industrial sites mostly require:

- Heating in a wide range starting from 100 °C up to 400 °C and even to very high temperatures close to 1,000 °C,
- Cooling in the range 20 °C to 50 °C, and chilling in the range 0 °C to 10 °C,
- Refrigeration to temperatures reaching −100 °C and lower.

A further classification of the required temperature ranges for process heating and cooling can be found in the design book (Towler and Sinnott, 2008), as well as in the comprehensive handbook (Green and Perry, 2008). A special class of applications is farming and agriculture production. There are various examples of using the low potential waste and renewables heat for supplying greenhouse demands, for example, (Kondili and Kaldellis, 2006).

Residential sites (residential dwellings and their complexes in the case of district heating) feature demands for:

- Moderate-temperature heating of space and hot water
- Air conditioning
- Direct electricity consumption for lighting, cooking, refrigerators and other household appliances
- Electricity for heat pumps

The energy demands for the service industry and for building complexes (hotels, hospitals, schools and universities, banks, entertainment premises and governmental complexes) are generally similar in structure to residential sites. Some specific features are:

- A part of the heating demand can be at a temperature in the range 90 °C to 150 °C. For example, steam can be used in hotels for cooking and in hospitals for sterilising bedding and other appliances.
- The share of air conditioning may be significantly higher compared with residential homes.
- The specific resource consumption per person in the service industry (as hotels and hospitals) is generally higher than in residential homes because of the overheads for running additional services, infrastructures and facilities such as restaurants, bars and entertainment facilities.

11.2.2.3 Variability of the demands

The time variations of energy demands have been subject to research in both industrial and residential contexts. An example is a study investigating the variation of residential energy consumption for heating, electricity and hot water (Varbanov and Klemeš, 2011). The results show two types of trends: hourly variations during each day, and seasonal variations during the year. For the hourly variations in residential energy consumption, there are nearly steady periods during the usual office hours and two consumption peak intervals in the morning and in the evening. Figure 11.12 shows a typical trace for electricity consumption, but hot-water consumption variation follows a close trend. The seasonal variations are relatively smooth, with more substantial space heating demands from October until April.

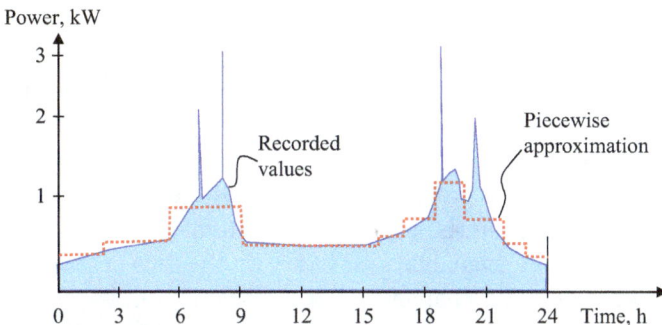

Fig. 11.12: Typical residential electricity demands within a 24 h cycle (after Varbanov and Klemeš, 2011).

Demand variations are mostly predictable and feature minor uncertainties – mainly in the timing of the consumption. The picture may differ among buildings, industrial sites and farms. A similar situation occurs in service buildings such as hotels and hospitals, where the demand levels will also depend on the occupancy rate and some less predictable features. Table 11.5 shows the types of temporal variations in energy demands typical for the various users.

Tab. 11.5: Demand variability types.

	House/ dwelling	Industrial site	Service building complex	Farms/ agriculture
Electricity	Peak/off-peak	Main shift/other sifts – campaign/off campaign	Day/night – sometimes less	Summer/winter
Heating	Winter/summer	Winter/summer – campaign/off campaign	Winter/summer	Winter/summer
Cooling	Summer/winter	Winter/summer – campaign/off campaign	Summer/winter	Summer/winter
Air conditioning	Summer mainly	Summer mainly	Summer mainly sometimes less predictable	Not used too widely
Direct shaft power	Not used	Widespread, using steam or gas turbines	Not typical	Not typical

11.2.2.4 Variability of renewable resources

For their efficient exploitation, it is necessary to assess an overall availability of renewables and variability with time. Some of them are close to the performance of fossil fuels and can be well stored for continuous energy generation. An example is biomass, where the supply varies by year seasons and by biowaste availability. However, sufficient storage could be made available. The availability of other renewable sources such as wind and solar varies more rapidly – in hours and even minutes.

Two examples of more detailed studies on the variability of renewable energy sources can be given. One is a report for the International Energy Agency (IEA, 2005), which deals with the intermittency of wind power generation as well as that from other renewable energy sources. It provides an analysis of the time scales of variation of the availability of those sources, based on the natural cycles of variation. It can be seen that solar energy availability may vary on the time scale from minutes to years. A second example is a chapter by Von Bremen (2010), which describes the variations of wind and solar power availability, also providing visual examples in terms of plots. The chapter also discusses the implications of these variations and strategies for performing power balancing in order to keep the demand satisfied.

These types of variation present an integration challenge where the time horizons of the changes are diverse. From the given examples, for biomass, the Time Slices to model the problem would last on the order of months and at smallest – weeks. For wind and solar energy, the Time Slice durations will obviously be much shorter (Nemet et al., 2012). This section describes the extensions of Total Site methodology to deal with such variations.

11.2.3 Feature requirements of the tools and solutions

From the above discussion, the need for several important requirements to energy system design and operation can be identified:
- It is necessary to cater to a diverse set of energy users. The most notable customers are large and small industrial plants, residential, commercial and service buildings, as well as potentially farm complexes.
- Integration of renewables is of strategic importance for reducing CO_2 emissions and for a great number of countries – to also lessen external energy dependence.
- It is vital to account for the variations in energy supplies and demands. A number of tools can be used, all centred around heat storage.
- A uniform framework for modelling the site heat sources and sinks, combined with heat storage and non-isothermal utilities is needed.

11.2.4 Thermal energy storage

The idea of using thermal storage for smoothing variations in energy supply and demand is not new. It allows delayed use of available heat or cooling utility. Heat can be stored using different principles. Based on that criterion, there can be physical, chemical and physic-chemical facilities for thermal energy storage (Dinçer and Ezan, 2018).

A simple but effective heat storage practice applied in industry is a stratified water tank used within a heat recovery loop. Atkins et al. (2010) have presented an evaluation of such an arrangement for a milk powder plant. Another review of the various forms of heat storage can be found in Huggins (2010). The source discusses several storage types, among which are sensible heat storage and latent heat storage. The latter is based on various phase-change materials (salts, paraffin-based materials and glycol polymers). Heat storage based on reversible chemical reactions is also described, emphasising salt hydration/dehydration.

The use of energy storage for industrial Heat Integration has been discussed in the book by Kemp (2007). In turn, this philosophy has been extended and applied to Total Sites first by Perry et al. (2008) who introduced other than industrial process types into the Total Site concept and after that by Varbanov and Klemeš (2011) who provided a full-fledged framework for modelling and selecting the levels and capacities of heat storage facilities located in the utility system of a Total Site.

Despite the variety of heat storage methods and devices, for the purpose of heat integration, they can be all classified into two types Fig. 11.13 (Nemet et al., 2012b) – facilities with temperature change during charge/discharge and others where each of the two operating modes takes place at a virtually constant temperature.

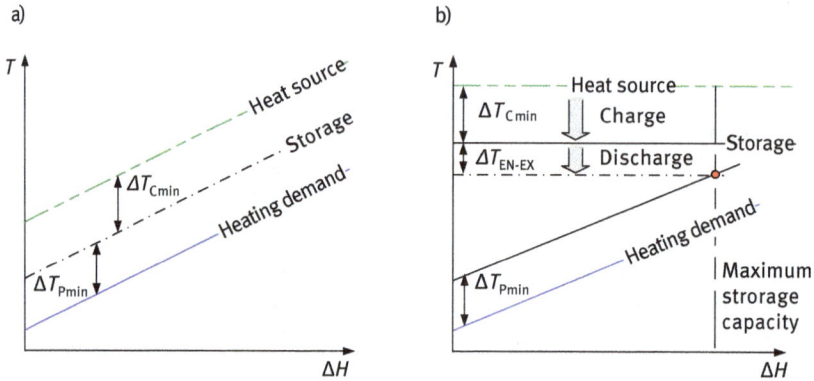

Fig. 11.13: Storage types by temperature change: (a) non-isothermal charge/discharge and (b) isothermal charge and discharge (after Nemet et al., 2012b).

11.2.5 Integration architecture

To maximise the energy efficiency of heat recovery under the conditions of variable supplies and demands (e.g. renewables), it is necessary to apply Total Site Integration combined with heat storage evaluation. Traditional Total Site models (see Section 11.1) include only industrial processes with their demands for heating, cooling and power, connected via the steam mains of the site utility system (Fig. 11.1). This understanding has served the industry for a couple of decades and proved its usefulness in a number of applications. However, a more extended model for the integration of energy flows is needed to fit the current context.

Perry et al. (2008) extended the Total Site concept by proposing to add also residential and service buildings (hospitals, hotels and offices) as site processes, as well as to consider alternative energy sources (biomass, wind, solar, ground heat) and conversion technologies (small-scale boilers, Stirling Engines, fuel cells, micro-turbines, solar thermal collectors and solar photovoltaic panels). For denoting the re-sulting architecture, they have coined the term Locally Integrated Energy Sector (LIES) (Fig. 11.14).

This conceptual extension allows applying the Total Site Integration strategy to local communities involving a larger number of smaller-scale entities and is referred

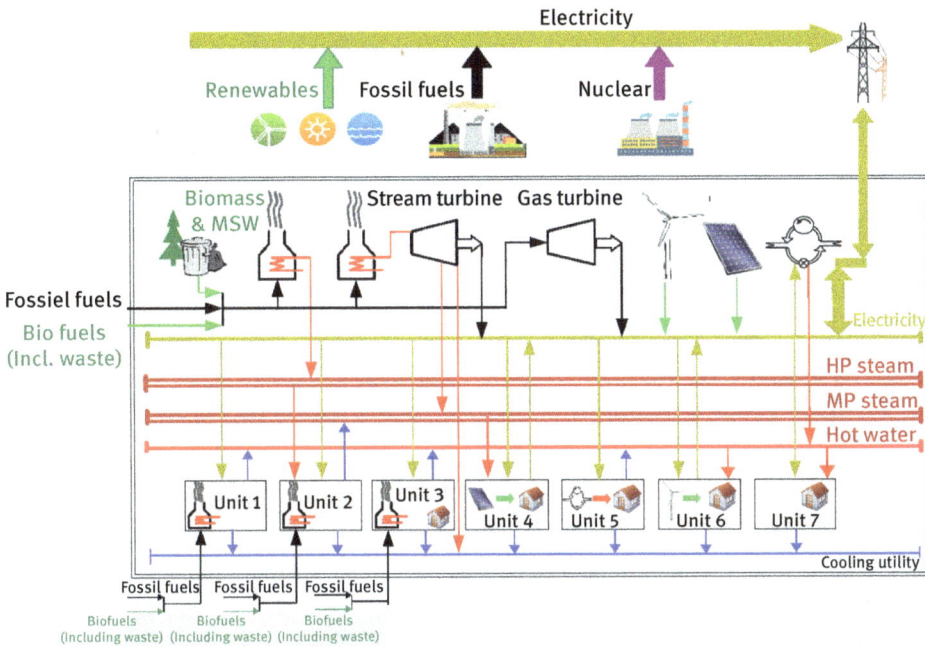

Fig. 11.14: Locally Integrated Energy Sector (after Perry et al., 2008).

to as LIES. Adding the energy user and generator types and conversion technologies allow to combine the temporal patterns of different user and generator types and to potentially smooth their inherent variations. Also, using distributed power and heat generation at or close to the locations of use enables a direct trade-off between centralised and distributed generation. The latter is an important element allowing to explore the most advantageous arrangement accounting for energy losses during transportation of fuel and the energy products (heat and power).

11.2.6 Terminology update on process streams and utilities

From the Heat Integration basics (Klemeš et al., 2018), it is known that hot utilities can be consumed for process heating and/or generated for process cooling. To establish an efficient procedure for dynamic energy targeting of Total Sites, the classification of the various heat sources and sinks on the site need to be updated. The heating and cooling demands of individual processes, remaining after internal Heat Integration, are used to create the Heat Source Profile and the Heat Sink Profile for an industrial site (Section 11.1). Further, the available utilities are placed using the Site Composite Curves.

From the viewpoint of Heat Integration, steam, hot oil, flue gas, cooling water and so on are clearly utilities. When integrating renewables and dealing with temporal variations in their availability as well as demand variations, it is necessary to decide how to treat the intermittent heat sources and sinks associated with the renewable energy supply and heat storage. Capturing renewable energy thermally always results in a heat source. On the other hand, the heat storage can be charged acting as a heat sink or discharged acting as a heat source.

An important issue is how to treat these heat sources and sinks – as additional processes and process streams or as utilities? The answer lies in analysing the role and the meaning of process heating and cooling demands (process heat sinks and sources) on the one hand, and that of the utilities on the other. Process heating and cooling demands are part of the problem specification. As such, they have to be satisfied and cannot be varied or manipulated by the user. On the other hand, the utilities are tools providing degrees of freedom to satisfy the heating and cooling demands with or without heat recovery. They may be used, but do not impose requirements.

Consider solar thermal energy capture. Solar irradiation falls on the capture area and as such cannot be controlled by the process operators and managers. Having a specification of the maximum potential capture area defines an upper limit on the solar heat flow that can be captured. Within this limit, any capture rate is possible, and this is a degree of freedom to be used. Similarly, heat storage also has a certain capacity and the storage may be charged or discharged according to this capacity, also providing a degree of freedom to the system. If the balance of process heat sources and sinks does not require it, neither solar heat should be captured, nor heat from storage needs to be used. As a result of the above analysis, renewables and heat storage are classified as utilities.

11.2.7 Identification of the Time Slices (periods)

Time Slices are defined as modelling time intervals, within which the rates of heat demand and supply streams can be modelled as constant. They are obtained by combining the starting and ending times of process streams activity. A key issue in applying this approach is the identification of the Time Slices. A procedure for identifying the number and durations of Time Slices for a problem featuring variable renewable energy supply has been formulated and developed for solar energy utilisation by Nemet et al. (2012).

11.2.7.1 Time Slices for the variable energy sources
The variation in the availability of renewable in time can be described as a series of continuous trends interrupted by weather changes, which creates discontinuous switch points. While the switch points are obvious candidates for Time Slice boundaries, it is

not so straightforward to identify Time Slices within a continuously varying energy flow profile. An example of the latter is shown in Fig. 11.15, where the availability of solar irradiation during a day is approximated by a parabolic curve, assuming that no clouds appear.

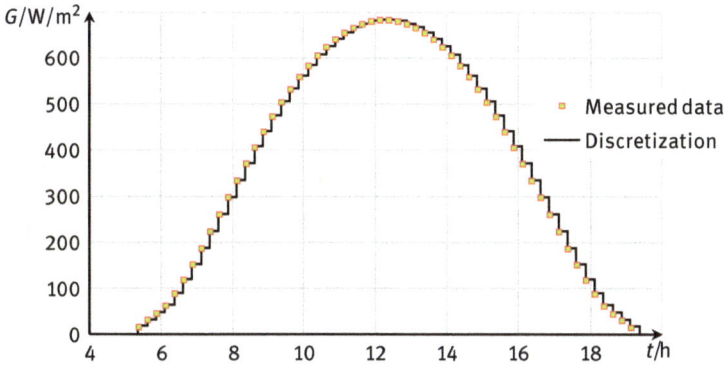

Fig. 11.15: Approximation of solar irradiation availability during a day (after Nemet et al., 2012b).

The profile is expressed as a series of sampling points connected with a piecewise-constant discretisation curve. The latter suggests all sampling points as potential Time Slice boundaries. Considering these as a superset, an optimal subset of sampling points has to be selected, which represents the continuous solar profile is a sufficiently small number of Time Slices at minimum inaccuracy. This task is performed by solving a MILP problem Nemet et al. 2012b), where each candidate Time Slice (Fig. 11.16) has a single level of the resource availability approximation, which is allowed to vary. These are accompanied by defining binary selection variables for each candidate for a Time Slice boundary where if the candidate boundary is selected, the availability level approximation for that interval is allowed to vary, while in the opposite case, the level is set equal to the one from the previous candidate slice.

The overall inaccuracy of the approximation using the selection of Time Slices is minimised subject to a set of constraints ensuring the formulation consistency. The overall inaccuracy itself is defined as a sum of the absolute deviations between the levels of the sampling points and the approximation variables. Several options for solving this optimisation problem exist and Nemet et al. (2012b) provide a discussion on their merits and weaknesses. The optimal selection of Time Slices depends on the required accuracy. Figure 11.17 shows a typical trend of resulting numbers of Time Slices at various tolerances.

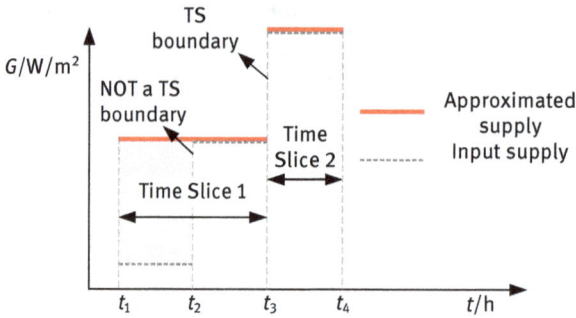

Fig. 11.16: Acceptance/rejection of the candidate time period boundary as a Time Slice boundary (after Nemet et al., 2012b).

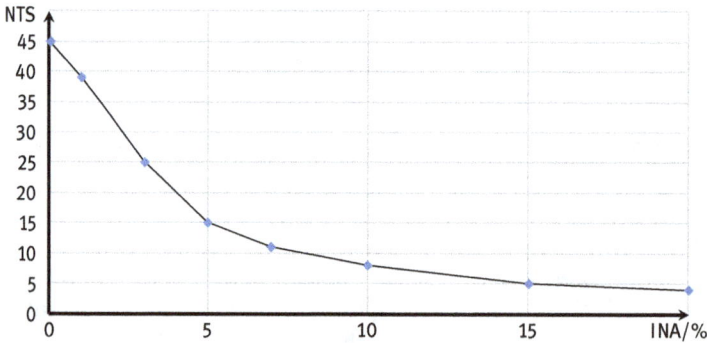

Fig. 11.17: Dependence on the Number of Time Slices (NTS) on the allowed tolerance (INA) value (after Nemet et al., 2012b).

11.2.7.2 Time Slices for the energy demands

For continuous industrial processes and for residential activities, the energy demands usually vary smoothly for most of the time and change discontinuously when the operating mode changes. This pattern results in a similar set of continuous trends and switching points as in the case of characterising the availability of renewable energy sources. Therefore, that procedure can be directly applied to this case too.

There is also the case of batch and semicontinuous processes in the industry, which are characterised with transient operations. This results in energy demands with periods of changing activity and inactivity. Each energy demand can usually be assigned clear starting and ending time points, which then become Time Slice boundaries.

11.2.7.3 Combining the Time Slices

Having identified the Time Slices for supply and demand sides of the problem allows combining them, which is necessary for treating the system states uniformly. The combination can be performed by merging the Time Slice boundaries for the supply and the demand sides, as shown in Fig. 11.18.

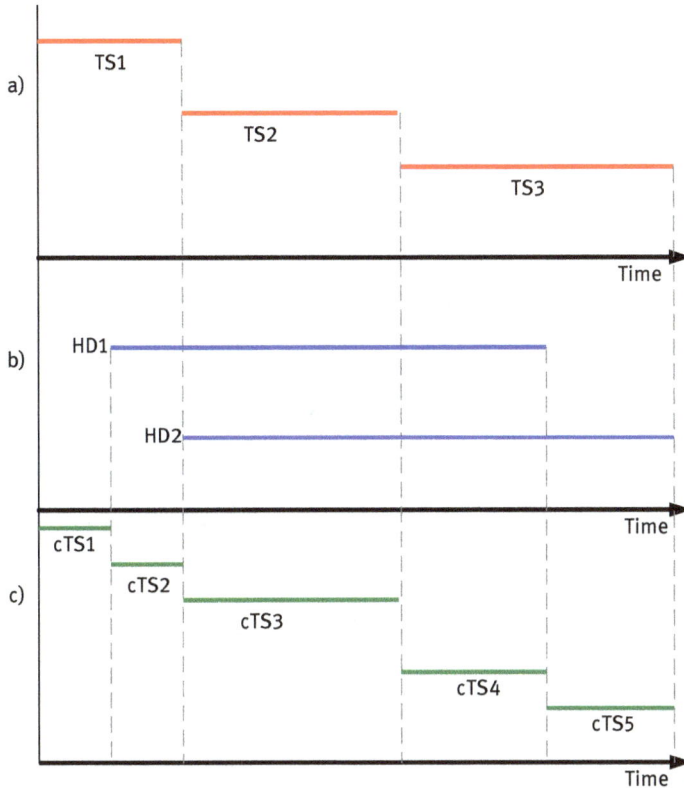

Fig. 11.18: Merging TS boundaries for supply and demand for obtaining combined Time Slices: (a) Time Slices for solar thermal energy; (b) Time Slices for the varying heat demand; (c) combined Time Slices (after Nemet et al., 2012b).

11.2.8 Targeting cascade for Total Site

Heat cascades have been developed previously for Total Sites, mainly for the purpose of identifying the specifications for utility system synthesis. The first relevant version is by Shang and Kokossis (2004) for performing initial targeting, and the second version has been formulated by Varbanov et al. (2005) for choosing steam header pressure levels. These constructions (Fig. 11.19) generally consist of

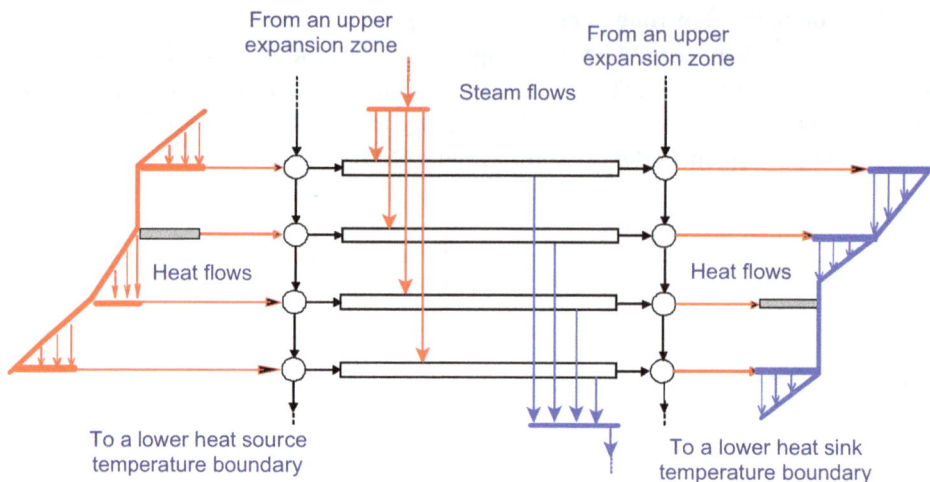

Fig. 11.19: Total Site heat cascade for building steam network superstructure during synthesis (after Varbanov et al., 2005).

three parts for expressing utility (mostly steam) generation from process heat sources, the utility (steam) cascading and distribution via steam mains, as well as utility use by process heat sinks.

The same cascade type can be constructed for describing numerically Total Site targets. A variation of this has been the work by Liew et al. (2012), which uses the same cascading principle expressed in a table form, termed the Total Site Problem Table. For the purpose of evaluating Total Site targets for varying energy supplies and demands, this technique has been revisited and adapted in Varbanov and Klemeš (2011) and is discussed below.

11.2.8.1 TSP and cascading of utilities

When utilities are placed against targeting profiles, it is necessary to account for the generation of each utility carrier from process heat sources or its consumption by process heat sinks. The potential for utility generation on a site is represented by the Site Source Composite Curve and that for consumption – by the Site Sink Composite Curve (Fig. 11.20a).

If, for a given utility, the generation from processes is insufficient to cover the demands, the two curves are displaced, and part of the demands for the current utility is left to be covered by a higher-temperature utility. For the example in Fig. 11.20b, the utility placement starts from hot water (the lowest-temperature utility), where the generation from processes is smaller than the process use of hot water. As a result, some of the generated steam should be used to cover the needs for hot water. Continuing up the temperature scale, the remaining unused LP steam from process generation is, in

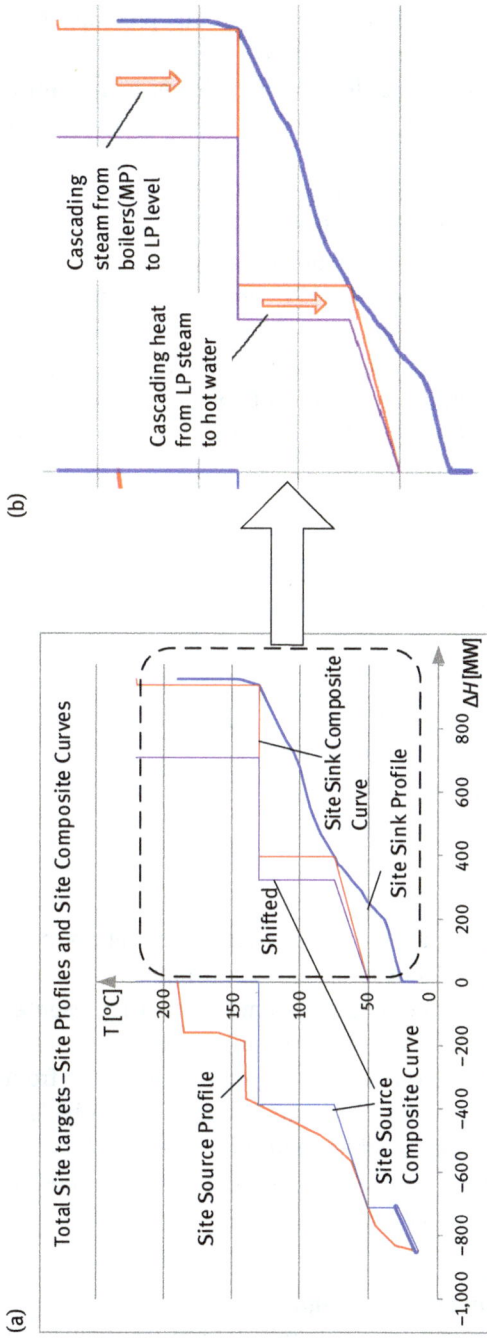

Fig. 11.20: Mapping of the heat cascading flows on the Site Composite Curves: (a) Total Site Profiles and Site Composite Curves; (b) cascading heat between utility levels (after Varbanov and Klemeš, 2011).

turn, insufficient to cover the LP steam demands and some medium-pressure (MP) steam from the site boilers is used to satisfy the remaining need.

This kind of use of higher-temperature utilities for satisfying lower-temperature utility needs is referred to as utility cascading. It can be implemented in various ways such as MP to LP steam let-down or direct use of the MP steam instead of LP steam. Utility cascading is similar to cascading heat in Pinch Analysis using the Problem Table Algorithm (Linnhoff and Hindmarsh, 1983). For the purposes of Total Site utility targeting and placement for continuous processes and without heat storage, the utility cascading is represented as in Fig. 11.21, where a part of the Site Source Profile, is allocated for generating a utility (e.g. steam) and feeding it to a mains vessel. Another input to this vessel is the heat cascaded from a higher-temperature utility. There are also two outputs – the use of the current utility by transferring heat to a part of the Site Sink Profile and the heat cascading to a lower-temperature utility.

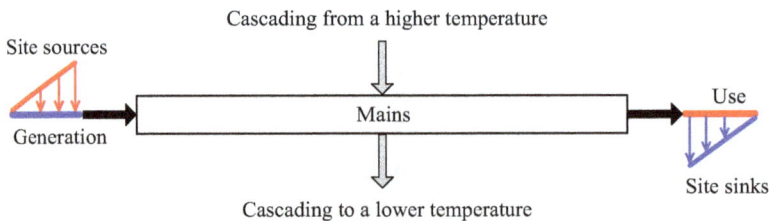

Fig. 11.21: Utility cascading without storage (after Varbanov and Klemeš, 2011).

The flowsheet-style diagram of utility cascading from Fig. 11.21 can be represented in a more compact way. This would allow incorporating it as a fragment in more comprehensive diagrams providing views of complete systems. Such a representation is shown in Fig. 11.22a, which also refers to a key term: utility interval. The latter is an analogue of the process heat cascade and the temperature intervals from the original Pinch Design Method (Linnhoff and Hindmarsh, 1983). An equivalent Sankey diagram in Fig. 11.22b illustrates the possible contributions of the heat inputs to the utility interval and the split between the outputs. This cascade representation is similar to the one by Shang and Kokossis (2004).

11.2.8.2 Accounting for non-isothermal heat sources and sinks

As discussed above, integrating the renewables uses an extended concept for Total Site modelling, called LIESs (Perry et al., 2008). Such an extended model involves entities and processes of different types – industrial, commercial (e.g. hotels), service (e.g. hospitals) and office buildings. The main reasons are the distributed

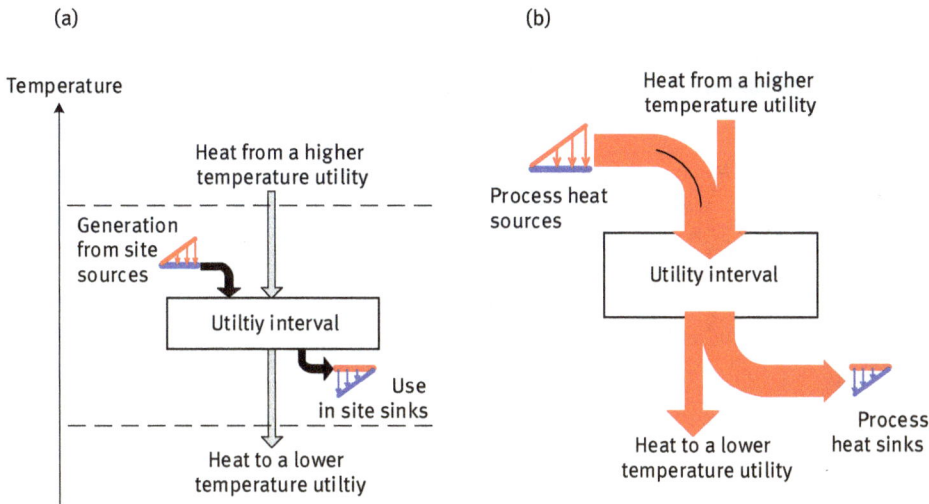

Fig. 11.22: Heat cascade cell for a Total Site with continuous processes: (a) flows configuration and (b) Sankey diagram (after Varbanov and Klemeš, 2011).

availability of the renewables and the need to maximise the reuse of any available waste heat in a given area. These necessitate minimising the transportation of the captured energy and limiting the scale of the capture facilities. As a result, some of the processes to be integrated feature lower-temperature heating demands, for which using hot water is sufficient. Hot water is a non-isothermal utility, and it is important that the Total Site targeting methodology supports its representation.

Non-isothermal utilities include hot water, cooling water, flue gas and hot oil, among others. The profile of a non-isothermal utility in a temperature–enthalpy plot is represented by one or more segments with slope, as can be seen in the lower temperature range of Fig. 11.20. This is in contrast to representing isothermal utilities (such as steam) by horizontal segments, approximating their heat transfer flows with latent heat exchange only.

It is important to stress that the slopes of the non-isothermal utility segments in the Site Composite Curves are provisional at the time of constructing the curves. It is possible later to break the Site Composite Curves into non-continuous shapes to accommodate, for instance, smaller CPs of utility generation by changing the shape of the utility use and the CP value in the corresponding Site Sink Composite Curve. The opposite situation is also possible – heat supply from process generation or solar heat capture may suggest large CP resulting in flatter segment slopes. This is usually accompanied by an excess of the generated and/or captured heat compared with the process demand for the particular utility (e.g. hot water). In such cases, it is possible to split the utility generation and/or solar heat capture stream (the T–H diagram at the top of Fig. 11.23) and produce two or more segments with steeper

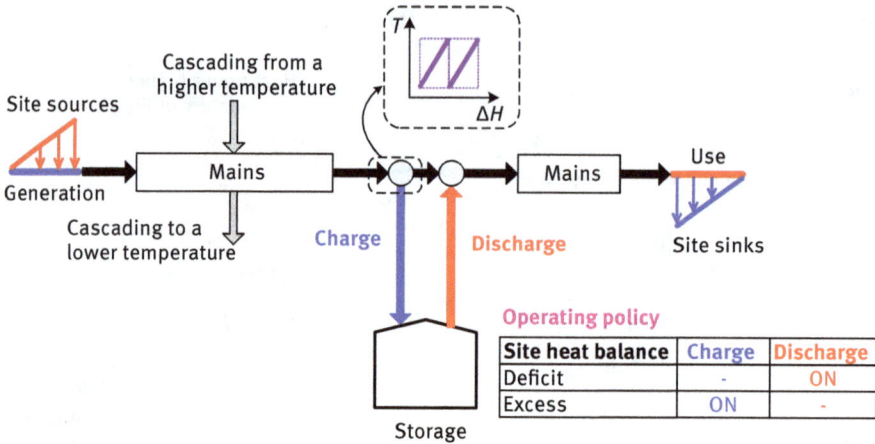

Fig. 11.23: Utility cascading with storage: (a) flows configuration and (b) Sankey diagram (after Varbanov and Klemeš, 2011).

slopes (smaller CP values), matching one of the segments with the utility use (Site Sink Composite Curve) and sending the remainder to heat storage. These techniques are demonstrated in the illustrative example in Section 11.2.9.

11.2.8.3 Handling the variability

11.2.8.3.1 General approach

Heating and cooling demands are imposed on the energy conversion and supply systems. They do not belong to the degrees of freedom, but to the system constraints. However, in some cases shifting certain demands in time can be suggested and, if sufficiently supported by economic gains, even achieved. Short-term fluctuations are usually modelled using time-differential equations. For the longer interval variations, considered here, piecewise constant approximations of the demands are used.

The piecewise representation of the demands can be embedded within a Total Site formulation to model the changes in the demand over longer periods. Typical examples are campaigns in the sugar industry, and first shift and the other shifts in the industrial plants, where the second and third shift could feature considerably lower energy consumption or might not be covered at all. Approximating winter and summer demands, especially for the residential buildings, is another example.

11.2.8.3.2 Heat cascading for intermittent processes and sites

The approach of representing the site integration parameters in a piecewise-constant form leads to the need for setting up a time horizon for the model and further partitioning it into a set of Time Slices. The partitioning can be performed using the procedure described in section 11.2.7 (Nemet et al., 2012b). The Total Site targeting is then performed within each Time Slice. An important feature of this type of problems is that some Time Slices may feature heat excess in one or several temperature intervals or deficit in other intervals. In such cases, heat can be transferred to Time Slices within the same temperature intervals or to lower temperatures, by means of thermal energy storage facilities.

The heat transfers from an earlier Time Slice to a later one can be implemented using heat storage based on a chemical or physical principle. These positions heat storage among the most important parts of the system related to the optimal design of intermittent processes. If storage is available at the required capacity with feasible cost, it can considerably increase the system heat recovery and efficiency.

Energy storage is a complicated and demanding issue, which is still waiting for a major breakthrough (Dinçer and Ezan, 2018). The Heat Integration methodology could contribute to this problem solution by providing targets that could be achieved.

The next step is to identify the possible degrees of freedom. Important ones are:
(i)   Integration of several unit processes and/or consumers into Total Sites. Integrating many users with different temporal energy consumption patterns provides the opportunity for more efficient utilisation of the primary resources as well as to exchange heat for better recovery.
(ii) Selection of the degree of utilisation of the available renewable resources – solar, wind, biomass (including waste), hydropower, geothermal and ground heat pumping.
(iii) Storing excess waste heat and its utilisation in a future Time Slice.

To ensure the transfer of heat from one Time Slice to another for a particular utility, it is necessary to use heat storage. One possibility is, starting from the utility use-generation pattern shown in Fig. 11.21, to separate the utility mains into two parts and add the heat storage facility between them, as it is illustrated in Fig. 11.23. The latter arrangement allows for several storage operating policies to be applied, including storage charge, discharge or lack of heat transfer, depending on the situation at any given moment.

Switching the view from process flowsheet to heat cascade, as shown in Fig. 11.24a allows showing the time dimension of the problem explicitly. The diagram shows a heat cascade cell in two dimensions – temperature (vertical) and time (horizontal). An equivalent Sankey diagram in Fig. 11.24b illustrates the possible contributions of the cell heat inputs and the split between its outputs.

(a)

(b)

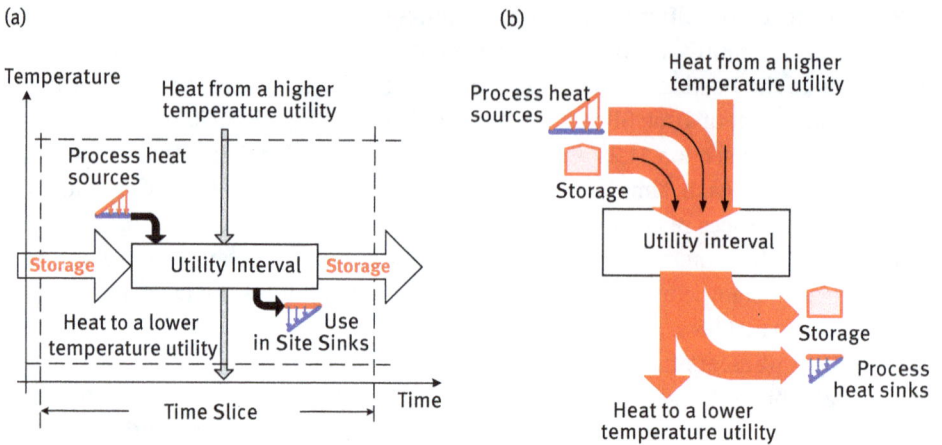

Fig. 11.24: Heat cascade cell for a Total Site with intermittent processes. (after Varbanov and Klemeš, 2011).

11.2.8.4 Heat storage model for Heat Integration

There is another important issue, which should be optimised: the size and location of heat storage facilities. They are three obvious options for implementing heat storage: centralised, distributed and also their combination. This optimisation has its own economic background and is related to specific features of heat storage. However, it is also closely interconnected with Total Site Heat Integration.

11.2.8.4.1 Storage scope

For Total Site targeting, a central heat storage facility per each utility is assumed to be available without pre-determining the installation of local storage capacities for the processes. One heuristic for deciding on the installation of local storage can be to evaluate the variability of the heating and cooling demands of the site processes and, if they are clearly concentrated within one or two processes and not typical for the remaining ones, then installing local storages may be justified and should be evaluated in detail at the stage of system design.

11.2.8.4.2 Heat losses

From any kind of heat storage, there are heat losses. They may be in the form of load loss, temperature loss or both. The following cases illustrate some of the possible storage options:

– If steam is stored in a large drum (steam accumulator) (Prieto et al., 2018) or another vessel without heat replenishment, part of the steam would condense as a result of casing heat losses. Until there is some steam in the vessel, the temperature would not deteriorate. Therefore, assuming reasonably short

storage durations and/or sufficient insulation, heat loss from steam storage can be assumed to involve only load loss and no temperature loss.
- If cooling capacity is stored in the form of ice, it would have similar properties as steam storage due to the fact that the main cooling effect comes from ice melting, that is, there is again a phase change.
- In the case of sensible heat storage – for instance, storing heat in the form of hot water or storing cooling capacity as cold/chilled water (storing "cold"), the load loss is always accompanied by an equivalent loss of temperature potential – temperature reduction for hot water and temperature increase for cooling water storage.
- There is also the possibility that heat is stored in a device based on a chemical or physic-chemical reaction. Storing heat in a reactor using a hydration–dehydration (endothermic/exothermic) reaction is a popular choice presented by Masruroh et al. (2006) and later in a different type of the process by Weber and Dorer (2008). In any case, charging the storage with heat takes place at a higher temperature than the discharge operation. There is always a temperature loss in addition to any load loss due to heat exchange with the ambient.

For energy targeting, the heat load losses can easily be accounted for. In a heat cascade (e.g. Fig. 11.24), the cascading flows between the Time Slices represent the heat storage. Adding another flow accounting for the load loss (Fig. 11.25) is straightforward.

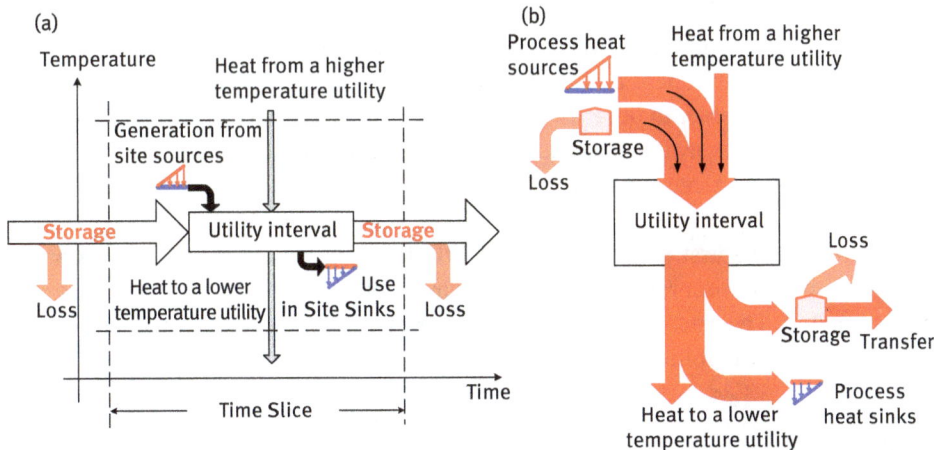

Fig. 11.25: Accounting partially for heat deterioration: (a) flows configuration and (b) Sankey diagram (after Varbanov and Klemeš, 2011).

However, accounting for temperature deterioration is not as simple. To complete it successfully, it is necessary to know specifics of the employed storage technology and make additional decisions on the storage operating policy, which do not belong to the targeting activity, or for making which at the stage of targeting the information is insufficient. Such a decision can involve whether to simply accept the losses or to provide some auxiliary heating to the storage maintaining its load and temperature. At the current state of the art in terms of Total Site targeting the heat losses during storage are generally neglected.

11.2.8.5 Operating policy and management of the heat storage

The energy storage may be used in three modes: charge, discharge and idle (dormant). For the purpose of targeting, if the storage exchanges heat, it is normally assumed that in a given Time Slice, it may operate only in one direction – either charge or discharge mode. Although it is technically possible to devise storage facilities, which can handle two-directional operation, this would tend to complicate the storage and increase its investment and operating costs at no added benefit for the site. Therefore, the storage is assumed to be exclusively in only one of its three operating modes during a given Time Slice.

11.2.8.6 The priority of utility placement and use

Both renewables and heat from storage have to be used before the constant-availability utilities (coming mainly from fuels) in order to maximise the energy efficiency and save fuel. The policy of unidirectional heat transfer from/to the energy storage has further implication on the priority with which the utilities are used. When both renewable heat and heat from the storage are available, precedence is given to the renewable, in order to guarantee that heat will be transferred only from the storage if the renewable source capacity is insufficient, or to the storage, in case that the sum of the renewable and stored heat exceeds the process needs for a given Time Slice.

Therefore, the first priority for using utilities is to maximise the utilisation of the heat captured from renewables. As a second option, the heat from storage facilities is applied. The remaining demands are covered by other non-transient utilities (with constant availability).

11.2.8.7 Total Site Heat Cascade

After considering the main issues and establishing the targeting and cascading rules, the Total Site Heat Cascade can be defined. Figure 11.26 provides a generalised view of the Total Site Heat Cascade where the temperature intervals are allocated to specific utilities –MP steam, LP steam and hot water. The diagram shows the options for heat cascading. After evaluating a particular operating scenario for

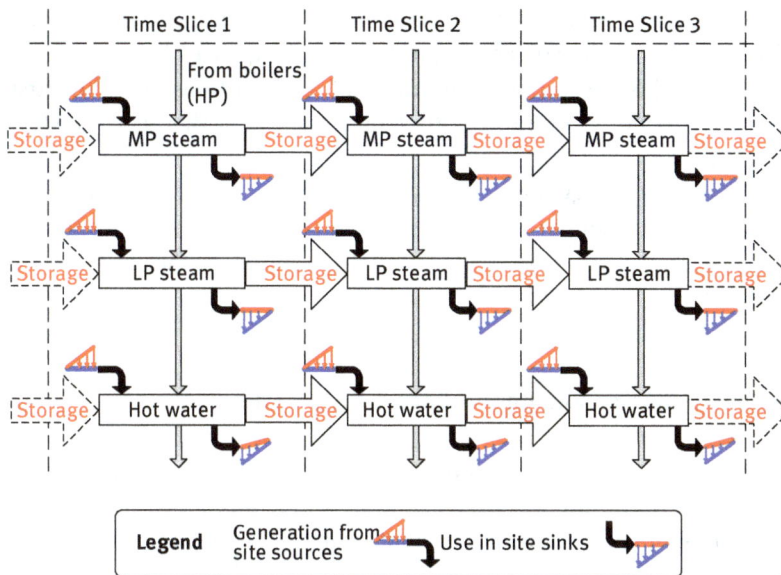

Fig. 11.26: Total Site Heat Cascade with Time Slices (after Varbanov and Klemeš, 2011).

the site, some of the flows shown may become zero and in such cases could be omitted from the diagram for enhancing the clarity. For instance, the heat storage transfer flows from the left of Time Slice 1 will exist only in the case of cyclic operation and also the heat storage transfers for corresponding utilities from the last Time Slice (No. 3 in Fig. 11.26) are positive.

11.2.8.8 Overall procedure
The overall procedure for targeting Total Sites with temporal variations and renewables is then defined as follows.
(i) Analyse the temporal variations in the heating and cooling demands of the site processes (plants, buildings and farms) and partition the modelling horizon into Time Slices.
(ii) Assume initially that no heat is stored before the first Time Slice for all utility types.

Steps (iii) to (iv) are performed for each Time Slice.
(iii) Obtain the Heat Integration targets for each process unit.
(iv) Obtain the Total Site utility targets using the Total Site Heat Cascade, TSP and Site Composite Curves. The targets calculation should follow the utility placement priorities given in Section 11.2.8.6.

(v) For cyclic operation, the heat transfer flows from the last Time Slice in the Total Site Heat Cascade should be fed/recycled to the first Time Slice. At this point, a reiteration of step (iv) is needed until the values of the heat transfers to the first Time Slice become equal to or sufficiently close to those of the heat transfers from the last Time Slice.

(vi) Identify the target for energy storage capacity for each utility type using the heat transfer flows between the Time Slices in the Total Site Heat Cascade. For each utility type, the heat storage capacity will be equal to the largest heat transfer flow between Time Slices.

11.2.9 Example: Integration of solar thermal energy into a LIES

The case study is based on a configuration first introduced by Perry et al. (2008) and adapted by Varbanov and Klemeš (2011). Four areas are integrated into a Total Site – two industrial plants, a hotel and a residential area (Fig. 11.27). Each of these areas is referred to as a process. Each process features a number of streams – hot and/or cold. It is possible to establish a connection to a district heating plant.

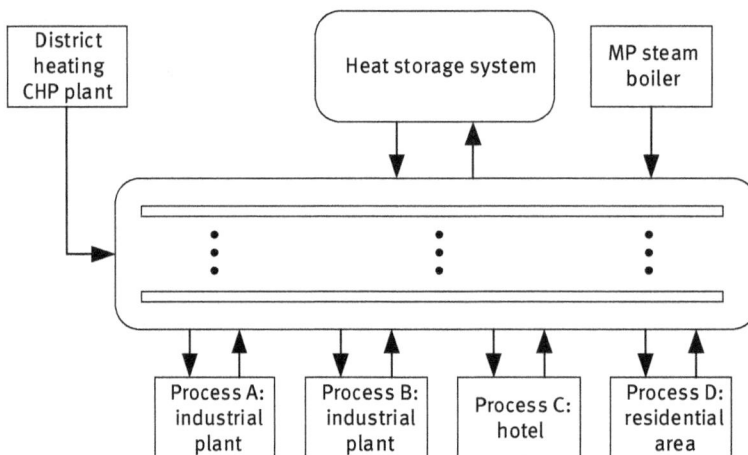

Fig. 11.27: Configuration of the considered LIES (derived from Perry et al., 2008).

The process streams with their main properties and periods of activity are given in Tabs. 11.6–11.9. The time intervals are expressed for a 24 h cycle. The minimum allowed temperature difference for all processes is $\Delta T_{min} = 10$ °C. The residential area has a number of solar thermal collector cells for generating domestic hot water and space heating. The utilities available at the Total Site are listed in Tab. 11.10. An assumed storage facility uses hot water. The operating temperature of the storage facility is assumed to be 75 °C, and for the current example, the heat losses are neglected.

Tab. 11.6: Heating and cooling demands for process A.

No.	Stream	Temperature [°C]		ΔH [kW]	Type	CP [kW/°C]	Time interval [h]	
		Supply	Target				From	To
1	A2	170	80	120.0	Hot	1.333	0	24
2	A1	150	149	180.0	Hot	180.000	0	24
3	A5-1	50	135	104.4	Cold	1.228	0	24
4	A5-2	85	100	82.3	Cold	5.487	0	24
5	A6	62	100	130.0	Cold	3.421	0	24
6	A7	72	55	130.0	Hot	7.647	0	24

Tab. 11.7: Heating and cooling demands for process B.

No.	Stream	Temperature [°C]		ΔH [kW]	Type	CP [kW/°C]	Time interval [h]	
		Supply	Target				From	To
1	B1	200	195	160.0	Hot	32.000	6	20
2	B2	20	54.7	10.0	Cold	0.288	6	20
3	B3	50.5	85	107.3	Cold	3.109	20	6
4	B4	100	120	130.0	Cold	6.500	6	20
5	B5	150	40	83.5	Hot	0.759	6	17
6	B6	80	95	48.0	Cold	3.200	6	20
7	B7	95	25	80.0	Hot	1.143	6	17

After analysing the process stream data, three switching time points over the 24 h horizon have been identified: 6 h, 17 h and 20 h. They define three Time Slices as listed in Tab. 11.11. In the residential area, there are also solar thermal collectors, which capture solar heat. The available solar heat flows are shown in the last column of Tab. 11.11.

The energy targets have been evaluated for the described Total Site, applying the procedure from Section 11.2.8.8. For Time Slice 1, the TSP and Utility Composite Curves (Fig. 11.28) have been obtained. Several observations can be made:

– The generation of hot water by processes (Shifted Site Source CC) can completely satisfy the process heat sinks in its temperature interval, and 11.0 kW of hot water (121.4 kWh for the whole slice duration) remains unused.

Tab. 11.8: Heating and cooling demands for process C.

No.	Stream	Temperature [°C]		ΔH [kW]	Type	CP [kW/°C]	Time interval [h]	
		Supply	Target				From	To
1	Soapy water	85	40	20.0	Hot	−0.444	6	17
2	Condensate	80	40	75.0	Hot	−1.875	6	17
3	Sanitary water	25	55	17.3	Cold	0.576	20	6
4	Laundry	55	85	18.0	Cold	0.600	20	6
5	BFW	33	60	12.0	Cold	0.446	0	24
6	Sanitary water	25	60	15.0	Cold	0.429	6	17
7	Sterilisation	82	121	34.1	Cold	0.874	20	6
8	Swimming pool water	25	28	23.1	Cold	7.700	6	17
9	Cooking	80	100	32.0	Cold	1.600	6	17
10	Heating	18	25	41.1	Cold	5.864	0	24
11	Bedpan washers	21	121	5.0	Cold	0.050	6	17

Tab. 11.9: Heating and cooling demands for process D.

No.	Stream	Temperature [°C]		ΔH [kW]	Type	CP [kW/°C]	Time interval, h	
		Supply	Target				From	To
1	Space heating	15	25	88.0	Cold	8.800	0	24
2	Hot-water base	15	45	25.0	Cold	0.833	0	24
3	Hot-water day	15	45	65.0	Cold	2.167	6	20

Tab. 11.10: Site utility specifications.

Name	Type	Temperature level(s)
Cooling water	Cold	15 °C to 30 °C
Solar hot water	Hot	80 °C to 50 °C
District hot water	Hot	75 °C to 50 °C
MP steam	Hot	at 220 °C
LP steam	Hot	at 130 °C

- The 112.9 kW (1,241.9 kWh) captured solar heat also cannot be immediately utilised.
- The site needs 113.2 kW (1,245.4 kWh) MP steam, of which 94.7 kW (1,042.2 kWh) is the MP steam cascaded to satisfy LP steam demands.

Tab. 11.11: Time Slices for the example.

	From [h]	To [h]	Duration [h]	Share [%]	Solar capture [kW]
Time Slice 1	6	17	11	46	112.9
Time Slice 2	17	20	3	12	40
Time Slice 3	20	6*	10	42	0

- The demand for cooling water is 139.8 kW (1,537.9 kWh).
- As a result, the excess heat available from solar hot water and the excess of process-generated hot water is directed to the storage, with the overall amount 1,363.3 kWh and are made available as a utility for Time Slice 2 at 75 °C.

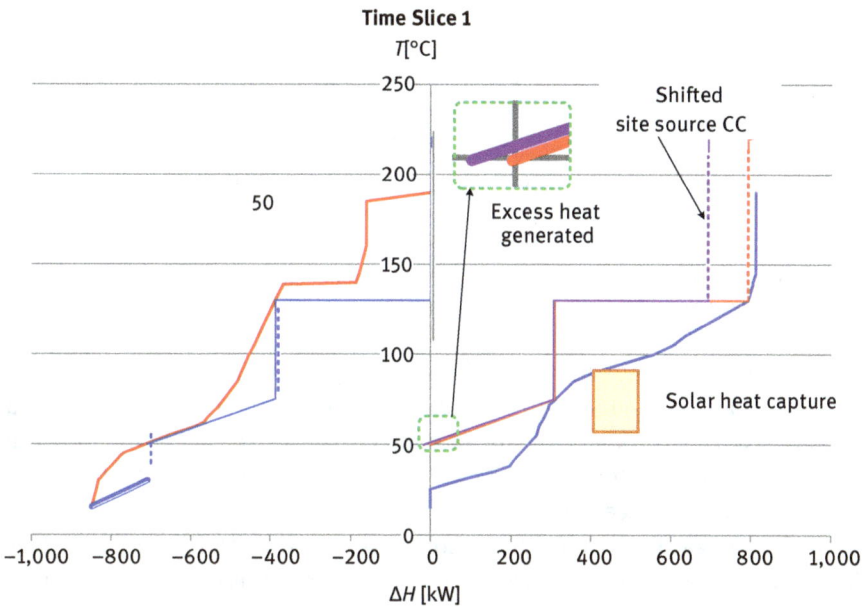

Fig. 11.28: Total Site targets for Time Slice 1 (after Varbanov and Klemeš, 2011).

For Time Slice 2, the active process streams can be supplemented with the heat available from the storage with the availability of 1,363.3 kWh. The targets for Time Slice 2 are shown in Fig. 11.29. It can be seen that 145.2 kW (435.5 kWh) heat is drawn from the storage, leaving another 927.8 kWh remaining stored. Also, the full extent of the solar capture for this Time Slice is used amounting to 40 kW (120 kWh). The remaining utility targets are 114.3 kW (342.8 kWh) MP steam and 56.2 kW (168.5 kWh) cooling water.

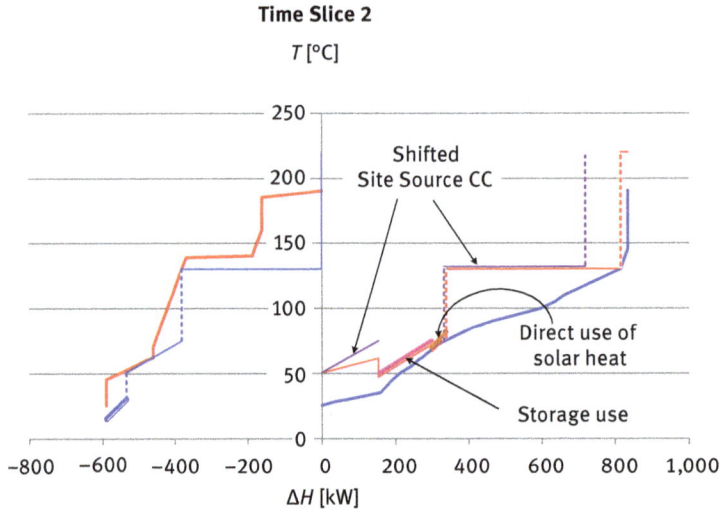

Fig. 11.29: Total Site targets for Time Slice 2 (after Varbanov and Klemeš, 2011).

Finally, the targets for Time Slice 3 are shown in Fig. 11.30. The remaining stored heat from Time Slice 2, amounting to 927.8 kWh, is completely used. There is also a need for 136.7 kW (1,366.7 kWh) MP steam and cooling water is not required. The fact that all the remaining stored heat is used in this Time Slice means that recalculation of the targets is not needed.

The Total Site heat cascade for the whole case is shown in Fig. 11.31. Besides providing a single view of the analysis, two more important results can be inferred from the diagram. One is the total of the utility targets for the overall process cycle of 24 h: the MP steam target is 2,954.9 kWh and for cooling water is 1,706.3 kWh. The other important result is the storage capacity required. In the current example, only storage for hot water is needed. It has to be taken as the largest heat transfer flow from the cascade between the Time Slices, which in this case is 1,363.3 kWh, representing 46 % of the MP steam target.

It is important to mention that the overall MP steam target for the same stream data but without applying heat storage would be 4,438.2 kWh, meaning that the

Time Slice 3

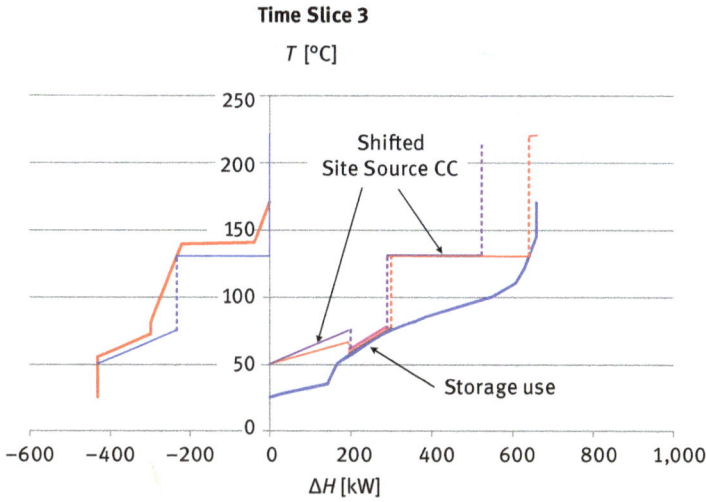

Fig. 11.30: Total Site targets for Time Slice 3 (after Varbanov and Klemeš, 2011).

All heat transfers are in [kWh]

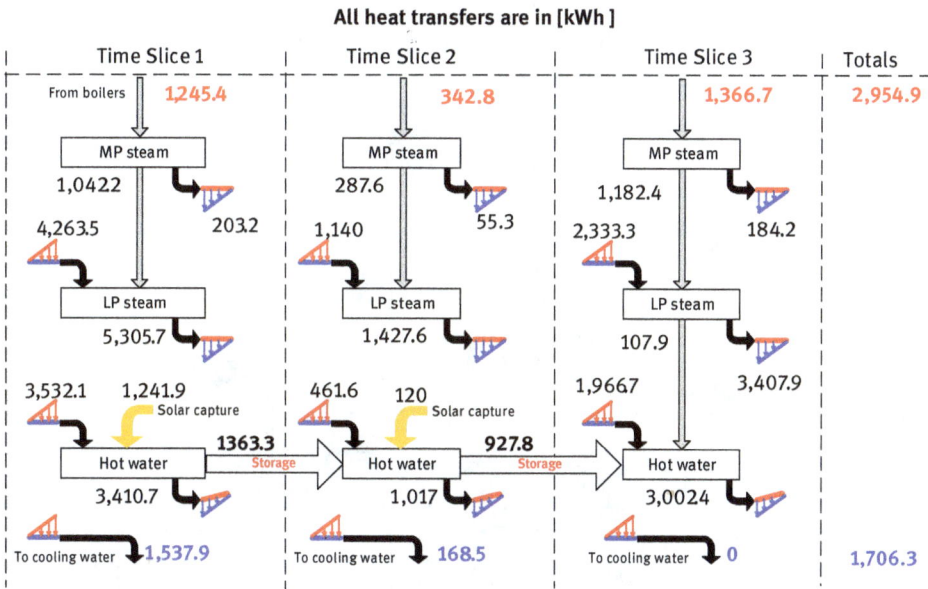

Fig. 11.31: Total Site heat cascade for the initial case (after Varbanov and Klemeš, 2011).

benefit from the heat storage is saving 1,483.3 kWh MP steam or 33 % of the no-storage case.

An important advantage of using the Total Site Heat Cascade is the ability to trace the routes for possible improvement of the site targets by rescheduling some of the heating or cooling demands. The benefits to looking for the rescheduling actions could involve reduction of the required storage capacity and also reduction in the final utility targets. For instance, in Fig. 11.31 it can be seen that the likely improvements in the utility and storage targets should be sought by rescheduling streams intersecting the interval of the hot water. Looking back into the process, heating and cooling demands, stream B3 from Tab. 11.7 (Process B) as well as streams "3: Sanitary water" and "7: Sterilisation" from Tab. 11.8 (process C) are present in the desired temperature range. The next step is to map the Time Slices during which these streams are active, and in which Time Slice there is most local excess of the required heat. All the identified streams are active during Time Slice 3. Also, the most heat excess for the hot water utility is in Time Slice 1. As a result, the targeting procedure has been performed, after moving the streams activity to Time Slice 1 and keeping their total heating demands as amounts (kWh) the same. As a result, the cascade shown in Fig. 11.32 has been obtained. Here, the required storage capacity takes approximately 15 % of the MP steam target.

The comparison of the diagrams from Figs. 11.31 and 11.32 shows that moving the identified streams activity would be beneficial for reducing both the requirements for MP steam and for heat storage, but would slightly increase the need for cooling water. Some simple dependencies can be inferred. For example, moving the

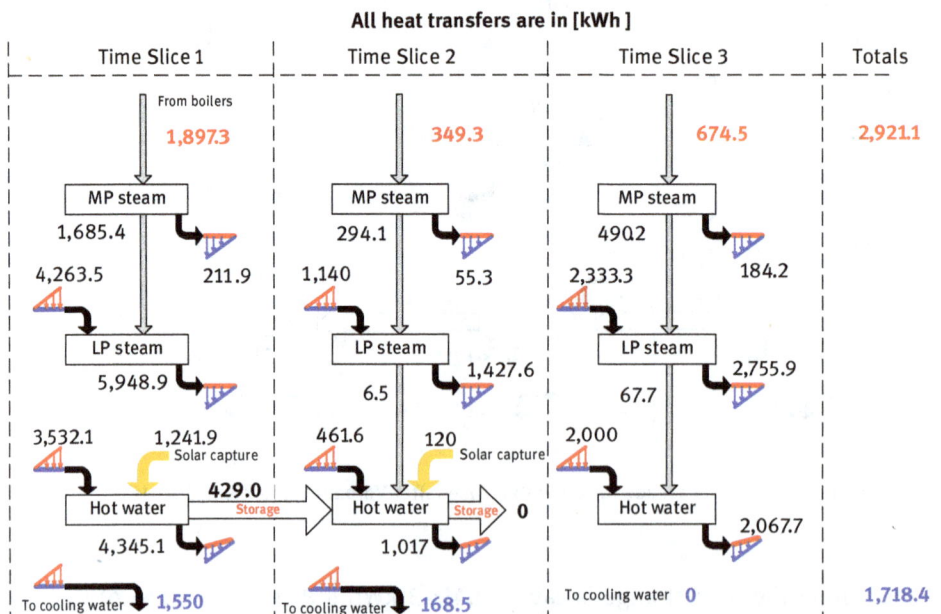

Fig. 11.32: Total Site Heat Cascade for the modified case (after Varbanov and Klemeš, 2011).

streams "3: Sanitary water" and "7: Sterilisation" in Process C from the night shift to the day time may introduce the need to increase the inventory of sanitary materials, beddings, bedsheets and so on. Increased inventory may also be needed for various tools which are sterilised, in order to have such items available while the other items await their processing.

11.3 Total Site targeting with process specific minimum temperature difference (ΔT_{min})

The section presents an extension of Total Site Integration to LIESs (Perry et al., 2008) producing more realistic utility and heat recovery targets in cases with significantly varying heat transfer coefficients between the site processes. Process-level Heat Integration, based on Pinch Analysis, aims to minimise the energy utility demand of individual industrial processes. The core Total Site targeting method takes a single specification for ΔT_{min}. Such an assumption is too simplistic in some cases and may lead to inadequate results due to imprecise estimation of the overall Total Site heat recovery targets. The extended Total Site targeting procedure, described in this section, has been developed in (Varbanov et al., 2012). It allows for obtaining more realistic heat recovery targets for Total Sites. The procedure takes specifications of different values for the minimum allowed temperature differences (ΔT_{min}) for each process on the site. It is illustrated with a case study for LIESs, also providing a comparison with the traditional targeting procedure and the advantages offered by the extended procedure.

11.3.1 Analysis of the Total Site method interfaces – single versus multiple ΔT_{min} specifications

Energy targeting methodologies such as Pinch Analysis have been developed and used in the past with proven and useful results (Klemeš et al., 2018). Transferring and recovering heat requires heat exchangers of reasonable size. The main driving force in heat transfer is derived from the temperature differences of the participating streams (Eq. 11.1), and the heat transfer area is inversely proportional to the driving force:

$$A = \frac{Q_{HE}}{U \times \Delta T_{LM}} \tag{11.1}$$

To ensure that excessively large heat transfer area is not selected, the temperature differences between the hot and the cold streams in each heat exchanger are kept larger than a certain specified lower bound. In Heat Exchanger Network design, the concept of the minimum allowed temperature difference (ΔT_{min}) is introduced, representing this lower bound. ΔT_{min} is a design parameter, determined by exploring the trade-offs between larger heat recovery and the larger heat transfer area requirement (Klemeš et al., 2018).

The core Total Site targeting procedure, introduced by Dhole and Linnhoff (1993) and summarised at the beginning of this chapter, does not provide flexibility in the specifications. Representative previous studies of Total Site Integration include the pivotal work on Total Site cogeneration (Klemeš et al., 1997), the Hui and Ahmad article in 1994 on heat recovery through the utility system (Hiu and Ahmad, 1994), and the 2008 extension to LIESs (Perry et al., 2008). They all assume a single uniform ΔT_{min} specification for all processes on a site and for exchanging heat with the utilities. For the adequate design of interprocess heat recovery systems, ΔT_{min} specifications are needed for several types of heat transfer: process-to-process, hot streams to cold utility and hot utility to cold streams. Moreover, these types of heat transfer are within the context of each process on the site. All these specifications should be available as degrees of freedom to the system designers. Specifying only one value for all of these means using only one of the degrees of freedom and eliminating the rest. This can be rather limiting for real problems, as explained next.

Process streams with different heat transfer characteristics require different ΔT_{min} values for a process to process heat exchange. For instance, two gas streams exchanging heat would need higher temperature difference than two liquid streams for achieving the same heat load and same heat transfer area. Realistic ΔT_{min} values derived for utility and heat exchanger costs in the year 2007 have been discussed in (Towler and Sinnott, 2008). The source indicates values of 10 °C to 30 °C as suitable for refinery and chemical processes, 50 °C to 80 °C for furnaces, and 3 °C to 5 °C for plate-fin heat exchangers in the food industry.

Although more than a decade has passed since then and the exact values have changed, the important message is that the magnitude of ΔT_{min} depends significantly on the type and context of heat transfer. For heat exchange between the processes and the utility system, the ΔT_{min} can also be different. If different types of plants are integrated on a site – for instance, oil refinery, food processing plant, brewery, hospital, football ground and stadium – different ΔT_{min} values could be the optimum for each plant.

When using a uniform ΔT_{min} specification, there can be two cases:
- If the ΔT_{min} is overestimated for some of the processes, the achievable heat recovery targets for those processes would be underestimated, resulting in higher energy cost estimates. This can have economic implications in the form of oversizing the utility system components – most notably the steam boilers, cooling circuit elements, heaters and coolers. This can further lead to excessive resource use and cost, as well as to inefficient use of capital for the utility system.

- In the opposite case, if ΔT_{min} is set smaller than the optimal for the corresponding heat transfer type, the heat recovery targets would be overestimated leading to strive for more heat recovery site-wide and excessive capital costs for heat transfer area. In addition, the underestimated hot and cold utility demands may result in undersized utility facilities – for instance, boilers and cooling towers. This, in turn, can potentially lead to either infeasible operational situations or at the very least reduced reliability of the utility system by having to operate near or it at its maximum capacity.

The presented reasoning has led to the development of modelling concepts and a procedure that enable the optimisation of the site condition and ΔT_{min} specifications via the minimisation of the steam generation from fuel and the use of utility cooling. This is done by introducing individual specifications for the ΔT_{min} values, for performing the heat exchange within each process, plus ΔT_{min} specifications for the heat exchange between each process and the relevant utilities. This provides the opportunity for a more realistic evaluation of the energy-capital trade-offs. Such a change in the interface of the targeting procedure has led to modifications in the algorithm for constructing the TSP and utility placement allowing more appropriate Process Integration. The modified procedure is described next.

11.3.2 Modified procedure for Total Site targeting with flexible ΔT_{min} specifications

The energy efficiency on a site can be improved within various scopes. The considered system boundary can be set within individual processes, whole plants, heterogeneous areas of energy users and also at the regional level. For easier understanding, the definitions of isothermal and non-isothermal utilities are revisited.

Isothermal is a **utility**, which does not change the temperature while being generated or used. For example, in most cases of targeting steam is considered isothermal by neglecting all other parts of its generation or use, except the evaporation and condensation. If this targeting approach is taken, then steam generation and use are termed as "isothermal utilities."

Similarly, if there is a temperature change during the generation or use of a utility, it is referred to as non-isothermal. Flue gas from a furnace is a typical non-isothermal hot utility, whose temperature change cannot be neglected Similar is the case with the generation and use of hot water. As can be seen from a recent extension to the Total Site targeting model (Liew et al., 2014), accounting for the segment of boiler feedwater preheating before evaporation, leads to more robust targets of fuel use and cooling water at the site level. In such a case, the steam generation is considered as a "non-isothermal utility."

A typical hot utility used in industry is steam. Choosing the pressure levels of steam mains affects the site utility demand significantly. Any heat excess in one process, after internal recovery, could be reused in another via the steam system. To achieve this, each steam pressure level has to allow both generation and use in sufficient amounts. For example, if at a given pressure level 10 MW of steam can be generated from process heat sources, but only 5 MW could be utilised by process heat sinks, this would result in only 5 MW of heat recovery through the utility system. On the other hand, if the steam pressure level would be reduced, it may be possible to achieve higher heat recovery via that particular steam main. Any steam raised from process cooling should be utilised as much as possible to maximise the site-level heat recovery. Traditionally, in Total Site targeting, the steam representation is simplified, showing only the constant-temperature segments, not considering preheat, superheat and subcooling segments for the purposes of targeting. In some cases, cooling with water can also be considered virtually isothermal, resulting in horizontal plots. This effect can be achieved by assuming very large cooling water flowrates.

Non-isothermal utilities, such as hot water, should be treated similarly. The only exception is that the generation and use temperatures are different. Therefore, each utility has an upper and a lower temperature. For non-isothermal utilities, these temperatures assume different values, while for isothermal utilities (e.g. saturated steam) the values for upper and lower temperatures are identical.

Heat exchangers have heat transfer areas, inversely proportional to the temperature differences between streams. This can be easily seen in Eq. 11.1. As a result, lower ΔT_{min} values result in larger heat transfer area requirements but also in potentially higher rates of heat recovery. Essentially, this creates a trade-off between equipment cost and heat recovery rate within Heat Exchanger Networks. As discussed in the previous section, this trade-off becomes even more complicated when the capital and operating costs related to the utility system are accounted for explicitly.

The traditional methodology (Section 11.1) proceeds as follows. Process stream data for heating and cooling demands are first extracted from the general process datasets (flowsheets). They are further used to obtain heat recovery targets for each process, most often represented by GCCs. In the process of GCC construction, the ΔT_{min} specification for the particular process is used to shift the process stream temperatures and enable a uniform temperature representation for both cold and hot streams. From the GCCs for all site processes, only the net heat sink and source segments are extracted, assuming that heat recovery is maximised within each plant (process) separately. The extracted segments are then combined to form a Site Heat Source Profile, and a Site Heat Sink Profile collectively referred to as TSP. Using the TSP, the Total Site Composite Curves for utility generation and use are constructed, using the specifications for the utility temperatures.

The traditional methodology – initiated in (Dhole and Linnhoff, 1993) and further matured in (Klemeš et al., 1997), proceeds as follows:

Step 1. Process stream data for heating and cooling demands are first extracted from the general process datasets (flowsheets)

Step 2. The data are further used to **obtain heat recovery targets for each process**, most often represented by GCCs. In the process of GCC construction, the ΔT_{min} specification for the particular process is used to shift the process stream temperatures and enable a uniform temperature representation for both cold and hot streams. The ΔT_{min} specification is set the same for all site processes.

Step 3. From the GCCs for all site processes, only the **net heat sink and source segments are extracted**, assuming that heat recovery is maximised within each process separately.

Step 4. The extracted segments are then combined to **form a Site Heat Source Profile and a Site Heat Sink Profile** collectively referred to as TSP.

Step 5. Using the TSP, the **Total Site Composite Curves** for utility generation and use are constructed, using specifications for the utility temperatures, provided by the users.

In this chapter, a modification to the traditional Total Site targeting procedure is presented referred to as the "modified" procedure. It employs an altered TSP construction algorithm, allowing separate ΔT_{min} specifications for heat exchange within each process and also between each process and each utility (Figure 11.1). It is important to note that ΔT_{min} limits can also be specified individually for each stream. This has the potential to allow even more thorough estimation of the heat recovery potentials over the Total Site, but also bears the potential of much higher complexity, comparable with detailed system design or optimisation.

The modified Total Site targeting procedure (Fig. 11.33) is formulated as follows:

Step 1: Parameter specification. The main parameters of process streams are defined similarly to the basic methodology. In addition, for each process, an individual ΔT_{min} value is specified for heat exchange between the process streams – Fig. 11.33. Separate ΔT_{min} values are specified for heat transfer from each hot utility to each process (PHU) and from each process to each cold utility (PCU). This can be done in the form of a ΔT_{min} matrix with the utilities forming the rows and the processes forming the columns. An example of such a matrix is given in the case study within this section (Tab. 11.17).

Step 2: Process-level Pinch Analysis. For each site process, the heat recovery targets are obtained using Pinch Analysis (Klemeš et al., 2018). The results are the process heat cascade, the GCC, the Pinch location and the overall minimum utility heating and minimum utility cooling demands of each process.

Step 3: Extraction of the heat source and heat sink segments from the GCCs for all processes. This is performed in the same way as in the basic procedure – see Klemeš et al. (2018) and Section 11.1 in this chapter. First, the GCC pockets are removed, then the remaining segments are extracted as heat sources and

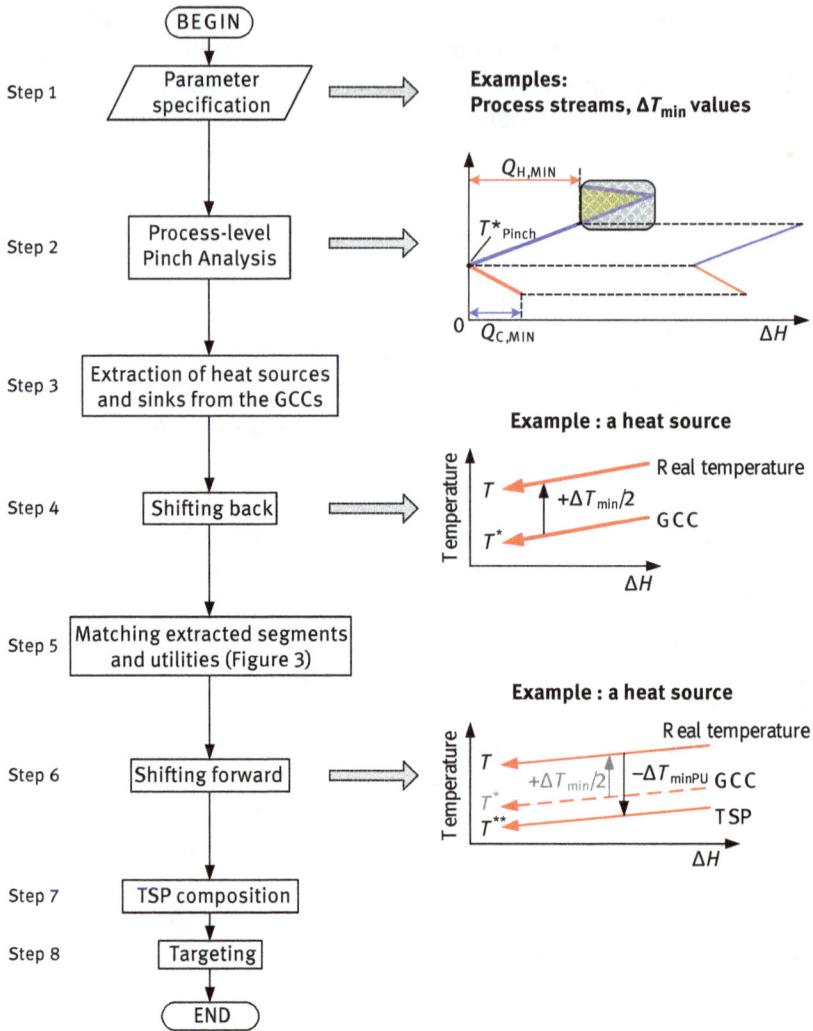

Fig. 11.33: Modified Total Site targeting procedure allowing flexible ΔT_{min} specifications (Varbanov et al., 2012).

heat sinks at the site level. As a result, at the Total Site level, only the net utility demands (heating and cooling) of the various processes are considered. When targeting for cogeneration, however, the pockets may also be extracted, if they bear significant power cogeneration potential.

Steps 4 and 5, discussed next, implement the idea of applying individual ΔT_{min} values to different heat exchange types.

Step 4: Shifting back. A shift of the extracted segments back to the temperature levels of the process streams. The heat source and sink segments are shifted back to the levels of the real stream temperatures by $\Delta T_{min,PP}/2$ (Eq. (11.2)). For each process the extracted heat source segments are shifted up by $\Delta T_{min,PP}/2$ and the heat sink segments are shifted down by $\Delta T_{min,PP}/2$ as follows:

$$T_{BACK} = \begin{cases} \text{sources: } T^* + \frac{1}{2} \times \Delta T_{min,PP} \\ \text{sinks: } T^* - \frac{1}{2} \times \Delta T_{min,PP} \end{cases} \tag{11.2}$$

Step 5: Matching the extracted segments and the utilities. This step aims at the selection of the appropriate ΔT_{min} specification for each extracted heat source and sink segment from the **ΔT_{min} matrix**. To do this, the following substeps are performed.

5.1. Projection of utility temperatures on heat sources. For the heat source segments in each process, to the lower temperature of each utility, which can be generated, is added the ΔT_{min} specification for each process and all heat sources from that process are checked whether they contain a point at the projected temperature (Fig. 11.34). Any such heat source segment is then split into two or more parts, depending on how many utility projections it crosses. This is repeated for all combinations of processes and appropriate utilities. The reason for this operation is that above the projected temperature the currently considered utility can be generated, while below that, only a lower-temperature utility can be generated.

5.2. Projection of utility temperatures on heat sinks. For evaluating the scope for utility use, from the upper temperature of each utility which can be used by process heat sinks, is subtracted the ΔT_{min} specification for each process and all heat sinks from that process are checked whether they contain a point at the projected temperature (Fig. 11.34). Any such heat sink segment is then split into two or more parts, depending on how many utility projections it crosses. This is repeated for all combinations of processes and appropriate utilities.

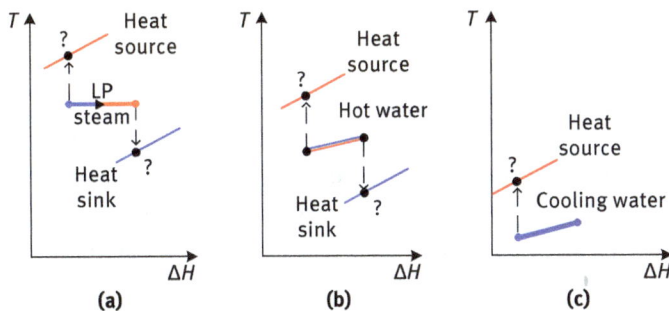

Fig. 11.34: Selection of ΔT_{min} specifications.

5.3. Selection of ΔT_{min} specifications. Completion of the previous two steps ensures that all resulting heat source and sink segments can be matched to exactly one utility type for generation or use. This is exploited to match each heat source segment to the hottest hot utility to whose generation it can contribute. Using the matched pair, the corresponding ΔT_{min} specifications are then selected from the ΔT_{min} matrix. Similarly, each heat sink segment is matched to the lowest-temperature hot utility it can use, selecting the ΔT_{min} specifications for the heat sink segments. An example of the ΔT_{min} matrix and its use is provided in the case study.

Step 6: Shifting forward. Shift the extracted segments to the temperature scale of the utilities by using the relevant individual specifications for minimum allowed temperature differences – Eq. 11.2 ($\Delta T_{min,\ PHU}$ and $\Delta T_{min,\ PCU}$). For each process the extracted heat source segments are shifted down by $\Delta T_{min,PCU}$ and the heat sink segments are shifted up by $\Delta T_{min,PHU}$ as follows:

$$T^{**} = \begin{cases} \text{sources: } T_{BACK} - \Delta T_{min,\ PCU} \\ \text{sinks: } T_{BACK} + \Delta T_{min,\ PHU} \end{cases} \tag{11.3}$$

Step 7: TSP composition. Combination of the extracted heat source segments into a **Heat Source Profile** and of the heat sink segments into a **Heat Sink Profile** (the TSP). This is performed in a way similar to the construction of the Composite Curves (Klemeš et al., 2018).

Step 8: Targeting – identification of the utility generation and use. This step is performed in the same way as in the traditional procedure (Klemeš et al., 1997), but using the modified heat source and sink segments, according to Steps 5 and 6. The construction of the Utility Generation Composite Curve starts from the lowest-temperature hot utility and moves toward the hottest one, maximising the utility generation at each utility level. Symmetrically, the construction of the Utility Use Composite Curve starts at the hottest hot utility and proceeds to the lower temperatures maximising the utility use at each level.

11.3.3 Demonstration case study

This section introduces a case study to illustrate the application of the modified targeting procedure and its potential advantages. The case study is also adapted from the example in (Perry et al., 2008), where four process areas are integrated into a Total Site. The considered site has the same topology as the example from Section 11.2.9 (Fig. 11.27) consists of the following processes: a chemical plant (Process A), a food industry plant (Process B), a hospital (Process C) and a residential area (Process D). The arrows in Fig. 11.27 represent the connectivity between the processes and the utility system via the energy carriers. It is possible to establish a connection to a district heating plant for supplying hot water.

For performing the traditional targeting procedure, a uniform $\Delta T_{min} = 12$ °C is used for heat exchange within all four processes as well as between the processes and all utilities. This value represents the average of the desired individual ΔT_{min} specifications for the individual processes ($\Delta T_{min,PP,A} = 20$ °C, $\Delta T_{min,PP,B} = 5$ °C, $\Delta T_{min,PP,C} = 10$ °C and $\Delta T_{min,PP,D} = 12$ °C).

Applying the modified targeting procedure from Section 11.3.2, the following individual ΔT_{min} values are used for each process to process heat exchange: $\Delta T_{min,PP,A} = 20$ °C, $\Delta T_{min,PP,B} = 5$ °C, $\Delta T_{min,PP,C} = 10$ °C and $\Delta T_{min,PP,D} = 12$ °C. For process B the minimum driving force of 5 °C is more appropriate for this process type – food industry. The ΔT_{min} matrix mapping the processes to utilities is shown in Tab. 11.17.

Each process has several hot and/or cold streams. Acquiring appropriate data from a process flowsheet for input is performed by "Data Extraction" – see Klemeš et al. (2018). Following the data extraction steps, the process streams with their main properties are given in Tabs. 11.12–11.15. Several points in Tab. 11.15 should be discussed. The process streams listed there refer to residential energy needs. The space heating refers to the air in the building, where the higher temperature is the one to be maintained in the rooms. The hot-water generation has two parts – "base" referring to the common parts of a building and a second part "Hot water 2" referring to the hot-water demands for use by residents. The utility options for the presented case study are shown in Tab. 11.16. These are sufficient to cover the required temperature ranges.

Tab. 11.12: Heating and cooling demands for chemical plant (Process A).

No.	Stream	Type	T_S [°C]	T_T [°C]	ΔH [MW]	CP [MW/°C]
1	A1	Hot	110	80	40.0	1.333
2	A2	Hot	150	149	180.0	180.000
3	A3	Cold	50	135	104.4	1.228
4	A4	Cold	85	100	82.3	5.487
5	A5	Cold	62	100	130.0	3.421
6	A6	Hot	92	55	130.0	7.647

Tab. 11.13: Heating and cooling demands for food plant (Process B).

No.	Stream	Type	T_S [°C]	T_T [°C]	ΔH [MW]	CP [MW/°C]
1	B1	Hot	200	195	160.0	32.000
2	B2	Cold	20	54	10.0	0.294
3	B3	Cold	50	85	107.3	3.066
4	B4	Cold	100	120	130.0	6.500
5	B5	Hot	150	40	83.5	0.759
6	B6	Cold	80	95	48.0	3.200
7	B7	Hot	95	25	80.0	1.143

Tab. 11.14: Heating and cooling demands for hospital (Process C).

No.	Stream	Type	T_S [°C]	T_T [°C]	ΔH [MW]	CP [MW/°C]
1	Soapy water	Hot	85	40	23.85	0.53
2	Condensate	Hot	80	40	96.4	2.41
3	Sanitary water	Cold	25	55	17.3	0.576
4	Laundry	Cold	55	85	18.0	0.600
5	BFW	Cold	33	60	12.0	0.446
6	Sanitary water	Cold	25	60	15.0	0.429
7	Sterilisation	Cold	82	121	34.1	0.874
8	Swimming pool water	Cold	25	28	23.1	7.700
9	Cooking	Cold	80	100	32.0	1.600
10	Heating	Cold	18	25	41.1	5.871
11	Bedpan washers	Cold	21	121	5.0	0.050

Tab. 11.15: Heating and cooling demands for residential area (Process D).

No.	Stream	Type	T_S [°C]	T_T [°C]	ΔH [MW]	CP [MW/°C]
1	Space heating	Cold	15	25	88.0	8.800
2	Hot water base	Cold	15	45	25.0	0.833
3	Hot water 2	Cold	15	45	65.0	2.167

Tab. 11.16: Cost and operating parameters.

Type	Temperature range	Unit cost [$/(kWy)]
Cooling water	From 13 °C to 25 °C	150
LP steam	150 °C	260
Hot water	From 40 to 70 °C	–

11.3.3.1 Process-level targeting using the traditional procedure

The problem tables, heat cascades and GCCs are constructed for each process using the uniform $\Delta T_{min} = 12$ °C. As an example for process A, Fig. 11.35 shows the Problem Table and Fig. 11.36 shows the GCC. For the food plant, hospital and residential area (processes B, C and D) the external utility requirements are also calculated and the GCC constructed in the same way as for the chemical plant (process A). The obtained targets are listed in Tab. 11.18, where they are also compared with the ones from the modified procedure.

T*[°C]	Stream population CP [MW/°C]	ΔT$_{interval}$ [°C]	ΣCP [MW/°C]	ΔH$_{interval}$ [MW]	Surplus/deficit
144		1	180	180	Surplus
143		2	0	0	-
141		35	−1.2	−43	Deficit
106		2	−10.2	−20.3	Deficit
104		13	−8.8	−114.4	Deficit
91		5	−3.3	−16.6	Deficit
86		12	4.3	52	Surplus
74		6	3	18	Surplus
68		12	6.4	77	Surplus
56		7	7.6	53.5	Surplus
49					

Stream population CP values: 2, CP = 180, CP = 3.421, CP = 1.333, CP = 7.647, 1, 4, CP = 1.228, CP = 5.487, 6, 5, 3

Fig. 11.35: Problem Table for process A for the uniform ΔT_{min} = 12 °C.

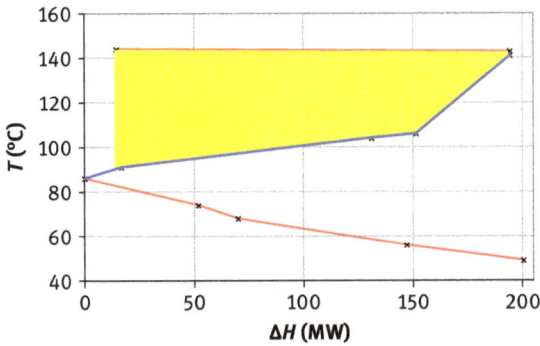

Fig. 11.36: Utility targets for process A, uniform ΔT_{min} = 12 °C.

Tab. 11.17: ΔT_{min} matrix for the case study, all values in °C.

	Process A	Process B	Process C	Process D
LP Steam	5	4	5	5
HW	10	6	8	10
CW	10	6	5	10

11.3.3.2 Process-level targeting using the modified ΔT_{min} procedure

The heat recovery targets have been obtained using the individual ΔT_{min} specifications provided at the case study setup. Figures 11.37 and 11.38 show the targeting results for the Food Plant (Process B).

Table 11.18 shows the utility targets for all processes obtained with both procedures, comparing them. The last column shows the absolute difference in the hot utility demands derived as the difference between the value obtained by the modified procedure and the one by the traditional procedure. This difference can be used as

Tab. 11.18: Process utility requirements for the traditional and the modified procedures.

Process	$\Delta T_{min,PP}$ [°C]		Minimum hot utility [MW]		Minimum cold utility [MW]		Hot utility difference [MW]
	(a)	(b)	(a)	(b)	(a)	(b)	
A	12	20	14.3	51.5	200.5	337.7	37.2
B	12	5	22.0	8.7	50.2	36.9	−13.3
C	12	10	77.4	77.4	0	0	0
D	12	12	178.0	178.0	0	0	0

Legend: (a) Traditional procedure; **(b)** Modified procedure

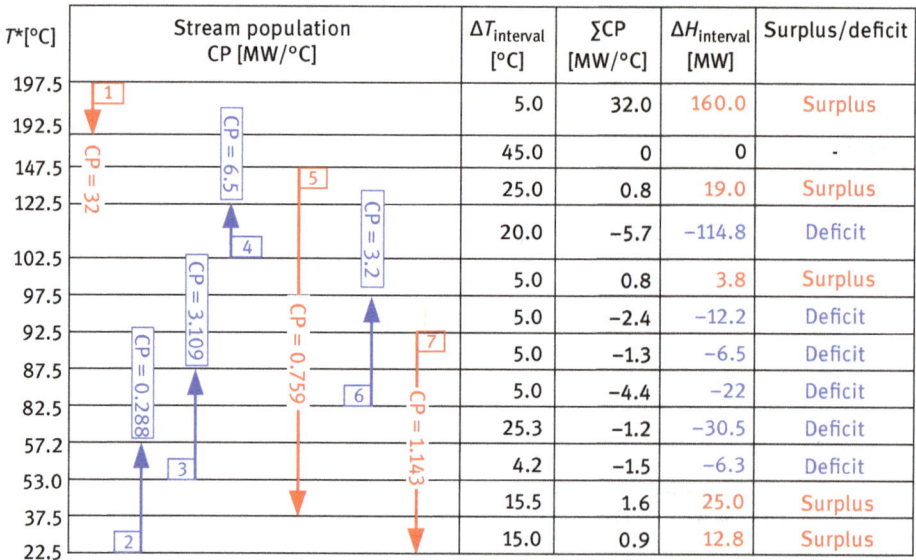

T^*[°C]	Stream population CP [MW/°C]	$\Delta T_{interval}$ [°C]	ΣCP [MW/°C]	$\Delta H_{interval}$ [MW]	Surplus/deficit
197.5		5.0	32.0	160.0	Surplus
192.5					
147.5		45.0	0	0	-
122.5		25.0	0.8	19.0	Surplus
		20.0	−5.7	−114.8	Deficit
102.5					
97.5		5.0	0.8	3.8	Surplus
92.5		5.0	−2.4	−12.2	Deficit
87.5		5.0	−1.3	−6.5	Deficit
82.5		5.0	−4.4	−22	Deficit
57.2		25.3	−1.2	−30.5	Deficit
53.0		4.2	−1.5	−6.3	Deficit
37.5		15.5	1.6	25.0	Surplus
22.5		15.0	0.9	12.8	Surplus

Fig. 11.37: Problem Table for process B, individual $\Delta T_{min,PP,B} = 5$ °C.

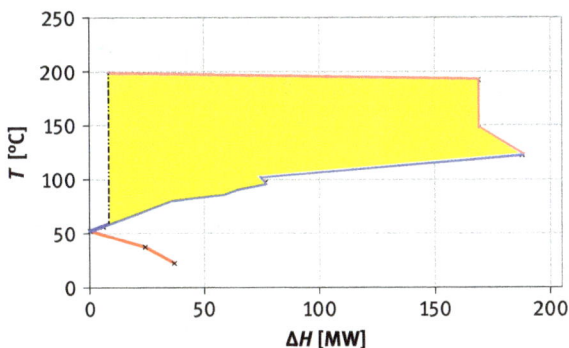

Fig. 11.38: Utility targets for process B for $\Delta T_{min,PP,B} = 5$ °C.

an approximate preliminary indicator of the under- or overestimation of the heat exchanger area and the heat recovery. Both processes C and D pose threshold problems for the given minimum driving forces. Also, for process D the ΔT_{min} specification is the same for both cases. The utility cost has also been estimated for both procedures and shown in Tab. 11.19.

Tab. 11.19: Process utility cost comparison.

Process	Hot utility [M$/y]		Cold utility [M$/y]		Total [M$/y]	
	(a)	(b)	(a)	(b)	(a)	(b)
A	3.7	13.3	30.1	35.7	33.8	49
B	5.7	2.2	7.5	5.5	13.2	7.7
C	20.1	20.1	0	0	20.1	20.1
D	46.3	46.3	0	0	46.3	46.3

Legend (a) Traditional procedure; **(b)** Modified procedure

11.3.3.3 Total Site targeting comparing the traditional and modified ΔT_{min} methodology

Having the process-level utility targets, the Total Site calculation can be performed. For the two procedures the pockets are removed to maximise the heat recovery inside each process. The resulting targets for external utility requirements of the processes feature only the non-overlapping segments of the GCCs, as illustrated in Fig. 11.38. The segments that require cooling (below the Pinch and in red in Fig. 11.38) form heat sources and the segments that require heating (above the Pinch and in blue in Fig. 11.38) form the heat sinks.

The TSP are constructed next, following the traditional and the modified procedures. In both cases, the Source Profile is formed by segments coming from the chemical plant (process A) and the food plant (process B). The modified procedure has been applied to the GCCs of the four processes. Steps 1 to 3 are identical for the modified and the traditional procedures. The application of Steps 4 and 5 of the modified procedure to the heat source segments is illustrated in Tab. 11.20. The first heat source segment from process B (source B1) has been split into two by the projection of the lower temperature of the hot-water utility. As a result, the hotter part (source B1-1) is matched with the hot water, while the colder part (source B1-2) – with the cooling water. As a result of applying these steps, the segment populations for heat sources and have been obtained. Figure 11.39 shows a view for the heat source segments. The heat sinks are obtained in a similar way, see Fig. 11.40.

Tab. 11.20: Steps 4 and 5 of the modified procedure applied to processes A and B.

Labels		T^* [°C]	T_{BACK}, [°C]	Process – utility	$\Delta T_{min,P-U}$ [°C]	T^{**} [°C]	Load [MW]
Source A1	T_H	82	92	A – HW	10	82	43.31
	T_L	72	82		10	72	
Source A2	T_H	72	82	A – HW	10	72	15.504
	T_L	70	80		10	70	
Source A3	T_H	70	80	A – HW	10	70	64.19
	T_L	60	70		10	60	
Source A4	T_H	60	70	A – HW	10	60	114.705
	T_L	45	55		10	45	
Source B1-1	T_H	52.5	55	B – HW	6	49	14.472
	T_L	–	46		6	40	
Source B1-2	T_H	–	46	B – CW	6	40	9.648
	T_L	37.5	40		6	34	
Source B2	T_H	37.5	40	B – CW	6	34	12.73
	T_L	22.5	25		6	19	

Legend HW: Hot-water generation; CW: Cooling water

The TSPs are constructed by combining all heat sources in one curve and all sinks in another following the standard algorithm from the traditional procedure. The TSPs are shown in Fig. 11.41, together with the Utility Composite Curves and the utility targets. Table 11.21 summarises the targets obtained via the modified procedure – compared with the ones obtained with the traditional procedure.

11.3.3.4 Discussion of the results from the two procedures
The results, shown in the previous section, have two parts. First, the process-level targets (Tab. 11.19) reveal changes in opposite directions. For Process A,

ΔH	MW	43.31	15.504	64.19	114.705	14.472	9.648	12.73
CP	MW/°C	4.331	7.752	6.419	7.647	1.608	1.608	0.849
TB [° C]		Source A1	Source A2	Source A3	Source A4	Source B1-1	Source B1-2	Source B2
82								
80		4.331						
77		4.331						
75		4.331						
72		4.331						
70			7.752					
62.4				6.419				
60				6.419				
56					7.647			
55					7.647			
49					7.647			
45					7.647	1.608		
40						1.608		
35							1.608	
34							1.608	
27.4								0.849
26								0.849
25								0.849
19								0.849

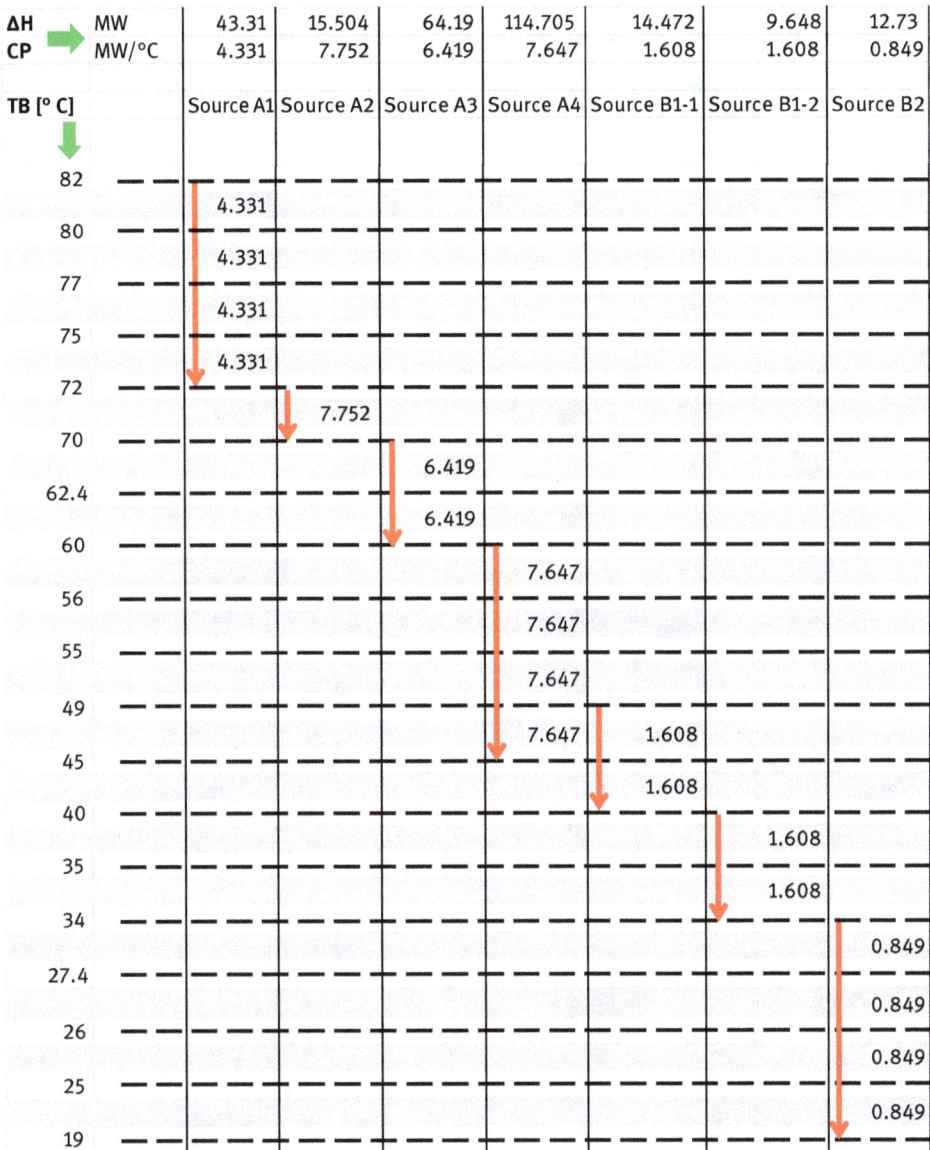

Fig. 11.39: Heat source segments for the modified procedure.

using the average ΔT_{min} specification significantly overestimates the possible heat recovery by 37.2 MW, which may result in higher capital costs for its HEN design. The case for process B is the opposite. There the uniform ΔT_{min} specification leads to an underestimation of the possible intra-process heat recovery by 13.3 MW, which may result in higher utility costs for the process. Processes D and

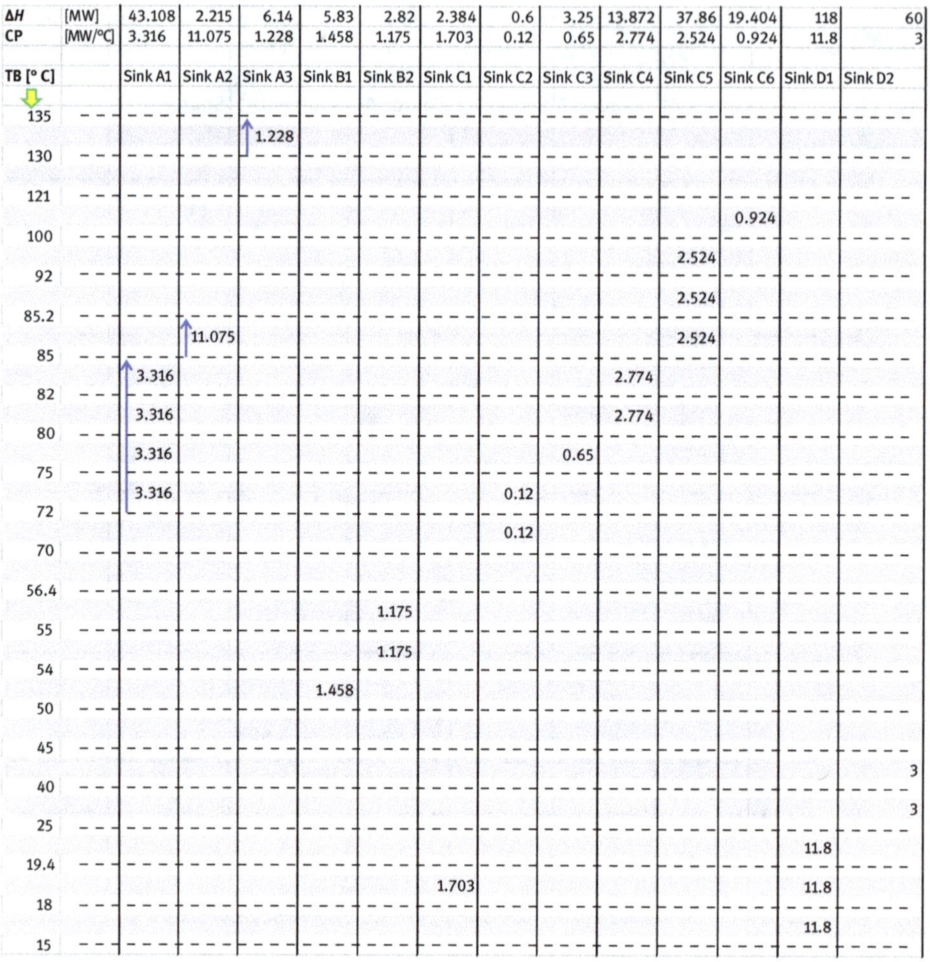

ΔH [MW]	43.108	2.215	6.14	5.83	2.82	2.384	0.6	3.25	13.872	37.86	19.404	118	60
CP [MW/°C]	3.316	11.075	1.228	1.458	1.175	1.703	0.12	0.65	2.774	2.524	0.924	11.8	3
TB [°C]	Sink A1	Sink A2	Sink A3	Sink B1	Sink B2	Sink C1	Sink C2	Sink C3	Sink C4	Sink C5	Sink C6	Sink D1	Sink D2
135			1.228										
130													
121											0.924		
100										2.524			
92										2.524			
85.2										2.524			
85		11.075											
82	3.316								2.774				
80	3.316								2.774				
75	3.316							0.65					
72	3.316						0.12						
70							0.12						
56.4													
55					1.175								
54					1.175								
50				1.458									
45													3
40													3
25												11.8	
19.4												11.8	
18						1.703						11.8	
15													

Fig. 11.40: Heat sink segments for the modified procedure.

C pose threshold problems as there is no cooling utility requirement. There is only 2 °C change in the used ΔT_{min} for Process C and this change is insufficient to also alter the process utility targets.

Second, at the site level, the targets are also quite different. The targets obtained with the modified procedure feature larger utility demands. The cooling water target difference is by 19.8 MW (30 %), and the difference for the LP Steam targets is 19.3 MW (18 %). Figure 11.42 shows a visual comparison of the obtained results. It can be noticed that the increased targets for both LP steam and cooling water are due to the shift in the TSP (source and sink). Further analysis can readily trace this to the smaller heat recovery in process A as a result of using the individual $\Delta T_{min,PP,A} = 20$ °C.

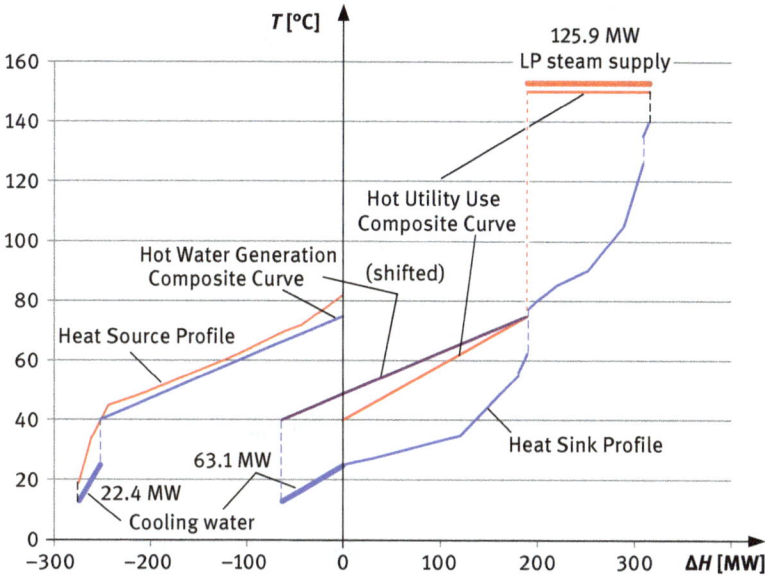

Fig. 11.41: Total Site targets obtained with the modified procedure.

Tab. 11.21: Comparison of the targets after identifying utility recovery.

	Load [MW]		Cost [M$/y]	
	Traditional	**Modified**	**Traditional**	**Modified**
Cooling water	65.7	85.5	27.7	32.7
LP steam	106.6	125.9	9.9	12.8
Total	–	–	37.6	45.5

11.3.4 Summary of the method of individual ΔT_{min} for Total Site targeting

This method offers more flexibility and site utility targets, which are more appropriate to the individual heat transfer properties of the various site processes. The described procedure modification results in several potential benefits for site energy managers.

In terms of methodology, the modified procedure offers a clear definition of the heat transfer contexts allowing the users to specify ΔT_{min} values for all important types of heat exchange on a site, intraprocess, process-to-utility and utility-to-process. This can be done using the ΔT_{min} specification matrix and implemented through the algorithmic improvements (steps 4 and 5) in the modified procedure.

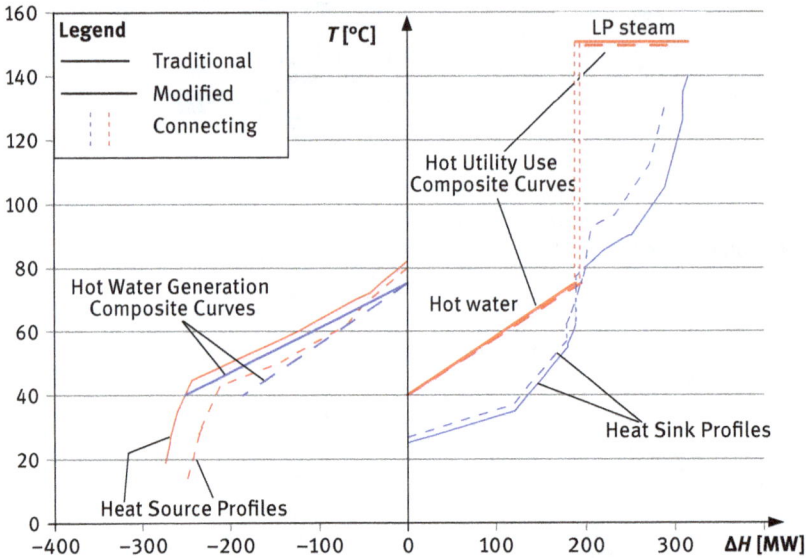

Fig. 11.42: Visual comparison of the targets for the traditional and the modified procedure.

The case study illustrates that the targeting discrepancies, introduced by the assumption for uniform ΔT_{min} in the traditional procedure, appear both at the process and at site levels. At the process level, the possible discrepancies are in opposite directions. This is due to the averaging the ΔT_{min} specifications for the traditional methodology, revealing that the traditional methodology may in many cases lead to underestimation of the heat recovery targets at process level and consequent utility cost losses. The opposite can also be the case – overestimation of the heat recovery potential, which may lead to higher capital costs for process and utility heat exchangers. The magnitude of the discrepancies between the targets in (30 % to 200 %) points to the significance of using the proper ΔT_{min} specifications.

At the site level, the modified methodology produces utility targets that are higher. For the case study, these are 18 % and 30 % for the hot and cold utilities, clearly traceable to the ΔT_{min} specifications. This also points to the significance of determining and using the correct ΔT_{min} specifications for the heat exchanges inside processes as well as between the processes and the utilities.

11.4 Summary

This chapter provides a systematic discussion of Total Site Integration methods. It starts with an overview of the fundamentals of the methodology and provides references for deeper learning on the fundamentals. The various extensions to the methodology are also analysed. Two of these extensions are presented in detail – for the integration of variable supply and demand streams, as well as for allowing more accurate targeting by using individual ΔT_{min} specifications.

As has been shown, applying Total Site heat integrated storage can significantly reduce the targets for fossil-fuel as well as solar-based utilities, when applied to the integration of renewables with time-varying availability. The methodology can be used for simultaneously obtaining several important targets for Total Sites integrating renewables. These are the maximum amounts (upper bound) of renewables capture, the minimum consumption (lower bound) of fossil fuels and the maximum capacity (upper bound) of energy storage. The identified targets can be further used for more detailed system design. The illustrated procedure shows the potential to employ the developed Total Site Heat Cascade as a tool for seeking energy efficiency improvement by demand-side management of the process energy demands and potential operation rescheduling.

The flexible targeting with process-specific ΔT_{min} specifications provides better targets, which can serve as more appropriate bounds to the detailed design and optimisation studies. As such, it has the potential to prevent the waste of investment, operating costs and resources, contributing significantly to the sustainability and economic performance of industrial and urban sites.

Nomenclature

Acronyms

Acronym	Description
BFW	Boiler feedwater
CC	Composite Curve – see Chapter 1.2
CHP	Combined Heat and Power (generation)
cTS	Combined Time Slices
CW	Cooling water
ChW	Chilled water
GCC	Grand Composite Curve
HD (HD1, HD2)	Time Slices for the varying heat demands
HEN	Heat Exchanger Network
HP (steam)	High Pressure (steam)
HW	Hot water
LIES	Locally Integrated Energy Sector
LP (steam)	Low Pressure (steam)
MILP	Mixed-Integer Linear Programming
MP (steam)	Medium Pressure (steam)
SUGCC	Site Utility Grand Composite Curve
TB	Temperature Boundary

(continued)

Acronym	Description
TS, TS1, TS2, TS3	Time Slice
TSP	Total Site Profiles
VHP (steam)	Very High Pressure (steam)

Symbols

Symbol	Unit	Description
CP	MW/°C	Heat-capacity flowrate of a process stream
$\sum CP_{interval}$	MW/°C	Sum of the CP values of heat source or sink segments present in a heat cascade interval
A	m^2	Heat transfer area
G	W/m^2	Solar irradiation
INA	%	Inaccuracy (allowed tolerance)
NTS	[1]	Number of Time Slices
$Q_{C,MIN}$ or Q_{Cmin}	MW	Minimum cooling demand
$Q_{H,MIN}$ or Q_{Hmin}	MW	Minimum heating demand
Q_{HE}	MW	Heat load
t	h	Time
T	°C	Temperature
T^*	°C	Shifted temperature at the scale of the GCC (GCC)
T^*_{start}	°C	Starting shifted temperature of a heat source/sink segment, at the scale of the GCC (GCC)
T^*_{end}	°C	Starting target temperature of a heat source/sink segment, at the scale of the GCC (GCC)
TB**	°C	Temperature boundary of the Total Site Heat Cascade of sources (Fig. 11.5) and sinks (Fig. 11.6)
T^{**}	°C	Shifted temperature at the scale of the utilities
T^{**}_{start}	°C	Starting shifted temperature of a heat source/sink segment, at the scale of the utilities
T^{**}_{end}	°C	Starting target temperature of a heat source/sink segment, at the scale of the utilities
T^*_{Pinch}	°C	Pinch Point at the shifted temperature scale
T_{BACK}	°C	Intermediate temperature level for a heat source or heat sink segment at the scale of the process stream temperatures – two instances per segment
T_H	°C	Higher temperature of a heat source/sink segment
T_L	°C	Lower temperature of a heat source/sink segment
T_S	°C	Starting temperature of a process stream or curve segment
T_T	°C	Ending (target) temperature of a process stream or curve segment

(continued)

Symbol	Unit	Description
$T_{S<\#>}$, $T_{T<\#>}$	°C	Start (S) and target (T) temperature coordinates on the T–H plot (e.g. Fig. 11.2)
U	MW/m^2/°C	Overall heat transfer coefficient
ΔH	MW	Enthalpy difference (enthalpy flow)
$\Delta H_{S<\#>}$, $\Delta H_{T<\#>}$	MW	Enthalpy Flow coordinates on the T-H plot (e.g. Fig. 11.2), indices mean start (S) and target (T)
$\Delta H_{Interval}$	MW	Enthalpyflow within a heat cascade interval
ΔT_{Cmin} (Fig. 11.13)	°C	Minimum allowed temperature difference between a process heat source and the heat storage
ΔT_{EN-EX} (Fig. 11.13)	°C	Temperature difference between the charge and the discharge processes of a facility for chemical energy storage
$\Delta T_{Interval}$	°C	Temperature span of a heat cascade interval
ΔT_{LM}	°C	Logarithmic mean temperature difference for a heat exchanger
ΔT_{min}	°C	Minimum allowed temperature difference
$\Delta T_{min, PCU}$	°C	Minimum allowed temperature difference for heat exchange between a process and a cold utility
$\Delta T_{min, PHU}$	°C	Minimum allowed temperature difference for heat exchange between a process and a hot utility
$\Delta T_{min,PP}$ $\Delta T_{min,PP,\{Process\ ID\}}$	°C	Minimum allowed temperature difference for process-to-process heat exchange
$\Delta T_{min,PU}$	°C	Minimum allowed temperature difference for heat exchange between a process and a utility
ΔT_{Pmin} (Fig. 11.13)		Minimum allowed temperature difference between the heat storage and a process heat sink

Subscripts

PP	Process-to-process
PCU	Process to cold utility
PHU	Process to hot utility
BACK [Eq. (11.2)]	Shifted back to the orginal temperature

References

Atkins, M.J., Michael, R.W.W., and Neale, J.R. 2010a. The Challenge of Integrating Non-Continuous Processes – Milk Powder Plant Case Study. Journal of Cleaner Production 18(9): 927–934. https://doi.org/10.1016/j.jclepro.2009.12.008.

Atkins, M.J., Walmsley, M.R.W., and Morrison, A.S. 2010b. Integration of Solar Thermal for Improved Energy Efficiency in Low-Temperature-Pinch Industrial Processes. Energy 35(5): 1867–1873. https://doi.org/10.1016/j.energy.2009.06.039.

Botros, B.B. and Brisson, J.G. 2011. Targeting the Optimum Steam System for Power Generation with Increased Flexibility in the Steam Power Island Design. Energy 36(8): 4625–4632. https://doi.org/10.1016/j.energy.2011.03.045.

Čuček, L., Varbanov, P.S., Klemeš, J.J., and Kravanja, Z. 2012. Total Footprints-Based Multi-Criteria Optimisation of Regional Biomass Energy Supply Chains. Energy 44(1): 135–145. https://doi.org/10.1016/j.energy.2012.01.040.

Dhole, V.R. and Linnhoff, B. 1993. Total Site Targets for Fuel, Co-Generation, Emissions, and Cooling. Computers & Chemical Engineering 17 (January): S101–9. https://doi.org/10.1016/0098-1354(93)80214-8.

Dinçer, İ. and Ezan, M.A. 2018. Heat Storage: A Unique Solution for Energy Systems. Cham, Switzerland: Springer International Publishing.

Faramarzi, S., Tahouni, N., and Hassan Panjeshahi, M. 2019. Total Site Heat Integration Considering Optimum Pressure Drops. Chemical Engineering Transactions 76 (October): 1225–1230. https://doi.org/10.3303/CET1976205.

Green, D.W. and Perry, R.H. 2008. Perry's Chemical Engineers Handbook. 8th ed. New York, USA: McGraw-Hill.

Hackl, R., Andersson, E., and Harvey, S. 2011. Targeting for Energy Efficiency and Improved Energy Collaboration between Different Companies Using Total Site Analysis (TSA). Energy 36(8): 4609–4615. https://doi.org/10.1016/j.energy.2011.03.023.

Hackl, R. and Harvey, S. 2015. From Heat Integration Targets toward Implementation – A TSA (Total Site Analysis)-Based Design Approach for Heat Recovery Systems in Industrial Clusters. Energy 90 (October): 163–172. https://doi.org/10.1016/j.energy.2015.05.135.

Hiu, C.W. and Ahmad, S. 1994. Total Site Integration Using the Utility System. Computers & Chemical Engineering 18(8): 729–742.

Huggins, R.A. 2010. Energy Storage. Boston, MA, USA: Springer US. https://doi.org/10.1007/978-1-4419-1024-0.

IEA. 2005. Variability of Wind Power and Other Renewables. Management Options and Strategies. https://refman.energytransitionmodel.com/publications/1629/download, accessed 14/11/2019.

Kemp, I.C. 2007. Pinch Analysis and Process Integration: A User Guide on Process Integration for the Efficient Use of Energy. 2nd ed. Amsterdam, The Netherlands; Boston, USA: Butterworth-Heinemann/Elsevier.

Klemeš, J.J., Varbanov, P.S., Wan Alwi, S.R., and Abdul Manan, Z. 2018. Sustainable Process Integration and Intensification. Saving Energy, Water and Resources. 2nd ed. Berlin, Germany: Walter de Gruyter GmbH.

Klemeš, J.J., Dhole, V.R., Raissi, K., Perry, S.J., and Puigjaner, L. 1997. Targeting and Design Methodology for Reduction of Fuel, Power and CO_2 on Total Sites. Applied Thermal Engineering 17(8–10): 993–1003. https://doi.org/10.1016/S1359-4311(96)00087-7.

Kondili, E. and Kaldellis, J.K. 2006. Optimal Design of Geothermal–Solar Greenhouses for the Minimisation of Fossil Fuel Consumption. Applied Thermal Engineering 26(8): 905–915. https://doi.org/10.1016/j.applthermaleng.2005.09.015.

Liew, P.Y., Wan Alwi, S.R., Ho, W.S., Manan, Z.A., Varbanov, P.S., and Klemeš, J.J. 2018. Multi-Period Energy Targeting for Total Site and Locally Integrated Energy Sectors with Cascade Pinch Analysis. Energy 155 (July): 370–380. https://doi.org/10.1016/j.energy.2018.04.184.

Liew, P.Y., Theo, W.L., Wan Alwi, S.R., Lim, J.S., Abdul Manan, Z., Klemeš, J.J., and Varbanov, P.S. 2017. Total Site Heat Integration Planning and Design for Industrial, Urban and Renewable Systems. Renewable and Sustainable Energy Reviews 68: 964–985. https://doi.org/10.1016/j.rser.2016.05.086.

Liew, P.Y., Wan Alwi, S.R., Lim, J.S., Varbanov, P.S., Klemeš, J.J., and Abdul Manan, Z. 2014. Total Site Heat Integration Incorporating the Water Sensible Heat. Journal of Cleaner Production 77: 94–104. https://doi.org/10.1016/j.jclepro.2013.12.047.

Liew, P.Y., Wan Alwi, S.R., Varbanov, P.S., Manan, Z.A., and Klemeš, J.J. 2012. A Numerical Technique for Total Site Sensitivity Analysis. Applied Thermal Engineering 40: 397–408. https://doi.org/10.1016/j.applthermaleng.2012.02.026.

Linnhoff, B. and Hindmarsh, E. 1983. The Pinch Design Method for Heat Exchanger Networks. Chemical Engineering Science 38(5): 745–763.

Masruroh, N.A., Li, B., and Klemeš., J. 2006. Life Cycle Analysis of a Solar Thermal System with Thermochemical Storage Process. Renewable Energy 31(4): 537–548. https://doi.org/10.1016/j.renene.2005.03.008.

Matsuda, K., Hirochi, Y., Tatsumi, H., and Shire, T. 2009. Applying Heat Integration Total Site Based Pinch Technology to a Large Industrial Area in Japan to Further Improve Performance of Highly Efficient Process Plants. Energy 34(10): 1687–1692.

Morandin, M., Toffolo, A., Lazzaretto, A., Maréchal, F., Ensinas, A.V., and Nebra, S.A. 2011. Synthesis and Parameter Optimization of a Combined Sugar and Ethanol Production Process Integrated with a CHP System. Energy 36(6): 3675–3690. https://doi.org/10.1016/j.energy.2010.10.063.

Nemet, A., Klemeš, J.J., Varbanov, P.S., and Kravanja, Z. 2012a. Methodology for Maximising the Use of Renewables with Variable Availability. Energy 44(1): 29–37. https://doi.org/10.1016/j.energy.2011.12.036.

Nemet, A. and Kravanja, Z. 2017. Enhanced Procedure for Simultaneous Synthesis of an Entire Total Site. In: Computer Aided Chemical Engineering, Vol. 40: 427–432. Elsevier. https://doi.org/10.1016/B978-0-444-63965-3.50073-8.

Nemet, A., Kravanja, Z., and Klemeš, J.J. 2012b. Integration of Solar Thermal Energy into Processes with Heat Demand. Clean Technologies and Environmental Policy 14(3): 453–463.

Perry, S., Klemeš, J., and Bulatov, I. 2008. Integrating Waste and Renewable Energy to Reduce the Carbon Footprint of Locally Integrated Energy Sectors'. Energy 33(10): 1489–1497. https://doi.org/10.1016/j.energy.2008.03.008.

Prieto, C., Rodríguez, A., Patiño, D., and Cabeza, L.F. 2018. Thermal Energy Storage Evaluation in Direct Steam Generation Solar Plants. Solar Energy 159 (January): 501–509. https://doi.org/10.1016/j.solener.2017.11.006.

Ren, X.-Y., Jia, -X.-X., Varbanov, P.S., Klemeš, J.J., and Liu, Z.-Y. 2017. Calculation of Cogeneration Potential of Total Site Utility Systems with Commercial Simulator. Chemical Engineering Transactions 61: 1231–1236. https://doi.org/10.3303/CET1761203.

Shang, Z. and Kokossis, A. 2004. A Transhipment Model for the Optimisation of Steam Levels of Total Site Utility System for Multiperiod Operation. Computers & Chemical Engineering 28(9): 1673–1688. https://doi.org/10.1016/j.compchemeng.2004.01.010.

Stijepovic, M.Z. and Linke, P. 2011. Optimal Waste Heat Recovery and Reuse in Industrial Zones. Energy 36(7): 4019–4031. https://doi.org/10.1016/j.energy.2011.04.048.

Tarighaleslami, A.H., Walmsley, T.G., Atkins, M.J., Walmsley, M.R.W., and Neale, J.R. 2018. Utility Exchanger Network Synthesis for Total Site Heat Integration. Energy 153 (June): 1000–1015. https://doi.org/10.1016/j.energy.2018.04.111.

Towler, G. and Sinnott, R. 2008. Chemical Engineering Design, Principles, Practice and Economics of Plant and Process Design. Amsterdam, The Netherlands: Elsevier.

Varbanov, P., Perry, S., Klemeš, J., and Smith, R. 2005. Synthesis of Industrial Utility Systems: Cost-Effective de-Carbonisation. Applied Thermal Engineering 25(7): 985–1001. https://doi.org/10.1016/j.applthermaleng.2004.06.023.

Varbanov, P.S. and Klemeš, J.J. 2011. Integration and Management of Renewables into Total Sites with Variable Supply and Demand. Computers and Chemical Engineering 35(9): 1815–1826. https://doi.org/10.1016/j.compchemeng.2011.02.009.

Varbanov, P.S., Fodor, Z., and Klemeš, J.J. 2012. Total Site Targeting with Process Specific Minimum Temperature Difference (ΔT_{min}). Energy 44(1): 20–28. https://doi.org/10.1016/j.energy.2011.12.025.

Varbanov, P.S., Walmsley, T.G., Klemes, J.J., Wang, Y., and Jia, -X.-X. 2018. Footprint Reduction Strategy for Industrial Site Operation. Chemical Engineering Transactions 67 (September): 607–612. https://doi.org/10.3303/CET1867102.

Von Bremen, L. 2010. Large-Scale Variability of Weather Dependent Renewable Energy Sources. In: Troccoli, A. (Ed.), Management of Weather and Climate Risk in the Energy Industry, Proceedings of the NATO Advanced Re-Search Workshop on Weather/Climate Risk Management for the Energy Sector, Santa Maria Di Leuca, Italy, 6–10 October 2008, Management of Weather and Climate Risk in the Energy Industry, 189–206. Dordrecht, The Netherlands: Springer.

Weber, R. and Dorer, V. 2008. Long-Term Heat Storage with NaOH. 82(7): 708–716. https://doi.org/10.1016/j.vacuum.2007.10.018.

12 Software for Total Site and utility system modelling

12.1 Introduction

Since Total Site Integration (TSI) was introduced by Dhole and Linnhoff (1993), mainly for addressing the heat recovery among industrial processes at the level of whole sites, there have been further developments. Klemeš et al. (1997) added targets for power cogeneration. The further developments include process-specific $\Delta T_{\min}$ values (Varbanov et al., 2012), as discussed in Chapter 11, allowing for appropriate consideration of the heat transfer properties inside each process. Also, other developments included accounting for the boiler feed water preheating (Liew et al., 2014b), utility system planning (Liew et al., 2013) and market and environment variations – including uncertainty over the entire lifetime of the sites (Nemet et al., 2015). Klemeš et al. (2018b) provided an extensive and in-depth review of the methods for resource conservation, analysing the trends and challenges stemming from the extensive growth of the methods, also paying attention to Total Site and utility system optimisation.

During all these developments, various tools were used. Initially, they were purely graphical tools – plotting on paper, supported by simple calculations. Dedicated applications and mathematical programming software were also used, which led to the development of the software tools by the University of Manchester Institute of Science and Technology, now The University of Manchester (UMIST) group.

For research and industrial application of TSI and utility system modelling, including site-level heat recovery and cogeneration optimisation, software packages are used as computing tools supporting the design and operation decision-making. The contemporary market offers a few tools capable of supporting various parts of the overall spectrum of the required features. This chapter provides an overview of the main methodologies and the software packages, mapping the feature requirements of the methods to the tools.

12.2 Total Site and utility optimisation developments

TSI considers an industrial site as a set of processes exchanging utilities – process heating, process cooling, refrigeration, power and water. The exchange is performed via the site utility system, playing the role of a marketplace, as discussed in Chapter 11. The methodology for modelling, simulation and optimisation of site utility systems as hubs interlinking site processes has developed over the years to account for many of the system aspects. There have also been some extensions dealing with broader resource management, which are closely interlinked with providing heat and power on industrial sites.

https://doi.org/10.1515/9783110630091-012

12.2.1 Total Site Analysis interfaces

Total Site Analysis involves identification of the process heating and cooling requirements, estimation of the targets for minimum fuel use, power cogeneration and power import or export. Heat sinks and heat sources represent the process of heating and cooling demands at the site level. They are characterised by starting and final temperatures and enthalpy changes. The Total Site Data Extraction (Chapter 11) was initially performed graphically using the process Grand Composite Curves, following the procedure laid out by Dhole and Linnhoff (1993). This is a manual procedure, involving a lot of plotting and carrying the potential for a lot of mistakes and repeated work. There have been proprietary software solutions, some of them still in developments and offered on the market, without published algorithms. They are summarised later in this chapter.

After the introduction of the Total Site Heat Cascade in the works of Varbanov and Klemeš (2011) and the Total Site Problem Table Algorithm by Liew et al. (2012), this calculation can be also performed numerically using any spreadsheet software – including Microsoft Excel (Microsoft, 2020) or LibreOffice Calc (The Document Foundation, 2020). Those research works accounted for the variability and uncertainties in the utility demands, as well as those on the supply side when renewable resources are used. The Total Site Heat Cascade and the Total Site Problem Table Algorithm have built-in support for evaluating variability scenarios, placing appropriate storage facilities and maximising the efficiency. These numerical models reveal important requirements to the software tools for supporting data transfer between the various stages: simulation, synthesis, simulation and optimisation and site-level studies.

12.2.2 Site-level heat recovery

Another notable development of the methodology is the method incorporating process-specific minimum allowed temperature differences into the Total Site targeting procedure (Varbanov et al., 2012), which has been presented in Chapter 11. The proposed concepts, including the matrix for providing the ΔT_{min} specifications, introduced new data structures that need to be explicitly implemented by software tools that would support the method. Those concepts have been refined by Liew et al. (2012), also adding an algorithm for obtaining the targets numerically. The introduced modelling constructs include the Problem Table Algorithm and evaluation of the system sensitivity.

A further method within this family includes the optimal design of Total Site Heat Integration systems, accounting for the utility price variations and their uncertainty over the entire lifetime of the sites (Nemet et al., 2015). That exposes to the potential implementation of the data structures related to long-term asset management and economic forecasts. The physical layout of the involved operating units of the utility system at the site level has also been modelled (Wang et al., 2017), setting out the layout data as candidates for the overall process and site databases.

There have been numerous applications of the Total Site methodology to real industrial cases. The recent review on the Pinch Analysis developments (Klemeš et al., 2018b) provides a summary of the most prominent implementations – including large chemical sites, a steel site, bromine production, food, dairy, pulp and paper and pharmaceuticals. This poses additional requirements to the software tools to use – for storing and inter-linking data about project setup, management and assets management. These should be complemented with models and data constructs, allowing the evaluation and optimisation of capital costs and environmental footprints, as illustrated in Ren et al., (2018).

12.2.3 Cogeneration within the site utility system

The cogeneration targeting using the Site Utility Grand Composite Curve, introduced in Klemeš et al. (1997), uses the so-called *T–H* model for evaluating power cogeneration. The model is based on considering equivalent heat flows through the utility system that represents the actual steam flows. The power generation model developed by Mavromatis and Kokossis (1998) for backpressure turbines was extended (Shang and Kokossis, 2004) to condensing steam turbines and steam boilers. These works provided complete hardware models capable of modelling all equipment types in utility systems. The models are discussed in Chapters 5 to 7.

Varbanov et al. (2004b) presented a critical correction to the steam turbine model and added models for gas turbines and the complete utility networks. These models are presented in detail in Chapter 6. The models have been then also used for utility system synthesis (Varbanov et al., 2005). Luo et al. (2011) have modelled utility systems containing multiple extraction steam turbines in General Algebraic Modelling System (GAMS) (GAMS, 2019) using models derived from Varbanov et al. (2004b). Rad et al. (2016) studied the design and optimisation of utility systems accounting for reliability. These developments clearly illustrate the need to combine features of component modelling with system-level simulation and optimisation.

12.2.4 Utility system synthesis, analysis, retrofit and planning

This aspect has been researched for a few decades using mainly mathematical programming. Among the numerous works, there has been a multi-period formulation by Iyer and Grossmann (1998) and complete utility system synthesis (Varbanov et al., 2005). Luo et al. (2016) treat the utility system and process heat recovery simultaneously.

Top-Level Analysis (Varbanov et al., 2004a) is used to select the processes for most profitable heat recovery retrofit by building a set of curves for the marginal steam prices. Liew et al. (2014a) presented the framework for targeting retrofits of the utility system and process changes employing algorithmic estimates of the thermal efficiency.

Planning of utility systems has been addressed by Liew et al. (2013), accounting for the variation and uncertainty in the supply and demand for utilities. They considered potential storage of utilities, variation scenarios, sizing of the main facilities and backup generators.

12.2.5 Challenges posed by biorefineries and waste-to-energy

Regional biorefinery synthesis has been performed by Halasz et al. (2005). This is an interesting work bridging process synthesis and regional resource optimisation. While it has accounted for the variety of resources, potential products and processing routes, it has not included consideration of the spatial dimension of regional resource management problems. This issue has been incorporated into the picture by the optimisation of regional energy supply chains with renewables by Lam et al. (2010). That work recognised that harvesting and utilising renewable bio-resources define a spatial development problem (Fig. 12.1). It applies a two-level procedure for the design of regional biomass utilisation networks, minimising the environmental impact. The proposed procedure successfully manages the complexity of the network problem by simplifying the infrastructure links and their design and tasks.

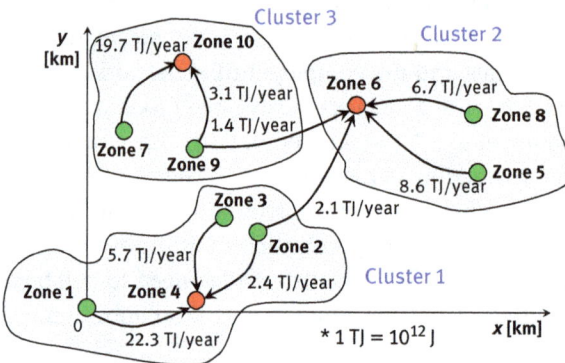

Fig. 12.1: Renewables as a spatial problem (after Lam et al., 2010).

The concept of regional supply chains has been developed further, by formulating a comprehensive mixed-integer programming model while considering additional products besides energy (Čuček et al., 2010). The method addresses the main challenges presented by the biomass resources. The first one is the distributed and varying availability of biomass by both location and time. The second is to maximise the economically viable utilisation of the resources, accounting for the competition between energy and food production. A four-layer supply chain architecture has been developed (Fig. 12.2), including harvesting, preparation, core processing and

Harvesting
and supply

Collection and
preprocessing

Main
processing

Use

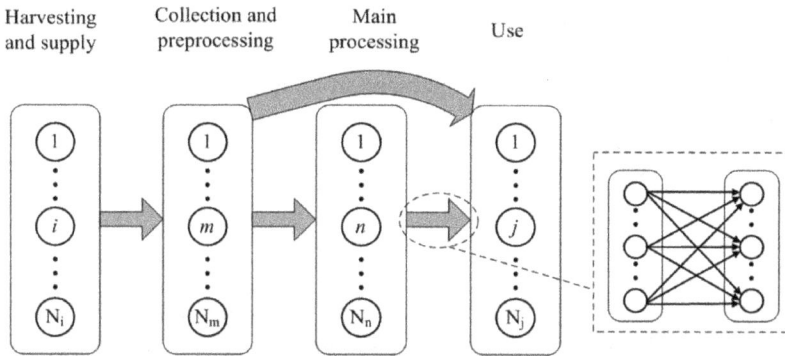

Fig. 12.2: The layer structure of the networks for renewable energy production and consumption (after Čuček et al., 2010).

products distribution. Profit maximisation is used as the optimisation criterion. The environmental impact is evaluated by the carbon footprint of the network.

A follow-up development was published (Čuček et al., 2012), embedding all relevant footprints – including water and nitrogen footprints. The model performs multi-criteria optimisation of the networks by simultaneously maximising the economic performance and minimising the environmental and social footprints.

An interesting contribution to resource management is related to waste minimisation. Halim and Srinivasan (2002a) have formulated a hierarchical decision-making procedure embedding P-graph and its application in the overall workflow. Using a cause and effect-based analysis, the work provides a complete algorithm and an illustrative case study. That was later followed by extensions developing an intelligent decision support system (Halim and Srinivasan, 2002b) and adaptation to batch processes (Halim and Srinivasan, 2006).

12.2.6 Power planning for Total Sites

The power management on industrial sites has been under intensive development in the last decade. It has already been discussed in detail in Klemeš et al. (2018a). The main developments stem from the work by Wan Alwi et al. (2012) who provided a power (electricity) planning method for systems comprising hybrid energy sources determining the targets for minimum outsourced power storage. Further development also includes optimising process modifications and load shifting (Alwi et al., 2013) and accounting for power losses (Rozali et al., 2013). Zahboune et al. (2016) presented sizing autonomous power management system, harvesting wind and solar sources. A study combining Total Site heat recovery with power and CO_2 emissions management has also been published (Aziz et al., 2017).

12.3 Most applicable software tools

Software packages supporting site energy optimisation include process optimisation and retrofit analysis tools, general mathematical modelling and flowsheet simulators. There are also solutions providing software functionality as a package of services.

12.3.1 Flowsheet process simulators and specialised packages

A variety of process simulators is available, as discussed by Lam et al. (2011). Aspen Plus (2019) stands out as the industry standard. The simulator provides a wide palette of tools around the core simulation environment. Most notable are the local operational optimisation facility and the Heat Integration analysis functionality. Aspen Plus has been successfully used in Ren et al. (2018) for obtaining heat and power integration targets for site utility systems (Fig. 12.3).

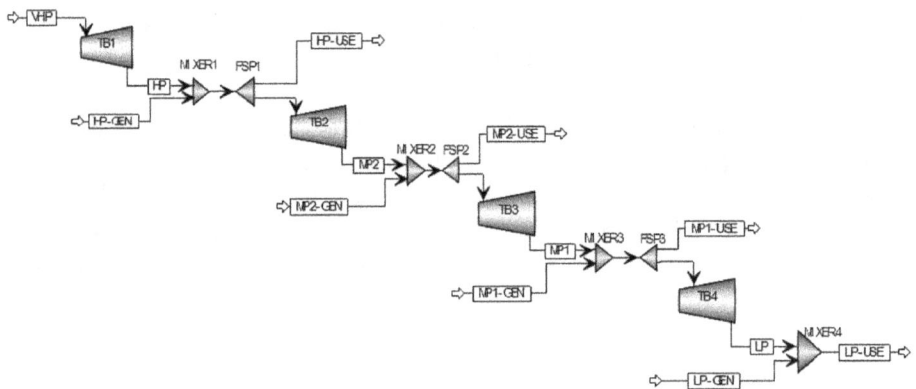

Fig. 12.3: Aspen Plus simulation flowsheet for calculating utility system cogeneration potential (after Ren et al., 2018).

It is worth looking further into the software tool packages offered by AspenTech – branded as the AspenONE Product Portfolio (2019). Besides the process simulation and analysis, the portfolio also includes the access to Aspen Energy Analyzer, Aspen Utilities Planner and many other tools related to data processing and supply chain decision-making. Aspen Energy Analyzer is a tool for process heat recovery design and planning with an emphasis on heat exchangers networks that can be linked to the Aspen Plus simulator for data extraction. There have been a couple of successful software tools implementing the Total Site energy optimisation. One such tool is the research software application STAR (UNIMAN, 2019). A similar software application, for commercial uses, is Steam (PIL, 2019). Both tools have overlapping functionality. They offer features such as steam system simulation and operational optimisation, obtaining the targets for Total Site Heat Integration. Specifically, i-Steam is described as also offering utility system design functionality, hydraulic modelling of the steam headers, pressure drop and heat losses.

Petro-SIM, supplemented with SuperTarget (KBC, 2019), is a comprehensive software package for process evaluation. The simulator part (Petro-SIM) is the KBC's fork of HYSYS. Besides the standard simulation features, it also has components for modelling steam systems – such as gas turbines, steam turbines, steam mains and deaerators. SuperTarget is mainly aimed at optimising process-level Heat Exchanger Networks, but it is described as having a Total Site module for evaluating and planning site-wide heat recovery. SuperTarget can exchange data with the most popular process simulators programmes – including Aspen Plus. Chapter 9 provides a complete example of steam system simulation, described step by step.

12.3.2 General-purpose optimisation tools

If suitable ways of calculating the fluid properties are established for explicit mathematical models, it is possible to apply general-purpose optimisation software. Prominent tools are the GAMS (2019) and LINGO (2019). GAMS, as a standard, offers a dedicated integrated development environment (IDE) in two variants, Excel and command-line interfaces for algebraic modelling. A screen showing a snapshot from modelling a utility system is given in Fig. 12.4. LINGO, on the other hand, emphasises

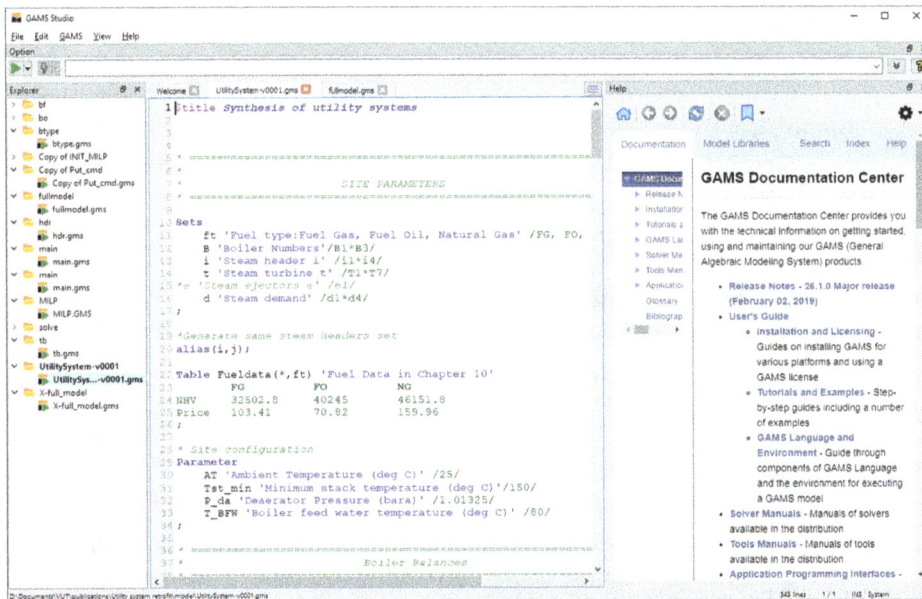

Fig. 12.4: GAMS Studio screenshot for modelling utility systems.

the Excel-based interface and modelling capability, while also offering an IDE and an application programming interface. While such general optimisers provide a lot of flexibility, the generality also prevents users from focusing on the core concepts. For industrial energy optimisation, they are best suited to prototyping and back-end tasks. In this regard, a good example of a front-end to GAMS is mixed-integer process synthesizer (Kravanja, 2010) – a modular package for the synthesis of industrial processes.

A good way for modelling and solving combinatorial problems is by using the P-graph framework (Friedler et al., 1992), based on a thorough mathematical foundation, currently implemented in a freeware software tool named "P-Graph Studio" (P-graph, 2019). The tool integrates all activities necessary for representing a process network problem as a P-graph and solving the optimisation problem. The specific implementation of the P-graph framework is based on a simplified modelling engine, which allows the performance of the operating units to be scaled proportionally with the unit size, while the operating and the investment costs allow full linear equations to be specified.

12.4 Mapping the requirements and tools

Identification of software requirements posed by an activity domain is a complex undertaking, referred to as "requirements engineering" (Sommerville, 2011). Requirements can be functional (what software should do) and non-functional (mainly constraints) requirements. The most used classification distinguishes user requirements and system requirements, defined more precisely, also including important system constraints. Here, software tools are analysed at the level of user requirements. Each workflow poses its own feature requirements to the potential tools to be used. The workflows differ from one type of activity to another and even for separate studies within the same type of activity when performed by different experts.

A general workflow for performing heat recovery studies can be found in Kemp (2007) – consisting of steps for describing the processes, their connectivity and flowsheets, their mass and energy balances, data extraction, energy targeting, energy optimisation and potential process modifications. The workflow can be summarised as follows:

1. The first step is to obtain a description of the process under analysis and develop a good general understanding of the flowsheet and the involved unit operations. This implies close interaction with the process designer and/or plant manager, especially if the plant is already operating.
2. Develop mass and heat balances. These should be based on process flowsheet data from a current design, calculations and/or on measurements taken from the operating plant.

3. Selection of Heat Integration Process Streams. This is a critical activity, known as Data Extraction. It applies the analysis of the unit operations and their heating and cooling demands. A detailed presentation of the data extraction principles and procedures is provided in Klemeš et al. (2018a).
4. Removal of all existing units related to the analysis. For Heat Integration, this means removing all heat exchangers – both recovery and utility exchangers. The corresponding heating and cooling demands are lumped.
5. Extraction of the Heat Integration process stream data. This is done using the result of Step 4.
6. Make an initial educated guess for the ΔT_{min} value appropriate for each analysed process. This value can be adjusted later at various stages of the design optimisation.
7. Pinch Analysis, which identifies the Pinch point temperatures and the utility targets.
8. Design of the initial Heat Exchanger Network using the criterion of maximising energy recovery.
9. Check for Cross-Pinch heat transfer and for inappropriate placement of heating and cooling utilities.
10. Check for proper placement of reactors, separation columns, heat engines and heat pumps.
11. Identification of the potential for further modifying the process in order to minimise energy consumption and reduce capital costs. Evaluation of the potential benefits of applying the plus–minus principle – see Chapter 4 in Klemeš et al. (2010) and the process modification rules derived from the Heat Integration.
12. Evaluation of the potential for integration between processes, stepping to Total Site Analysis.
13. Incorporation into the model of implementation aspects. One such aspect includes the implications of pressure drop and the trade-offs between heat savings and extra energy for pumping. It is also necessary to account for the physical layout, determining the capital cost of heat exchangers and for piping.
14. Pre-selection of heat exchangers and preliminary costing. The costs should account for the likely variations in the future price of energy.
15. First optimisation of the plant/site design and adjustments to the appropriate ΔT_{min} values.
16. Based on the optimisation from Step 15, adjusted data is extracted and return to step 7. Perform an additional loop (or loops) while screening and scoping for potential simplifications. Once the variations in the key design properties achieve convergence, the loops are stopped.
17. Consideration of real plant constraints. These include safety, technology limitations, controllability, operability, flexibility, availability and maintainability.
18. Incorporation of provisions for process start-up and shutdown of the process.

19. Repeated process optimisation for tuning the design. If necessary, a return to any appropriate previous step for adjustment can be performed, and iterations are repeated until convergence.
20. Detailed design.

One possible implementation of that general workflow has been presented by Bohnenstaedt et al. (2014). They combined several software tools – including process data acquisition packages, a flowsheet simulator, an energy optimisation package and a project data bank. A workflow (Fig. 12.5) based on decades of experience was proposed by Chew et al. (2013) in their analysis of Total Site industrial implementation issues. Although having common elements with the previous references, the workflow in Fig. 12.5 follows different steps and accounts for various parts of the processes, indicating the need for flexible tools. In practice, complete project workflows are more complex and difficult to foresee in advance.

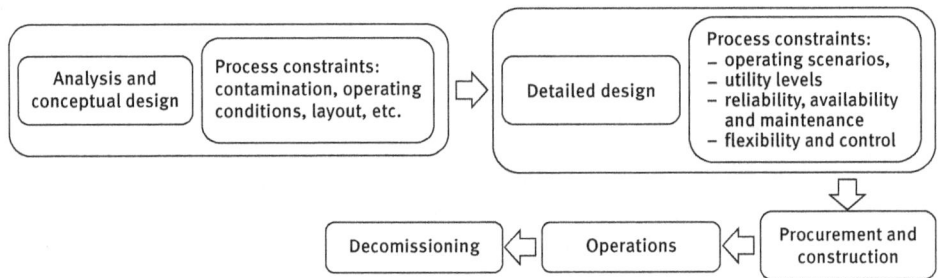

Fig. 12.5: Workflow for TS optimisation (after Chew et al. 2013).

The provided workflow examples can be used as good initial templates, reflecting the main evaluation principles and project stages. The specific procedures to be followed are determined by the implementers. This nature of process design and optimisation – including that of utility systems, requires modelling flexibility, which can only partially be accommodated by commercial tools such as Aspen Custom Modeler (Aspentech, 2019). To some extent, the Modelica implementations (Modelica, 2019) provide such flexibility. However, all language implementations specialise in control applications and discrete event simulation, which makes them unsuitable for modelling and optimising site utility systems.

A key obstacle to building a flexible modelling platform for utility systems and chemical processes is also the licencing cost. While academic licences for the main process simulators can be provided at a lower cost not exceeding several thousand USD/y, the licence terms usually do not allow distribution of the models and their wider use beside teaching.

An important requirement, which is becoming apparent, is the holistic evaluation of the site energy systems and networks. In addition to the economic performance, the environmental and social footprints also have to be accounted for, as has been done in Ren et al. (2018) using Aspen Plus and data retrieved using GaBi® LCA (ThinkStep, 2019). Proper integration of the environmental performance indicators into commercial simulators is still pending. This problem features mainly licencing and administrative obstacles. Documentation is a standard requirement for complex models and software systems. For site energy optimisation, it is essential to provide capabilities for documenting modelling, algorithmic and case study specific facts, providing continuity.

12.5 Summary

Software can be used for Process Systems Engineering and optimisation of utility systems. These tools provide mainly fixed implementations of previously developed methodologies. They are very useful to industrial users and to students for learning. However, for technology developers, researchers, and consultants, they are only partially helpful. Because of the fixed functionality, they are difficult to extend and adapt to research cases.

A possible alternative is an open computational and development environment for performing custom studies and adding extensions. Features to seek to include the following:

- Modelling and procedural flexibility and documentation capability. Barna et al. (2015) presented an initial step in this direction by proposing an open XML format for Heat Exchanger Network data. While this is a good development allowing the exchange of data among different tools, the need for an integrated modelling environment is apparent. This would allow to reuse and customise tools and to build their own in-house tools.
- Dataset storage and holistic use, preserving data integrity, reasoning and semantics for modelling process systems and energy integration. This has been discussed at length in Varbanov et al. (2019).

Another important issue is the need for seamless integration of environmental performance databases into the used software tools. The main difficulties in implementing this seem to be related mainly to the licencing cost, but also to the lack of appropriate application interfaces.

References

'Aspen Plus'. 2019. https://www.aspentech.com/en/products/engineering/aspen-plus, accessed 21/03/2019.

'AspenONE Product Portfolio'. 2019. https://www.aspentech.com/en/products/full-product-listing, accessed 21/11/2019.

Abdul Aziz, E., Alwi, S.R.W., Lim, J.S., Manan, Z.A., and Klemeš, J.J. 2017. An Integrated Pinch Analysis Framework for Low CO_2 Emissions Industrial Site Planning. Journal of Cleaner Production 146 (March): 125–138. https://doi.org/10.1016/j.jclepro.2016.07.175.

Aspentech. 2019. Aspen Custom Modeler. Build and Share Your Own Custom Models. https://www.aspentech.com/en/products/pages/aspen-custom-modeler, accessed 24/11/2019.

Barna, B., Varbanov, P.S., Klemeš, J.J., and Yong, J.Y. 2015. Object-Oriented Model and File Format for Heat Exchanger Network Computations. Chemical Engineering Transactions 45: 319–324. https://doi.org/10.3303/CET1545054.

Bohnenstaedt, T., Brandt, C., Fieg, G., and Dietrich, W. 2014. Energy Integration Manager: A Workflow for Long Term Validity of Total Site Analysis and Heat Recovery Strategies. Computer Aided Chemical Engineering 33: 1819–1824. Elsevier. https://doi.org/10.1016/B978-0-444-63455-9.50138-0.

Chew, K.H., Klemeš, J.J., Alwi, S.R.W., and Manan, Z.A. 2013. Industrial Implementation Issues of Total Site Heat Integration. Applied Thermal Engineering 61(1): 17–25. https://doi.org/10.1016/j.applthermaleng.2013.03.014.

Čuček, L., Lam, H.L., Klemeš, J.J., Varbanov, P.S., and Kravanja, Z. 2010. Synthesis of Regional Networks for the Supply of Energy and Bioproducts. Clean Technologies and Environmental Policy 12(6): 635–645. https://doi.org/10.1007/s10098-010-0312-6.

Čuček, L., Varbanov, P.S., Klemeš, J.J., and Kravanja, Z. 2012. Total Footprints-Based Multi-Criteria Optimisation of Regional Biomass Energy Supply Chains. Energy 44(1): 135–145. https://doi.org/10.1016/j.energy.2012.01.040.

Dhole, V.R. and Linnhoff, B. 1993. Total Site Targets for Fuel, Co-Generation, Emissions, and Cooling. Computers & Chemical Engineering 17 (January): S101–9. https://doi.org/10.1016/0098-1354(93)80214-8.

The Document Foundation. 2020. Calc | LibreOffice – Free Office Suite – Fun Project – Fantastic People. https://www.libreoffice.org/discover/calc/, accessed 02/04/2020.

Friedler, F., Tarján, K., Huang, Y.W., and Fan, L.T. 1992. Graph-Theoretic Approach to Process Synthesis: Axioms and Theorems. Chemical Engineering Science 47(8): 1973–1988. https://doi.org/10.1016/0009-2509(92)80315-4.

GAMS. 2019. General Algebraic Modeling System. https://www.gams.com/, accessed 20/08/2019.

Halasz, L., Povoden, G., and Narodoslawsky, M. 2005. Sustainable Processes Synthesis for Renewable Resources. Resources, Conservation and Recycling 44(3): 293–307. https://doi.org/10.1016/j.resconrec.2005.01.009.

Halim, I. and Srinivasan, R. 2002a. Systematic Waste Minimization in Chemical Processes. 1. Methodology. Industrial & Engineering Chemistry Research 41(2): 196–207. https://doi.org/10.1021/ie010207g.

Halim, I. and Srinivasan, R. 2002b. Systematic Waste Minimization in Chemical Processes. 2. Intelligent Decision Support System. Industrial & Engineering Chemistry Research 41(2): 208–219. https://doi.org/10.1021/ie0102089.

Halim, I. and Srinivasan, R. 2006. Systematic Waste Minimization in Chemical Processes. 3. Batch Operations. Industrial & Engineering Chemistry Research 45(13): 4693–4705. https://doi.org/10.1021/ie050792b.

Iyer, R.R. and Grossmann, I.E. 1998. Synthesis and Operational Planning of Utility Systems for Multiperiod Operation. Computers & Chemical Engineering 22(7): 979–993. https://doi.org/10.1016/S0098-1354(97)00270-6.

KBC. 2019. Petro-SIM | Process Simulation Software | KBC. https://www.kbc.global/software/process-simulation-software, accessed 19/08/2019.

Kemp, I.C. 2007. Pinch Analysis and Process Integration: A User Guide on Process Integration for the Efficient Use of Energy. 2nd ed. Amsterdam, The Netherlands; Boston, USA: Butterworth-Heinemann.

Klemeš, J., Varbanov, P.S., Alwi, S.R.W., and Manan, Z.A. 2018a. Sustainable Process Integration and Intensification: Saving Energy, Water and Resources. 2nd ed. Berlin, Germany: Walter de Gruyter GmbH.

Klemeš, J.J., Varbanov, P.S., Walmsley, T.G., and Jia, X. 2018b. New Directions in the Implementation of Pinch Methodology (PM). Renewable and Sustainable Energy Reviews 98 (December): 439–468. https://doi.org/10.1016/j.rser.2018.09.030.

Klemeš, J.J., Dhole, V.R., Raissi, K., Perry, S.J., and Puigjaner, L. 1997. Targeting and Design Methodology for Reduction of Fuel, Power and CO_2 on Total Sites. Applied Thermal Engineering 17(8–10): 993–1003. https://doi.org/10.1016/S1359-4311(96)00087-7.

Klemeš, J.J., Friedler, F., Bulatov, I., and Varbanov, P.S. 2010. Sustainability in the Process Industry: Integration and Optimization. New York, USA: McGraw-Hill Education.

Kravanja, Z. 2010. Challenges in Sustainable Integrated Process Synthesis and the Capabilities of an MINLP Process Synthesizer MipSyn. Computers & Chemical Engineering 34(11): 1831–1848. https://doi.org/10.1016/j.compchemeng.2010.04.017.

Lam, H.L., Klemeš, J.J., Kravanja, Z., and Varbanov, P.S. 2011. Software Tools Overview: Process Integration, Modelling and Optimisation for Energy Saving and Pollution Reduction. Asia-Pacific Journal of Chemical Engineering 6(5): 696–712. https://doi.org/10.1002/apj.469.

Lam, H.L., Varbanov, P.S., and Klemeš, J.J. 2010. Optimisation of Regional Energy Supply Chains Utilising Renewables: P-Graph Approach. Computers and Chemical Engineering 34(5): 782–792. https://doi.org/10.1016/j.compchemeng.2009.11.020.

Liew, P.Y., Lim, J.S., Wan Alwi, S.R., Abdul Manan, Z., Varbanov, P.S., and Klemeš, J.J. 2014a. A Retrofit Framework for Total Site Heat Recovery Systems. Applied Energy 135: 778–790. https://doi.org/10.1016/j.apenergy.2014.03.090.

Liew, P.Y., Wan Alwi, S.R., Lim, J.S., Varbanov, P.S., Klemeš, J.J., and Abdul Manan, Z. 2014b. Total Site Heat Integration Incorporating the Water Sensible Heat. Journal of Cleaner Production 77: 94–104. https://doi.org/10.1016/j.jclepro.2013.12.047.

Liew, P.Y., Wan Alwi, S.R., Varbanov, P.S., Manan, Z.A., and Klemeš, J.J. 2012. A Numerical Technique for Total Site Sensitivity Analysis. Applied Thermal Engineering 40: 397–408. https://doi.org/10.1016/j.applthermaleng.2012.02.026.

Liew, P.Y., Wan Alwi, S.R., Varbanov, P.S., Manan, Z.A., and Klemeš, J.J. 2013. Centralised Utility System Planning for a Total Site Heat Integration Network. Computers and Chemical Engineering 57: 104–111. https://doi.org/10.1016/j.compchemeng.2013.02.007.

LINGO. 2019. LINGO and Optimization Modeling. https://www.lindo.com/index.php/products/lingo-and-optimization-modeling, accessed 20/08/2019.

Luo, X., Huang, X., El-Halwagi, M.M., Ponce-Ortega, J.M., and Chen, Y. 2016. Simultaneous Synthesis of Utility System and Heat Exchanger Network Incorporating Steam Condensate and Boiler Feedwater. Energy 113 (October): 875–893. https://doi.org/10.1016/j.energy.2016.07.109.

Luo, X., Zhang, B., Chen, Y., and Songping, M. 2011. Modeling and Optimization of a Utility System Containing Multiple Extractions Steam Turbines. Energy 36(5): 3501–3512. https://doi.org/10.1016/j.energy.2011.03.056.

Mavromatis, S.P. and Kokossis, A.C. 1998. Conceptual Optimisation of Utility Networks for Operational Variations – I. Targets and Level Optimisation. Chemical Engineering Science 53(8): 1585–1608. https://doi.org/10.1016/S0009-2509(97)00431-4.

Microsoft. 2020. Excel Help & Learning – Office Support. https://support.office.com/en-us/excel, accessed 02/04/2020.

Modelica. 2019. Modelica Tools – Modelica Association. https://www.modelica.org/tools, accessed 24/11/2019.

Mohammad Rozali, N.E., Alwi, S.R.W., Klemeš, J., and Hassan, M.Y. 2013. Process Integration of Hybrid Power Systems with Energy Losses Considerations. Energy 55 (June): 38–45. https://doi.org/10.1016/j.energy.2013.02.053.

Nemet, A., Klemeš, J.J., and Kravanja, Z. 2015. Designing a Total Site for an Entire Lifetime under Fluctuating Utility Prices. Computers & Chemical Engineering 72 (Supplement C): 159–182. https://doi.org/10.1016/j.compchemeng.2014.07.004.

P-graph. 2019. Download | P-Graph. http://p-graph.com/download/, accessed 21/11/2019.

PIL. 2019. I-Steam™ – Process Integration Limited. Process Integration Limited | Chemical Engineering Consultancy (blog). https://www.processint.com/software/i-steam/, accessed 21/11/2019.

Rad, M.P., Khoshgoftar Manesh, M.H., Rosen, M.A., Amidpour, M., and Hamedi, M.H. 2016. New Procedure for Design and Exergoeconomic Optimization of Site Utility System Considering Reliability. Applied Thermal Engineering 94 (February): 478–490. https://doi.org/10.1016/j.applthermaleng.2015.10.091.

Ren, X.-Y., Jia, -X.-X., Varbanov, P.S., Klemeš, J.J., and Liu, Z.-Y. 2018. Targeting the Cogeneration Potential for Total Site Utility Systems. Journal of Cleaner Production 170: 625–635. https://doi.org/10.1016/j.jclepro.2017.09.170.

Shang, Z. and Kokossis, A. 2004. A Transhipment Model for the Optimisation of Steam Levels of Total Site Utility System for Multiperiod Operation. Computers & Chemical Engineering 28(9): 1673–1688. https://doi.org/10.1016/j.compchemeng.2004.01.010.

Sommerville, I. 2011. Software Engineering. 9th ed. Boston, MA, USA: Pearson.

ThinkStep. 2019. GaBi: Life Cycle Assessment LCA Software. http://www.gabi-software.com/ce-eu-english/index/, accessed 24/11/2019.

UNIMAN. 2019. STAR Overview, Centre for Process Integration, CEAS, The University of Manchester. http://documents.manchester.ac.uk/display.aspx?DocID=35772, accessed 21/11/2019.

Varbanov, P., Perry, S., Klemeš, J., and Smith, R. 2005. Synthesis of Industrial Utility Systems: Cost-Effective de-Carbonisation. Applied Thermal Engineering 25(7): 985–1001. https://doi.org/10.1016/j.applthermaleng.2004.06.023.

Varbanov, P., Perry, S., Makwana, Y., Zhu, X.X., and Smith, R. 2004a. Top-Level Analysis of Site Utility Systems. Chemical Engineering Research and Design 82(6): 784–795. https://doi.org/10.1205/026387604774196064.

Varbanov, P.S., Doyle, S., and Smith, R. 2004b. Modelling and Optimization of Utility Systems. Chemical Engineering Research and Design 82(5): 561–578. https://doi.org/10.1205/026387604323142603.

Varbanov, P.S. and Klemeš, J.J. 2011. Integration and Management of Renewables into Total Sites with Variable Supply and Demand. Computers and Chemical Engineering 35(9): 1815–1826. https://doi.org/10.1016/j.compchemeng.2011.02.009.

Varbanov, P.S., Yong, J.Y., Klemeš, J.J., and Chin, H.H. 2019. Data Extraction for Heat Integration and Total Site Analysis: A Review. Chemical Engineering Transactions 76 (October): 67–72. https://doi.org/10.3303/CET1976012.

Varbanov, P.S., Fodor, Z., and Klemeš, J.J. 2012. Total Site Targeting with Process Specific Minimum Temperature Difference (ΔT_{min}). Energy 44(1): 20–28. https://doi.org/10.1016/j.energy.2011.12.025.

Wan Alwi, S.R., Rozali, N.E.M., Abdul-Manan, Z., and Klemeš, J.J. 2012. A Process Integration Targeting Method for Hybrid Power Systems. Energy 44(1): 6–10. https://doi.org/10.1016/j.energy.2012.01.005.

Wan Alwi, S.R., Tin, O.S., Rozali, N.E.M., Manan, Z.A., and Klemeš, J.J. 2013. New Graphical Tools for Process Changes via Load Shifting for Hybrid Power Systems Based on Power Pinch Analysis. Clean Technologies and Environmental Policy 15(3): 459–472. https://doi.org/10.1007/s10098-013-0605-7.

Wang, R., Wu, Y., Wang, Y., and Feng, X. 2017. An Industrial Area Layout Design Methodology Considering Piping and Safety Using Genetic Algorithm. Journal of Cleaner Production 167 (November): 23–31. https://doi.org/10.1016/j.jclepro.2017.08.147.

Zahboune, H., Zouggar, S., Yong, J.Y., Varbanov, P.S., Elhafyani, M., Ziani, E., and Zarhloule, Y. 2016. Modified Electric System Cascade Analysis for Optimal Sizing of an Autonomous Hybrid Energy System. Energy 116: 1374–1384. https://doi.org/10.1016/j.energy.2016.07.101.

Part 4: **Conclusion**

13 Conclusions and sources of further information

This chapter summarises the book content and provides pointers to further information as well as on closely related topics. The main sources can be grouped into books and research publications and are discussed next.

13.1 Summary of the material presented in the book

This book discusses the core of the subjects related to industrial utility systems. The focus is on the site utility system as a supplier of heating, cooling and power, and on its role as a marketplace for exchanging steam as the main utility used in industry for heating.

Chapter 1 discusses the main issues related to energy use, supply and sourcing – related to the growing user demands, the resulting energy waste and emissions to the environment. The chapter then proceeds to the definitions of the fundamental concepts of the industrial site and the typical process energy interfaces – heating, cooling, power and water demands. The main decision items for a utility system are outlined – fuel consumption, power import/export, selection and sizing of equipment items (steam turbines, gas turbines and steam generators) and their arrangement into an integrated network for utility supply and exchange. The chapter links the utility system design and operation to the issues of sustainability – environmental, societal and economical and provides a concise historical overview of the modelling developments.

Chapter 2 introduces the fundamental concepts needed to model heat and power utility systems, providing a host of references for the readers to use if they would be interested in a more detailed treatment of the concepts. It starts with an overview of the typical utility heating and utility cooling options. Given that context, the fundamentals of the thermodynamic concepts used in the other book chapters are introduced – the laws of Thermodynamics, the concepts of temperature, work, internal energy, enthalpy and equations of the state relating the state variables and the modelling and description of thermodynamic state transitions. The most common thermodynamic cycles are introduced – polytropic process definition and Carnot cycle. The estimation of the thermodynamic properties of substances is reviewed, giving examples of the potential tools capable of performing this task.

Chapter 3 provides an overview of site utility systems. The departure point is how the utility system is viewed from the perspective of the served processes including the typical utilities and their topology arrangements. The chapter proceeds to overview the typical architecture and items in a heat-and-power utility system – the steam mains (headers), steam generators, steam turbines, gas turbines and letdown stations. The concepts of combined heat and power (CHP) generation and combined

https://doi.org/10.1515/9783110630091-013

cycle are defined and clearly distinguished from one another to help the readers avoid a typical misconception equating these concepts. The chapter discusses the life cycle of the utility system facilities and concludes with an overview of the developments in the field of utility system modelling.

Chapter 4 turns the readers' attention to the primary resources used for energy supply. It starts with the main physical principles and mechanisms for energy transfer and storage. The global energy flows supplying energy to the Earth are outlined, providing a context for distinguishing between renewable and non-renewable primary energy sources. Given that, the most used primary energy sources are discussed – solar energy, biomass, wind, hydropower, geothermal and fossil fuels. The relation of fuel combustion to the main environmental impacts is analysed, introducing the concept of environmental footprints and the impact of fuel use on human health.

Chapter 5 discusses steam generators, also known as steam boilers. The main principle of steam generation in the industry is presented, on the example of a water-tube drum boiler, referring the readers to more detailed sources dedicated to boiler design and operation. The main boiler classifications are introduced – water-tube versus fire-tube, field-erected versus packaged and dedicated boilers versus heat recovery steam generators. The main principles and procedures of boiler operation are introduced, explaining the need for boiler blowdown and the ways to minimise the energy losses associated with this practice. The operation patterns and issues are overviewed – efficiency, heat losses, flue gas and its acid dew point, the related construction materials, the minimum stack temperature limit. The main factors determining boiler efficiency are discussed, accompanied by examples of boiler efficiency trends. The models of boilers and heat recovery steam generators used in optimising utility systems are presented.

Chapter 6 is dedicated to steam turbines. It introduces the Rankine–Clausius cycle. The steam turbine typical construction is presented, discussing the operation and the main relationships – specific internal work, energy balances and simplified thermodynamic models. The classifications of steam turbines are overviewed – by service type (generation versus direct drive) and by their arrangement in the utility system (backpressure, condensing and extraction). Examples of steam turbine power plants are provided for illustration of the typical steam turbine installations and auxiliary equipment. The connection and governance of steam turbines are explained, also including the typical efficiency trends and steam turbine maps. The suppliers of steam turbines are overviewed, providing references to dedicated market studies. The remainder of the chapter presents the steam turbine model based on the Willans Line concept, tailored for using in large-scale optimisation models of steam networks.

Chapter 7 presents gas turbines and their use in industrial utility systems. The principle of operation of gas turbines is shown on the example of a single-shaft engine. The Brayton cycle, which is implemented in gas turbines, is introduced with the main cycle variations – open and closed. The connection and use of gas turbines as process units are explained followed by an analysis of the typical performance trends. As with steam turbines, the main gas turbine suppliers and market trends are outlined in this chapter. The application contexts for gas turbine are summarised. The chapter then presents the gas turbine model developed for embedding in models for utility network simulation and optimisation. The procedure for the model execution is given, and the derivation of its main parameters is illustrated in a literature example using real gas turbine data from a manufacturer.

Chapter 8 discusses the modelling of steam networks as a whole. It starts with a discussion of the main contribution aspects of utility system to sustainability and the problem description. The further discussion concerns the main degrees of freedom in site utility systems by subsystem groups – firing machines (boilers and gas turbines) and the steam distribution part (steam mains, steam turbines, letdown stations and vents). The network performance is discussed, concerning the environmental emissions and economic performance, leading to the objective functions used for the optimisation formulations. The model features are discussed in the context of their implications on the optimality and numerical convergence for the case of implementing the model using direct algebraic models and mathematical programming. It is shown how to model power import–export balance of the site to incorporate into the mathematical model. The chapter concludes with an illustrative example of utility system optimisation based on industrial data.

Chapter 9 gives a step-by-step utility system modelling example. It has two parts – first, the network is modelled explicitly in MS Excel, applying all the equations from the previous chapters, and in the second part, the same network is modelled using KBC's Petro-SIM process simulator. The chapter illustrates the importance of applying engineering judgement in performing such studies and illustrates one possible approach to use the independent modelling tools for obtaining higher confidence in the modelling results.

Chapter 10 presents two types of macroanalyses that can be performed using utility system optimisation models and tools. These are the Top-Level Analysis and R-curve analysis. The former allows focusing data collection and process modification efforts on the most important parts of an industrial site giving the highest returns, while the latter allows analysing and improving the thermodynamic efficiency of the system in relation to the power-to-heat ratios of the process demands and the utility system facilities.

Chapter 11 delivers the Total Site Integration developments and their relation to the utility system. The discussion starts with presenting the fundamentals of the method, referring the readers for more details to the relevant sources. The chapter then proceeds to present two essential extensions of the method. One is the Total Site heat cascade with Time Slices, which allows modelling variable energy supplies and demands. The other is the Total Site targeting with process-specific minimum allowed temperature difference (ΔT_{min}). This extension allows energy engineers to reflect more accurately the heat transfer properties of the equipment and streams on the site and obtain more realistic heat recovery and utility supply targets.

Chapter 12 provides a review of the features of utility system models, the possible applications and matching them to the software tools available to complete the task. The available tools are classified into categories of general computation, flowsheet simulators and dedicated proprietary tools for energy optimisation. The overall workflow for Total Site modelling and optimisation is reviewed, suggesting the possible applications of the discussed tools to accomplishing tasks in the workflow.

Chapter 13 provides conclusions and sources of further information – this chapter.

13.2 Recommended use

This book is intended as a reference for students (under- and postgraduate) and their lecturers. Industrial process engineers at varies stages of their professional career would also benefit from the information provided. The book can be used as a direct learning tool as well as the reference material for the recommended reading of dedicated courses on utility system modelling and optimisation.

The step-by-step case study from Chapter 9 is accompanied by two data files which can be freely used by the book readers for learning and teaching. One is an MS-Excel workbook containing all the modelling steps explicitly, following the discussion in the first part of Chapter 9. The second data file is specific to the Petro-SIM modelling tool and can be used by holders of Petro-SIM licences. This case study can be the basis of a working session inside a dedicated steam network modelling course.

13.3 Sources of further information

The sources of further information specific to boilers, steam turbines and gas turbines are discussed in their dedicated chapters. The content of this section is intended to provide the readers with obtaining further information on general energy modelling and optimisation in industry, as well as on utility and steam systems as a whole.

13.3.1 Books

There are many books on the topics of energy systems modelling and optimisation – including general thermodynamics (Balmer, 2011) or more specialised texts based on the concept of exergy (Dinçer and Rosen, 2007). This section reviews other books, treating the energy efficiency and utilities at industrial sites, from the viewpoint of the overall system.

13.3.1.1 General Process and Heat Integration

A cornerstone of Process Integration literature has been the "red book" – the first comprehensive manual on Heat Integration, which has been developed under the leadership of Bodo Linnhoff. The first version (Linnhoff et al., 1982) has started the wide application of the methodology and its spread towards other application areas. The book provides a complete reference on Heat Integration of industrial processes, starting from fundamental concepts, and discussing Heat Exchanger Network targeting, synthesis and follow-up optimisation for energy conservation. The book is suitable for teaching, as well as for self-study. The user guide has been updated 12 years later (Linnhoff et al., 1994) to include the overall Process Integration methodology formulation (e.g. the Onion Diagram of process design hierarchy) and the first Water Integration and Total Site developments. Although that book has been published a very long time ago, it is an excellent teaching and learning material on Heat Integration and several more Process Integration applications, providing the reader with a smooth and understandable learning path.

Valuable references for engineers, teachers and students are the books published by the successor of Bodo Linnhoff as a Head of Department of Process Integration at UMIST and after the merge with the University of Manchester, the Director of Centre for Process Integration – Robin Smith. He published three books. The first was "Chemical Process Design" (Smith, 1995), which was followed by an extended version which included more recent development – "Chemical Process Design and Integration", which has two editions (Smith, 2005 and 2016). This series of books discusses chemical process design in all its stages, from the perspective of Process Integration. It starts with the conception of potential chemical products and the main reaction pathways of obtaining them, moving to reactor design, separation system design, heat and energy recovery and utilities – following the famous "Onion Diagram" model.

Ian Kemp undertook the task to develop the second edition of the *User Guide* from 1994 (Linnhoff et al., 1994) and created an updated and extended version (Kemp, 2007). He has been widely using the UMIST course materials [CPI (Centre for Process Integration), 2005]. This book has also an accompanying Excel spreadsheet available as a web annex.

Another key source of information on the related topics is the *Handbook of Process Integration* – with its first edition (Klemeš, 2013), whose second edition is expected in 2022. This is a comprehensive collection of all Process Integration methods from various processing domains. These include chemical and food industries, heat and power, supply chains and water network optimisation, to name a few. The book contains several chapters related to process and site level heat recovery and CHP generation.

Another book, already briefly mentioned in the previous chapters, is worth a special emphasis. This is the book on *Sustainable Process Integration and Intensification* (Klemeš et al., 2018b), published by "Walter de Gruyter GmbH". This book is authored by Klemeš, Varbanov, Wan Alwi and Manan and is in its second edition. The materials presented there provide a smooth learning experience on the fundamental parts of Process Integration – Heat Integration, Water Integration, basics of Total Site Heat Integration (TSHI), CO_2 emissions targeting and minimisation. In this context, this book is a continuation focusing on the utility systems and both books form a series that can be beneficially used in combined courses on heat and Process Integration.

On Total Sites and between them, beside energy flows, discussed in this book, some utilities come in the form of materials and a number of energy-intensive processes are used for material supply and regeneration. Within this context, the minimisation of energy demands of industrial processes should start from the reuse and minimisation of materials and other resources besides energy – including water. For this purpose, a comprehensive Process Integration text for Mass Integration and waste material reuse is given in (Foo, 2012). Another source, useful in this direction, is the Process Integration textbook *Sustainable Design Through Process Integration* (El-Halwagi, 2017), which treats key areas as sustainable design, Heat Integration, material recycling, mass exchange networks and separators. It also contains a chapter on the main thermodynamic cycles and their thermal interactions with industrial processes.

13.3.1.2 Books specialising in steam and utility systems

A good reference on the fundamentals and details of the thermodynamic cycles, as well on the equipment design details, is the book by Boyce (2010). In addition to the main utility system equipment types considered in the current book (steam/gas turbines, boilers), an extended selection of equipment is discussed there, used in other process types, for instance, gasifiers, which are key for building integrated gasification combined cycle power plants running on coal.

The book by Smith (2016) has to be mentioned here too. It has a chapter dedicated to part of the topics discussed in this book, including the modelling of boilers and turbines.

Other two books on industrial steam systems deserve mentioning and using jointly with the current book. One of them covers design fundamentals and practices (Sabet, 2016), focusing on the design codes and regulations, system setup and operation. The other book (Merritt, 2016) covers in detail a host of topics laying out the fundamentals of steam system operation and design: steam as a heat transfer fluid; steam generation, accumulation, transportation and use; heat transfer fundamentals, details of boiler design and operation; the condensate system; operating practices – start-up, shutdown, maintenance.

13.3.2 Scientific journals

For more up-to-date information on the discussed topics, one can refer to the available scientific journals. It has to be stressed that all published journals are impossible to enumerate. Here are summarised only the journals, which the authors of the current book have published in or from which they have sourced information for their research.

The *Journal of Cleaner Production* (Elsevier, 2020d), $IF_{2019} = 7.246$, is an international, transdisciplinary journal focusing on cleaner production, environmental, and sustainability research and practice. It publishes research on a host of topics, grouped into seven topic clusters, where the energy issues, Total Site Integration and utility systems are an integral part of several of the groups:
1. Cleaner production and technical processes
2. Sustainable development and sustainability
3. Sustainable consumption
4. Environmental and sustainability assessment
5. Sustainable products and services
6. Corporate sustainability and corporate social responsibility
7. Education for sustainable development

Renewable and Sustainable Energy Reviews (Elsevier, 2020f), $IF_{2019} = 12.11$, is another journal related to energy issues, specialising in reviews or research articles with a substantial review component and a topical focus on renewables and sustainability. The main topical groups of the journal include:
1. Energy resources – this includes biomass – dedicated crops and waste, nuclear, fossil fuels, hydropower, as well as less used sources as geothermal, hydrogen, marine and ocean energy, solar and wind.
2. Applications – including buildings, industry and transport
3. Energy conversion and storage
4. Environment impacts – related to the atmosphere, climate issues, meteorology, mitigation technologies
5. Relation of the technology to the socio-economic aspects
6. System-level issues and sustainability in general

Energy (Elsevier, 2020b), $IF_{2019} = 6.082$, is an international, multidisciplinary journal, publishing research in energy engineering areas. The journal covers research in mechanical engineering and thermal sciences, energy analysis, energy modelling and prediction. This includes integrated energy systems, energy planning and energy management. Parts of the scope are also energy conservation, energy efficiency, biomass and bioenergy, renewable energy, electricity supply and demand, energy storage, energy in buildings, economic and policy issues.

Journal of Sustainable Energy Engineering (De Gruyter, 2020), is an open-access (since March 2019) electronic journal. Its scope includes research on sustainable methods of energy generation. The list of subtopics includes renewable energy sources, energy efficiency, energy planning, energy policy and economics, sustainable development, thermodynamics, security of energy supply, waste management and environmental issues.

For more information on assessment and minimisation of the environmental impact of utility systems, the readers may refer to the *Journal of Environmental Management* (Elsevier, 2020e). This is a peer reviewed journal (IF_{2019} = 5.647) publishing original research on managing environmental systems and improving environmental quality. The scope includes the following topics:

1. Technological developments related to environmental ecosystems
2. Management of natural resources
3. Ecological conservation
4. Management and valorisation of waste
5. Environmental quality management
6. Environmental analysis and assessment
7. Social, economic and policy aspects

Clean Technologies and Environmental Policy (Springer, 2020), IF_{2019} = 2.429, is another journal, which publishes research related to site utility systems and energy issues. This journal publishes papers that treat subjects related to the development, demonstration, and commercialisation of cleaner products and processes. The topics of interest also include effective environmental policy strategies. Many of the research papers treat heat recovery in industry at process and site levels. Further energy-related extensions concern energy sustainability, security of supply, as well as municipal and regional studies, all aimed at minimising emissions and environmental impacts.

There are many further related scientific journals. From those, a partial list is as follows:

- *Energy Conversion and Management* (Elsevier, 2020c), IF_{2019} = 7.181, 8.208 focused on the technology developments
- *Applied Energy* (Elsevier, 2020a), IF_{2019} = 8.848, publishing on the topics of energy conversion and conservation, the optimal use of energy resources, analysis and optimisation of energy processes, mitigation of environmental pollutants, and sustainable energy systems
- *Energies* (MDPI, 2020), IF_{2019} = 2.702, an open-access journal, publishing scientific research, technology development, engineering, and the studies in policy and management related to energy issues, including in industrial context.
- *International Journal of Energy Research* (John Wiley & Sons, 2020), IF_{2019} = 3.741, having among its topics energy conversion/conservation and management, energy storage, integrated energy systems for multi-generation

- *Energy and Environment* (Sagepub, 2020), $IF_{2019} = 1.775$, aims to be a bridging platform of research and policy discussions on the topics related to energy conversion, supply and use, as well as environmental issues.

13.3.3 Further developments in related fields

This section overviews some of the developments in fields related to utility system modelling and optimisation. During the past two decades, the development and extension of Total Site Pinch Methodology has accelerated. The research group located in 2007–2015 at the University of Pannonia, Hungary led by Jiří Jaromír Klemeš, developed several extensions making Total Site Methodology more implementable in real life, in collaboration with the Centre for Process Integration at The University of Manchester as well as with the University of Maribor.

An important basis for extending the scope of Total Site considerations was set up by Perry et al. (2008). The study added to the scope further possible systems and activities – including business units, large shopping centres, hospital complexes, schools and universities, government complexes and residential buildings, expanding the types of processes considered for Total Site Integration. Varbanov et al. (2012) proposed an enhanced model and algorithm for Total Site Integration (TSI), accounting for the heat transfer properties within each site process. The integration of renewables with variable supply and demand has been tackled in Varbanov and Klemeš (2011). These developments have been discussed in detail in Chapter 11.

In 2017 the core of the Centre for Process Integration and Intensification CPI^2 won a prestigious EU-supported project Sustainable Process Integration Laboratory (SPIL), funded as project no. CZ.02.1.01/0.0/0.0/15_003/0000456, by Czech Republic Operational Programme Research and Development, Education, Priority 1: Strengthening capacity for quality research. This has given the start of the new laboratory, which has been established at the Brno University of Technology (VUT Brno), in the Czech Republic. The laboratory has enveloped as a centre of excellence from where a number of new developments originated. They have been summarised and analysed by recent review publications – on heat recovery systems retrofit (Klemeš et al., 2020) and on the development of the Pinch Methodology in service of process optimisation (Klemeš et al., 2018a).

SPIL has been proactively collaborating with a number of world-leading universities, and many of them developed progressive and high-quality research works in the field of this book. They include
- The University of Manchester, UK
- University of Maribor, Slovenia
- Universiti Teknologi Malaysia, Johor Bahru/Kuala Lumpur, Malaysia
- University Technology Petronas, Malaysia
- Hebei University of Technology, Tianjin, China

- Fudan University, Shanghai, China
- Pázmány Péter Catholic University, Budapest, Hungary
- Cracow University of Technology, Poland
- Russian Mendeleev University of Chemical Technology, Moscow
- The University of Waikato, Hamilton, New Zeland
- Xi'an Jiaotong University, Xi'an, China
- De La Salle, Manila, Philippines
- D. Serikbayev East Kazakhstan State Technical University, Kazakhstan
- Kharkiv Polytechnic Institute, Ukraine
- Beijing Normal University, Beijing, China
- South China University of Technology, Guangzhou
- Tianjin University in China

Zdravko Kravanja from the University of Maribor, Slovenia collaborated on optimising Total Sites for an entire lifetime under fluctuating utility prices Nemet et al. (2015). Čuček et al. (2012) studied the potential of Total Site Process Integration for balancing and decreasing the key environmental footprints and also joint research with the University of Maribor and the University of Waikato (Nemet et al., 2012) considering the Total Site methodology as a tool for planning and strategic decisions on Total Sites.

An increasing number of developments have been originating from University Technology Malaysia led by Zainuddin Abdul. Manan and Sharifah Rafidah Wan Alwi are in collaboration with Hungarian universities and presently the SPIL at the Brno University of Technology. One important extension dealt with Power (Electricity) Pinch (Rozali et al., 2018). A series of works were presented by Chew and coworkers, bringing the analogy from single-process modification to Total Site to reduce the energy demand (Chew et al., 2015a) and to reduce the capital cost (Chew et al., 2015b). From the UTM collaboration came the extension dealing with the sensitivity analysis (Liew et al., 2012) and a contribution that dealt with layout issues (Liew et al., 2014). The scope of Pinch Methodology was also extended by the UTM group to carbon emissions (Greenhouse Gas emissions) planning and minimisation (Munir et al., 2012).

A number of developments have also come from India led by Santanu Bandyopadhyay. Their team revisited the cogeneration potential targeting (Bandyopadhyay et al., 2010), which was further studied by Manesh et al. (2013) who also made an attempt to again use the R-curve approach (Karimkashi and Amidpour, 2012), originally introduced by Kimura and Zhu (2000) and further elaborated by Varbanov et al. (2004).

The research group around Xiao Feng published several works dealing with Total Site layout and implementation issues. Wang and Feng (2017) dealt with the distance factor, and some recent works have been applicable to Total Sites. They used Mathematical Programming for optimising the layout (Wang et al., 2017) or heuristic algorithm dealing mainly with piping length and related cost (Wu et al., 2016).

The Total Site methodology has been very widely implemented in a variety of settings, starting with industrial complexes and later extended into locally integrated energy sectors and even sustainable regions using renewable energy (Liew et al., 2017). The most common focus of TSI studies has been TSHI, including CHP.

TSHI involving both heat and power has been successfully implemented by Matsuda et al. (2009), who presented the comprehensive TSI study of a large chemical and petrochemical complexes in Kashima, Japan and later the Chiyoda Corp. team again – as described by Matsuda et al. (2012) – for a large steel site in Japan. Isaksson et al. (2011) made a successful implementation in the steel industry as well. These implementations clearly showed that the applicability of Total Site Methods is wider than the oil and petrochemicals industry, for which the TSHI was originally developed. TSHI applications to the steel industry have also broken another stereotype – that the steel industry is not suitable for Heat and Energy Integration.

The variety of applications has been even wider. Boldyryev and Varbanov (2015) showed the optimal heat recovery on a site for bromine production by applying TSI. Building on the earlier work on integrated evaporation system design of Westphalen and Maciel (2000) in combination with TSHI, Walmsley (2016) placed evaporation systems, which are commonly needed to concentrate multicomponent liquid solutions, suspensions, and emulsions in the food and dairy, pulp and paper, petrochemical, chemical, and pharmaceutical industries, in the TSHI context and presented an extension including vapour recompression.

The work of Axelsson et al. (2006) was one of the first industrial implementations in the pulp and paper industry. Walmsley et al. (2017) succeeded with TSHI of multieffect evaporators with vapour recompression for older Kraft Mills in the pulp and paper industry, where a number of previous implementations were also successfully completed. Walmsley et al. presented an ultralow-energy milk powder plant design using TSHI (Walmsley et al., 2018), which combined previous works that focused on the integrated designs of the milk evaporation system (Walmsley et al., 2015a) and milk spray dryer system (Walmsley et al., 2015b). Lundberg et al. (2014) applied TSHI within the context of a case study for overall performance improvement and conversion of a Kraft pulp mill into a biorefinery. The energy analysis in this work identified several options for steam and fuel savings, which included increasing the return temperature of the process steam condensate and preheating of the make-up water before mixing with the condensate. The identified energy savings of up to 5–6 %, combined with other process improvements, allowed 23 % increase of the site processing capacity. The Total Site idea has also been further extended in a recent implementation (Chang et al., 2017), to provide multi-plant Heat Integration, on an example of a 5-plant site, which has also accounted for the necessary piping cost.

One option for reduction of the GHG footprint from power generation, discussed in the literature, is CO_2 capture and sequestration (CCS). CCS involves substantial consumption of heat and power – for both the capture and transportation, which have to be supplied by the fuel spent by the process. An efficient way of reducing this energy penalty is TSHI. By analysing the power plant flowsheets and applying Pinch Analysis for HI, Harkin et al. (2010) reported a possible reduction of the energy penalty by up to 50 %.

Emphasising the power generation aspect of utility systems, Matsuda et al. (2015) presented the use of the Total Site Analysis in the Map Ta Phut industrial area in Thailand. The authors describe the application of the R-curve concept for triggering collaborative energy-saving projects. They reported the achievement of a 28 % reduction in the overall energy consumption of the area, in addition to any previous energy-saving measures already implemented by the individual sites.

13.3.4 Promising directions for future development

Several recent works provide an indication of the possibly advantageous directions of development of the TSI methodology. Tarighaleslami et al. (2017b) bridged the gap between high- and low-Pinch processing site applications through the development of a unified method. Tarighaleslami et al. (2017a) then further developed utility selection and optimisation using cost and exergy derivative analysis as well as the design of the utility exchanger network (Tarighaleslami et al., 2018).

Aziz et al. (2017) presented an integrated Pinch Analysis framework for low CO_2 emissions industrial site planning, emphasising the emissions minimisation as the objective function. Ong et al. (2017) developed Total Site Mass, Heat and Power Integration using Process Integration and P-graph, while Hassiba et al. (2017) integrated CO_2 and heat on industrial parks.

Chang et al. (2016) contributed with an optimisation algorithm for using waste heat by integration between two plants, using a heat recovery loop. From the same research group came several other contributions. Chang et al. (2017) optimised multi-plant Heat Integration using intermediate fluid circles and Chang et al. (2015) using hot water circuits. The most recent contribution came from Chang et al. (2018) for stepwise optimisation of indirect interplant heat recovery among plans, dealing with the issue of allocation between the plants of the costs and benefits from energy efficiency projects.

Wang et al. (2014) analysed the trade-off between energy and distance-related cost for Heat Integration across plants. Another contribution from Wang et al. (2015) presented a study on the use of combined direct and indirect Heat Integration modes on multiplant sites. Song et al. (2016) suggested Interplant Shifted Composite Curves. It is an interesting idea. However, it remains to be seen whether it will be further developed and widely used. Song et al. (2018) again presented an interesting contribution suggesting a parallel connection pattern among three plants (Stamp and Majozi, 2017).

Key issues in Total Site targeting have been addressed by Ren et al. (2018). These include the automation of the targeting procedure, ensuring a better approach to the underlying steam network fundamentals and the need to obtain derived targets on the system performance – including investment costs and environmental impact indicators. The work also illustrated how to perform the targeting within the process simulator environment, allowing seamless integration with modelling work performed prior to the utility system design.

Interesting ideas, based on TSHI, have been recently presented by several authors in relation to industrial symbiosis of industrial parks. El Massah (2018) published a paper expanding the TSHI considerations including not only heat but energy and water as well as integrating products and waste. This extension, dealing with waste-to-energy, waste-to-wealth and potentially closing the loop by implementing circular economy principles, presented a novel avenue where Pinch Methodology can be implemented to achieve long-term sustainability (El Massah, 2018).

One further area that needs attention is the retrofit of TSHI networks (Liew et al., 2014b), including the utility system and its structure. Careful consideration for future energy supply systems should also be recognised as part of embedding new technology during retrofits (Philipp et al., 2018).

13.3.5 Scientific conferences

Another key driver for development of the field of utility systems modelling and for sourcing information are scientific conferences. There are many energy-related and process systems engineering conferences, which are available. They can be even searched on conference platforms – such as "10times" (Tentimes Online Private Limited, 2020). However, this platform – as many others – usually lists many events and of different types, including fairs and society meetings. This section mentions scientific conferences, with which the authors have had personal experience.

To closely follow the most recent developments, the *Conferences on Process Integration, Modelling and Optimisation for Energy Saving and Pollution Reduction – PRES* (PRES, 2020) are naturally the best option. In 2017, it celebrated 20 years since its foundation (Klemeš et al., 2017). The series of conferences PRES is a recognised platform for bringing together scientists, engineers and decision-makers, to discuss and innovate on solutions and technologies for improved efficiency and sustainability of industrial and regional systems. This series of conferences is dedicated to Process Integration and related methods with a very strong presence of energy, Total Site and related issues in the conference topics. As of April 2020, the *Chemical Engineering Transactions Journal* (AIDIC, 2020), which publishes special issue selections from the conference PRES, features more than 150 articles on "utility system". A search on the term "Total Site" returns even more – close to 400 entries.

The conferences have a 23-year tradition of knowledge dissemination and creativity with a focus on addressing global challenges in energy, water, pollution and sustainability using classical chemical engineering approaches, enhanced by numerous interdisciplinary collaborations. The 16th Conference on Process Integration, Modelling and Optimisation for Energy Saving and Pollution Reduction – PRES'13 took place in Rhodes, Greece; the 17th Conference PRES 2014 (Varbanov et al., 2014) in Prague, the Czech Republic and the 18th PRES'15 (Varbanov et al., 2015) in Malaysia, Sarawak, Borneo. The 19th PRES 2016 (Varbanov et al., 2016) was held traditionally in Prague and the jubilee 20th PRES'17 (Varbanov et al., 2017) again in Asia – in the Chinese city of Tianjin. After PRES 2018 in Prague, the most recent PRES'19 returned again to Greece to the beautiful island of Crete. Despite the fact that the difficult situation connected to world-widespread of #covid-19, which created a situation when many conferences are cancelled or postponed, PRES'20 was successfully organized as a fully virtual conference in Xi'an, a former emperor's seat in China. This has come with a new technologically advanced format as a virtual conference.

There are other international conferences, worth mentioning
- The International Conference on Low Carbon Asia (ICLCA, 2020), which addresses the challenges related to the UN Sustainable Development Goals. The conference topics include urban planning and smart cities, sustainable energy, waste and water management, sustainable agriculture and urban food production, nexus and circular economy, clean technologies and other topics relevant to sustainable development. The coming 2020 edition of the conference will be held in Shanghai – China.
- The conferences organised by the International Centre for Sustainable Development of Energy, Water and Environment Systems (SDEWES, 2020) are dedicated to a spectrum of topics related to sustainable development. Within that spectrum, the issues of energy supply, conversion and use take a central place.
- The Global Conference on Global Warming – coming in its 9th edition in 2020 (GCGW-20, 2020) – is a multi-disciplinary international conference on all aspects of global warming. It provides a forum for the exchange of knowledge and technical information for a more sustainable development and energy security, which includes research and developments on utility systems.
- The scientific and technical conference "Modern Power Systems and Units" (MPSU, 2020) is organised periodically by the Cracow University of Technology. The 5th edition of the conference will be held in Cracow, 16–15 October 2020. The conference topics include fuels as energy sources, thermal energy conversion, energy use and societal/economic aspects.

- Another potential source of information is the series of conferences ECOS – "International Conference on Efficiency, Cost, Optimization, Simulation and Environmental Impact of Energy Systems" (ECOS, 2020). This is an annual events specialising on the environmental aspects of the energy systems analysis.

Nomenclature

Symbol	Measurement unit	Description
ΔT_{min}	°C	Minimum allowed temperature difference
CHP	–	Combined Heat and Power generation
TSHI	–	Total Site Heat Integration
GHG	–	Greenhouse gas
CCS	–	CO_2 capture and sequestration
HI	–	Heat Integration
TSI	–	Total Site Integration

References

Abdul Aziz, E., Alwi, S.R.W., Lim, J.S., Manan, Z.A., and Klemeš, J.J. 2017. An Integrated Pinch Analysis Framework for Low CO_2 Emissions Industrial Site Planning. Journal of Cleaner Production 146 (March): 125–138. https://doi.org/10.1016/j.jclepro.2016.07.175.

AIDIC. 2020. Chemical Engineering Transactions. https://www.cetjournal.it/index.php/cet, accessed 06/04/2020.

Axelsson, E., Olsson, M.R., and Berntsson, T. 2006. Heat Integration Opportunities in Average Scandinavian Kraft Pulp Mills: Pinch Analyses of Model Mills. Nordic Pulp and Paper Research Journal 21(4): 466–475.

Balmer, R.T. 2011. Modern Engineering Thermodynamics. Amsterdam; Boston, USA: Academic Press.

Bandyopadhyay, S., Varghese, J., and Bansal, V. 2010. Targeting for Cogeneration Potential through Total Site Integration. Applied Thermal Engineering. 30(1): 6–14. https://doi.org/10.1016/j.applthermaleng.2009.03.007.

Boldyryev, S. and Varbanov, P.S. 2015. Low Potential Heat Utilization of Bromine Plant via Integration on Process and Total Site Levels. Energy. 90 (October): 47–55. https://doi.org/10.1016/j.energy.2015.05.071.

Boyce, M.P. 2010. Handbook for Cogeneration and Combined Cycle Power Plants. 2nd ed. New York, USA: ASME Press.

Chang, C., Chen, X., Wang, Y., and Feng, X. 2016. An Efficient Optimization Algorithm for Waste Heat Integration Using a Heat Recovery Loop between Two Plants. Applied Thermal Engineering 105 (July): 799–806. https://doi.org/10.1016/j.applthermaleng.2016.04.079.

Chang, C., Chen, X., Wang, Y., and Feng, X. 2017. Simultaneous Optimization of Multi-Plant Heat Integration Using Intermediate Fluid Circles. Energy 121 (February): 306–317. https://doi.org/10.1016/j.energy.2016.12.116.

Chang, C., Wang, Y., and Feng, X. 2015. Indirect Heat Integration across Plants Using Hot Water Circles. Chinese Journal of Chemical Engineering 23(6): 992–997. https://doi.org/10.1016/j.cjche.2015.01.010.

Chang, H.H., Chang, C.T., and Li, B.H. 2018. Game-Theory Based Optimization Strategies for Stepwise Development of Indirect Interplant Heat Integration Plans. Energy. 148 (April): 90–111. https://doi.org/10.1016/j.energy.2018.01.106.

Chew, K.H., Klemeš, J.J., Alwi, S.R.W., and Manan, Z.A. 2015a. Process Modifications to Maximise Energy Savings in Total Site Heat Integration. Applied Thermal Engineering 78: 731–739. https://doi.org/10.1016/j.applthermaleng.2014.04.044.

Chew, K.H., Klemeš, J.J., Alwi, S.R.W., and Manan, Z.A. 2015b. Process Modification of Total Site Heat Integration Profile for Capital Cost Reduction. Applied Thermal Engineering 89 (October): 1023–1032. https://doi.org/10.1016/j.applthermaleng.2015.02.064.

CPI (Centre for Process Integration). 2005. Heat Integration and Energy Systems, MSc Course. Manchester, UK: School of Engineering and Analytical Science, The University of Manchester.

Čuček, L., Varbanov, P.S., Klemeš, J.J., and Kravanja, Z. 2012. Potential of Total Site Process Integration for Balancing and Decreasing the Key Environmental Footprints. Chemical Engineering Transactions 29: 61–66. https://doi.org/10.3303/CET1229011.

Dinçer, İ. and Rosen, M. 2007. Exergy: Energy, Environment, and Sustainable Development. Amsterdam, The Netherlands; Boston, USA: Elsevier.

ECOS. 2020. 33rd International Conference on Efficiency, Cost, Optimization, Simulation and Environmental Impact of Energy Systems. https://ecos2020.org/, accessed 06/04/2020.

El-Halwagi, M.M. 2017. Sustainable Design Through Process Integration. 2nd ed. New York, USA: Elsevier.

ElMassah, S. 2018. Industrial Symbiosis within Eco-industrial Parks: Sustainable Development for Borg El-Arab in Egypt. Business Strategy and the Environment March 27(7): 884–892. https://doi.org/10.1002/bse.2039.

Elsevier. 2020a. Applied Energy. https://www.journals.elsevier.com/applied-energy, accessed 04/04/2020.

Elsevier. 2020b. Energy. https://www.journals.elsevier.com/energy, accessed 04/04/2020.

Elsevier. 2020c. Energy Conversion and Management. https://www.journals.elsevier.com/energy-conversion-and-management, accessed 04/04/2020.

Elsevier. 2020d. Journal of Cleaner Production. https://www.journals.elsevier.com/journal-of-cleaner-production, accessed 04/04/2020.

Elsevier. 2020e. Journal of Environmental Management. https://www.journals.elsevier.com/journal-of-environmental-management, accessed 04/04/2020.

Elsevier. 2020f. Renewable & Sustainable Energy Reviews. https://www.journals.elsevier.com/renewable-and-sustainable-energy-reviews, accessed 04/04/2020.

Foo, D.C.Y. 2012. Process Integration for Resource Conservation. Boca Raton, FL, USA: CRC Press, Taylor & Francis Group.

GCGW-20. 2020. GCGW-20 – 9th Global Conference on Global Warming. http://gcgw.org/gcgw2020/, accessed 06/04/2020.

De Gruyter 2020. Journal of Sustainable Energy Engineering. De Gruyter. https://www.degruyter.com/view/journals/josee/josee-overview.xml, accessed 04/04/2020.

Harkin, T., Hoadley, A., and Hooper, B. 2010. Reducing the Energy Penalty of CO_2 Capture and Compression Using Pinch Analysis. Journal of Cleaner Production 18(9): 857–866. https://doi.org/10.1016/j.jclepro.2010.02.011.

Hassiba, R.J., Al-Mohannadi, D.M., and Linke, P. 2017. Carbon Dioxide and Heat Integration of Industrial Parks. Journal of Cleaner Production 155 (July): 47–56. https://doi.org/10.1016/j.jclepro.2016.09.094.

ICLCA. 2020. ICLCA | International Conference on Low Carbon Asia and Beyond. https://iclcaconf. com/, accessed 06/04/2020.

Isaksson, J., Harvey, S., Grip, C.-E., and Karlsson, J. 2011. Possibilities to Implement Pinch Analysis in the Steel Industry – A Case Study at SSAB EMEA in Luleå. Industrial Energy Efficiency November: 1660–1667. https://doi.org/10.3384/ecp110571660.

John Wiley & Sons. 2020. International Journal of Energy Research. Wiley Online Library. 4 April 2020. https://doi.org/10.1002/(ISSN)1099-114X.

Karimkashi, S. and Amidpour, M. 2012. Total Site Energy Improvement Using R-Curve Concept. Energy 40(1): 329–340. https://doi.org/10.1016/j.energy.2012.01.067.

Kemp, I.C. 2007. Pinch Analysis and Process Integration: A User Guide on Process Integration for the Efficient Use of Energy. 2nd ed. Amsterdam, The Netherlands; Boston, USA: Butterworth-Heinemann/Elsevier.

Khoshgoftar Manesh, M.H., Amidpour, M., Khamis Abadi, S., and Hamedi, M.H. 2013. A New Cogeneration Targeting Procedure for Total Site Utility System. Applied Thermal Engineering 54(1): 272–280. https://doi.org/10.1016/j.applthermaleng.2013.01.043.

Kimura, H. and Zhu, X.X. 2000. R – Curve Concept and Its Application for Industrial Energy Management. Industrial & Engineering Chemistry Research 39(7): 2315–2335. https://doi.org/10.1021/ie9905916.

Klemeš, J.J., Varbanov, P.S., Alwi, S.R.W., and Manan, Z.A. 2018b. Sustainable Process Integration and Intensification. Saving Energy, Water and Resources. 2nd ed. Berlin, Germany: Walter de Gruyter GmbH.

Klemeš J.J., Varbanov, P.S., Van Fan, Y., and Lam, H.L. 2017. Twenty Years of PRES: Past, Present and Future Process Integration towards Sustainability. Chemical Engineering Transactions 61: 1–24. https://doi.org/10.3303/CET1761001.

Klemeš, J.J., Varbanov, P.S., Walmsley, T.G., and Jia, X. 2018a. New Directions in the Implementation of Pinch Methodology (PM). Renewable and Sustainable Energy Reviews 98 (December): 439–468. https://doi.org/10.1016/j.rser.2018.09.030.

Klemeš, J.J., Wang, Q.-W., Varbanov, P.S., Zeng, M., Chin, H.H., Lal, N.S., Li, N.-Q., Wang, B., Wang, X.-C., and Walmsley, T.G. 2020. Heat Transfer Enhancement, Intensification and Optimisation in Heat Exchanger Network Retrofit and Operation. Renewable and Sustainable Energy Reviews 120 (March): 109644. https://doi.org/10.1016/j.rser.2019.109644.

Klemeš, J.J., ed. 2013. Handbook of Process Integration (PI): Minimisation of Energy and Water Use, Waste and Emissions. Cambridge, UK: Woodhead / Elsevier.

Liew, P.Y., Alwi, S.R.W., and Klemeš, J.J. 2014a. Total Site Heat Integration Targeting Algorithm Incorporating Plant Layout Issues. Computer Aided Chemical Engineering 33 (January): 1801–1806. https://doi.org/10.1016/B978-0-444-63455-9.50135-5.

Liew, P.Y., Alwi, S.R.W., Varbanov, P.S., Manan, Z.A., and Klemeš, J.J. 2012. A Numerical Technique for Total Site Sensitivity Analysis. Applied Thermal Engineering 40 (July): 397–408. https://doi.org/10.1016/j.applthermaleng.2012.02.026.

Liew, P.Y., Lim, J.S., Alwi, S.R.W., Varbanov, P.S., and Klemeš, J.J. 2014b. A Retrofit Framework for Total Site Heat Recovery Systems. Applied Energy 135 (December): 778–790. https://doi.org/10.1016/j.apenergy.2014.03.090.

Liew, P.Y., Theo, W.L., Alwi, S.R.W., Lim, J.S., Manan, Z.A., Klemeš, J.J., and Varbanov, P.S. 2017. Total Site Heat Integration Planning and Design for Industrial, Urban and Renewable Systems. Renewable and Sustainable Energy Reviews 68: 964–985. https://doi.org/10.1016/j. rser.2016.05.086.

Linnhoff, B., Townsend, D.W., Boland, D., Thomas, B.E.A., Guy, A.R., and Marsland, R.H. 1982. A User Guide on Process Integration for the Efficient Use of Energy. Rugby, UK: Institution of Chemical Engineers.

Linnhoff, B., Townsend, D.W., Boland, D., Thomas, B.E.A., Guy, A.R., and Marsland, R.H. 1994. A User Guide on Process Integration for the Efficient Use of Energy. Revised First Edition. Rugby, UK: Institution of Chemical Engineers.

Lundberg, V., Bood, J., Nilsson, L., Axelsson, E., Berntsson, T., and Svensson, E. 2014. Converting a Kraft Pulp Mill into a Multi-Product Biorefinery: Techno-Economic Analysis of a Case Mill. Clean Technologies and Environmental Policy 16(7): 1411–1422. https://doi.org/10.1007/s10098-014-0741-8.

Matsuda, K., Hirochi, Y., Tatsumi, H., and Shire, T. 2009. Applying Heat Integration Total Site Based Pinch Technology to a Large Industrial Area in Japan to Further Improve Performance of Highly Efficient Process Plants. Energy 34(10): 1687–1692.

Matsuda, K., Hirochi, Y., Kurosaki, D., and Kado, Y. 2015. Area-Wide Energy Saving Program in a Large Industrial Area. Energy 90 (October): 89–94. https://doi.org/10.1016/j.energy.2015.05.058.

Matsuda, K., Tanaka, S., Endou, M., and Iiyoshi, T. 2012. Energy Saving Study on a Large Steel Plant by Total Site Based Pinch Technology. Applied Thermal Engineering 43 (October): 14–19. https://doi.org/10.1016/j.applthermaleng.2011.11.043.

MDPI. 2020. Energies – Open Access Journal. https://www.mdpi.com/journal/energies, accessed 04/04/2020.

Merritt, C. 2016. Process Steam Systems | Wiley Online Books. Hoboken, NJ, USA: John Wiley & Sons, Inc. https://doi.org/10.1002/9781119085454.

Mohammad Rozali, N.E., Ho, W.S., Alwi, S.R.W., Klemeš, J.J., Yunus, M.N.S.M., and Zaki, S.A.A.S. M. 2018. Peak-off-Peak Load Shifting for Optimal Storage Sizing in Hybrid Power Systems Using Power Pinch Analysis Considering Energy Losses. Energy 156 (August): 299–310. https://doi.org/10.1016/j.energy.2018.05.020.

MPSU. 2020. V Scientific and Technical Conference "Modern Power Systems and Units". https://wtiue.conrego.pl/en/home, accessed 06/04/2020.

Munir, S.M., Abdul Manan, Z., and Wan Alwi, S.R. 2012. Holistic Carbon Planning for Industrial Parks: A Waste-to-Resources Process Integration Approach. Journal of Cleaner Production 33 (September): 74–85. https://doi.org/10.1016/j.jclepro.2012.05.026.

Nemet, A., Klemeš, J.J., Varbanov, P.S., Walmsley, M.R.W., and Atkins, M.J. 2012. Total Site Methodology as a Tool for Planning and Strategic Decisions. Chemical Engineering Transactions 29 (September): 115–120. https://doi.org/10.3303/CET1229020.

Nemet, A., Klemeš, J.J., and Kravanja, Z. 2015. Designing a Total Site for an Entire Lifetime under Fluctuating Utility Prices. Computers & Chemical Engineering 72 (Supplement C): 159–182. https://doi.org/10.1016/j.compchemeng.2014.07.004.

Ong, B.H.Y., Walmsley, T.G., Atkins, M.J., and Walmsley, M.R.W. 2017. Total Site Mass, Heat and Power Integration Using Process Integration and Process Graph. Journal of Cleaner Production 167: 32–43. https://doi.org/10.1016/j.jclepro.2017.08.035.

Perry, S., Klemeš, J., and Bulatov, I. 2008. Integrating Waste and Renewable Energy to Reduce the Carbon Footprint of Locally Integrated Energy Sectors. Energy PRES '07 10th Conference on Process Integration, Modelling and Optimisation for Energy Saving and Pollution Reduction, 33(10): 1489–1497. https://doi.org/10.1016/j.energy.2008.03.008.

Philipp, M., Schumm, G., Peesel, R.-H., Walmsley, T.G., Atkins, M.J., Schlosser, F., and Hesselbach, J. 2018. Optimal Energy Supply Structures for Industrial Food Processing Sites in Different Countries Considering Energy Transitions. Energy 146 (March): 112–123. https://doi.org/10.1016/j.energy.2017.05.062.

PRES'20 2020. Conference on Process Integration, Modelling and Optimisation for Energy Saving and Pollution Reduction – PRES. 6 April 2020. http://conferencepres.site/index.php/PRES.

Ren, X.-Y., Jia, X.-X., Varbanov, P.S., Klemeš, J.J., and Liu, Z.-Y. 2018. Targeting the Cogeneration Potential for Total Site Utility Systems. Journal of Cleaner Production 170 (January): 625–635. https://doi.org/10.1016/j.jclepro.2017.09.170.

Sabet, M. 2016. Industrial Steam Systems: Fundamentals and Best Design Practices. Boca Raton, FL, USA: CRC Press, Taylor & Francis Group.

Sagepub. 2020. Energy & Environment. SAGE Journals. https://journals.sagepub.com/home/eae, accessed 04/04/2020.

SDEWES. 2020. The International Centre for Sustainable Development of Energy, Water and Environment Systems. https://www.sdewes.org/, accessed 06/04/2020.

Smith, R. 1995. Chemical Process Design. New York, USA: McGraw-Hill, Inc.

Smith, R. 2005. Chemical Process Design and Integration. 1st ed. Chichester, West Sussex, UK; Hoboken, NJ, USA: Wiley.

Smith, R. 2016. Chemical Process Design and Integration. 2nd ed. Chichester, West Sussex, United Kingdom: Wiley.

Song, R., Feng, X., and Wang, Y. 2016. Feasible Heat Recovery of Interplant Heat Integration between Two Plants via an Intermediate Medium Analyzed by Interplant Shifted Composite Curves. Applied Thermal Engineering 94 (February): 90–98. https://doi.org/10.1016/j.applthermaleng.2015.10.125.

Song, R., Wang, Y., Panu, M., El-Halwagi, M.M., and Feng, X. 2018. Improved Targeting Procedure To Determine the Indirect Interplant Heat Integration with Parallel Connection Pattern among Three Plants. Industrial & Engineering Chemistry Research 57(5): 1569–1580. https://doi.org/10.1021/acs.iecr.7b04327.

Springer. 2020. Clean Technologies and Environmental Policy. Springer. https://www.springer.com/journal/10098, 04/04/2020.

Stamp, J.D. and Majozi, T. 2017. Long-Term Heat Integration in Multipurpose Batch Plants Using Heat Storage. Journal of Cleaner Production 142 (January): 1492–1509. https://doi.org/10.1016/j.jclepro.2016.11.155.

Tarighaleslami, A.H., Walmsley, T.G., Atkins, M.J., Walmsley, M.R.W., and Neale, J.R. 2017a. Total Site Heat Integration: Utility Selection and Optimisation Using Cost and Exergy Derivative Analysis. Energy 141: 949–963. https://doi.org/10.1016/j.energy.2017.09.148.

Tarighaleslami, A.H., Walmsley, T.G., Atkins, M.J., Walmsley, M.R.W., Liew, P.Y., and Neale, J.R. 2017b. A Unified Total Site Heat Integration Targeting Method for Isothermal and Non-Isothermal Utilities. Energy 119: 10–25. https://doi.org/10.1016/j.energy.2016.12.071.

Tarighaleslami, A.H., Walmsley, T.G., Atkins, M.J., Walmsley, M.R.W., and Neale, J.R. 2018. Utility Exchanger Network Synthesis for Total Site Heat Integration. Energy 153 (June): 1000–1015. https://doi.org/10.1016/j.energy.2018.04.111.

Tentimes Online Private Limited. 2020. Top Power & Energy Events, Trade Fairs, Conferences to Attend – Global Ranking 2020'. https://10times.com/top100/power-energy, accessed 06/04/2020.

Varbanov, P., Perry, S., Makwana, Y., Zhu, X.X., and Smith, R. 2004. Top-Level Analysis of Site Utility Systems. Chemical Engineering Research and Design 82(6): 784–795. https://doi.org/10.1205/026387604774196064.

Varbanov, P.S., Fodor, Z., and Klemeš, J.J. 2012. Total Site Targeting with Process Specific Minimum Temperature Difference (ΔT_{min}). Energy Integration and Energy System Engineering, European Symposium on Computer-Aided Process Engineering 2011, 44(1): 20–28. https://doi.org/10.1016/j.energy.2011.12.025.

Varbanov, P.S. and Klemeš, J.J. 2011. Integration and Management of Renewables into Total Sites with Variable Supply and Demand. Computers & Chemical Engineering Energy Systems Engineering, 35(9): 1815–1826. https://doi.org/10.1016/j.compchemeng.2011.02.009.

Varbanov P.S., Klemeš J.J., Liew P.Y., Yong J.Y., 2014. Chemical Engineering Transactions, Vol.39, AIDIC – The Italian Association of Chemical Engineering, Milano, Italy.

Varbanov P.S., Klemeš J.J., Wan Alwi S.R., Yong J.Y., Liu X., 2015. Chemical Engineering Transactions, Vol.45, AIDIC – The Italian Association of Chemical Engineering, Milano, Italy.

Varbanov P.S., Liew P.Y., Yong J.Y., Klemeš J.J., Lam H.L., 2016. Chemical Engineering Transactions, Vol.52, AIDIC – The Italian Association of Chemical Engineering, Milano, Italy.

Varbanov P.S., Su R., Lam H.L., Klemeš J.J., 2017. Chemical Engineering Transactions, Vol.61, AIDIC – The Italian Association of Chemical Engineering, Milano, Italy.

Walmsley, T.G., Atkins, M.J., Ong, B.H.Y., Klemeš, J.J., Walmsley, M.R.W., and Varbanov, P.S. 2017. Total Site Heat Integration of Multi-Effect Evaporators with Vapour Recompression for Older Kraft Mills. Chemical Engineering Transactions 61 (October): 265–270. https://doi.org/10.3303/CET1761042.

Walmsley, T.G., Walmsley, M.R.W., Atkins, M.J., Neale, J.R., and Tarighaleslami, A.H. 2015b. Thermo-Economic Optimisation of Industrial Milk Spray Dryer Exhaust to Inlet Air Heat Recovery. Energy 90 (October): 95–104. https://doi.org/10.1016/j.energy.2015.03.102.

Walmsley, T.G., Walmsley, M.R.W., Neale, J.R., and Atkins, M.J. 2015a. Pinch Analysis of an Industrial Milk Evaporator with Vapour Recompression Technologies. Chemical Engineering Transactions 45: 7–12. https://doi.org/10.3303/CET1545002.

Walmsley, T.G. 2016. A Total Site Heat Integration Design Method for Integrated Evaporation Systems Including Vapour Recompression. Journal of Cleaner Production 136 (November): 111–118. https://doi.org/10.1016/j.jclepro.2016.06.044.

Walmsley, T.G., Atkins, M.J., Walmsley, M.R.W., Philipp, M., and Peesel, R.-H. 2018. Process and Utility Systems Integration and Optimisation for Ultra-Low Energy Milk Powder Production. Energy 146 (March): 67–81. https://doi.org/10.1016/j.energy.2017.04.142.

Wang, R., Wu, Y., Wang, Y., and Feng, X. 2017. An Industrial Area Layout Design Methodology Considering Piping and Safety Using Genetic Algorithm. Journal of Cleaner Production 167 (November): 23–31. https://doi.org/10.1016/j.jclepro.2017.08.147.

Wang, Y., Chang, C., and Feng, X. 2015. A Systematic Framework for Multi-Plants Heat Integration Combining Direct and Indirect Heat Integration Methods. Energy 90 (October): 56–67. https://doi.org/10.1016/j.energy.2015.04.015.

Wang, Y. and Feng, X. 2017. Heat Integration Across Plants Considering Distance Factor. In: Advances in Energy Systems Engineering, edited by, Kopanos, G.M., Liu, P., and Georgiadis, M.C., 621–648. Cham, Switzerland: Springer International Publishing. https://doi.org/10.1007/978-3-319-42803-1_21.

Wang, Y., Feng, X., and Chu, K.H. 2014. Trade-off between Energy and Distance Related Costs for Different Connection Patterns in Heat Integration across Plants. Applied Thermal Engineering 70(1): 857–866. https://doi.org/10.1016/j.applthermaleng.2014.06.012.

Westphalen, D.L. and Wolf Maciel, M.R. 2000. Pinch Analysis of Evaporation Systems. Brazilian Journal of Chemical Engineering 17(4–7): 525–538. https://doi.org/10.1590/S0104-66322000000400017.

Wu, Y., Wang, Y., and Feng, X. 2016. A Heuristic Approach for Petrochemical Plant Layout Considering Steam Pipeline Length. Chinese Journal of Chemical Engineering 24(8): 1032–1037. https://doi.org/10.1016/j.cjche.2016.04.043.

Index

https://doi.org/10.1515/9783110630091-014

www.ingramcontent.com/pod-product-compliance
Lightning Source LLC
Chambersburg PA
CBHW080129220326
41598CB00032B/5009